High-Order Methods for Incompressible Fluid Flow

High-order numerical methods provide an efficient approach to simulating many physical problems. This book considers the range of mathematical, engineering, and computer science topics that form the foundation of high-order numerical methods for the simulation of incompressible fluid flows in complex domains. Introductory chapters present high-order spatial and temporal discretizations for one-dimensional problems. These are extended to multiple space dimensions with a detailed discussion of tensor-product forms, multidomain methods, and preconditioners for iterative solution techniques. Numerous discretizations of the steady and unsteady Stokes and Navier–Stokes equations are presented, with particular attention given to enforcement of incompressibility. Advanced discretizations, implementation issues, and parallel and vector performance are considered in the closing sections. Numerous examples are provided throughout to illustrate the capabilities of high-order methods in actual applications.

Computer scientists, engineers, and applied mathematicians interested in developing software for solving flow problems will find this book a valuable reference.

M. O. Deville is a professor of numerical fluid mechanics at the Ecole Polytechnique Fédérale de Lausanne in Switzerland. His work concerns spectral methods, turbulence modelling, and large-eddy simulation. He has applied these techniques to a wide variety of applications related to biomedical engineering, direct numerical simulation, and, more recently, non-Newtonian fluids. He is the coauthor of other books on numerical modeling. He is also consultant for ONERA/DSNA (France).

P. F. Fischer is a research scientist in the Mathematics and Computer Science Division of Argonne National Laboratory. He has published extensively in the areas of computational fluid dynamics, spectral methods, iterative methods, and large-scale parallel algorithms and is currently working on applications in biofluid dynamics. He was a recipient of the Gordon Bell Prize for high-performance computing in 1999.

Since 1967, E. H. Mund has worked as a research scientist at the (Belgian) Fonds National de la Recherche Scientifique (FNRS). He has published extensively in the areas of nuclear reactor physics and computation, radiation transport, and applied mathematics. Dr. Mund is also part-time professor of nuclear engineering at the Université Catholique de Louvain and a member of the scientific advisory committee of the SCK.CEN nuclear research center in Mol (Belgium).

**CAMBRIDGE MONOGRAPHS ON
APPLIED AND COMPUTATIONAL
MATHEMATICS**

Series Editors
P. G. CIARLET, A. ISERLES, R. V. KOHN, M. H. WRIGHT

---

9   High-Order Methods for Incompressible Fluid Flow

The *Cambridge Monographs on Applied and Computational Mathematics* reflect the crucial role of mathematical and computational techniques in contemporary science. The series presents expositions on all aspects of applicable and numerical mathematics, with an emphasis on new developments in this fast-moving area of research.

State-of-the-art methods and algorithms as well as modern mathematical descriptions of physical and mechanical ideas are presented in a manner suited to graduate research students and professionals alike. Sound pedagogical presentation is a prerequisite. It is intended that books in the series will serve to inform a new generation of researchers.

Also in this series:

A Practical Guide to Pseudospectral Methods, *Bengt Fornberg*

Dynamical Systems and Numerical Analysis, *A. M. Stuart and A. R. Humphries*

Level Set Methods and Fast Marching Methods, *J. A. Sethian*

The Numerical Solution of Integral Equations of the Second Kind, *Kendall E. Atkinson*

Orthogonal Rational Functions, *Adhemar Bultheel, Pablo González-Vera, Erik Hendriksen, and Olav Njåstad*

Theory of Composites, *Graeme W. Milton*

Geometry and Topology for Mesh Generation, *Herbert Edelsbrunner*

Schwarz–Christoffel Mapping, *Tobin A. Driscoll and Lloyd N. Trefethen*

High-Order Methods for Incompressible Fluid Flow

M. O. DEVILLE
Ecole Polytechnique Fédérale de Lausanne

P. F. FISCHER
Argonne National Laboratory

E. H. MUND
*Université Libre de Bruxelles and
Université Catholique de Louvain*

PUBLISHED BY THE PRESS SYNDICATE OF THE UNIVERSITY OF CAMBRIDGE
The Pitt Building, Trumpington Street, Cambridge, United Kingdom

CAMBRIDGE UNIVERSITY PRESS
The Edinburgh Building, Cambridge CB2 2RU, UK
40 West 20th Street, New York, NY 10011-4211, USA
477 Williamstown Road, Port Melbourne, VIC 3207, Australia
Ruiz de Alarcón 13, 28014 Madrid, Spain
Dock House, The Waterfront, Cape Town 8001, South Africa

http://www.cambridge.org

© Cambridge University Press 2002

This book is in copyright. Subject to statutory exception
and to the provisions of relevant collective licensing agreements,
no reproduction of any part may take place without
the written permission of Cambridge University Press.

First published 2002

Printed in the United Kingdom at the University Press, Cambridge

Typeface Times Roman 10/13 pt.   *System* LATEX 2_ε [TB]

A catalog record for this book is available from the British Library.

Library of Congress Cataloging in Publication Data
Deville, M. O. (Michel O.)
High-order methods for incompressible fluid flow / M. O. Deville, P. F. Fischer, E. H. Mund.
p. cm. – (Cambridge monographs on applied and computational mathematics ; 9)
Includes bibliographical references and index.
ISBN 0-521-45309-7
1. Fluid dynamics.   I. Fischer, P. F. (Paul F.)   II. Mund, E. H. (Ernest H.)
III. Title.   IV. Series.
QA929 .D48   2002

532′.051 – dc21   2001052969

ISBN 0 521 45309 7 hardback

*To our wives Christina, Dimpy, Monique,
and our families*

Je cherche un h.o.m.
Diogenes

Contents

List of Figures	*page* xvii
Preface	xxv

1 Fluid Mechanics and Computation: An Introduction   1

1.1	Viscous Fluid Flows	1
1.2	Mass Conservation	3
1.3	Momentum Equations	5
	1.3.1 Linear Momentum	5
	1.3.2 Angular Momentum	6
1.4	Energy Conservation	6
1.5	Thermodynamics and Constitutive Equations	7
1.6	Fluid Flow Equations and Boundary Conditions	8
	1.6.1 Isothermal Incompressible Flow	8
	1.6.2 Thermal Convection: The Boussinesq Approximation	9
	1.6.3 Boundary and Initial Conditions	10
1.7	Dimensional Analysis and Reduced Equations	11
1.8	Vorticity Equation	15
1.9	Simplified Models	16
1.10	Turbulence and Challenges	17
1.11	Numerical Simulation	22
	1.11.1 Hardware Issues	22
	1.11.2 Software Issues	24
	1.11.3 Algorithms	26
	1.11.4 Advantages of High-Order Methods	28

2 Approximation Methods for Elliptic Problems — 33
- 2.1 Variational Form of Boundary-Value Problems — 34
  - 2.1.1 Variational Functionals — 34
  - 2.1.2 Boundary Conditions — 39
  - 2.1.3 Sobolev Spaces and the Lax-Milgram Theorem — 40
- 2.2 An Approximation Framework — 46
  - 2.2.1 Galerkin Approximations — 47
  - 2.2.2 Collocation Approximation — 51
- 2.3 Finite-Element Methods — 53
  - 2.3.1 The h-Version of Finite Elements — 54
  - 2.3.2 The p-Version of Finite Elements — 60
- 2.4 Spectral-Element Methods — 62
- 2.5 Orthogonal Collocation — 67
  - 2.5.1 Orthogonal Collocation in a Monodomain — 67
  - 2.5.2 Orthogonal Collocation in a Multidomain — 69
- 2.6 Error Estimation — 71
- 2.7 Solution Techniques — 73
  - 2.7.1 The Conditioning of a Matrix — 74
  - 2.7.2 Basic Iterative Methods — 81
  - 2.7.3 Preconditioning Schemes of High-Order Methods — 83
  - 2.7.4 Iterative Methods Based on Projection — 86
- 2.8 A Numerical Example — 92

3 Parabolic and Hyperbolic Problems — 98
- 3.1 Introduction — 98
- 3.2 Time Discretization Schemes — 99
  - 3.2.1 Linear Multistep Methods — 100
  - 3.2.2 Predictor–Corrector Methods — 110
  - 3.2.3 Runge–Kutta Methods — 113
- 3.3 Splitting Methods — 119
  - 3.3.1 The Operator-Integration-Factor Splitting Method — 121
  - 3.3.2 OIFS Example: The BDF3/RK4 Scheme — 123
- 3.4 The Parabolic Case: Unsteady Diffusion — 124
  - 3.4.1 Spatial Discretization — 126
  - 3.4.2 Time Advancement — 127
- 3.5 The Hyperbolic Case: Linear Convection — 129
  - 3.5.1 Spatial Discretization — 130
  - 3.5.2 Eigenvalues of the Discrete Problem and CFL Number — 131
  - 3.5.3 Example of Temporal and Spatial Accuracy — 135
  - 3.5.4 Inflow–Outflow Boundary Conditions — 137

	3.6	Steady Advection–Diffusion Problems	137
		3.6.1 Spectral Elements and Bubble Stabilization	138
		3.6.2 Collocation and Staggered Grids	141
	3.7	Unsteady Advection–Diffusion Problems	145
		3.7.1 Spatial Discretization	146
		3.7.2 Temporal Discretization	149
		3.7.3 Outflow Conditions and Filter-Based Stabilization	149
	3.8	The Burgers Equation	151
		3.8.1 Space and Time Discretization	151
		3.8.2 Numerical Results	153
	3.9	The OIFS Method and Subcycling	155
	3.10	Taylor–Galerkin Time Integration	158
		3.10.1 Nonlinear Pure Advection	159
		3.10.2 Taylor–Galerkin and OIFS Methods	161
4	**Multidimensional Problems**		**162**
	4.1	Introduction	162
	4.2	Tensor Products	162
	4.3	Elliptic Problems	169
		4.3.1 Weak Formulation and Sobolev Spaces	170
		4.3.2 A Constant-Coefficient Case	172
		4.3.3 The Variable-Coefficient Case	177
	4.4	Deformed Geometries	178
		4.4.1 Generation of Geometric Deformation	183
		4.4.2 Surface Integrals and Robin Boundary Conditions	186
	4.5	Spectral-Element Discretizations	188
		4.5.1 Continuity and Direct Stiffness Summation	191
		4.5.2 Spectral–Element Operators	194
		4.5.3 Inhomogeneous Dirichlet Problems	196
		4.5.4 Iterative Solution Techniques	197
		4.5.5 Two-Dimensional Examples	198
	4.6	Collocation Discretizations	202
		4.6.1 The Diffusion Case	202
		4.6.2 The Advection–Diffusion Case	216
	4.7	Parabolic Problems	220
		4.7.1 Time-Dependent Projection	222
		4.7.2 Other Diffusion Systems	224
	4.8	Hyperbolic Problems	226
	4.9	Unsteady Advection–Diffusion Problems	230
	4.10	Further Reading	232

5	**Steady Stokes and Navier–Stokes Equations**	**234**
	5.1 Steady Velocity–Pressure Formulation	234
	5.2 Stokes Equations	236
	5.2.1 The Weak Formulation	236
	5.2.2 The Spectral-Element Method	238
	5.2.3 Collocation Methods on Single and Staggered Grids	245
	5.3 Linear Systems, Algorithms, and Preconditioners	253
	5.3.1 Spectral-Element Methods and Uzawa Algorithm	253
	5.3.2 Collocation Methods	257
	5.4 Poisson Pressure Solver and Green's-Function Technique	259
	5.4.1 General Considerations	259
	5.4.2 The Green's-Function Method	260
	5.4.3 Implementation	263
	5.5 Divergence-Free Bases	264
	5.6 Stabilization of the $\mathbb{P}_N$–$\mathbb{P}_N$ Approximation by Bubble Functions	269
	5.7 hp-Methods for Stokes Problems	272
	5.8 Steady Navier–Stokes Equations	273
	5.8.1 Weak Formulation	274
	5.8.2 Collocation Approximation of the Navier–Stokes Equations	275
	5.8.3 Solution Algorithms: Iterative and Newton Methods	277
	5.9 Applications	278
	5.9.1 Stokes Problems	278
	5.9.2 Navier–Stokes Problems	283
	5.10 Complements and Engineering Considerations	288
6	**Unsteady Stokes and Navier–Stokes Equations**	**291**
	6.1 Unsteady Velocity–Pressure Formulation	291
	6.2 Unsteady Stokes Equations	293
	6.2.1 The Weak Formulation	293
	6.2.2 Uzawa Algorithm	295
	6.2.3 Splitting and Decoupling Algorithms	296
	6.3 Pressure Preconditioning	300
	6.4 Unsteady Navier–Stokes Equations	303
	6.4.1 Weak Formulation	303
	6.4.2 Advection Treatment	304
	6.5 Projection Methods	309
	6.5.1 Fractional-Step Method	310
	6.5.2 Pressure Correction Method	313

6.6	Stabilizing Unsteady Flows		315
6.7	Arbitrary Lagrangian–Eulerian Formulation and Free-Surface Flows		318
	6.7.1	ALE Formulation	319
	6.7.2	Free-Surface Conditions	320
	6.7.3	Variational Formulation of Free-Surface Flows	322
	6.7.4	Space and Time Discretization	325
6.8	Unsteady Applications		326
	6.8.1	Extrusion from a Die	326
	6.8.2	Vortex-Sheet Roll-Up	327
	6.8.3	Unsteady Flow in Arteriovenous Grafts	329
6.9	Further Reading and Engineering Considerations		329

7 Domain Decomposition — 333

7.1	Introduction		333
7.2	Preconditioning Methods		334
	7.2.1	Substructuring and the Steklov–Poincaré Operator	334
	7.2.2	Overlapping Schwarz Procedures	338
	7.2.3	Schwarz Preconditioners for High-Order Methods	345
	7.2.4	Spectral-Element Multigrid	348
7.3	The Mortar Element Method		352
	7.3.1	Elliptic Problems	357
	7.3.2	Implementation	358
	7.3.3	Steady Stokes Problems	363
	7.3.4	Applications	365
7.4	Adaptivity and Singularity Treatment		368
	7.4.1	Coupling between Finite and Spectral Elements	369
	7.4.2	Singularity Treatment	370
	7.4.3	Triangular and Tetrahedral Elements	371
	7.4.4	Error Estimates and Adaptivity	377
7.5	Further Reading		378

8 Vector and Parallel Implementations — 379

8.1	Introduction		379
8.2	Serial Architectures		380
	8.2.1	Pipelining	381
	8.2.2	Memory, Bandwidth, and Caches	382
8.3	Tensor-Product Operator Evaluation		384
	8.3.1	Tensor-Product Evaluation	385
	8.3.2	Other Operations	390

	8.4	Parallel Programming	391
		8.4.1	Communication Characteristics	393
		8.4.2	Vector Reductions	396
	8.5	Parallel Multidomain Methods	399
		8.5.1	Data Distribution and Operator Evaluation	399
		8.5.2	Direct Stiffness Summation	401
		8.5.3	Domain Partitioning	406
		8.5.4	Coarse-Grid Solves	407
	8.6	Applications	408
		8.6.1	Hairpin Vortices	408
		8.6.2	Driven Cavity	410
		8.6.3	Backward-Facing Step	412
	8.7	Further Reading	416
A	**Preliminary Mathematical Concepts**			**417**
	A.1	Metric Spaces	417
		A.1.1	Definition	417
		A.1.2	Open Set, Closed Set, Neighborhood	418
		A.1.3	Cauchy Sequence, Limit Points, Dense Sets	419
		A.1.4	Mapping, Domain, Range, Continuity	419
		A.1.5	Convergence, Completeness, Completion Process	420
	A.2	Normed Spaces	421
		A.2.1	Definition	421
		A.2.2	Banach Spaces	422
	A.3	Linear Operators and Functionals in Normed Spaces	423
		A.3.1	Linear Operator, Domain, Range, Nullspace	423
		A.3.2	The Inverse Operator	423
		A.3.3	Bounded Operators, Compact Operators	424
		A.3.4	Bounded Linear Functionals, Dual Spaces	425
		A.3.5	The Fréchet Derivative of an Operator	425
	A.4	Inner-Product Spaces	427
		A.4.1	Definition	427
		A.4.2	Hilbert Spaces	428
		A.4.3	Cauchy–Schwarz Inequality	429
		A.4.4	The Riesz Representation	430
		A.4.5	Orthogonality, Orthogonal Projection	431
		A.4.6	Separable Hilbert Spaces, Basis	432
		A.4.7	Gram–Schmidt Orthonormalization Process	433

	A.5	Distributions	434
		A.5.1 Definitions	434
		A.5.2 Basic Properties of Distributions	438
B	**Orthogonal Polynomials and Discrete Transforms**		**442**
	B.1	Systems of Orthogonal Polynomials	442
		B.1.1 Eigensolutions of Sturm–Liouville Problems	444
		B.1.2 The Legendre Polynomials	445
		B.1.3 The Chebyshev Polynomials	447
	B.2	Gaussian-Type Quadratures	448
		B.2.1 Fundamental Theorems	448
		B.2.2 Gaussian Rules Based on Legendre Polynomials	450
		B.2.3 Gaussian Rules Based on Chebyshev Polynomials	451
		B.2.4 Discrete Inner Products and Norms	452
	B.3	Spectral Approximation and Interpolation	454
		B.3.1 Preliminaries	454
		B.3.2 Discrete Spectral Transforms	455
		B.3.3 Approximate Evaluation of Derivatives	458
		B.3.4 Estimates for Truncation and Interpolation Errors	464
Bibliography			467
Index			489

List of Figures

1.11.1	Evolution of machines, algorithms, and their combination over the past five decades for the solution of a three-dimensional Poisson equation.	page 27
1.11.2	Time evolution of high-order methods for a 3D Poisson equation.	30
1.11.3	For a fixed accuracy, combination of best machine and algorithm using the best high-order method.	31
2.3.1	Piecewise linear and quadratic ($p = 1, 2$) finite elements.	56
2.3.2	Structures of (a) element stiffness matrix $\hat{K}$ and (b) element mass matrix $\hat{M}$ of a p-type FEM with $p = 11$.	61
2.4.1	An example of Legendre-spectral element basis functions (2.4.3) corresponding to $N = 10$.	63
2.4.2	Global stiffness matrix (a) and mass matrix (b) structures of a spectral-element approximation with three elements and $N = 5$.	65
2.7.1	Distribution of Gauss–Lobatto–Chebyshev points on $[-1, 1]$.	79
2.7.2	Comparison of eigenvalues for second-order diffusion operator, $-u_{xx} = \lambda u$, $u(0) = u(1) = 0$: (a) linear finite elements with tridiagonal (M) and lumped ($\tilde{M}$) mass matrices, spectral elements with $(E, N) = (64, 4)$, and analytical spectrum; (b) Legendre SEM with $E = 1$ and N ranging from 4 to 256. [Exact and FEM ($\tilde{M}$) curves are also shown.]	81
2.8.1	The L^2 error norms for the Q_1 FE ($\star$), Chebyshev collocation ($\triangle$), and Legendre spectral-element ($\diamond$) solutions of the model problem (2.8.1)–(2.8.3), as a function of the number of degrees of freedom, N.	95

xviii   *List of Figures*

2.8.2   Spectral condition number of three matrix operators involved in a Legendre collocation scheme with Q_1 FE preconditioning: (a) $\kappa(L_{FE})$, (b) $\kappa(L_C)$, and (c) $\kappa(L_{FE}^{-1} \cdot L_C)$. The parameter N is the polynomial degree of the approximation.   96

2.8.3   Decay of the L^2 error norm for FD ($\star$) and FE ($\diamond$) preconditioned solutions of a Chebyshev collocation calculation, as a function of the iteration index k. The problem solved is (2.8.1)–(2.8.3): (a) calculations with the relaxation factor $\alpha = 1$, and (b) with $\alpha = \alpha_{opt}$. The factors α_{opt} are equal to 0.590617 (FD) and 1.16086 (FE).   96

3.2.1   Stability regions for ABk schemes with $1 \leq k \leq 4$.   108
3.2.2   Stability regions for AMk schemes with $0 \leq k \leq 4$.   108
3.2.3   Stability regions for BDFk schemes with $1 \leq k \leq 4$.   110
3.2.4   Stability regions for the $P(EC)^m E$ scheme (3.2.32) based on the AB2 and AM2 methods. The figure displays the stability contours of the PC scheme with $m = 1, 2$, as well as of the AB2 and AM2 schemes.   112
3.2.5   Stability regions for the Runge–Kutta schemes.   116
3.3.1   Stability regions for two second-order implicit–explicit schemes: (a) the ABCN scheme and (b) the MABCN scheme. Both set of curves depend on a parameter μ, with $0 \leq \mu \leq 6$ [see (3.3.7)].   121
3.3.2   An example of OIFS integration grids with time steps h_t for the diffusion component (BDF3) and h_s for the advection component (ERK4). In this example ($h_t = 5\, h_s$).   124
3.5.1   Comparison of eigenvalues for periodic convection operator, $u_x = \lambda u$, $u(0) = u(1)$: (a) exact, Q_1 linear finite elements with tridiagonal mass matix M (FE), and Q_1 linear finite elements with lumped mass matix $\tilde{M}$ (FD); (b) eigenvalues for Legendre spectral element method with $E = 1$ and N ranging from 4 to 256.   132
3.5.2   (a) Scale factors S for spectral-element discretizations. The other parts show SEM/AB3 solutions to $u_t + u_x = 0$, $u(0, t) = u(1, t)$ with (b) $(E, N) = (256, 1)$, CFL $= 0.72$; (c) $(E, N) = (256, 1)$, CFL $= 0.77$; (d) $(E, N) = (82, 3)$, CFL $= 0.72362/1.5$. The vertical axis in (b)–(d) is $t + u/15$.   134
3.5.3   Initial condition (left) and final solution (right) for $(E, N) = (256, 1)$ results of Table 3.5.1.   135

List of Figures

3.6.1 Collocation solutions of advection-dominated diffusion problems on (a) Legendre and (b) Chebyshev grids. — 142

3.6.2 An example of a staggered grid (bullets) for the advection–diffusion problem $-\nu d^2/dx^2 + xd/dx$ on $\hat{\Omega}$ (with $\nu = 10^{-2}$ and $N = 10$). The upper and lower parts of the figure display the GLL and Gauss–Legendre grids, respectively. — 144

3.7.1 Spectra for spectral-element discretization of the advection–diffusion problem with $(E, N) = (1, 64)$. A few isolated negative eigenvalues are off the scale in (b)–(f). — 148

3.7.2 Advection–diffusion results. — 150

3.8.1 Solution to Burgers equation on $[-1, 0]$ using the Legendre spectral method, $(E, N) = (1, 64)$, with AB3CN. — 154

4.2.1 Clockwise from upper left: GLL nodal point distribution on $\hat{\Omega}$ for $M = N = 4$, and Lagrangian basis functions for $M = N = 10$: $\pi_{10}\pi_2$, $\pi_2\pi_9$, and $\pi_3\pi_4$. — 163

4.4.1 Sketch of coordinate transformation from physical domain Ω to computational domain $\hat{\Omega}$. — 179

4.4.2 Application of the Gordon–Hall algorithm in $\mathbb{R}^3$. Subscripts i, j, and k range from 0 to N. Summations involving $\hat{i}$, $\hat{j}$, and $\hat{k}$ range from 0 to 1, accounting for contributions from opposing faces in each of the three coordinate directions. — 184

4.4.3 Description of surface geometry in $\mathbb{R}^3$. — 187

4.4.4 Computational domains for Hamel flow. — 188

4.4.5 Influence of mesh distortion D on convergence for Hamel flow example (from [345]). — 189

4.5.1 Example of spectral-element discretization in $\mathbb{R}^2$, showing GLL nodal lines for $(E, N) = (3, 4)$. — 190

4.5.2 Example of spectral-element mesh for $(E, N) = (2, 2)$: (a) global values are mapped to (b) local values by (c) the operation $\underline{u}_L = Q\underline{u}$. — 192

4.5.3 Pseudocode for serial implementation of Q and Q^T in $\mathbb{R}^2$. — 194

4.5.4 Global (left) and local (right) mask arrays. Open circles imply $(\mathcal{M})_{\hat{i}} = (\mathcal{M}_L)_{ij}^e = 0$. Solid circles imply $(\mathcal{M})_{\hat{i}} = (\mathcal{M}_L)_{ij}^e = 1$. — 196

4.5.5 Global (left) and local (right) node numbering for spectral-element example with $(E, N) = (2, 2)$. Dirichlet boundary segments are $\partial\Omega_{D_1}$ at $r = 0.5$ and $\partial\Omega_{D_2}$ at $y = 1$. — 199

4.5.6 Diagonal preconditioning results for $(E, N) = (3, 8)$. — 201

4.6.1 (a) Topological structure of the collocation matrix L_C for the elliptic problem (4.6.1). In this example $N = 7$. (b) Topological structure of the Q_1 Lagrangian FE matrix L_P for the same problem on the same grid. ... 204

4.6.2 Spectrum of $I - L_P^{-1} \cdot L_C$ for Q_1 FE preconditioning of two diffusion operators $\mathcal{L}u := -\nabla \cdot (p\nabla u) + u$, with Dirichlet conditions on the reference square $\hat{\Omega}$, and collocation on the Chebyshev grid with $N = 16$: (a) $p = 1$, and (b) $p = 1 + 10\,x^2 y^2$. ... 207

4.6.3 Spectrum of $I - L_P^{-1} \cdot L_C$ for Q_1 FE preconditioning of $\mathcal{L}u := -\nabla \cdot [(1 + x^2 y^2)\nabla u] + u$, with Dirichlet conditions on the reference square $\hat{\Omega}$, and collocation on the Chebyshev grid with $N = 16$. In this case, the preconditioner L_P results from the FE approximation of $\mathcal{L}u := -\Delta u + u$. ... 207

4.6.4 Spectrum of the operator $I - L_P^{-1} \cdot L_C$ for FD preconditioning of $\mathcal{L}u := -\nabla \cdot (p\nabla u) + u$, with Dirichlet conditions on the reference square $\hat{\Omega}$, and collocation on the Chebyshev grid with $N = 16$: (a) $p = 1$, (b) $p = 1 + x^2 y^2$, (c) $p = 1 + 10\,x^2 y^2$, and (d) $p = 1 + x^2 y^2$, the preconditioner L_P being such that $L_P u$ is the FD approximation of $\mathcal{L}u := -\Delta u + u$. ... 208

4.6.5 Convergence histories for the Chebyshev collocation solution of (4.6.17) with $N = 17$: (a) the preconditioner results from the FE ($\star$) or FD ($\square$) approximation of (4.6.17), (b) the preconditioner results from the FE or FD approximations of $\mathcal{L}u := -\Delta u + u$. ... 209

4.6.6 (a) The quadrangular integration domain Ω and (b) its partition into four nonoverlapping subdomains Ω^k. ... 213

4.6.7 Two topological structures of L_C, the orthogonal collocation matrix on a 2×2 multidomain, depending on the ordering of the unknowns (see text). In this example, the polynomial expansion in each subdomain has $N = 5$. ... 215

4.6.8 (a) The Gauss–Lobatto–Legendre grid with $N_x = N_y = 8$, and (b) the corresponding upwind grid for the problem $-\frac{1}{100}\Delta u + x\frac{\partial u}{\partial x} - y\frac{\partial u}{\partial y} = f$. ... 218

4.7.1 Jacobi–PCG iteration counts with and without projection for an unsteady conduction problem in the unit square with $(E, N) = (16, 8)$. ... 224

4.8.1 Convected-cone problem on a 32×32 grid comprising E_1^2 elements of order N. ... 229

List of Figures

5.2.1	Staggered spectral element for $N = 6$ (left) and $N = 7$ (right).	242
5.2.2	Legendre staggered collocation grid for $N_x = N_y = 7$.	251
5.5.1	Countours of computed scalar function $\rho(\mathbf{x})$ for the grooved channel and eddy-promoter geometries. The spectral-element discretization is based on $N_x = N_y = 13$.	268
5.9.1	Square-cavity problem: velocity and pressure error with and without special treatment for top-corner singularities. Crosses indicate special treatment; circles indicate no special treatment.	279
5.9.2	Creeping flow in a wedge [331]. Velocity boundary conditions are no slip on the sides and unit horizontal velocity on the top. The $E = 30$, $N = 8$ spectral-element discretization is shown in (a), and the sequence of decaying eddies, in the form of streamlines, in (b). The vortex strength decays by a factor of roughly 400 from one eddy to the next.	280
5.9.3	Grooved-channel flow: velocity-vector plots of the first four singular Stokes eigenfunctions. In each spectral element, $N_x = N_y = 13$. The first, second, third, and fourth eigenfunctions are shown in (a)–(d), respectively. In (e)–(j) are given the velocity-vector plots of eigenfunctions 10, 20, 30, 40, 50, and 60, respectively.	281
5.9.4	Wannier–Stokes flow: domain decomposition (top) and corresponding mesh with $N = 9$ in each subdomain (bottom).	282
5.9.5	Isolines for the Wannier–Stokes flow of the velocity components (a, b), pressure (c), and vorticity (d).	284
5.9.6	Evolution of the pressure along horizontal lines located under the cylinder.	284
5.9.7	Kovasznay flow: spectral-element convergence (a) and streamlines (b) for $Re = 40$.	285
5.9.8	Vertical velocity profile for the grooved channel at $Re = 100$: (a) $x = 1.3$; (b) $x = 3.1$.	286
5.9.9	The cooled-hot-cylinder problem: geometry.	287
5.9.10	The cooled-hot-cylinder problem: isothermal lines.	287
6.8.1	Free-surface die problem (due to Rønquist [335]).	327
6.8.2	Vorticity contours for different (E, N) pairings: $(a-d)$ thick shear layer, $\rho = 30$, $Re = 10^5$, contours from -70 to 70 by $140/15$; $(e-f)$ thin shear layer, $\rho = 100$, $Re = 40,000$, contours from -36 to 36 by $72/13$ (cf. Figure 3c in [56]).	328
6.8.3	Flow in an arteriovenous graft model.	330

7.2.1	Subdomain decomposition of a bounded domain $\Omega \subset \mathbb{R}^2$.	335
7.2.2	Domain Ω decomposed into overlapping domains Ω^1 and Ω^2 having boundaries $\partial \Omega^1$ and $\partial \Omega^2$.	339
7.2.3	Error behavior for the one-dimensional alternating Schwarz method on $\Omega = (-1, 1)$ with overlap $\delta = 0.25$ with respect to $x = 0$.	340
7.2.4	Overlapping subdomain operators $K_s = R_s^T K R_s$ corresponding to principal submatrices of K.	342
7.2.5	Overlapping Schwarz degrees of freedom (closed circles) for fine and coarse subproblems with $E = 9$ subdomains. The local system (left) is built on the Gauss points, using a tensor product of linear FE functions. The global coarse-grid problem (right) is based on linear triangles, with homogenous Dirichlet boundary conditions along the outflow boundary on the left.	346
7.2.6	(a) Convergence histories for $N_0 = 1$ case of Table 7.1 with $E = 93, 372,$ and $1488.$ (b) Contours of $\underline{p}^{\text{final}} - \underline{p}^{25}$ for $E = 1488$.	347
7.2.7	Multigrid V-cycle with m_d and m_u smoothings on the downward and upward sweeps, respectively.	350
7.3.1	Close-up of mesh near a blowing or suction slot (a) reveals how propagating (conforming) refinement leads to undesirable (unnecessary, high-aspect-ratio) elements in the far field (b).	352
7.3.2	Functionally (a) and geometrically (b) varying refinement.	353
7.3.3	Example of nonconforming domain and mortar decomposition in $\mathbb{R}^2$.	354
7.3.4	Example of geometrically conforming domain decomposition in $\mathbb{R}^2$.	356
7.3.5	Example of nonconforming domain decomposition in $\mathbb{R}^2$.	357
7.3.6	Nonconforming connectivity matrix for $(E, N) = (2, 2)$ spectral-element configuration of Figure 4.5.2.	359
7.3.7	Mortar decompositions in a geometrically nonconforming case.	361
7.3.8	Rotating-impeller problem.	366
7.3.9	Close-up of vorticity contours near a blowing or suction slot (a) for the resonator problem with nonpropagating mesh refinement [cf. Figure 7.3.1(a)]. The far-field view (b) reveals the absence of high-aspect-ratio elements that can degrade iterative solver performance.	366
7.3.10	Clearance-gap problem: geometry and boundary conditions.	367

7.3.11	Clearance-gap problem: polynomial degree distribution in the mortar elements around the blade.	368
7.3.12	Clearance-gap problem: isobars.	369
7.4.1	The reference square and triangle.	372
8.2.1	Pipelined execution of $c_i = a_i + b_i$ for an s-stage pipe. Result c_1 is ready as a_{s+1} and b_{s+1} enter the pipe.	381
8.2.2	In-cache performance for $\underline{a} = \underline{a} + \underline{b}$.	383
8.3.1	Interpretations of 3D data array as (*a*) consecutive elements in memory; (*b*) an $n \times n^2$ matrix ($n := N + 1$); (*c*) a sequence of n $n \times n$ matrices; and (*d*) an $n^2 \times n$ matrix.	386
8.4.1	Network of $P = 10$ distributed processor/memory units.	392
8.4.2	Measured communication times from ping-pong test for three distributed-memory computers.	394
8.4.3	Performance parameter trends in distributed-memory computing. The straight line follows Moore's law.	395
8.4.4	(*a*) First three stages of a contention-free binary-tree fan-in for vector reduction on a one-dimensional network. (*b*) Recursive doubling (butterfly) exchange for four-processor network.	397
8.5.1	Decomposition of five-element mesh into two subdomains.	399
8.5.2	Direct stiffness communication patterns: (*a*) physical grid for $(E, N) = (4, 4)$ in $d = 2$ space dimensions; (*b*) edge–vertex exchange algorithm requiring three messages per element (processor); (*c*) synchronized edge–edge exchange algorithm requiring d messages per element.	402
8.5.3	(*a*) Global numbering and (*b*) partition onto two processors.	405
8.5.4	Schematic of $Q_m^T = \hat{Q}_m J$ applied to (interpolation-based) nonconforming interface variables.	406
8.6.1	(*a*) Computational domain showing inlet velocity profile, flat plate, hemisphere, and isolated hairpin vortex. For clarity, the vortex has been reflected about the symmetry plane. (*b*) Strouhal number versus Reynolds number for the $E = 1021$ and $E = 1535$ meshes.	409
8.6.2	Profile (top) and planform (bottom) views of hairpin and secondary vortices generated in a boundary layer by a hemispherical roughness element for $Re_k = 922$. The spectral-element parameters are $(E, N) = (1021, 11)$.	410
8.6.3	Three-dimensional driven cavity at $Re = 5000$ (due to Leriche [239]).	411

8.6.4 Geometry of the backward-facing step with a 1 : 1.94 expansion ratio. For the Reynolds number considered here, two recirculation zones are of interest. Their locations are indicated in the symmetry plane. 413

8.6.5 Spectral-element distribution. True aspect ratio. 413

8.6.6 Comparison of the length of the first recirculation zone as a function of the Reynolds number. 414

8.6.7 $Re = 343$. Streamwise velocity component along the three observation lines in the lower part of the geometry, downstream of the recirculation zone. Results for $N = 9$ and $N = 11$ are shown but are virtually indistinguishable. 414

8.6.8 $Re = 343$. Streamwise velocity component along the three observation lines in the upper part of the geometry, downstream of the recirculation zone. The results for $N = 9$ and $N = 11$ are given, but are difficult to distinguish. 415

8.6.9 $Re = 343$. Spanwise velocity component along the three observation lines in the lower part of the geometry, downstream of the recirculation zone. 415

A.5.1 Two examples of distributions: (a) a piecewise linear function; (b) the Heaviside step function. 436

Preface

High-order methods have gained increasing attention in recent years. Their theoretical development has reached a high level of sophistication, and at the same time the range of applications has been broadening, including such diverse topics as global atmospheric modeling, aerodynamics, oceanography, thermal convection, and theoretical chemistry. Specialized conferences on the subject like the International Conference on Spectral Applications and High-Order Methods (ICOSAHOM) have been launched to bring mathematicians, engineers, and computer scientists together in order to stimulate further work in the field and to prospect new areas: high-order time schemes, treatment of singularities, complex geometries, mixed discretization techniques, domain decomposition, and parallelism. These topics were once considered as the stumbling block of spectral methods. As time goes on, this is no longer true, and high-order methods apply more and more to real-life engineering problems.

The monograph by Gottlieb and Orszag [163] and the book by Canuto et al. [64] remain milestones in the subject. They are cited in almost every paper written on the topic. Gottlieb and Orszag's monograph was the first on the subject and contains very little about applications. Moreover, it is silent on the topics mentioned above. Most of the developments covered by Canuto et al. are devoted to simple geometries, but a last chapter entitled "Domain Decomposition Methods" introduces extensions to more complex geometries. Recent achievements in the field of high-order methods have far-reaching consequences for geometrically complex configurations. As these constitute the basic ingredients of engineering design problems, they may be embodied in the formulation of the problem itself and in the related algorithms.

This book is intended to be an advanced textbook on high-order methods applied to incompressible fluid flow problems. Its intended audience includes engineers, numerical analysts, computational scientists, and students with a background in numerical methods and fluid mechanics. The text is not aimed at mathematical proofs of convergence, but rather at practical applications and results achievable with large-scale computing capabilities. The goal is to show the realm of feasibility of high-order methods: accuracy versus efficiency, tractable problems, nonlinearities, complex geometrical configurations, and the influence of computer architectures. Moreover, as the previous texts cover much of the spectral approaches either in Fourier or in Chebyshev space, we focus in this book on high-order methods in physical space, namely, collocation and spectral elements.

The introductory Chapter 1 presents the basic equations of incompressible Newtonian flows, including natural convection through the Boussinesq approximation. Several simplified models are derived on the basis of dimensional analysis, which will be examined in further chapters in order to build the appropriate algorithms. The chapter concludes with considerations on numerical simulation.

Chapter 2 is devoted to the numerical solution of one-dimensional elliptic problems. One recalls the variational weak formulation, because this leads to physical interpretation. The framework of Galerkin methods is introduced in the context of polynomial approximations. High-order methods are closely associated with Gaussian quadrature rules, and the resulting algebraic equations are intrinsically of large size and numerically ill-conditioned. Therefore, their implementation requires preconditioning techniques.

Chapter 3 analyzes hyperbolic and parabolic problems in the same fashion as elliptic problems in the preceding chapter. The first part of the chapter deals with numerical schemes for the solution of nonlinear systems of ordinary differential equations. Steady-state advection and/or advection–diffusion problems are treated in the second part, as well as evolutionary problems. Finally, the Burgers equation is introduced as an example of a nonlinear model.

Chapter 4 generalizes the introductory concepts to multiple space dimensions. The key element lies in a tensor-product formulation of the one-dimensional space approximation that is at the heart of algorithmic developments. Elliptic, parabolic, and hyperbolic problems are systematically studied. Multidimensional implementation issues are discussed in detail. This chapter makes a thorough analysis of orthogonal collocation methods and spectral element discretizations.

Chapter 5 deals with the velocity–pressure formulation of the steady Stokes and Navier–Stokes equations. The key issue of compatible discretizations of velocity and pressure is fully considered. The Babuška–Brezzi or inf–sup condition is recalled for the right functional spaces, and a decoupling of the pressure computation is presented in the Uzawa algorithm. The chapter ends with the analysis of multidimensional Navier–Stokes problems.

In Chapter 6, we treat the unsteady Stokes and Navier–Stokes equations presenting the state-of-the-art splitting and projection methods. The arbitrary Lagrangian–Eulerian method is also introduced to deal with free surface flows. A new filtering technique is described that enables simulations at higher Reynolds number values. The chapter ends with a few relevant applications.

Chapter 7 is devoted to the domain decomposition methods and adaptivity techniques. Substructuring and overlapping Schwarz methods are introduced as building blocks for efficient iterative solution of elliptic problems. Discretization developments include the mortar element method and unstructured approaches based on high-order simplicial mesh elements.

Chapter 8 considers the issues of vectorization and parallelism through study cases that include spectral element and collocation techniques.

The book may be used in two different ways for teaching purposes. The first consists in using it at the undergraduate level as an introductory course in high-order methods devoted to applied mathematics problems. Chapters 2, 3, and 4 would present the material of space and time discretizations and related algorithmic considerations. The second path builds upon the previous material and covers the incompressible Navier–Stokes equations treated by spectral element and collocation methods for direct numerical simulation. Each chapter describes the theoretical and implementation considerations and illustrates these by several examples of current applications. This second course is more advanced and would require as prerequisites courses in scientific computing or numerical methods and some knowledge of the basic concepts of incompressible fluid mechanics.

The book has been written through an effort spread over several years. Throughout the course of writing, we have benefited from fruitful discussions with numerous researchers active in the field. These discussions allowed us to synthesize some thoughts about high-order methods and their application to physical and engineering problems. We are solely responsible for these points of view and hope that they will be of use to the community at large.

Given the rapid expansion of the field, it is difficult to provide a complete bibliography. We have tried to be as comprehensive as possible, and we apologize to authors whose work we may have overlooked.

We would like to thank David Gottlieb for his support and initiative in encouraging this effort. We gratefully acknowledge the careful feedback and constructive comments of our colleagues Mejdi Azaïez, Robert Beauwens, Cédric Chauvière, Emmanuel Leriche, Anthony T. Patera, Etienne Perchat, Alfio Quarteroni, Einar Rønquist, Henry Tufo, and Olof Widlund. We would like to thank Gail Pieper for her attention to detail in editing the manuscript.

1
Fluid Mechanics and Computation: An Introduction

According to the Greek philosopher Heraclitus, who used to say "$\pi\alpha\nu\tau\alpha$ $\rho\epsilon\iota\ldots$,"[1] daily life is concerned with the flow of ordinary fluids: water, air, blood, and so forth, in very common situations like breathing, coffee drinking, and hand washing.

Most flows are generated by nature (e.g., oceans, winds, rivers) and by human industrial activity (e.g., planes, cars, materials processing, biomedical engineering). There is a need to model fluid flow problems in order to improve the basic understanding of these complex phenomena and to increase the design quality of technological applications. With the advent of large and powerful computational tools, modeling has become more and more a substitute for direct experimentation. In some circumstances, experimentation may be too expensive – particularly if it leads to the destruction of the facility – or even impossible to perform, so that modeling is the only reasonable way to get answers and to study a range of parameters for optimal design.

1.1 Viscous Fluid Flows

We know from experience that many flows are set into motion by shear forces, and hence viscous effects play a vital role in fluids. In general, the viscosity depends on the shear rate (roughly speaking, the velocity gradient), as is explained by non-Newtonian theory. In this book, however, we will restrict ourselves mainly to the case of viscous Newtonian incompressible fluids in isothermal situations or under the influence of thermal convection as described by the Boussinesq approximation.

---

[1] Everything flows....

The conceptual framework through which the governing equations will be derived is the axiomatic presentation based on the principles of the mechanics of continuous media (see, e.g., Gurtin [179], Truesdell [385], Truesdell and Toupin [387], and, more recently, Truesdell and Rajagopal [386]), although the equations also can be obtained from the principles of statistical mechanics (Cercignani [77]). This process will deliver the well-known Navier–Stokes equations, a set of nonlinear partial differential equations (PDEs) subject to both initial and boundary conditions.

A major parameter characterizing the flow features is given by the Reynolds number, which is a nondimensional measure of the importance of inertial forces relative to viscous forces. Starting from a zero value (creeping Stokes flow), where the physics is straightforward because the model is basically linear, the sweep to increasing values leads from laminar flow to the transition to turbulence. Later stages are weak turbulence and fully developed turbulence; correspondingly the Reynolds number goes up from a few hundred to several million. In this latter situation, the physics becomes quite complicated, even in simple geometrical configurations, because the flow structure is highly nonlinear and time-dependent, and involves spatial scales spread over several orders of magnitude. These circumstances render the use of computers essential.

The advent of computers led to the emergence of numerical fluid mechanics, also known as computational fluid dynamics (CFD). At the beginning, this new discipline brought some hope that computation would open the door to simulations of transient, three-dimensional (3D) flows in or around industrial (complex) geometries at high Reynolds numbers. These flows correspond to internal or external flows, respectively. Reality has disappointed the expectations of this naive picture. We will analyze the difficulties faced by computational scientists at the present time and will study carefully the advantages (and drawbacks) of applying high-order methods to CFD. Despite spectacular breakthroughs in software and hardware, the present tools are still insufficient to cope with direct numerical simulation (DNS) at moderate ($\approx 10^4$–10^5) and high ($\approx 10^6$–10^7) Reynolds numbers. Therefore, large-eddy simulation (LES), where the gross structures of the flow are resolved spatially and temporally and where small scales are modeled through a subgrid-scale approximation, constitutes a viable compromise between the expensive DNS and the first- or second-order moment closure models applied to the Reynolds-averaged Navier–Stokes equations (RANS). Fortunately, even though it is more complicated to model, LES retains the same computational complexity as the classical Navier–Stokes equations.

1.2. Mass Conservation

Before turning to mathematical models, we mention examples of flows arising in nature and industrial processing. The following list is by no means exhaustive:

- Creeping flows with thermal convection: plate tectonics, glass flow in a furnace.
- Materials processing: crystal growth, mold filling, thermal convection in a molten tin bath of a float glass process.
- Biofluid mechanics: blood flow in stenosed arteries (see, for example, Tu et al. [388]) through artificial valves.
- Hydraulic machines: flow through a Francis runner.
- Turbulent flows with thermal convection: liquid-metal processing, planetary sciences (Jupiter's red spot; see, for example, Marcus [266]).

As examples of DNS, we mention the computation of a flow in a square duct (e.g., Gavrilakis [150]), in a 3D parallelepipedic cavity (e.g., Deville et al. [98]), and in a cubical cavity (e.g., Leriche and Gavrilakis [240]).

In this chapter, we review the general principles of fluid mechanics: mass conservation (Section 1.2), momentum and angular momentum conservation (Section 1.3), and the first and second principles of thermodynamics (Sections 1.4 and 1.5, respectively). Then, we discuss the fluid flow equations (Section 1.6) and the relevant dimensionless numbers characterizing the physical situations (Section 1.7). Section 1.8 presents the vorticity equations. In Section 1.9, we discuss simplified models that can be derived as limit cases of the general equations when some dimensionless numbers go to extreme values. The challenge raised by turbulence then is examined. Several turbulence models are given, and large-eddy simulation is also summarized (Section 1.10). Because these last problems lead to large-scale, computation-intensive applications, the questions coming from scientific computing are evoked in a broad sense, that is, in the context of software and hardware interactions (Section 1.11). In particular, parallel computing and second-generation languages are scrutinized.

1.2 Mass Conservation

The principle of mass conservation states: *In a given deforming material volume $\Omega(t)$, the mass $M(t)$ is constant with respect to time t.* Translated into mathematical terms, this principle yields the relation

$$\frac{d}{dt} M(t) = 0, \qquad (1.2.1)$$

where the mass $M(t)$ is expressed as

$$M(t) = \int_{\Omega(t)} \rho \, dV, \tag{1.2.2}$$

with ρ the volumetric mass of the fluid. Generally speaking, the Reynolds transport theorem is employed to evaluate the rate of change of volume integrals of a material property, say f, when the volume is changing in time. Defining the integral

$$I(t) = \int_{\Omega(t)} f(\mathbf{x}, t) \, dV, \tag{1.2.3}$$

where f is a function of position $\mathbf{x}$ and time, one obtains

$$\frac{dI}{dt} = \int_{\Omega(t)} \left(\frac{Df}{Dt} + f \, \text{div} \, \mathbf{v} \right) dV. \tag{1.2.4}$$

In Equation (1.2.4), the symbol D/Dt denotes the material time derivative

$$\frac{D}{Dt} = \frac{\partial}{\partial t} + \mathbf{v} \cdot \nabla, \tag{1.2.5}$$

and $\mathbf{v}$ is the velocity field in the Eulerian representation. Combining Equations (1.2.1), (1.2.2), and (1.2.4) gives

$$\int_{\Omega(t)} \left(\frac{D\rho}{Dt} + \rho \, \text{div} \, \mathbf{v} \right) dV = 0. \tag{1.2.6}$$

The relationship is valid for any material volume inside the fluid. Consequently, this equation holds only if the integrand vanishes everywhere in the fluid. This gives rise to the local equation of mass conservation:

$$\frac{D\rho}{Dt} + \rho \, \text{div} \, \mathbf{v} = 0, \tag{1.2.7}$$

which can be rewritten as

$$\frac{\partial \rho}{\partial t} + \text{div}(\rho \, \mathbf{v}) = 0. \tag{1.2.8}$$

A fluid shows the property of incompressibility if ρ does not depend on the pressure and temperature. For an incompressible fluid, ρ is a constant, and

$$\frac{D\rho}{Dt} = 0, \tag{1.2.9}$$

1.3. Momentum Equations

which implies by (1.2.7) that

$$\text{div } \mathbf{v} = 0. \tag{1.2.10}$$

This is the incompressibility constraint, requiring that the velocity field be solenoidal.

1.3 Momentum Equations

In this section, we review the generalization of Newton's law for continuous media, which brings into consideration also the equilibrium equation and the angular momentum conservation law.

1.3.1 Linear Momentum

The principle of linear momentum conservation generalizes Newton's law of rational mechanics to continuous media: *The time rate of change of the momentum of a deforming material region embedded in $\Omega(t)$ is equal to the sum of the forces applied to that region.* These forces can be decomposed into a body force $\mathbf{f}$ and surface contact forces $\mathbf{t}$. The former is a volume force (gravity, for example), while the latter are exerted across the surface $\partial\Omega(t)$ surrounding the material. Mathematically, one obtains

$$\frac{d}{dt}\int_{\Omega(t)} \rho\mathbf{v}\,dV = \int_{\partial\Omega(t)} \mathbf{t}\,dS + \int_{\Omega(t)} \rho\mathbf{f}\,dV. \tag{1.3.1}$$

The Cauchy principle implies that the density of contact forces represented by the stress vector $\mathbf{t}$ depends on the oriented normal $\mathbf{n}$ to $\partial\Omega(t)$ at $\mathbf{x}$. More precisely, there is a tensor field such that

$$\mathbf{t}(\mathbf{x}, \mathbf{n}) = \sigma(\mathbf{x})\mathbf{n}, \tag{1.3.2}$$

where σ denotes the stress tensor. In indicial notation, (1.3.2) becomes

$$t_i = \sigma_{ij} n_j, \tag{1.3.3}$$

with the assumption of the Einstein summation convention for repeated indices. The stress component σ_{ij} gives the stress in the ith direction of the coordinate system on a plane surface element with a normal that is positively oriented in direction j.

With the help of the transport theorem (1.2.4), the mass conservation property (1.2.8), the Cauchy principle (1.3.2), and the divergence theorem, Equation (1.3.1) provides the equation of motion:

$$\rho\frac{D\mathbf{v}}{Dt} = \text{div }\sigma + \rho\mathbf{f}. \tag{1.3.4}$$

1.3.2 Angular Momentum

The principle of angular momentum conservation states that *the rate of change of the angular momentum of a deforming material region $\Omega(t)$ is equal to the sum of the applied torques*. After some algebra [387], omitted for the sake of conciseness, we arrive at the symmetry condition of the stress tensor,

$$\sigma = \sigma^T, \tag{1.3.5}$$

the superscript T indicating the transpose.

1.4 Energy Conservation

The general principle of energy conservation, also known as the first law of thermodynamics, states that *the rate of change of the total energy within a deforming material region $\Omega(t)$ is equal to the rate at which this energy is received through heat and work transfer*. One writes

$$\frac{d}{dt} \int_{\Omega(t)} \rho \left[\frac{1}{2} \mathbf{v} \cdot \mathbf{v} + U \right] dV = \int_{\partial\Omega(t)} \mathbf{t} \cdot \mathbf{v} \, dS + \int_{\Omega(t)} \rho \mathbf{f} \cdot \mathbf{v} \, dV$$
$$+ \int_{\partial\Omega(t)} (-\mathbf{q} \cdot \mathbf{n}) \, dS + \int_{\Omega(t)} \rho r \, dV. \tag{1.4.1}$$

In (1.4.1), U is the internal energy per unit mass, $\mathbf{q}$ the heat flux, and r the heat source per unit mass. This last term represents, for example, the heat supply (or consumption) through chemical reactions in the fluid mass. Invoking again the transport theorem, the mass conservation principle, the Cauchy principle, and the divergence theorem, one can rewrite (1.4.1) as follows:

$$\rho \frac{D}{Dt} \left(\frac{1}{2} \mathbf{v}^2 + U \right) = \text{div}(\sigma \cdot \mathbf{v}) + \rho \mathbf{f} \cdot \mathbf{v} - \text{div} \, \mathbf{q} + \rho r. \tag{1.4.2}$$

Subtraction of the equation of motion (1.3.4) multiplied by $\mathbf{v}$ from (1.4.2) yields the relation for the rate of change of the internal energy

$$\rho \frac{DU}{Dt} = \text{tr}(\sigma \nabla \mathbf{v}) + \rho r - \text{div} \, \mathbf{q}, \tag{1.4.3}$$

where tr denotes the trace operator. In the right-hand side (r.h.s.) of (1.4.3), the first term is the rate of mechanical energy dissipation.

1.5 Thermodynamics and Constitutive Equations

The irreversibility of thermodynamic processes implies that *the time rate of change of the entropy inside the material region $\Omega(t)$ is always bounded from below by the heat supply*. Designating by S the entropy per unit mass and by T the absolute temperature, the second law of thermodynamics postulates

$$\frac{d}{dt}\int_{\Omega(t)} \rho S\, dV \geq \int_{\Omega(t)} \rho \frac{r}{T}\, dV - \int_{\partial\Omega(t)} \frac{\mathbf{q}\cdot\mathbf{n}}{T}\, d\Omega. \tag{1.5.1}$$

Again, the application of the Reynolds transport theorem, the mass conservation principle, and the divergence theorem to (1.5.1) leads to the local form

$$\rho \frac{DS}{Dt} \geq \rho \frac{r}{T} - \operatorname{div}\left(\frac{\mathbf{q}}{T}\right). \tag{1.5.2}$$

Combining (1.4.3) and (1.5.2) to eliminate r, one obtains the Clausius–Duhem inequality

$$\rho \frac{DS}{Dt} \geq \frac{1}{T}\left(\rho \frac{DU}{Dt} - \operatorname{tr}(\sigma \nabla \mathbf{v})\right) + \frac{1}{T^2}\mathbf{q}\nabla T. \tag{1.5.3}$$

This inequality, which has to be satisfied by viscous Newtonian fluids, needs the introduction of the constitutive relations for the stress tensor and the heat conduction flux. The constitutive equations relate the stress tensor σ to the pressure p and to the rate-of-deformation tensor $\mathbf{d}$, the symmetric part of the velocity gradient tensor, defined by the expression

$$\mathbf{d} := \frac{1}{2}[\nabla\mathbf{v} + (\nabla\mathbf{v})^T]. \tag{1.5.4}$$

These general material constitutive relations must be independent of the frame of reference used to describe position and time. For a Newtonian fluid, the frame-indifferent stress tensor σ is given by

$$\sigma = (-p + \lambda \operatorname{tr}\mathbf{d})I + 2\mu\mathbf{d}, \tag{1.5.5}$$

where λ is the bulk viscosity and μ is the dynamic shear viscosity. The quantity $\operatorname{tr}\mathbf{d}$ is the first invariant of the tensor $\mathbf{d}$ and is equal to $\operatorname{div}\mathbf{v}$. For an incompressible fluid, $\operatorname{tr}\mathbf{d} = 0$; see (1.2.10).

The heat flux $\mathbf{q}$ in an isotropic heat-conducting medium obeys Fourier's law, which expresses a first-order approximation of $\mathbf{q}$ with respect to the temperature gradient,

$$\mathbf{q} = -k\nabla T, \tag{1.5.6}$$

where k is the coefficient of thermal conductivity. From the Clausius–Duhem inequality (1.5.3), the material coefficients in (1.5.5) and (1.5.6) satisfy the classical inequalities

$$\mu \geq 0, \qquad 3\lambda + 2\mu \geq 0, \qquad k \geq 0. \tag{1.5.7}$$

For an incompressible fluid where $U = U(\mathcal{S}, T)$ and $c = \partial U/\partial T$, the heat capacity c is unique because the difference between heat capacities at constant pressure and constant volume, c_p and c_v, respectively, vanishes [293]. With the help of (1.5.5) and (1.5.6), the energy equation (1.4.3) is now

$$\rho c \frac{DT}{Dt} = \text{div}(k\,\nabla T) + \rho r + \Phi, \tag{1.5.8}$$

where Φ is the dissipation function defined by the relation

$$\Phi = 2\mu\,\mathbf{d} : \mathbf{d}. \tag{1.5.9}$$

The product $\mathbf{d} : \mathbf{d}$ is the dyadic product $d_{ij}d_{ij}$. To obtain this last expression, we used the identities

$$\nabla \mathbf{v} = \mathbf{d} + \mathbf{w}, \tag{1.5.10}$$

with $\mathbf{w}$ the skew-symmetric rate of rotation tensor and

$$\mathbf{d} : \mathbf{w} \equiv 0. \tag{1.5.11}$$

1.6 Fluid Flow Equations and Boundary Conditions

In this section, we present the incompressible Navier–Stokes equations, which will be the major cornerstone of our mathematical model. To include thermal effects, we describe the Boussinesq approximation where the fluid is still considered as incompressible. The section ends with the initial and boundary conditions in various cases.

1.6.1 Isothermal Incompressible Flow

A Newtonian incompressible fluid has the constitutive relation

$$\sigma = -p\mathbf{I} + 2\mu\mathbf{d}. \tag{1.6.1}$$

Substituting this equation into the momentum equation (1.3.4), one obtains

$$\rho \frac{D\mathbf{v}}{Dt} = -\nabla p + \nabla \cdot \{\mu[\nabla \mathbf{v} + (\nabla \mathbf{v})^T]\} + \rho\mathbf{f}. \tag{1.6.2}$$

1.6. Fluid Flow Equations and Boundary Conditions

If the viscosity μ is constant in (1.6.2), one establishes the Navier–Stokes equations for incompressible fluids,

$$\rho \frac{D\mathbf{v}}{Dt} = -\nabla p + \mu \Delta \mathbf{v} + \rho \mathbf{f}, \qquad (1.6.3)$$

which cannot be separated from the divergence-free condition

$$\text{div } \mathbf{v} = 0. \qquad (1.6.4)$$

The symbol Δ represents the Laplacian operator ($\Delta = \partial^2/\partial x_j \partial x_j$). This set of nonlinear equations is to be solved subject to boundary and initial conditions elaborated on later in this section.

1.6.2 Thermal Convection: The Boussinesq Approximation

If the thermal effects are such that the incompressibility assumption is no longer valid, one must resort to the general Navier–Stokes equations for a compressible fluid. However, in some nonisothermal problems such as materials processing (glass flow, polymer or plastics processing), the fluid can still be considered as incompressible. The coupling between fluid dynamics and temperature evolution is made through the Boussinesq approximation. In this approximation, the volumetric mass is taken as a constant everywhere in the momentum and energy equations except in the body-force term. There, the volumetric mass varies according to the following equation of state:

$$\rho = \rho_0[1 - \alpha(T - T_0)], \qquad (1.6.5)$$

where α is the volume expansion coefficient and $\rho_0 = \rho(T_0)$ with T_0 a reference temperature. For natural convection, the body-force term $\mathbf{f}$ is equal to the gravity acceleration $\mathbf{g}$. The so-called Boussinesq equations are

$$\text{div } \mathbf{v} = 0, \qquad (1.6.6)$$

$$\rho_0 \frac{D\mathbf{v}}{Dt} = -\nabla p + \mu \Delta \mathbf{v} + \rho_0 \mathbf{g}[1 - \alpha(T - T_0)], \qquad (1.6.7)$$

$$\rho_0 c \frac{DT}{Dt} = \text{div}(k \nabla T) + \rho_0 r + \Phi. \qquad (1.6.8)$$

The last term in the r.h.s. of (1.6.7) is the buoyancy force that drives the motion by the temperature differences in the physical domain of the flow. In most applications, the volume source term $\rho_0 r$ in (1.6.8) does not appear, and we will discard it in the sequel.

1.6.3 Boundary and Initial Conditions

The transient Navier–Stokes equations (1.6.3) are of mixed type. In the classical sense, they are parabolic because of the ellipticity of the viscous terms. However, for high-Reynolds-number flows, the nonlinear part of the material time derivative dominates, emphasizing the hyperbolic character of the operator. Close to the walls, the viscosity effects remain important, and hyperbolic and elliptic operators are of the same order of magnitude.

The continuum mechanics hypothesis implies that a viscous fluid sticks to the wall. Therefore, one imposes the no-slip wall condition

$$\mathbf{v} = \mathbf{v}_w, \qquad (1.6.9)$$

$\mathbf{v}_w$ being the prescribed wall velocity. If the Reynolds number is high enough and if one assumes that the viscous influence may be neglected near the boundaries, the free-slip condition requires

$$\mathbf{v} \cdot \mathbf{n} = \mathbf{v}_w \cdot \mathbf{n}. \qquad (1.6.10)$$

An interesting class of fluid flow problems is concerned with free-surface flows, where a viscous fluid is directly in contact with a relatively inviscid fluid such as air. Some industrial examples are coating flows, extrusion, and crystal growth. These problems are highly nonlinear, because the shape of the free surface, where special conditions are to be applied, is also part of the solution itself. Since the free surface is in mechanical equilibrium, we have, from (1.3.2),

$$\mathbf{t}_{\text{fluid}} + \mathbf{t}_{\text{gas}} = 0. \qquad (1.6.11)$$

The projection of this relation onto the normal (n) with respect to the viscous fluid and tangent (τ) unit vectors of the surface produces the free-surface conditions ($\mathbf{n}_{\text{gas}} = -\mathbf{n}$)

$$\mathbf{t}_{\text{fluid}} \cdot \mathbf{n} = \mathbf{t}_{\text{gas}} \cdot \mathbf{n} = -p_{\text{gas}}, \qquad (1.6.12)$$

$$\mathbf{t}_{\text{fluid}} \cdot \boldsymbol{\tau} = 0. \qquad (1.6.13)$$

If the surface tension γ is taken into account, (1.6.12) and (1.6.13) are modified to be

$$\mathbf{t}_{\text{fluid}} \cdot \mathbf{n} = -p_{\text{gas}} + \gamma \langle R^{-1} \rangle, \qquad (1.6.14)$$

$$\mathbf{t}_{\text{fluid}} \cdot \boldsymbol{\tau} = \boldsymbol{\tau} \cdot \nabla \gamma, \qquad (1.6.15)$$

where $\langle R^{-1} \rangle$ represents the mean curvature of the surface. The r.h.s. of (1.6.15) models the surface tension gradient, known as the Marangoni effect, induced by temperature variations at the free surface [352].

1.7. Dimensional Analysis and Reduced Equations

For the temperature equation (1.6.8), the general boundary condition is

$$\mathbf{q} \cdot \mathbf{n} = \mathbf{q}_0 \cdot \mathbf{n} + h(T - T_e). \tag{1.6.16}$$

In this last equation, $\mathbf{q}_0$ is a constant prescribed heat flux (for example, the heat loss across the wall), h is a heat transfer coefficient, and T_e is an exterior reference temperature. The relation (1.6.16) may be rewritten as a Robin condition

$$aT + b\nabla T \cdot \mathbf{n} = d, \tag{1.6.17}$$

with a, b, and d given constants.

The initial condition requires that the velocity field satisfy

$$\mathbf{v}(\mathbf{x}, t=0) = \mathbf{v}^0(\mathbf{x}), \tag{1.6.18}$$

with

$$\operatorname{div} \mathbf{v}^0 = 0, \tag{1.6.19}$$

and the initial boundary conditions. The superscript indicates the time level.

1.7 Dimensional Analysis and Reduced Equations

From the physical or engineering point of view, it is always rewarding to write down the dimensionless set of equations. The immediate payoff is insight into the dominant phenomena and, consequently, the simplifications that the full model can endure to embody them.

The cornerstone of dimensional analysis, the celebrated Π theorem (see, for example, Panton [293]), indicates by linear algebra arguments that the process of nondimensionalization is not unique. We illustrate this point with the isothermal incompressible Navier–Stokes equations (1.6.3). Let us choose the dimensionless variables, denoted by primes, with the following basic scales: a reference length L for space, an inertial time L/U, and a reference velocity U. The pressure is scaled by the dynamic pressure ρU^2, and the body force by U^2/L. The variables are now $\mathbf{x}' = \mathbf{x}/L$, $t' = Ut/L$, $\mathbf{v}' = \mathbf{v}/U$, $p' = (p - p_0)/(\rho U^2)$, and $\mathbf{f}' = \mathbf{f} L/U^2$. Dropping the primes for ease of notation, we obtain

$$\frac{D\mathbf{v}}{Dt} = -\nabla p + (Re)^{-1} \Delta \mathbf{v} + \mathbf{f}. \tag{1.7.1}$$

The Reynolds number Re by definition yields the ratio of inertia to viscous forces,

$$Re := \frac{\rho UL}{\mu} = \frac{UL}{\nu}, \tag{1.7.2}$$

12   1. Fluid Mechanics and Computation: An Introduction

where ν is the kinematic viscosity. We note that if $\mathbf{f}$ is equal to the gravity acceleration $\mathbf{g}$, then the amplitude $\|\mathbf{f}'\|$ is equal to the reciprocal of the Froude number Fr:

$$Fr := \frac{U^2}{L \, \|\mathbf{g}\|}. \quad (1.7.3)$$

For advection-dominated flows characterized by high Reynolds numbers, the Navier–Stokes equations become a singular perturbation problem. The boundary layers develop and drive the dynamics. Special analytical methods have been designed to treat those problems (Bender and Orszag [30]). If we ignore the influence of $\mathbf{f}$ in (1.7.1), the limit form of the Navier–Stokes equations for $Re \to \infty$ is the Euler equation

$$\frac{D\mathbf{v}}{Dt} = -\nabla p. \quad (1.7.4)$$

If we look for the other limit (i.e., $Re \to 0$), we observe that (1.7.1) fails to produce the Stokes equations. The reasons for this failure are associated with the choice of the time and pressure scales that emphasize the inertial terms. To obtain the right model for a creeping flow, we must rely essentially on viscosity. Defining $t' := \nu t/L^2$ and $p' := (p - p_0)/(\mu U/L)$, and dropping the primes as before, we have for the dimensionless Navier–Stokes equations

$$\frac{\partial \mathbf{v}}{\partial t} + Re \, (\mathbf{v} \cdot \nabla \mathbf{v}) = -\nabla p + \Delta \mathbf{v} + Re \, \mathbf{f}. \quad (1.7.5)$$

For $Re \to 0$, we recover the Stokes equations

$$\frac{\partial \mathbf{v}}{\partial t} = -\nabla p + \Delta \mathbf{v}. \quad (1.7.6)$$

The Euler equations (1.7.4) are nonlinear and hyperbolic, with an elliptic term coming through the pressure gradient, while the Stokes relations (1.7.6) are linear and parabolic. A typical example of a fluid flow modeled by the Stokes equations is the flow of oil, whereas the Euler equations are used to study the water flow in hydraulic machines (pumps, turbines).

We now proceed with the Boussinesq equations (1.6.6)–(1.6.8). Defining again $\mathbf{x}' := \mathbf{x}/L$, $\mathbf{v}' := \mathbf{v}/U$, $t' := tU/L$, we introduce the quantities $p' := (p - \rho_0 \mathbf{g} \cdot \mathbf{x})/(\rho_0 U^2)$ and $T' := (T - T_0)/\delta T$. Here δT is the difference between two reference temperatures ($\delta T = T_1 - T_0$, say) and gives an order-of-magnitude estimate of the thermal gradient. We notice that the reduced pressure is modified by the presence of the hydrostatic pressure. Leaving out the primes,

1.7. Dimensional Analysis and Reduced Equations

we obtain

$$\text{div } \mathbf{v} = 0, \tag{1.7.7}$$

$$\frac{D\mathbf{v}}{Dt} = -\nabla p + (Re)^{-1} \Delta \mathbf{v} - Ri \frac{\mathbf{g}}{\|\mathbf{g}\|} T, \tag{1.7.8}$$

$$\frac{DT}{Dt} = (Pe)^{-1}(\Delta T + Br\, \Phi). \tag{1.7.9}$$

The dimensionless numbers in (1.7.8) and (1.7.9) are the Reynolds number, the Richardson number

$$Ri := \alpha g\, \delta T L / U^2, \tag{1.7.10}$$

and the Péclet number

$$Pe := UL/\kappa, \tag{1.7.11}$$

where $\kappa := k/\rho_0 c$ is the thermal diffusivity and $g = \|\mathbf{g}\|$. We observe that Pe is the analog of Re for the temperature equation, comparing convection with thermal diffusion (conduction), while Ri compares the buoyant force with the kinetic energy.

In the case of natural convection, the previous dimensional setting is a source of difficulty because it is not obvious how to choose the reference velocity U. The introduction of new nondimensional numbers will overcome this problem. The Brinkman number Br is defined by the equation

$$Br := \frac{\mu U^2}{k\, \delta T} \tag{1.7.12}$$

and compares the heat produced by viscous dissipation with thermal conduction. It usually has very low values (typically 10^{-3}), with some noticeable exceptions such as natural convection in the earth's mantle. The Prandtl number

$$Pr := \nu/\kappa = Pe/Re \tag{1.7.13}$$

is a material characteristic. Its value ranges from 10^{-2}–10^{-3} for a liquid metal to a few hundred for molten glass; it is approximately 1 for air. The Grashof number Gr represents the ratio of buoyant forces to viscosity effects:

$$Gr := \alpha g\, L^3\, \delta T / \nu^2. \tag{1.7.14}$$

The Rayleigh number

$$Ra := \alpha g\, L^3\, \delta T / \nu \kappa \tag{1.7.15}$$

is most used in transition problems because it compares the destabilizing buoyant term to stabilizing viscous and conductive mechanisms. With Equations (1.7.10) and (1.7.13)–(1.7.15), we have

$$Ra = Gr\, Pr, \qquad Ri = Gr/Re^2. \qquad (1.7.16)$$

If we select the reference velocity with respect to the dominant physical phenomenon, we are left with particular definitions of Re, Pe, and Ri and specific forms of the dimensionless equations (1.7.8) and (1.7.9). We obtain for a characteristic velocity based on the buoyancy term (Gray and Giorgini [165])

$$U_\alpha = (\alpha g\, \delta T\, L)^{1/2}, \qquad Re = \left(\frac{Ra}{Pr}\right)^{1/2}, \qquad Pe = (Ra\, Pr)^{1/2}, \qquad Ri = 1; \qquad (1.7.17)$$

for a viscous characteristic velocity (Chorin [78])

$$U_\nu = \nu/L, \qquad Re = 1, \qquad Pe = Pr, \qquad Ri = Gr; \qquad (1.7.18)$$

and for a thermal-diffusivity-based velocity (de Vahl Davis [93])

$$U_\kappa = \kappa/L, \qquad Re = (Pr)^{-1}, \qquad Pe = 1, \qquad Ri = Ra\, Pr. \qquad (1.7.19)$$

From the dimensional analysis, we have some guidelines to help answer the question: Which model should we use to investigate, say, the float glass process? This process involves use of two furnaces. The first one melts the raw material, which is basically sand and a few additional chemical products to help the fusion. The second furnace is aligned with respect to the previous one in such a way that the molten glass floats over a bath of molten tin. As the two fluids are immiscible, glass remains on top of the tin, cools down, and forms a plate. In the molten-glass furnace, where viscosity constitutes the main physical effect, Equation (1.7.18) determines the flow dynamics with respect to the viscous time scale ν/L^2, which is of the order of a few hours and is in good agreement with experimental data. Clearly, the nonlinear advection comes from the temperature that is transported by the glass, an insulating material (at room temperature). On the contrary, in the tin bath ($Pr \approx 10^{-2}$), heat conduction is of primary concern. The choice of (1.7.19) yields the conduction time scale κ/L^2. The nonlinearity is now on the hydrodynamics side, where boundary-layer dynamics will govern the flow.

1.8 Vorticity Equation

The velocity–pressure (primitive variables) formulation in the isothermal incompressible case (1.6.3)–(1.6.4) is the most general one, especially in the presence of free surfaces. One major drawback comes from the pressure calculation, because the pressure is not provided by a prognostic equation as the velocity is. The vorticity–velocity formulation is another possible choice to circumvent this difficulty. Taking the curl of (1.6.3) and defining the vorticity $\omega = \nabla \times \mathbf{v}$, we can describe the vorticity dynamics by the relation

$$\frac{D\omega}{Dt} = \omega \cdot \nabla \mathbf{v} + \nu \Delta \omega, \qquad (1.8.1)$$

while the velocity field may be obtained from the Poisson equation

$$\Delta \mathbf{v} = -\nabla \times \omega. \qquad (1.8.2)$$

The first term on the right-hand side of (1.8.1) has no counterpart in the classical Navier–Stokes equations. It is a three-dimensional effect connected with the extension or shortening of vortex filaments or with their angular deformation induced by the local flow [23]. We recall that vorticity is produced at the wall by viscosity and pressure-gradient effects, diffused away from the wall, and then advected in the interior of the flow. To integrate (1.8.1)–(1.8.2), we apply the kinematic boundary conditions on solid walls to the velocity $\mathbf{v}$ and (through the vorticity definition) to ω. Two-dimensional calculations where the term $\omega \cdot \nabla \mathbf{v}$ is canceled in (1.8.1) were performed recently by Daube [90]. Three-dimensional solutions were obtained by Gatski et al. [149].

Another way to look for solutions of (1.8.1) involves the use of the vector (3D) potential Ψ or the (2D) scalar stream function ψ. The latter technique has been widely used since the infancy of CFD because it avoids the pressure computation. The 2D dimensionless form of the equations (1.8.1)–(1.8.2) is

$$\frac{D\omega}{Dt} = \frac{1}{Re}\Delta\omega, \qquad (1.8.3)$$

$$\Delta\psi = -\omega. \qquad (1.8.4)$$

The major difficulty of this formulation is in determining the appropriate vorticity boundary conditions. The traditional approach (in finite differences, for example) couples ω_{BC}, where the subscript BC refers to boundary conditions, with adjacent ψ-values at points close to the walls (Roache [328]). A weak formulation in a finite-element treatment was proposed by Campion-Renson and Crochet [61], where the normal derivative of ψ is the natural boundary condition for the ψ Poisson equation, while the normal derivative of ω is the natural condition for the vorticity dynamics. In summary, because of the coupling between ψ and ω, the modern view of imposing boundary conditions for the two-dimensional vorticity–stream-function formulation involves the stream function itself and its normal derivative $\partial \psi / \partial n$.

The reader is referred to [166] for a complete review of the preceding topics.

1.9 Simplified Models

For the development and testing of numerical methods, it is often useful to consider subsets of the full Navier–Stokes equations. For example, if we neglect the pressure contribution and integrate (1.7.1) with an implicit Euler time treatment of the viscous term and an explicit scheme for the nonlinear term that will be transferred to the right-hand side, we obtain a Helmholtz equation, which reads in its scalar version (with u replacing $\mathbf{v}$)

$$\lambda u - \Delta u = f, \tag{1.9.1}$$

with $\lambda = (\Delta t)^{-1} Re$. If the Helmholtz parameter λ vanishes, we get a Poisson equation of the same type as (1.8.4):

$$-\Delta u = f. \tag{1.9.2}$$

Finally, for a zero source in (1.9.2), we obtain the Laplace equation

$$\Delta u = 0. \tag{1.9.3}$$

These last three relations are elliptic and do not involve time explicitly. The associated physics is stationary and linear, and a great deal of mathematical literature supplies existence and uniqueness proofs together with appropriate solution methods.

If we neglect the pressure contribution in (1.7.6), the one-dimensional version is the parabolic relation

$$\frac{\partial u}{\partial t} = \Delta u, \tag{1.9.4}$$

which is again a linear mixed initial–boundary-value problem.

The linearized Euler equation (1.7.4) with the pressure contribution discarded yields a one-dimensional hyperbolic wave equation of the form

$$\frac{\partial u}{\partial t} + c \frac{\partial u}{\partial x} = 0, \quad c > 0. \tag{1.9.5}$$

If we now incorporate viscous effects into (1.9.5), we get a linear advection–diffusion equation

$$\frac{\partial u}{\partial t} + c\frac{\partial u}{\partial x} = \nu\frac{\partial^2 u}{\partial x^2}. \qquad (1.9.6)$$

Finally, the one-dimensional counterpart of the Navier–Stokes equation with no pressure contribution consists of the nonlinear Burgers equation

$$\frac{\partial u}{\partial t} + u\frac{\partial u}{\partial x} = \nu\frac{\partial^2 u}{\partial x^2}, \qquad (1.9.7)$$

the solution of which may be obtained in closed form.

All the simplified equations (1.9.3)–(1.9.7) represent part of the analytical and numerical difficulties that must be considered in the design of Navier–Stokes solvers. In the first chapters of this book, we will review the basic numerical techniques to handle these reduced and more tractable problems efficiently. Once those building blocks have been well established, we will turn to the development of techniques directed to the difficulties encountered in numerical fluid dynamics per se.

1.10 Turbulence and Challenges

Above a certain critical Reynolds number, flows enter a transition stage and follow a route to chaos, weak turbulence, and, eventually, developed turbulence. Turbulent flows are unsteady, rotational, dissipative, and highly random.

The critical Reynolds number typically ranges from a few hundred to a few thousand. It depends very much on the geometry (internal or external flow, complex shapes, etc.) and the surface conditions (smooth or rough). Turbulent flows at $Re \approx 10^4$ are within the realm of feasibility of present direct numerical simulations. However, they involve discretizations of the order of several million points and require approximately one thousand hours on a state-of-the-art computer to perform a single calculation. Clearly, direct simulations for $Re \gg 10^4$ are beyond the reach of most present supercomputers.

As an example, we consider the flow of a river in its estuary ($Re \approx 10^6$). The space scales are in the range from 100 m for the large vortex structures to 10^{-2} m for the small eddies where viscosity dissipates the mechanical energy into heat. Using a uniform mesh (a bad idea in this case!) with $\Delta x = 10^{-2}$ m, we are led to 10^4 grid points in each spatial direction in order to resolve all length scales. The most pertinent dynamical time scale is of the order of one minute for large eddies. If an explicit time-marching algorithm is designed, the Courant–Friedrichs–Lewy (CFL) stability condition coming from the dynamics of the small-scale events imposes a time step of the order of the mesh size, typically 10^{-2} s. Consequently, $60/10^{-2} \approx 10^4$ steps will be needed to simulate one revolution of a typical large vortex. This represents a total of 4×10^{16} unknowns (three velocities and one pressure per grid point and per time step). If we assume that the program coding involves 100 floating-point operations per unknown, we are left with the prospect of 4×10^{18} operations. If we assume a computational speed of 2 Gflops (gigaflops ≡ one billion floating-point operations per second), we have to execute this program for about ten years, a decision that is not going to elate our sponsors, not to mention the skyrocketing amount of the final invoice.

The solution to this difficult challenge is twofold: (1) turbulence modeling and large-eddy simulation, a software improvement by methodological means; (2) computer architecture (i.e., hardware) redesign. We defer the examination of the second solution until the next section and concentrate here on the exploration of software and modeling issues.

Numericists tackle the computation of viscous turbulent flows from both ends of the Reynolds-number spectrum. On the one hand, for moderate to high Reynolds numbers, LES integrates unsteady filtered Navier–Stokes equations (see, e.g., Ferziger [122]) in conjunction with a subgrid scale model based on the hypothesis that unresolved small-scale motions have a universal homogeneous behavior. Mean statistical quantities are then obtained from the database results generated by LES. On the other hand, for very high Reynolds numbers aimed at developed turbulence, one addresses the Reynolds-averaged Navier–Stokes equations *closed* by a zero-, one-, or two-equation model (Tennekes and Lumley [377] or Reynolds [324]).

The basic philosophy behind the RANS approach consists in decomposing all quantities in terms of average or mean variables and fluctuations. The velocity field is the sum of two quantities

$$\mathbf{v} = \bar{\mathbf{v}} + \mathbf{v}', \tag{1.10.1}$$

where $\bar{\mathbf{v}}$ is the mean velocity and $\mathbf{v}'$ the fluctuation around that mean. The mean velocity field is usually defined as the time average over the interval $\mathcal{T}$:

$$\bar{\mathbf{v}} = \frac{1}{\mathcal{T}} \int_{t-\frac{\mathcal{T}}{2}}^{t+\frac{\mathcal{T}}{2}} \mathbf{v}(\tau) \, d\tau. \tag{1.10.2}$$

Applying the averaging procedure to (1.2.10) and (1.3.4), one obtains the Reynolds-averaged Navier–Stokes equations

$$\text{div } \bar{\mathbf{v}} = 0, \tag{1.10.3}$$

$$\rho \frac{\bar{D}\bar{\mathbf{v}}}{Dt} = \text{div}(\bar{\sigma} + \mathbf{R}) + \rho \bar{\mathbf{f}}, \tag{1.10.4}$$

1.10. Turbulence and Challenges

where $\bar{\sigma}$ denotes the Newtonian mean stress tensor and $\bar{\mathbf{f}}$ the mean body force. The components of the symmetric Reynolds stress tensor given by

$$\mathbf{R} = -\rho \overline{\mathbf{v}' \otimes \mathbf{v}'} \qquad (1.10.5)$$

provide the turbulence contribution to the flow. The mean material time derivative involves the mean velocity field in the transport term:

$$\frac{\bar{D}}{Dt} = \frac{\partial}{\partial t} + \bar{\mathbf{v}} \cdot \nabla. \qquad (1.10.6)$$

A review by Launder [236] of Reynolds stress closure emphasizes the fact that, in addition to Equations (1.10.3)–(1.10.4), six additional dynamical equations for $\mathbf{R}$ should be integrated. These equations involve new correlations that are difficult to model (especially those including pressure–rate-of-deformation tensor correlations) and to treat numerically, because these advection–diffusion-type relations have stiff source or sink terms. Let us consider the classical turbulence modeling, the so-called k–ϵ model, which handles only two additional quantities: the kinetic turbulent energy k defined by the relationship

$$k = \frac{1}{2}\overline{v_i' v_i'}, \qquad (1.10.7)$$

and the rate of energy dissipation,

$$\epsilon = \nu \overline{\frac{\partial v_i'}{\partial x_j}\frac{\partial v_i'}{\partial x_j}}. \qquad (1.10.8)$$

Following the line of reasoning of continuum mechanics, we can link the Reynolds stress tensor to the mean velocity field by the constitutive relation

$$\mathbf{R} = -\frac{2}{3}\rho k \mathbf{I} + 2\mu_T \bar{\mathbf{d}}, \qquad (1.10.9)$$

where μ_T is the turbulent viscosity introduced by Boussinesq in 1877 [45]. We notice that this approach "by analogy" is not completely justified, however, since turbulence is not a material characteristic but instead a flow property.

The zero-equation model, known also as the mixing-length theory (see Schlichting [344]), defines μ_T as

$$\mu_T = \rho l^2 (\mathrm{II}_{\bar{d}})^{1/2}, \qquad (1.10.10)$$

$$l = \kappa n^+, \qquad (1.10.11)$$

where $\mathrm{II}_{\bar{d}}$ is the second invariant of $\bar{\mathbf{d}}$, κ the Karman parameter ($\kappa = 0.4$) and n^+ the normal wall distance

$$n^+ = n v_*/\nu, \qquad (1.10.12)$$

with v_* the friction velocity obtained from the mean wall shear stress. The mixing-length concept applied to the turbulent boundary-layer profile in a plane channel flow yields the universal logarithmic law

$$\frac{\bar{u}}{v_*} = \frac{1}{\kappa}(\ln n^+ - \ln \beta), \tag{1.10.13}$$

where β is a constant evaluated from experimental data and is equal to 0.1.

The one-equation k–l model gets a turbulent velocity scale from integration of the k equation

$$\rho \frac{\bar{D}k}{Dt} = \text{div}\left[\left(\mu + \frac{\mu_T}{\sigma_k}\right)\nabla k\right] + 2\mu_T \, \bar{\mathbf{d}} : \bar{\mathbf{d}} - \rho \frac{k^{3/2}}{l_\epsilon}, \tag{1.10.14}$$

$$\mu_T = \rho l \sqrt{k}, \tag{1.10.15}$$

$$l_\epsilon = l/C_D. \tag{1.10.16}$$

In these equations, the empirical constants are set to $\sigma_k = 1$ and $C_D = 0.164$.

Finally, the k–ϵ two-equation model implies a dynamical equation for the rate of energy dissipation. The equations are

$$\rho \frac{\bar{D}k}{Dt} = \text{div}\left[\left(\mu + \frac{\mu_T}{\sigma_k}\right)\nabla k\right] + 2\mu_T \, \bar{\mathbf{d}} : \bar{\mathbf{d}} - \rho \frac{k^{3/2}}{l_\epsilon}, \tag{1.10.17}$$

$$\rho \frac{\bar{D}\epsilon}{Dt} = \text{div}\left[\left(\mu + \frac{\mu_T}{\sigma_\epsilon}\right)\nabla \epsilon\right] + 2\mu_T \, \bar{\mathbf{d}} : \bar{\mathbf{d}} \, C_{\epsilon_1} \frac{\epsilon}{k} - \rho C_{\epsilon_2} \frac{\epsilon^2}{k}, \tag{1.10.18}$$

$$\mu_T = \rho C_\mu k^2 / \epsilon. \tag{1.10.19}$$

The numerical values of the empirical constants are chosen as

$$C_\mu = 0.09, \quad \sigma_\epsilon = 1.3, \quad C_{\epsilon_1} = 1.35, \quad C_{\epsilon_2} = 1.8. \tag{1.10.20}$$

The linear model just described is well known to be defective in 3D computations. For example, it cannot generate secondary flows in the cross section of a square duct. A remedy is brought by the nonlinear k–ϵ model of Speziale [364, 365], which takes more turbulence memory effects into account by using a nonlinear constitutive equation. The generalization of (1.10.9) to quadratic terms in the mean velocity gradients includes the presence of the upper-convected derivative of $\bar{\mathbf{d}}$ as in the Oldroyd-B model in non-Newtonian fluids [216].

From the numerical point of view, it is fairly easy in a finite-element method to construct a mixed model as a combination of two classical models. Close to the solid walls, where shear flows dominate, the k–l model is used, and further

1.10. Turbulence and Challenges

inside the computational domain, the k–ϵ model produces a full turbulence description. This approach is known as the two-layer model.

Over the past decade, since the pioneering work of Yakhot and Orszag [416], renormalization group (RNG) theory has paved the way to new developments and generated extensive research to justify and complete the existing models. The idea is to perform a recursive elimination of infinitesimal strips of small scales (high frequencies) until the iteration converges to a fixed point at which the model does not change anymore. This procedure brings up a new term (not shown here) in Equation (1.10.18), which is definitely an emergence of the RNG.

With respect to LES, the goal is to deal with high-Reynolds-number unsteady flows by using coarse meshes in which the contributions of small, *subgrid-scale* modes are not resolved but are modeled. Large-scale quantities, designated by a hat, are obtained by a filtering process. The application of the filter to the function f yields the filtered quantity

$$\hat{f}(\mathbf{x}, t) = \int_\Omega f(\boldsymbol{\xi}, t) \, G(\mathbf{x}, \boldsymbol{\xi}, t) \, d\boldsymbol{\xi}, \qquad (1.10.21)$$

where G is a filter function such as a Gaussian, a sharp Fourier cutoff, or the automatic low-pass filter provided by a standard second-order-accurate finite-difference or finite-volume technique. Assuming the filter commutes with differentiation and applying the filter (1.10.21) to the Navier–Stokes equations (1.6.3) and (1.6.4), one obtains the following relations:

$$\operatorname{div} \hat{\mathbf{v}} = 0, \qquad (1.10.22)$$

$$\rho \frac{\hat{D}\hat{\mathbf{v}}}{Dt} = -\nabla \hat{p} + \mu \Delta \hat{\mathbf{v}} + \rho \operatorname{div} \hat{\boldsymbol{\tau}}. \qquad (1.10.23)$$

In (1.10.23), the material time derivative $\hat{D}\hat{\mathbf{v}}/Dt$ is defined in an analogous fashion to (1.10.6), with the filtered velocity replacing the mean velocity. The subgrid-scale (SGS) stress tensor $\hat{\boldsymbol{\tau}}$ picks up the small-scale effects and is given by

$$\hat{\tau}_{ij} = \widehat{v_i v_j} - \widehat{\hat{v}_i \hat{v}_j}. \qquad (1.10.24)$$

One of the most popular SGS models, due to Smagorinski [362], uses a turbulent viscosity. It is given by the relation

$$\hat{\boldsymbol{\tau}} = -\frac{1}{3} \operatorname{tr}(\hat{\boldsymbol{\tau}}) \, \mathbf{I} + 2\mu_{\text{LES}} \hat{\mathbf{d}}, \qquad (1.10.25)$$

where $\hat{\mathbf{d}}$ is the filtered rate of deformation tensor and

$$\mu_{\text{LES}} = \rho \, l_S^2 \, (\mathrm{II}_{\hat{\mathbf{d}}})^{1/2}, \qquad (1.10.26)$$

in which l_S is a characteristic length proportional to the filter width. More technical details on modeling issues and related numerics, boundary conditions, and engineering applications may be found in review papers (see, e.g., Piomelli [312], Reynolds [325], or Rogallo and Moin [330]) and in books (see, e.g., Lesieur [243] and Sagaut [341]).

1.11 Numerical Simulation

In this closing section, we leave for a while the mathematical models and evaluate the present status of CFD. Computational fluid dynamics is at the crossroads of mathematics, numerical discretization methods, and computer science techniques. We will discuss the interplay between hardware and software issues and suggest some prospects for future development, although astrologers are keener on this topic than ordinary scientists.

1.11.1 Hardware Issues

The landscape of computer hardware has evolved quite rapidly in the past decade. Although major supercomputing centers were equipped with vector mainframes (Cray, NEC, IBM, etc.), the workstation world experienced a technological revolution with the advent of RISC chips providing more computing power at a lower price. This technological achievement has had an impressive effect on parallel computer architectures, where several hundreds or thousands of interconnected processors are assembled.

The interconnection can be realized in a nonuniform memory access (NUMA) machine, in which the distributed memory is considered to be shared. Thus, the memory allocation in an application is handled by the underlying system; the user needs only to take care of the data exchange between the tasks running on the different processors. The data exchange is treated by threads (a shared-memory communication library) or even by automatic parallelization at the compiler level (e.g., in Fortran 90). Unfortunately, the scalability of these approaches is often limited by the fine granularity of the tasks; hence, NUMA machines have had no more than 128 processors, and scalability levels off before full use of even those processors.

Another architecture is clusters of distributed-memory computational boxes, each box of which can be a shared-memory multiprocessor (SMP) machine. Such clusters require that the user handle the memory distribution and, thus, dictate a message-passing library (MPI [171] has become a standard) that enables passing messages over the communication network. Cluster architecture include NUMA machines that, in most cases, show better performance when

1.11. Numerical Simulation

the messages are exchanged via MPI rather than by threading or automatic parallelization by the compiler. With the cluster concept the user can implement algorithms with minimal communication loads that scale up to thousands of processors. The efficiency of a single computational box depends on the number of shared-memory processors per box, on the cycle period of the processor, on the number of simultaneously active computational pipelines, on the memory access subsystem, and on the size and speed of the caches. These are discussed in detail in Chapter 8.

Price and adequacy of the hardware play an increasing role in the choice of parallel computer architecture. A few years ago, vector machines dominated the market, but they are now considered to be too expensive. Engineers develop and optimize their programs on their own PCs or workstations, and they are increasingly reluctant to adapt these programs to vector architectures. They are more concerned with the decision either to program for NUMA machines, which offer programming ease albeit with reduced flexibility and scalability, or to go for a flexible, scalable, and portable distributed-memory concept at the expense of an increased programming effort. The books by Hockney and Jesshope [209] and Dowd and Severance [110] constitute an excellent introduction to parallel machines.

Computer vendors expected to produce peak teraflops (10^{12} operations per second) capability by the mid-nineties, and sustained teraflops by the turn of the millennium. Indeed, this goal has been achieved. However, the market expectations, especially from industrial research groups, have shifted recently. Today, commodity technology seems to be the way to go and aims at producing installed peak gigaflops computers for costs ten to twenty times less than vector supercomputers. We note that the achievement of sustained teraflops performance will render the solution of Reynolds-averaged Navier–Stokes equations over a complete aircraft affordable. According to Karniadakis and Orszag, however [220], direct numerical simulation of turbulent flow around an aircraft body including the small structures will require an exaflop (10^{18} operations per second) architecture. That we are still far from such a goal underscores the present and future importance of turbulence models in industrial applications.

We could feel desperate about the computational speedup required. However, looking back over the history of computing since the late 1940s, we observe a 10^6–10^7 increase in hardware efficiency. Table 1.11.1, which lists a few landmark machines with the observed sustained computational speed on CFD codes, summarizes the hardware developments over the past five decades. Also worth mentioning is the ASCI (accelerated strategic computing initiative) project that was launched by the U.S. Department of Energy to change the slope in the speedup related to high-performance computing and promote teraflops

1. Fluid Mechanics and Computation: An Introduction

Table 1.11.1. *Hardware history.*

Machine	Year	Processing speed
Mark I	1947	200 flops
IBM 701	1954	3000
IBM Stretch	1960	0.3 Mflops
CDC 6600	1964	2
CDC 7600	1969	5
Cray 1	1976	50
Cray 2	1985	150
Cray YMP	1988	225
Cray C90	1992	500–750
Intel Paragon	1992	30 Mflops/proc
Thinking Machine CM-5	1992	60
Cray T3D	1993	50
Cray T3E	1996	150
NEC SX-5	1998	8 Gflops/proc

machines with thousands of processors. Petaflops (10^{15} flops) machines are expected by the end of the first decade of the millennium.

This spectacular improvement comes from technological breakthroughs. From tubes to transistors, then to very large-scale integrated (VLSI) circuits, the hardware designers went also from serial machines where instructions were performed sequentially to vector computers where pipelining and vector registers allowed array processing, achieving impressive speedups. We also mention the remarkable size reduction in the machine packaging. For example, the Cray-2 processor was contained in a volume of about 30 cm^3. In the mid-eighties, the first parallel machines came in with the prospect of linear speedups with respect to the number of processors. For some time, it seemed that commodity-based parallel and clustered massively parallel processors (MPPs) like the Cray T3D and SMP systems would replace vector machines. However, parallel machines have proven difficult to program. This is due partly to current trends toward a single-system image paradigm, which facilitates program porting and development. Scalable parallel architectures beyond thousands of processors raise fundamental questions in software design, programming languages, operating systems, communications, and synchronization.

1.11.2 Software Issues

Parallel computing would be ideal if a compiler could extract parallelism from existing serial or vector programs. This is unlikely to happen in the near future, however, and some of the software burden will continue to be placed on the shoulders of the developer.

1.11. Numerical Simulation

Most of the parallel codes for multiprocessors on distributed-memory architectures use message-passing libraries such as *parallel virtual machine* (PVM) or the *message-passing interface* (MPI). These enable the use of standard (Fortran or C) languages, with extensions, to explicitly control the parallel execution.

An emerging trend is to program in an entirely new language designed to make parallelism easier. Fortran 90 is a step in that direction in that it incorporates many new intrinsic statements aimed at array manipulations. A version of this language in the Fortran 90 context is HPF (high-performance Fortran); see, for example, Loveman [248].

Because of the complexity of large 3D transient fluid flow simulations in industrial applications, the need to obtain top computational performance from highly parallel algorithms at run time is of primary concern. The use of high-order methods such as preconditioned spectral collocation and spectral and mortar elements calls for sophisticated data structures that are beyond the scope of classical Fortran indexed arrays. Nonlinear data structures are especially needed when adaptive algorithms are designed. In addition, advanced software engineering concepts impose new programming criteria: software maintainability, ease of generalization, error-free design, reusability, and modularity. These criteria are met partially by Fortran 90: Machiels and Deville [252] describe a spectral-element solver designed in Fortran 90 for the solution of an elliptic equation. However, object-oriented concepts are finding increasing favor, in order to provide the user with general and elaborate tools (see Dubois-Pèlerin et al. [116, 115]). They are based essentially on data abstraction: data encapsulation, class inheritance, and polymorphism.

Parallel operating systems are under development for Unix-based platforms, as reported by Almasi and Gottlieb [4]. Some of them are in infancy and need strong testing to deliver a user-friendly environment. It is important that such operating systems embody dynamic scheduling mechanisms to balance the load between processors during the program execution itself.

Up to now, we have considered hardware and software needs general to a CFD practitioner using either high- or low-order techniques. With spectral discretizations, software interfaces between geometric modeling in CAD systems and mesh generation are at present inadequate. Indeed, if the geometry is described by high-order polynomials such as Bézier curves or nonuniform rational B-splines (NURBS) with some specifications on slopes and curvatures, there should be no loss of information going to the spectral representation. Here, however, effective tools are lacking.

Another area where developments are urgently needed is in data postprocessing. On parallel machines achieving tremendous processing speed (gigaflops and beyond), raw data are produced in huge amounts. Storage may be done on optical disks, even using compression algorithms such as wavelets (although, in some sense, spectral data are already compressed). From the viewpoint of visualization, parallel processing of graphics is desirable. Also needed is a way to handle spectral (highly accurate) results. Current graphics packages deal with standard low-order approximations.

1.11.3 Algorithms

While hardware has been progressing at a rapid pace, algorithms have moved along with the same impetus. In incompressible flows, the principal bottleneck for large simulations resides in the solution of the algebraic system of linear equations. As an example, let us suppose we want to solve the 3D Poisson equation (1.9.2) on a cube of unit side with homogeneous Dirichlet boundary conditions, using second-order centered finite differences giving rise to the seven-point discretization molecule. We will employ $N+1$ intervals of equal length in each space direction. If we use the natural ordering of the unknowns, the so-called lexicographic order (i.e., from left to right and from bottom to top, and then plane by plane), the matrix system is indeed symmetric, positive definite, and sparse and presents a bandwidth of N^2, typically the number of nodes in a 2D plane of the mesh.

At the time of the Mark I machine (1947), the straightforward algorithm was Gaussian elimination. According to Golub and Ortega [161], the operation count for the banded LU factorization is $O(N^7)$. Since the small memory space prevented solving systems of sufficient size, the methodology went from direct solver to iterative technique, with the intent of storing only the discrete values in core.

The successive overrelaxation (SOR) method was born in 1954 with suboptimal parameter (Young [419]). The operation count is directly proportional to the number of degrees of freedom per iteration [$O(8N^3)$]. The number of iterations depends strongly on the spectral radius of the iteration matrix. Here, estimates yield $O(N^2)$ iterations and a final operation count of $O(8N^5)$. When Young [420] proposed his optimal evaluation of the SOR parameter in the early 1960s, the operation count dropped to $O(8N^4 \log N)$.

Considering the tensor product form of the matrix system, in 1964 Lynch et al. [250] carried out a one-dimensional diagonalization procedure in each space direction, leading to an improved count of $O(2N^4)$ operations. Later, in 1970, Buzbee et al. [59] took full advantage of the special structure of the matrix to reduce it to tridiagonal form by cyclic reduction. The algorithm has a complexity of $O(8N^3 \log N)$ operations. Eventually, in 1977, Brandt [50] produced the optimal algorithm: the multigrid method that converges in $O(60N^3)$ operations. From that time on, parallel algorithms have been developed, with the multigrid method taking the lead.

1.11. Numerical Simulation

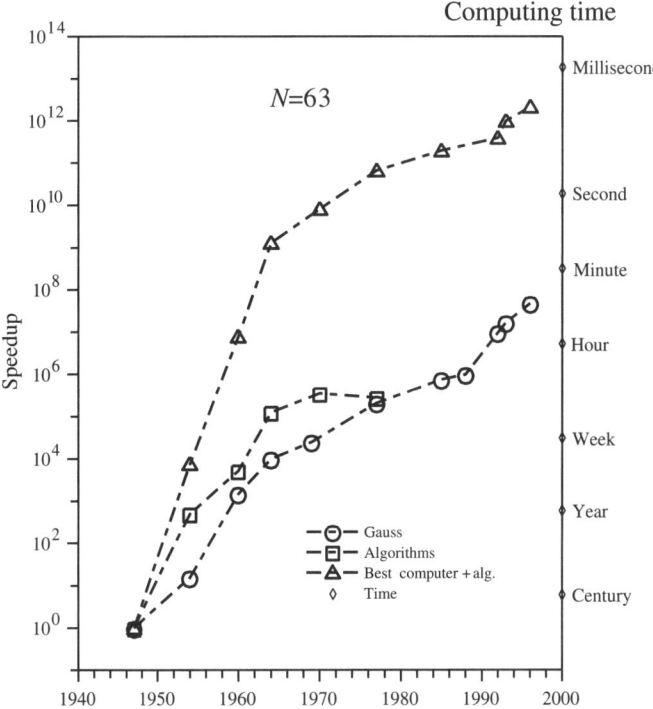

Figure 1.11.1. Evolution of machines, algorithms, and their combination over the past five decades for the solution of a three-dimensional Poisson equation.

The review written by McBrian and coauthors [271] summarizes the state of the art in the early 1990s and points out that classical multigrid performs poorly on MPPs because it is difficult to keep all processors busy when operating on the coarse grids. Nevertheless, McBrian et al. propose a parallel superconvergent multigrid technique where several grids (e.g., coexisting coarse grids with the even points and odd points coming from the fine grid) are solved concurrently and then assembled to give a more accurate fine-grid solution. The operation count is $O(30N^3(\log N)/P)$, where the factor $\log N$ is incurred by communication and P denotes the number of processors.

Let us now look at an instantiation of the previous example with an even number of intervals, namely, $N = 63$, in such a way that the multigrid method may be easily applied. Figure 1.11.1 shows three curves versus time over the past five decades. The curve with the circles corresponds to running Gaussian elimination on the best available computer at the moment. To solve the 63^3 problem on the Mark I of 1947 would have required a few centuries (6.5, to be precise – not to mention the added wait due to system failure), while today we get the answer in a few minutes. For the parallel machines, we have fixed the number of processors P equal to 32. The acceleration factor from hardware improvement is of the order of 10^7. The second curve, with the squares, results from running the best available algorithm on the Mark I machine. From Gaussian elimination to multigrid, the computing time decreases from centuries to twenty hours, a speedup factor of 10^6, showing that algorithmic creativity has not fallen behind that of hardware designers. Finally, if we combine both best machine and best algorithm, we obtain the curve with the triangles, which shows an acceleration factor of 10^{13}, an impressive figure indeed. The triangles after 1977 correspond to the multigrid algorithm run on the best available machine.

Of course, we have to remind ourselves, from time to time, that "the purpose of computing is insight, not numbers," as R. Hamming wrote [184], but this is achieved only if our calculations are correct, that is, are accurate enough. For a smooth exact solution made of the product of sines, for example, with the previous discretization, the error in maximum norm is about 10^{-4}. The question is: How do we improve the accuracy? One approach is to use more grid points by doubling, for example, the number of intervals in each space direction. This approach is not in favor with practitioners, however, because it is time-consuming. Another approach when dealing with structured grids – and this numerical trick has been, unfortunately, practically forgotten – is the extrapolation to the limit, or Richardson extrapolation. Combining the (second-order) solution on a coarse and a fine grid, we can reach fourth-order accuracy. Finally, we may attain the same level of precision by the use of higher-order discretization methods.

1.11.4 Advantages of High-Order Methods

The most obvious choice for solving the 3D Poisson equation (1.9.2) with a fourth-order accurate method is the five-point centered discrete formula for second-order one-dimensional derivatives. However, the fact that the formula is not diagonally dominant poses some difficulties for the convergence of iterative solvers. Furthermore, because of the spatial extension of the discrete formula, it cannot be used for grid points close to the boundaries. One has to degrade the accuracy of the scheme locally by second-order formulae. This strategy was acknowledged in the sixties, and in the mid-seventies fourth-order compact (designed on three consecutive discrete points) schemes were proposed for the Navier–Stokes equations (see Hirsh [204]). A 2D algorithm was proposed by Shaanan and Ferziger in 1978 [354], and Mercier and Deville [274] extended the algorithm to three dimensions in 1981. Using a fast one-dimensional diagonalization method, the cost of the algorithm is of the order of N^4.

1.11. Numerical Simulation

Spectral methods have been used since the late sixties and early seventies, largely due to Gottlieb and Orszag [163]. For the spectral tau method (carried out in spectral space), Haidvogel and Zang [181] applied the diagonalization technique to 2D problems in 1979. The algorithm was extended by Haldenwang et al. [182] in 1984 to 3D problems. The cost is proportional to $O(12N^4)$ operations, occurring mainly in the preprocessing stage of the one-dimensional fast diagonalization methods.

As the mid-eighties proceeded toward computation in real space, collocation schemes and spectral elements began to have a growing influence. Chebyshev collocation involves matrices for second-order operators such as the Poisson equation with a condition number of $O(N^4)$. The idea of preconditioning was proposed by Orszag [288] in 1980. The use of finite elements as preconditioners was independently worked out by Canuto and Quarteroni [69] and Deville and Mund [99] in 1985. This preconditioning technique is robust and efficient. It suffices to state here that any algorithm available for the preceding finite-difference formulation can still be applied. However, the convergence of iterative methods is very sensitive to the condition number, and a deeper analysis must be carried out. We defer this analysis until later in the book. If a multigrid technique is used as the preconditioner, the 3D problem may be solved in $O(N^4)$ operations.

Spectral (Chebyshev) elements also appeared in 1984, with the pioneering work of Patera [297]. From the beginning, conjugate gradient (CG) methods were suggested, since the system matrix is symmetric positive definite. It is well known that CG converges slowly, and thus the preconditioning issue is of vital importance. For one example, let us consider Legendre spectral elements as proposed by Maday and Patera [261] in 1989. When preconditioned by the diagonal, as is usually done, the system has a condition number $O(E_1^2 N^2)$, where E_1 is the number of spectral elements in a typical spatial direction and E is the total number of spectral elements in the 3D problem. Most of the time, E will be $O(E_1^3)$. Therefore, the number of iterations, which is proportional to the square root of the condition number, will be $O(E_1 N)$. This leads to an overall operation count of $O(EN^4 E_1 N)$ for the preconditioned conjugate gradient (PCG), since the computational heart of CG is basically a matrix–vector product involving EN^4 operations (because the tensor-product feature of the spectral method will be fully exploited). Another approach extends the fast diagonalization method to the spectral-element case when the elements are rectilinear (see Couzy [83]). The price of inverting the Poisson equation is two PCG iterations, leading to a computational cost of $O(EN^4)$ operations.

Figure 1.11.2 displays, over two decades (from 1976 to 1996), the curve obtained by the combination of best machine, best high-order method, and best solution algorithm. We observe that going from fourth-order compact (1981)

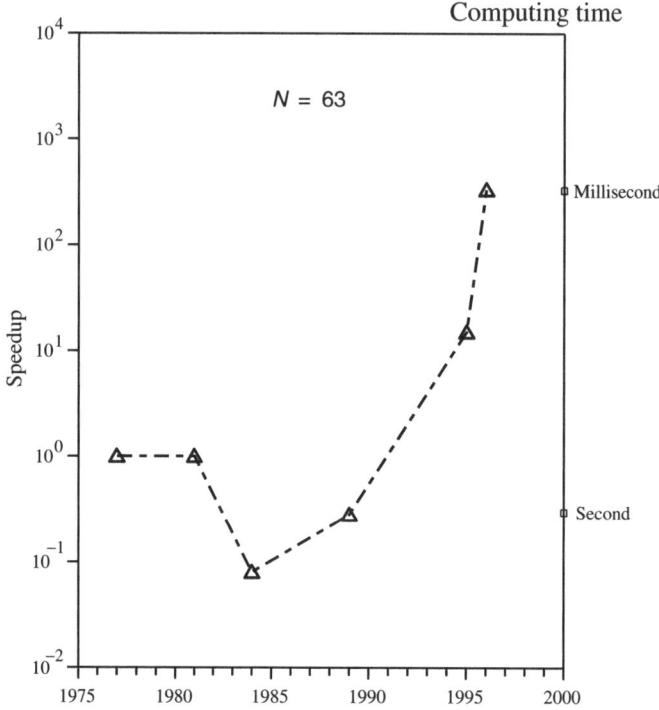

Figure 1.11.2. Time evolution of high-order methods for a 3D Poisson equation.

to spectral tau (1984) produces an increase of one order of magnitude in computing time. However, the error converges spectrally; that is, it yields machine accuracy. We have paid ten times more computing time, but we have gained seven significant digits. The spectral-element method yields machine accuracy with $E_1 = 4$ and $N = 16$, which amounts to 63 interior points per spatial direction. The last three triangles in the figure show the computing time required by the spectral-element method with the CG preconditioned by the diagonal coefficients (1989) and then by the fast diagonalization technique (1995, 1996).

Of course, in engineering practice, we do not need full accuracy. Therefore, the question is: If we want 10^{-3} accuracy, how many degrees of freedom do we need, and what is the cost of the computation? For this example, the answer is plotted in Figure 1.11.3. The number of points for the second-order method (1977) is $N = 55$; for the fourth-order compact scheme (1981), $N = 10$. The spectral tau method (1984) still needs $N = 8$, because at the beginning the spatial convergence is algebraic. The method converges faster than the fourth-order scheme, but not yet exponentially. Finally, for the spectral-element

1.11. Numerical Simulation

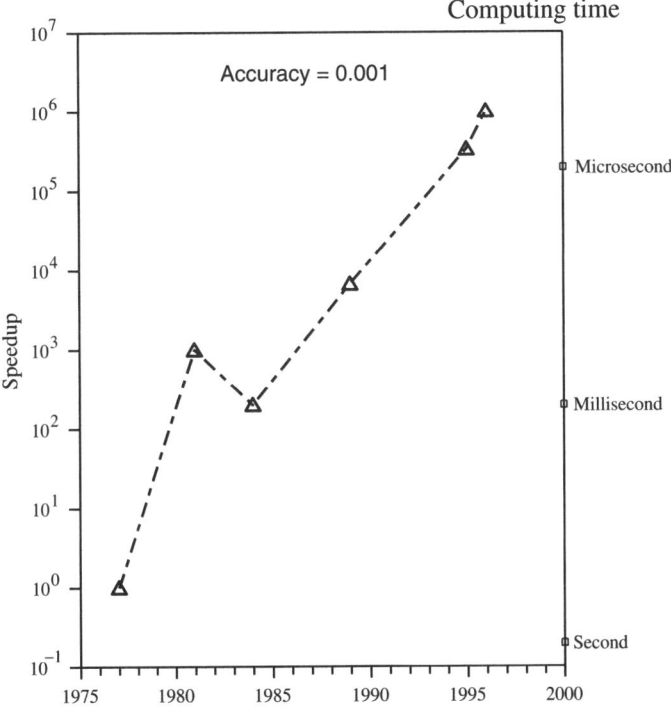

Figure 1.11.3. For a fixed accuracy, combination of best machine and algorithm using the best high-order method.

method (1989–1996) with $E = 64$ as before, we need only $N = 2$, the same accuracy as that provided by the quadratic finite-element method. In this case, it should be better to decrease E_1 and to increase N in order to converge spatially more quickly. This is one of the advantages of the spectral-element method. The acceleration factor is again striking: it reaches the value of 10^6 over twenty years.

Extrapolating our observations on a simple linear elliptic problem to 3D Navier–Stokes equations, we can deduce that the spectral convergence of the high-order methods will help to get solutions where we can test the error levels by looking, for example, at the spectra and to resolve the physics in a better fashion. The disadvantages of spectral methods are well known: geometric singularities, solution singularities, algorithmic efficiency, time-marching schemes with an accurate spatial discretization, and so forth. Later in the book, we will present high-order methods to cope with such disadvantages. Specifically, we will show that spectral methods designed in physical space (e.g., collocation or spectral elements) are able to deal with complex geometries and complex boundary conditions and to treat singularities by mortar elements or domain decomposition.

The use of high-order numerical techniques on massively parallel processing machines opens the door to many large-scale applications in computational science and engineering. Possibilities include biomedical engineering with heart flow simulation, a fascinating example of fluid–structure interaction; non-Newtonian flows with 3D time-dependent simulations; oceanography and meteorology; and heat-exchanger design. This activity will entail much feedback from numerical simulation to mathematical modeling. Again, without any pretense of exhaustion, we list some topics: turbulence, non-Newtonian constitutive relations, chemical processes, control, shape optimization, and inverse problems.

2
Approximation Methods for Elliptic Problems

The first chapter gave an overview of basic principles of fluid mechanics. In particular, it introduced the fundamental equations governing the motion of fluid flows. In this chapter we deal with approximation methods allowing these equations to be solved on a computer. We use a simple one-dimensional elliptic model problem in order to introduce the basic concepts. This problem will be carried throughout the chapter. Parabolic, hyperbolic, and multidimensional problems are treated in Chapters 3 and 4, respectively.

In Section 2.1 we begin with the derivation of a boundary-value problem (BVP) from a variational principle. Different types of boundary conditions are presented, with particular attention to the choice of test functions. The existence conditions for a solution are discussed within the Lax–Milgram theorem. Section 2.2 enlarges the scope to more general linear BVPs and introduces the approximation framework of Galerkin and collocation methods. In Section 2.3 we present various finite-element approximations, some of which have connections with spectral methods. Sections 2.4 and 2.5 discuss the spectral-element and the orthogonal collocation techniques, the two main forms of high-order methods studied in this book. Section 2.6 deals with the error estimation in connection with the various approximations introduced in the chapter. Section 2.7 is dedicated to some efficient solution techniques for the algebraic systems entailed by high-order methods. Finally, Section 2.8 discusses a numerical example.

In order to make the text reasonably self-contained, key mathematical notions have been gathered in two appendices. Appendix A deals with some fundamental concepts of functional analysis; Appendix B is devoted to orthogonal polynomials, Gaussian quadrature rules, and discrete transforms that play a central role in spectral approximations. Readers unfamiliar with these topics may first wish to review the material in the appendices. Further information can be found in the books by Kreyszig [228], Showalter [361], and Davis and Rabinowitz [92].

The bibliography covering the topics of this chapter is quite large. We single out the monographs of Axelsson and Barker [14], Carey and Oden [71], Ciarlet [79], Johnson [214], Oden and Reddy [286], and Quarteroni and Valli [319]. The book by Strang and Fix on finite-element techniques [369], although almost thirty years old, remains among the most comprehensive in the field. For information on weighted residual methods and spectral methods, interested readers should consult Finlayson [123], Fletcher [133, 134], Villadsen and Michelsen [400], Bernardi and Maday [38], Boyd [48], Canuto et al. [64], Funaro [142, 144], Gottlieb and Orszag [163], and Mercier [273].

2.1 Variational Form of Boundary-Value Problems

The equations governing fluid flow phenomena that were presented in Chapter 1 are all local PDEs, that is, functional relationships between physical variables (velocities, pressure, etc.) that must be satisfied at every point in a given domain. A large category of numerical methods to solve these equations is not based on the local form, but rather on an equivalent integral form known as the *variational formulation* of the problem.

2.1.1 Variational Functionals

Variational methods have played a major part in physical sciences and engineering since their introduction by Johann Bernoulli (1667–1748), who solved (1696) the *brachistochrone* problem of determining the shape of the curve along which a particle initially at rest falls under gravity with the smallest time of descent. Other problems in physics and engineering can be stated in a similar form: Determine a function u depending on $k \in \mathbb{N}^*$ variables and satisfying given boundary conditions that minimizes (or, maybe, maximizes) a functional $J : V \to \mathbb{R}$, where V is some set of admissible functions defined on the domain of the problem. $\mathbb{N}^*$ and $\mathbb{R}$ are the sets of positive integers and real numbers, respectively.

For convenience we consider a functional involving an element u depending on a single variable x (see [14]):

$$V := \{v \in C^2[a, b]; \ v(a) = 0\},$$
$$J(u) := \int_a^b \mathcal{J}(x, u(x), u'(x)) \, dx \quad \forall u \in V. \tag{2.1.1}$$

2.1. Variational Form of Boundary-Value Problems

The integrand $\mathcal{J}(x, u, u')$ is a real function defined on $a \leq x \leq b$ and on $-\infty < u, u' < \infty$, having continuous partial derivatives of order ≤ 2. This function determines the physical meaning of $J(u)$.

The set of admissible functions V in (2.1.1) includes all functions having continuous second-order derivatives in the closed interval $[a, b]$ and satisfying a zero boundary condition at $x = a$. Clearly, V is a linear space in the sense that for any two elements $u, v \in V$ and for any constants $\alpha, \beta \in \mathbb{R}$, the element $w = \alpha u + \beta v \in C^2[a, b]$ and is such that $w(a) = 0$.

The problem consists in finding an element $\hat{u} \in V$ that minimizes (or maximizes) $J(u)$, perhaps locally in V. Since we are interested mainly in physical problems, we will assume for the time being that the nature of these problems warrants the existence of, for instance, a minimizing element, the existence conditions being given later in the Lax–Milgram theorem. Then, in an open neighborhood B of $\hat{u}$ of radius ε (see Sections A.1.2 and A.2.1),

$$B(\hat{u}; \varepsilon) := \{v \in V; \ v = \hat{u} + \tau \eta, \eta \in V, \|\eta\| = 1, \tau \in [0, \varepsilon)\}, \qquad (2.1.2)$$

where the symbol $\|\cdot\|$ denotes some *norm* defined on the elements of V, one has

$$J(\hat{u}) = \min_u J(u), \qquad u \in B(\hat{u}; \varepsilon). \qquad (2.1.3)$$

The definition (2.1.2) of a neighborhood of $\hat{u}$ separates directions, represented by the elements η with unit norm, from distance, measured by τ. Therefore, by introducing a "test" element into J, one is able to probe the minimizing condition (2.1.3) in a constructive way. Assuming (as mentioned previously) the differentiability of $\mathcal{J}$, one defines the mth-order directional derivative of J at $\hat{u}$ in a given direction η by

$$J^{(m)}(\hat{u}; \eta) = \lim_{\tau \to 0} \frac{d^m}{d\tau^m} J(\hat{u} + \tau \eta).$$

By expanding $J(\hat{u} + \tau \eta)$ into a Taylor series of τ one gets, up to order two,

$$J(\hat{u} + \tau \eta) = J(\hat{u}) + \tau J^{(1)}(\hat{u}; \eta) + \frac{\tau^2}{2} J^{(2)}(\hat{u}; \eta) + o(\tau^2).$$

It turns out that, in order to minimize J, $\hat{u}$ must satisfy the condition

$$J^{(1)}(\hat{u}; \eta) = 0, \qquad \forall \eta \in V, \ \|\eta\| = 1. \qquad (2.1.4)$$

This is only a *necessary* condition, however. Nothing prevents the second-order directional derivative $J^{(2)}(\hat{u}; \eta)$ from taking a positive or negative sign, depending on η. If this is the case, $\hat{u}$ will simply ensure a stationary value of J. If, on the contrary, $J^{(2)}(\hat{u};\eta)$ keeps a positive sign independently of η, then $\hat{u}$ minimizes $J(u)$ in a neighborhood the size of which depends on high-order terms. If, additionally, it happens that high-order terms are absent, the functional being quadratic, then $\hat{u}$ is the unique minimizer of $J(u)$.

We turn back to the functional (2.1.1) and evaluate its first-order directional derivative at u:

$$J^{(1)}(u;\eta) = \int_a^b \left(\frac{\partial \mathcal{J}}{\partial u}\eta + \frac{\partial \mathcal{J}}{\partial u'}\eta'\right) dx.$$

By integrating the last term on the right-hand side by parts, one gets

$$J^{(1)}(u;\eta) = \int_a^b \left(\frac{\partial \mathcal{J}}{\partial u} - \frac{d}{dx}\left(\frac{\partial \mathcal{J}}{\partial u'}\right)\right)\eta\, dx + \left.\frac{\partial \mathcal{J}}{\partial u'}\eta\right|_{x=b}. \qquad (2.1.5)$$

Since η belongs to V, the boundary contribution in (2.1.5) involves only a single term at $x = b$. Applying the condition (2.1.4), one concludes that the element $\hat{u}$ is compelled to satisfy the relationships

$$\frac{\partial \mathcal{J}}{\partial \hat{u}} - \frac{d}{dx}\left(\frac{\partial \mathcal{J}}{\partial \hat{u}'}\right) = 0, \qquad a < x < b, \qquad (2.1.6)$$

$$\left.\frac{\partial \mathcal{J}}{\partial \hat{u}'}\right|_{x=b} = 0. \qquad (2.1.7)$$

Equation (2.1.6) is the Euler–Lagrange (or *local*) equation associated to the variational problem (2.1.1)–(2.1.3), while (2.1.7) is a boundary condition to be satisfied by $\hat{u}$, not included in the definition of the set of admissible functions. Boundary conditions of this type are called *natural*, while those satisfied by the elements of V are called *essential* boundary conditions.

Let us now be more specific and introduce as an example the model problem

$$V := \{v \in C^2[a,b];\ v(a) = 0\},$$
$$J(u) = \frac{1}{2}\int_a^b (pu'^2 + qu^2 - 2fu)\, dx \qquad \forall u \in V, \qquad (2.1.8)$$

where $p(x) \in C^1[a,b]$, $0 < p_0 \leq p(x) \leq p_1$ for $a \leq x \leq b$, and where $q(x)$, $f(x) \in C[a,b]$, $0 \leq q(x) \leq q_1$ for $a \leq x \leq b$. One can verify that (2.1.8) is a quadratic functional with Euler–Lagrange equation and boundary conditions

$$-\frac{d}{dx}\left(p\frac{du}{dx}\right) + qu = f, \qquad a < x < b,$$
$$u(a) = u'(b) = 0. \qquad (2.1.9)$$

The boundary condition involving the derivative comes from (2.1.7).

2.1. Variational Form of Boundary-Value Problems

The problem (2.1.9) is the canonical form of a second-order self-adjoint BVP that appears in various forms for the description of steady-state diffusion processes: molecular diffusion (mass conservation and Fick's law), thermal conduction [energy conservation and Fourier's law (1.5.6)], viscous flow [momentum conservation and constitutive Newtonian equations (1.5.4)–(1.5.5)], and the like. The diffusion term has a strictly positive coefficient $p(x)$ for both physical and mathematical reasons. The functional $J(u)$ in this example is a measure of the energy stored in the system. For convenience we will henceforth denote by $\mathcal{L}$ the differential operator in the left-hand side of (2.1.9). With the hypotheses on $p(x)$, $q(x)$, and the source term $f(x)$, a solution $\hat{u} \in V$ to (2.1.9) exists that is unique.

Following the earlier path, we introduce a test element $v = \hat{u} + \tau\eta$, $\eta \in V$, into $J(u)$, and we obtain

$$J(\hat{u} + \tau\eta) = J(\hat{u}) + \tau \int_a^b \left(p\frac{d\hat{u}}{dx}\frac{d\eta}{dx} + q\hat{u}\eta - f\eta \right) dx$$

$$+ \frac{\tau^2}{2} \int_a^b (p\eta'^2 + q\eta^2) dx \qquad \forall \eta \in V.$$

From this relationship we can draw two conclusions. First, $J(u)$ reaches a stationary value at $\hat{u} \in V$ if

$$\int_a^b \left(p\frac{d\hat{u}}{dx}\frac{d\eta}{dx} + q\hat{u}\eta \right) dx = \int_a^b f\eta \, dx, \qquad \forall \eta \in V. \qquad (2.1.10)$$

Second, since $J^{(2)}(u;\eta) > 0 \; \forall \eta \in V, \|\eta\| = 1$, $\hat{u}$ is indeed the unique minimizer of (2.1.8) in V.

At this point we introduce some convenient notation:

$$(u, v) := \int_a^b u(x)v(x) \, dx, \qquad u, v \in V, \qquad (2.1.11)$$

$$\mathcal{A}(u, v) := \int_a^b (pu'v' + quv) \, dx, \qquad u, v \in V, \qquad (2.1.12)$$

$$\mathcal{F}(v) := \int_a^b fv \, dx, \qquad v \in V. \qquad (2.1.13)$$

The quantity $(\cdot, \cdot)$ is the *inner product* of elements in V. With the hypotheses on $p(x)$ and $q(x)$, the mapping $\mathcal{A} : V \times V \to \mathbb{R}$ is a symmetric bilinear form [i.e., $\mathcal{A}(u, v) = \mathcal{A}(v, u)$, and $\mathcal{A}$ is linear with respect to both u and v] that satisfies the axioms (A.4.1)–(A.4.4) for an inner product, and $\mathcal{F} : V \to \mathbb{R}$ is a linear functional on the set of test functions V. The bilinear form $\mathcal{A}(u, v)$ is the *energy inner product* associated with $\mathcal{L}$. For notational convenience and in accordance with the conventions stated in Appendix A (see the Remark in Section A.4.2), we will also denote this inner product by $(u, v)_\mathcal{L}$ and the induced *energy norm* by $\|u\|_\mathcal{L}$:

$$(u, v)_\mathcal{L} := \mathcal{A}(u, v), \qquad (2.1.14)$$

$$\|u\|_\mathcal{L} := (u, u)_\mathcal{L}^{1/2} = \mathcal{A}^{1/2}(u, u). \qquad (2.1.15)$$

Introducing (2.1.12) and (2.1.13) into (2.1.8) and (2.1.10) gives

$$J(u) = \frac{1}{2}\mathcal{A}(u, u) - \mathcal{F}(u) \qquad \forall u \in V, \qquad (2.1.16)$$

and

$$J(\hat{u}) = \min_{u \in V} J(u), \qquad (2.1.17)$$

the stationary value $J(\hat{u})$ being attained at $\hat{u}$, the solution to the following problem: Find $\hat{u} \in V$ such that

$$\mathcal{A}(\hat{u}, v) = \mathcal{F}(v) \qquad \forall v \in V. \qquad (2.1.18)$$

Setting $u = \hat{u} + w$ in the right-hand side of (2.1.16) and taking (2.1.18) into account after expansion of the bilinear form $\mathcal{A}$, we get another useful relationship that clearly shows the characterization of $\hat{u}$ as the unique element minimizing J:

$$J(u) = J(\hat{u}) + \frac{1}{2}\mathcal{A}(u - \hat{u}, u - \hat{u}) \qquad \forall u \in V. \qquad (2.1.19)$$

Finally, proceeding backwards with an integration by parts in (2.1.12) and transferring anew all derivatives upon u, which poses no problem here because $p(x) \in C^1[a, b]$, we conclude that the solution to the Euler–Lagrange equation and boundary conditions (2.1.9) is $\hat{u} \in V$ such that

$$(\mathcal{L}\hat{u} - f, v) = 0 \qquad \forall v \in V. \qquad (2.1.20)$$

In the literature, the differential equation (2.1.9) is often called the *strong* (or classical) formulation of the problem, while (2.1.18) is its *weak* (or variational) formulation. We will meet (2.1.20) on several occasions later on, since this relationship is at the heart of the weighted-residual methods. The two integral relationships (2.1.18) and (2.1.20) correspond to the well-known principle of virtual work in classical mechanics (see, for example, [370]). Both are cornerstones of numerical fluid mechanics.

2.1. Variational Form of Boundary-Value Problems

2.1.2 Boundary Conditions

So far we have met two types of homogeneous boundary conditions: Dirichlet conditions, specifying values of the unknown function, and Neumann conditions, specifying its first derivative. While the former were included in the definition of the set of test functions V, the latter appeared only in the differential formulation of the problem. We clarify this difference further and examine the cases of inhomogeneous boundary conditions and mixed boundary conditions.

As seen in the preceding section, the weak formulation of a BVP may be viewed as the result of multiplying the differential equation by a test function, followed by integration by parts. For the diffusion process (2.1.9) and a test function $v(x) \in V$, where V is defined in (2.1.8), one gets the Green's formula

$$(\mathcal{L}u(x), v(x)) = \left. -p(x)\frac{du(x)}{dx}v(x)\right|_{x=a}^{x=b} + \mathcal{A}(u, v). \qquad (2.1.21)$$

The structure of the boundary term in (2.1.21) shows that u and du/dx play quite different roles in the boundary conditions, u being absent.[1] All constraints on u, whether homogeneous or inhomogeneous, should therefore be included in the definition of the set of admissible functions, whence their name "essential" boundary conditions. This is not the case for the derivative du/dx [or of any derivative $u^{(k)}(k = m, \ldots, 2m - 1)$, with a differential operator of order $2m$], which is present in the boundary term and is therefore an integral part of the computational process. Approximation techniques such as those developed later in this chapter ensure that natural boundary conditions are automatically satisfied in the limit, when the approximate solutions converge to the solution of the problem.

Let us give an example involving inhomogeneous boundary conditions:

$$-\frac{d}{dx}\left(p\frac{du}{dx}\right) + qu = f, \quad a < x < b,$$
$$u(a) = \mu_a, \quad u'(b) = \mu'_b, \qquad (2.1.22)$$

with $\mu_a, \mu'_b \in \mathbb{R}$. Because of the constraint at $x = a$, this problem requires two different sets of admissible functions. Let

$$V_{\mu_a} := \{v \in C^2[a, b];\ v(a) = \mu_a\},$$
$$V := \{v \in C^2[a, b];\ v(a) = 0\}$$

---

[1] More generally, in the case of a differential operator of order $2m$, any derivative $u^{(k)}$ ($k = 0, \ldots, m - 1$) of the solution would be absent in the boundary term.

denote sets of functions that satisfy the inhomogeneous and homogeneous essential conditions, respectively. The weak formulation of (2.1.22) is: Find $\hat{u} \in V_{\mu_a}$ such that

$$\mathcal{A}(\hat{u}, v) = \mathcal{F}(v) + p(b)\mu'_b v(b) \qquad \forall v \in V.$$

In this case, the solution to the problem is to be found inside V_{μ_a}, the set of *trial* functions, which differs from V, the linear space of *test* functions that handles the boundary term in (2.1.21). Incidentally, we remark that V_{μ_a} is *not* a linear space, but rather a translation of a linear space. Also, the inhomogeneous Neumann condition at $x = b$ has introduced an additional term in the weak formulation of the problem.

Boundary conditions may be neither Dirichlet nor Neumann, as is the case, for instance, with the problem

$$-\frac{d}{dx}\left(p\frac{du}{dx}\right) + qu = f, \qquad a < x < b,$$

$$u'(a) - \alpha u(a) = 0, \qquad u'(b) + \beta u(b) = 0,$$

where α and β are real positive constants. Such mixed homogeneous boundary conditions are called *Robin conditions*. Using the above analysis and eliminating the derivatives at both end points, one verifies that the weak formulation of this problem is: Find $\hat{u} \in V := \{v \in C^2[a, b]\}$ such that

$$\mathcal{A}(\hat{u}, v) + \alpha p(a)\hat{u}(a)v(a) + \beta p(b)\hat{u}(b)v(b) = \mathcal{F}(v), \qquad \forall v \in V. \quad (2.1.23)$$

Here, the admissible functions are totally unconstrained. Trial and test functions are chosen from the linear space of functions having continuous second-order derivatives.

2.1.3 Sobolev Spaces and the Lax-Milgram Theorem

The reader has certainly noticed that, up to now, trial and test functions in the variational problem (2.1.18) were selected in function spaces with elements having fairly smooth properties. This approach was in accordance with the hypotheses on $p(x)$, $q(x)$, and $f(x)$ in (2.1.8) ensuring the existence of a solution $\hat{u} \in \mathcal{D}(\mathcal{L})$ with $\mathcal{D}(\mathcal{L}) (= C^2[a, b])$ denoting the domain of $\mathcal{L}$. There are several reasons, however, why sets of admissible functions having less regularity should be identified. The first reason lies in the problem itself. One could be faced with a situation where some of the hypotheses in (2.1.8) would not be fulfilled. For instance, $p(x)$ might be only piecewise continuous, or the forcing term $f(x)$, instead of being continuous, might belong to $L^2(a, b)$, the

2.1. Variational Form of Boundary-Value Problems

space of square-integrable functions in (a, b). In the latter case, a solution to the variational problem (2.1.18) still exists, but $\hat{u} \notin \mathcal{D}(\mathcal{L})$. The second reason is that if one looks for an *approximation* of $\hat{u}$ in the weak sense, there are no compelling reasons to introduce C^2 regularity in the test elements, since (2.1.18) involves first-order derivatives only.

Identifying appropriate spaces of admissible functions for a BVP is always a delicate question. We use the problem (2.1.9) as an example to show the main ideas that establish the numerical treatment on firm theoretical ground (see Sections A.2, A.4, and A.5, as well as [14, 286]).

A proper choice of trial and test elements may be found in classes of functions that are "larger" than $\mathcal{D}(\mathcal{L})$. Two basic principles preside over the enlargement. All quadratures involved in the weak formulation should be well defined (i.e., should be finite), and every Cauchy sequence in the suitable norm [in our case, the energy norm (2.1.15)] should converge toward an element belonging to the same class of functions; that is, the set of functions should be complete. The function spaces resulting from this completion process are called *Sobolev spaces* [14, 3]. Examples of functions belonging to Sobolev spaces will be given in Section 2.3 with the description of finite-element methods.

We introduce three important mappings $V \times V \to \mathbb{R}$ and $V \to \mathbb{R}$, where V [$\equiv \mathcal{D}(\mathcal{L})$] is the linear space defined in (2.1.8):

$$(u, v)_{H^1} := \int_a^b (u'v' + uv)\,dx, \qquad u, v \in V, \tag{2.1.24}$$

$$\|u\|_{H^1} := (u, u)_{H^1}^{1/2} = \left\{ \int_a^b (u'^2 + u^2)\,dx \right\}^{1/2},$$

$$= (\|u'\|^2 + \|u\|^2)^{1/2}, \qquad u \in V, \tag{2.1.25}$$

$$|u|_{H^1} := \left\{ \int_a^b u'^2\,dx \right\}^{1/2} = \|u'\|, \qquad u \in V. \tag{2.1.26}$$

The subscript H^1 will be clarified below. One can verify that (2.1.24) and (2.1.25) satisfy the axioms for an inner product and of a norm, respectively. In particular, $\|u\|_{H^1} = 0$ implies that the L^2 norms $\|u\| = \|u'\| = 0$, and therefore that u is identically equal to zero in (a, b). The applications (2.1.24) and (2.1.25) are called the *Sobolev inner product* and the *Sobolev norm* of order one on V.

The quantity $|u|_{H^1}$ only partially satisfies the conditions of a norm, because, in general, there may exist elements of V not identically equal to zero for which $|u|_{H^1} = 0$. This quantity, known as the *Sobolev seminorm* of order one, will also be used later.

We need an inequality, due to Poincaré and Friedrichs, that sets a lower bound to the L^2 norm of a derivative (in the one-dimensional case) in terms of the L^2 norm of the function. A proof of this inequality is given in [14].

Lemma 2.1 (Poincaré–Friedrichs inequality)

$$(u', u') \geq \frac{2}{(b-a)^2}(u, u) \quad \forall u \in W := \{v \in C^1[a, b]; \ v(a) \cdot v(b) = 0\}. \tag{2.1.27}$$

Using (2.1.27), one may establish the equivalence between the energy norm and the Sobolev norm, that is, the existence of positive constants ρ and β such that

$$\rho \|u\|_{H^1}^2 \leq \|u\|_{\mathcal{L}}^2 \leq \beta \|u\|_{H^1}^2 \quad \forall u \in V. \tag{2.1.28}$$

With the conditions imposed on the coefficients $p(x)$ and $q(x)$ in the model problem (2.1.8) starting from (2.1.15), one gets

$$\|u\|_{\mathcal{L}}^2 \geq p_0 \int_a^b u'^2(x)\,dx \geq \rho \|u\|_{H^1}^2 \tag{2.1.29}$$

and

$$\|u\|_{\mathcal{L}}^2 \leq p_1 \int_a^b u'^2\,dx + q_1 \int_a^b u^2\,dx \leq \beta \|u\|_{H^1}^2, \tag{2.1.30}$$

with the constants $\rho := \min(p_0/2, p_0/(b-a)^2)$ and $\beta := \max(p_1, q_1)$. The inequality (2.1.29) involving ρ is very important and is known as the *coercivity* (or "ellipticity") condition, while the inequality (2.1.30) involving β is called the *boundedness* condition. We further remark that (2.1.29) implies $\mathcal{A}(u, u) \geq 0$ and also $\mathcal{A}(u, u) = 0 \Leftrightarrow u = 0$, legitimizing the use of $\mathcal{A}(u, v)$ as an inner product on V.

Coercivity plays a crucial role in the completion process of V. The argument goes as follows. Suppose $s := \{u_k\}_{k=1}^\infty$ is a Cauchy sequence, in the energy norm $\|\cdot\|_{\mathcal{L}}$, of functions belonging to V. Then, according to (A.1.9), one has

$$\lim_{k,l \to \infty} \|u_k - u_l\|_{\mathcal{L}} = 0. \tag{2.1.31}$$

The coercivity condition (2.1.29) and the relationship (2.1.25) imply that, for sequences satisfying (2.1.31), one has

$$\lim_{k,l \to \infty} \|u_k - u_l\| = 0 \quad \text{and} \quad \lim_{k,l \to \infty} \|u_k' - u_l'\| = 0.$$

2.1. Variational Form of Boundary-Value Problems

This result shows that the sequences s and $s' := \{u'_k\}_{k=1}^{\infty}$ are Cauchy sequences in the L^2 norm. Since $C^2[a, b]$ is not complete in the L^2 norm, one concludes that some Cauchy sequences s may not converge, in the energy norm $\|\cdot\|_\mathcal{L}$, to an element of V. But since $L^2(a, b)$ is complete, for all Cauchy sequences in $\|\cdot\|_\mathcal{L}$ there exist elements u and $\tilde{u}$ in $L^2(a, b)$ that are limits of s and s' in the L^2 norm:

$$\lim_{k \to \infty} \|u_k - u\| = 0 \quad \text{and} \quad \lim_{k \to \infty} \|u'_k - \tilde{u}\| = 0. \quad (2.1.32)$$

The element $\tilde{u}$ obtained in this way is uniquely determined by u and is called the *generalized derivative* of u in the sense of distributions (see the discussion in Section A.5). The reason is that whenever $u \in C^1[a, b]$, $\tilde{u}$ reduces to the classical derivative u', but for pathological functions like piecewise continuous functions, $\tilde{u}$ is a distribution.

Let $\hat{V}$ denote the space of functions that can be obtained as limits of Cauchy sequences in the energy norm $\|\cdot\|_\mathcal{L}$, of elements belonging to V defined in (2.1.8). Obviously $\hat{V}$ contains V as a proper dense subspace and is embedded into $L^2(a, b)$, $V \subset \hat{V} \subset L^2(a, b)$. With the help of the foregoing one concludes that $\hat{V}$ is made of functions belonging to $L^2(a, b)$ with first-order derivatives also in L^2, such functions having C^0 continuity in $[a, b]$ (see [276]). With the Sobolev inner product (2.1.24) and associated norm (2.1.25), $\hat{V}$ is a Hilbert space.

If V were unconstrained by an essential boundary condition, the resulting Hilbert space $\hat{V}$ would be the so-called Sobolev space $H^1(a, b)$ [14]:

$$H^1(a, b) := \left\{ u;\ u \in L^2(a, b), \frac{du}{dx} \in L^2(a, b) \right\}, \quad (2.1.33)$$

where du/dx is a distributional derivative. However, because of the essential boundary condition that is preserved in the completion process, $\hat{V}$ is a subspace of $H^1(a, b)$:

$$\hat{V} := \{u \in H^1(a, b);\ u(a) = 0\}.$$

In the sequel, a special notation will be used for Sobolev spaces satisfying homogeneous Dirichlet conditions on the boundary:

$$H^1_0(a, b) := \{u \in H^1(a, b);\ u(a) = u(b) = 0\}. \quad (2.1.34)$$

This Hilbert space is the closure in $H^1(a, b)$ of the set of functions belonging to $C^1(a, b)$ with compact support in (a, b). A definition of functions with compact support is given in Section A.5.1.

2. Approximation Methods for Elliptic Problems

The Sobolev space H^1 is the relevant space of admissible functions for the variational solution of second-order differential problems. It has less smoothness than the original space $C^2[a, b]$, and this calls for some remarks. First the reader should be convinced that the enlargement of the space of functions does not affect those problems having a solution in $\mathcal{D}(\mathcal{L})$. However, by recognizing that the range of solutions is larger than $C^2[a, b]$ and by identifying the appropriate function space, one is able to capture the set of solutions for all problems (2.1.9) with $f \in L^2(a, b)$. Another and perhaps more important consideration is that the enlargement to Sobolev spaces offers considerable practical advantages, as we will see in the next sections, where we encounter concrete examples of H^1 spaces. The overwhelming success of finite-element methods rests essentially with the ease of generating subspaces of H^1 with the simplest functions, namely, piecewise continuous polynomials.

Some of the numerical techniques developed in this book will require a larger category of Sobolev spaces, called *weighted Sobolev spaces*. These spaces may be obtained through the same completion process as above, except for the inner product and associated norm. Without entering into details we mention that if $w(x)$, $x \in (a, b)$, denotes a nonnegative, integrable function over the interval (a, b) (i.e., $\int_a^b w(x)\,dx < \infty$), satisfying all requirements for a weighting function, the weighted Sobolev space $H_w^1(a, b)$ is the completion of an inner-product space such that

$$H_w^1(a, b) := \left\{ u;\ u \in L_w^2(a, b), \frac{du}{dx} \in L_w^2(a, b) \right\},$$

where $L_w^2(a, b)$ denotes the Hilbert space of weighted square-integrable functions (see Example 2 in Section A.4.2).

Finally, we note that for differential problems of order $2k$, the process of identifying the proper set of admissible functions would be essentially the same, yielding the Sobolev space H_w^k (or $H_{0,w}^k$, depending on the boundary conditions):

$$H_w^k(a, b) := \left\{ u;\ \text{for } 0 \leq \ell \leq k, \frac{d^\ell u}{dx^\ell} \in L_w^2(a, b) \right\}, \qquad (2.1.35)$$

and

$$H_{0,w}^k := \left\{ u \in H_w^k(a, b);\ u^{(j)}(a) = u^{(j)}(b) = 0, 0 \leq j \leq (k-1) \right\}.$$

These spaces are endowed with an inner product $(u, v)_{H_w^k}$ and associated norms $\|u\|_{H_w^k}$ and seminorms $|u|_{H_w^k}$ that are straightforward generalizations of the

2.1. Variational Form of Boundary-Value Problems

expressions (2.1.24)–(2.1.26):

$$(u, v)_{H_w^k} := \int_a^b w(x) \left(\sum_{i=0}^k u^{(i)} v^{(i)} \right) dx, \qquad u, v \in H_w^k,$$

$$\|u\|_{H_w^k} := (u, u)_{H_w^k}^{1/2}, \qquad u \in H_w^k, \qquad (2.1.36)$$

$$|u|_{H_w^k} := \left\{ \int_a^b w(x) u^{(k)2} \, dx \right\}^{1/2}, \qquad u \in H_w^k.$$

The Sobolev spaces $H^k(a, b)$, for $k \geq 1$, form a hierarchy of Hilbert spaces with inclusion relationships. In the simple one-dimensional case, one has $H^k(a, b) \subset H^{k-1}(a, b) \subset \cdots \subset H^0(a, b) \equiv L^2(a, b)$. If $u \in C^k[a, b]$, then certainly $u \in H^k(a, b)$, and therefore $C^k[a, b] \subset H^k(a, b)$. Also, if $u \in H^k(a, b)$, it can be shown that $u \in C^{k-1}[a, b]$, and hence that $H^k(a, b) \subset C^{k-1}[a, b]$. All these inclusion relationships are continuous in the sense defined in Section A.3.3 [see Equation (A.3.1) and Theorem A.3]. Such relationships are called *Sobolev embedding theorems* in the literature (see [79, 286, 361]). Let us stress that some of these inclusion relationships are strictly valid for one-dimensional problems. We shall come back to this topic in Chapter 4.

Before leaving (temporarily) the topic of Sobolev spaces, we briefly extend their definition to *negative* values of k. In particular, we want to introduce the space H_w^{-1}, which will be needed in forthcoming chapters. In a quite consistent approach, negative exponents in this case refer to antiderivatives, and $H_w^{-1}(a, b)$ denotes the space of functions having their integral in $L_w^2(a, b)$. Without entering into the subtleties of the definition, which can be found in the literature, we mention that negative Sobolev spaces $H_w^{-k}(a, b)$ are the duals of $H_{0, w^{-1}}^k(a, b)$, or, more precisely, the spaces of continuous linear functionals defined on $H_{0, w^{-1}}^k(a, b)$ (see [71]).

Having identified suitable spaces of functions for the variational formulation of BVPs, we must answer one important question before setting the variational method to work. This question deals with the existence of the solution itself. In other words: What are the conditions to be satisfied by the various parameters of a problem in order to ensure that it has a *unique* solution? The answer to this question is given by the Lax–Milgram theorem:

Theorem 2.1 (Lax–Milgram) *Let V denote a Hilbert space with norm $\|\cdot\|_V$. Consider $\mathcal{A} : V \times V \to \mathbb{R}$, a bilinear form, and $\mathcal{F} : V \to \mathbb{R}$, a bounded linear functional. Assume the existence of positive constants β, ρ such that*

$$|\mathcal{A}(u, v)| \leq \beta \|u\|_V \|v\|_V \qquad \forall u, v \in V, \qquad (2.1.37)$$

$$\mathcal{A}(u, u) \geq \rho \|u\|_V^2 \qquad \forall u \in V. \qquad (2.1.38)$$

Then there exists a unique element $\hat{u} \in V$ such that

$$\mathcal{A}(\hat{u}, v) = \mathcal{F}(v) \qquad \forall v \in V. \tag{2.1.39}$$

Boundedness and coercivity were key elements to identify the Sobolev spaces of admissible functions for the variational solution of the problem (2.1.9). They are also of central importance for the existence and uniqueness of the solution to the problem. It is easy to show that the equivalence (2.1.28) between the Sobolev and energy norms of any element $u \in V$ corresponds to the conditions (2.1.37)–(2.1.38) of the Lax–Milgram theorem. Hence, all conditions are met to ensure that the solution to (2.1.9) exists and is, indeed, unique.

2.2 An Approximation Framework

Let us summarize the results obtained so far. We have seen that, provided the hypotheses of the Lax–Milgram theorem are fulfilled, the solution $\hat{u}$ to the BVP

$$\begin{aligned} \mathcal{L}u &= f, & x \in \Omega, \\ \mathcal{B}u &= 0, & x \in \partial\Omega, \end{aligned} \tag{2.2.1}$$

where $\mathcal{L}$ is a linear differential operator, $\mathcal{B}$ is a boundary operator, and Ω is an open domain in $\mathbb{R}$ with boundary $\partial\Omega$, exists and is unique. This solution may be characterized by one of the following requirements:

- $J(\hat{u}) = \min_{u \in V} J(u)$, where $J(u)$ is the variational functional with associated Euler–Lagrange equation $\mathcal{L}u = f$ [see (2.1.17)].
- Find $\hat{u} \in V$ such that $\mathcal{A}(\hat{u}, v) = \mathcal{F}(v) \, \forall v \in V$ [see (2.1.18)].
- Find $\hat{u} \in \mathcal{D}(\mathcal{L})$ such that $(\mathcal{L}\hat{u} - f, v) = 0 \, \forall v \in \mathcal{D}(\mathcal{L})$.

Here V denotes the Hilbert space of admissible functions (a Sobolev space) with inner product $(\cdot, \cdot)_V$ and induced norm $\|\cdot\|_V$; $\mathcal{A} : V \times V \to \mathbb{R}$ is a bilinear form associated to the problem; and $\mathcal{D}(\mathcal{L})$, the domain of the differential operator $\mathcal{L}$, is a dense subspace of V. Usually, $\hat{u}$ is a distribution, that is, an element of a space obtained by a completion process. In some cases it reduces to a continuous function.

Clearly, the minimization procedure (2.1.17) is meaningful only for symmetric problems. Although we will keep the model problem (2.1.9) throughout this chapter, we momentarily free ourselves from the existence of a (minimum or maximum) variational principle. Consequently, we characterize the solution $\hat{u}$ of the BVP (2.2.1) by (2.1.18) or by (2.1.20). We emphasize that these relationships are *formally* equivalent, that is, they are equivalent only if $\hat{u} \in \mathcal{D}(\mathcal{L})$.

2.2. An Approximation Framework

If the solution does not belong to $\mathcal{D}(\mathcal{L})$ (because the right-hand side f does not belong to the range of $\mathcal{L}$), the approach (2.1.20) is more delicate. In that case, one has to look for the closure of $\mathcal{D}(\mathcal{L})$. In conclusion, the variational (or weak) formulation (2.1.18) is the most general formulation of BVPs, and it will be used extensively in later chapters.

Once the space of admissible functions V has been identified and a unique solution $\hat{u}$ is known to exist, the remaining task is the determination of $\hat{u}$. Since this problem is infinite-dimensional in nature, only *approximations* of $\hat{u}$ can be found in general. There is a large collection of approximation schemes, depending on a variety of options. Among the most important families of schemes we have the Galerkin approximations, the collocation method, the method of subdomains, and the method of least squares. All these schemes are more or less based on the same principle. Let $\{V_N\}_{N=1}^{\infty}$ represent a family of finite-dimensional subspaces of V with

$$\lim_{N\to\infty} \inf_{u\in V_N} \|v - u\|_V = 0 \quad \forall v \in V, \qquad (2.2.2)$$

providing a dense coverage of V in the limit. Instead of looking for the solution $\hat{u} \in V$, one looks for an approximation $u_N \in V_N$. Each method has its own features in selecting the space V_N and (possibly) a numerical approximation to the equations (2.1.18) or (2.1.20).

Because of their importance in the numerical solution of incompressible fluid flows by high-order schemes, we will describe the Galerkin and collocation methods in greater detail. For a description of other methods, the reader is referred to the literature (see, for instance, [133, 400]).

2.2.1 Galerkin Approximations

All Galerkin approximations apply a projection technique. The differences between the approximations lie in the various possibilities of implementing the projection.

Galerkin Method

Consider the weak form of the BVP (2.2.1): Find $\hat{u} \in V$ such that

$$\mathcal{A}(\hat{u}, v) = \mathcal{F}(v) \quad \forall v \in V,$$

where $\mathcal{A} : V \times V \to \mathbb{R}$ is a bilinear form associated to the operator $\mathcal{L}$ by integration by parts, and $\mathcal{F} : V \to \mathbb{R}$ is a bounded linear functional. Given a family $\{V_N\}_{N=1}^{\infty}$ of finite-dimensional subspaces of the Hilbert space V, the Galerkin method consists in finding $u_N \in V_N$ such that

$$\mathcal{A}(u_N, v) = \mathcal{F}(v) \qquad \forall v \in V_N. \tag{2.2.3}$$

Let $\{\psi_n(x)\}_{n=0}^{N}$ denote a set of $N+1$ linearly independent functions of V, forming a basis for V_N. A potential candidate for $\hat{u}$ is

$$u_N(x) := \sum_{n=0}^{N} u_n \psi_n(x) = \underline{\psi}^T(x) \cdot \underline{u}, \qquad u_N \in V_N. \tag{2.2.4}$$

In this relationship $\{u_n\}_{n=0}^{N}$ is a set of coefficients to be determined. For brevity, the basis functions and the coefficients have been assembled into column vectors $\underline{\psi}(x)$ and $\underline{u}$, and $\underline{\psi}^T(x) \cdot \underline{u}$ is their inner product, with the superscript denoting the transpose. Introducing the expansion (2.2.4) into (2.2.3) and then taking the inner product with all basis functions ψ_k ($k = 0, \ldots, N$) leads to an algebraic system

$$H\underline{u} = M\underline{f}, \tag{2.2.5}$$

where H is an $(N+1) \times (N+1)$ matrix with elements

$$H_{mn} := \mathcal{A}(\psi_n, \psi_m), \qquad 0 \leq m, n \leq N. \tag{2.2.6}$$

For reasons that will be more transparent later, the right-hand side $M\underline{f}$ [where M is another, yet undefined, $(N+1) \times (N+1)$ matrix operator] denotes the $(N+1)$-vector with components

$$(M\underline{f})_m := \mathcal{F}(\psi_m), \qquad 0 \leq m \leq N. \tag{2.2.7}$$

An important issue of the computation is the method by which the algebraic system (2.2.5) is solved in practice. Various techniques will be presented later in Section 2.7. Here we solve (2.2.5) formally, assuming that H is nonsingular. Entering (2.2.5) in (2.2.4), we get $u_N(x) \in V_N$, approximation of $\hat{u}(x)$:

$$u_N(x) = \underline{\psi}^T(x) \cdot H^{-1} M \underline{f}. \tag{2.2.8}$$

For the Galerkin approximation scheme (2.2.3), one has the following theorem:

Theorem 2.2 *Provided the boundedness and coercivity assumptions of the Lax–Milgram theorem are fulfilled, the problem (2.2.3) has a unique solution (2.2.8) that satisfies the inequality*

$$\|u_N\|_V \leq C \|f\|, \tag{2.2.9}$$

2.2. An Approximation Framework

where $C > 0$ is a constant that does not depend on N. One has, moreover,

$$\|u_N - \hat{u}\|_V \leq \frac{\beta}{\rho} \inf_{v \in V_N} \|\hat{u} - v\|_V. \qquad (2.2.10)$$

The fact that, for a given value of N, coercivity implies the existence and uniqueness of $u_N(x)$ is easy to understand. The condition (2.1.38) means that the matrix H is positive definite and hence, nonsingular. The inequality (2.2.9) is very important and proves the *stability* of the Galerkin approximation. A numerical scheme is stable if the norm of any approximate solution is bounded, *independently* of the discretization parameter N. The theorem establishes also that, as a consequence of (2.2.2), $u_N(x)$ converges to $\hat{u}(x)$, with increasing dimensions of V_N. Of the three conditions, stability, consistency, and convergence, two are satisfied; hence, according to Lax's equivalence theorem [327], the third condition (in this case, consistency) is also verified.

The r.h.s. of (2.2.10) may be evaluated by making assumptions about the smoothness of $\hat{u}$ and by using the interpolant of $\hat{u}$ in V_N. This is at the root of all error estimates in finite-element approximations.

In the particular case where $V_N \subset \mathcal{D}(\mathcal{L})$, $N < \infty$, the substitution of (2.2.4) may also be based on the formally equivalent formulation (2.1.20) of the BVP. The approximation problem is then: Find $u_N \in V_N$ such that

$$(\mathcal{L}u_N - f, v) = 0 \qquad \forall v \in V_N. \qquad (2.2.11)$$

This equation has a simple and straightforward interpretation. Except in unusual circumstances, $\hat{u} \notin V_N$. Therefore the introduction of any trial element u_N in the equation $\mathcal{L}u = f$ leaves a residual

$$r_N(x) := \mathcal{L}u_N(x) - f(x). \qquad (2.2.12)$$

According to (2.2.11), the solution is obtained when the residual (2.2.12) is orthogonal to V_N with respect to the inner product $(\cdot, \cdot)$. Hence, the method is known in the engineering literature as the *method of weighted residuals* [123].

Remark 2.1 When the bilinear form is symmetric, $\mathcal{A}(u, v) = \mathcal{A}(v, u)$, and an approximate solution may be obtained by using the variational principle (2.1.16)–(2.1.17) as well. The finite-dimensional problem then becomes: Find $u_N \in V_N$ such that

$$J(u_N) = \min_{u \in V_N} J(u) = \min_{u \in V_N} \left\{ \frac{1}{2} \mathcal{A}(u, u) - \mathcal{F}(u) \right\}. \qquad (2.2.13)$$

This is known as the *Ritz method*. Putting the expansion (2.2.4) into (2.2.13) and then minimizing J over V_N leads to the same algebraic system as the Galerkin method (2.2.5)–(2.2.7). The interest of this formulation lies in the geometrical interpretation. Using (2.1.19) one may write also

$$\mathcal{A}(u_N - \hat{u}, u_N - \hat{u}) = \min_{v \in V_N} \mathcal{A}(v - \hat{u}, v - \hat{u}), \qquad (2.2.14)$$

which shows that of all elements belonging to V_N, $u_N(x)$ is at the shortest distance in energy norm from $\hat{u}(x)$. Subtraction of (2.2.3) from (2.1.18) shows indeed that $u_N(x)$ is the orthogonal projection of $\hat{u}(x)$ onto V_N in the energy inner product $(\cdot, \cdot)_\mathcal{L}$.

Petrov–Galerkin Methods

We will meet circumstances in forthcoming chapters where, for a given problem, the Galerkin approximation (2.2.3) offers poor numerical performance, although the problem satisfies the coercivity and boundedness requirements. A classical example is the advection-dominated diffusion equation where, in the absence of precautions, spurious oscillations (*wiggles*) may develop in the numerical solution. This example will be analyzed more closely in Chapter 3. A remedy to these wiggles is provided by Petrov–Galerkin methods.

We introduce two families of finite-dimensional subspaces of V, $\{V_N\}_{N=1}^\infty$ and $\{W_N\}_{N=1}^\infty$, satisfying the property (2.2.2). A *Petrov–Galerkin method* is applied when the approximation of (2.2.1) is as follows: Find $u_N \in V_N$ such that

$$\mathcal{A}(u_N, w) = \mathcal{F}(w) \qquad \forall w \in W_N. \qquad (2.2.15)$$

For example, let $V_N := \{\psi_k(x)\}_{k=0}^N$ and $W_N := \{\theta_k(x)\}_{k=0}^N$ denote two different subspaces of V. The problem (2.2.15) reduces to a linear system (2.2.5) with a matrix H having elements

$$H_{mn} := \mathcal{A}(\psi_n, \theta_m), \qquad 0 \le m, n \le N, \qquad (2.2.16)$$

and with r.h.s. components

$$(M\underline{f})_m := \mathcal{F}(\theta_m), \qquad 0 \le m \le N. \qquad (2.2.17)$$

The existence and convergence properties of $u_N(x)$, the solution to the Petrov–Galerkin approximation (2.2.15), are governed by the same assumptions as those of Theorem 2.2.

2.2. An Approximation Framework

Generalized Galerkin Method

So far we have assumed that the quadratures in $\mathcal{A}$ and $\mathcal{F}$ were performed *exactly*. We will develop families of numerical methods where these quadratures are performed *numerically*. Our interest in this book will be focused on high-order Gaussian quadrature schemes. In this case, a Galerkin approximation of the variational problem (2.1.18) will be written as follows: Find $u_N \in V_N$ such that

$$\mathcal{A}_N(u_N, v) = \mathcal{F}_N(v) \qquad \forall v \in V_N, \tag{2.2.18}$$

where $\mathcal{A}_N$ and $\mathcal{F}_N$ are approximations of $\mathcal{A}$ and $\mathcal{F}$ derived from Gaussian quadrature rules, defined on $V_N \times V_N$ and V_N, respectively. This procedure, called the *generalized Galerkin method* by Quarteroni and Valli [319], is at the heart of the spectral-element method (SEM) that we introduce in Section 2.4.

In this particular case, we have a stability and convergence result contained in the following theorem.

Theorem 2.3 *Assume that the bilinear form* $\mathcal{A} : V \times V \to \mathbb{R}$ *satisfies the boundedness and coercivity conditions of Theorem 2.1. Suppose further that these conditions are fulfilled when $\mathcal{A}$ and $\mathcal{F}$ are replaced by $\mathcal{A}_N$ and $\mathcal{F}_N$. Then, for a continuous f, there exists a unique solution $u_N \in V_N$ to (2.2.18) that satisfies the inequality*

$$\|u_N\|_V \leq C \|I_N f\|_{L^2}, \tag{2.2.19}$$

where $I_N f$ is the interpolation of f at the Gaussian quadrature nodes. Moreover, if $\hat{u}$ is the solution of (2.1.18), one has the error bound

$$\|\hat{u} - u_N\|_V \leq \frac{C}{\rho_N} \left[\inf_{v \in V_N} \left(\|\hat{u} - v\|_V + \sup_{\substack{z \in V_N \\ z \neq 0}} \frac{\mathcal{A}(v, z) - \mathcal{A}_N(v, z)}{\|z\|_V} \right) \right.$$
$$\left. + \sup_{\substack{z \in V_N \\ z \neq 0}} \frac{\mathcal{F}(z) - \mathcal{F}_N(z)}{\|z\|_V} \right], \tag{2.2.20}$$

where ρ_N is the ellipticity constant of the discrete bilinear form $\mathcal{A}_N$.

2.2.2 Collocation Approximation

Consider again the BVP (2.2.1) with Dirichlet conditions. We introduce an ordered set of nodes $\{x_i\}_{i=0}^N$ including the boundaries. Let V_N represent a finite-dimensional subspace of V spanned by the basis functions $\{\psi_n(x)\}_{n=0}^N$. We assume that these basis functions have preliminarily been bi-orthonormalized with respect to the set of pointwise interpolation functionals at the elements of $\{x_i\}_{i=0}^N$ (see Davis [91]) so that

$$\psi_k(x_l) = \delta_{kl}, \quad 0 \leq k, l \leq N, \tag{2.2.21}$$

where δ_{kl} denotes the Kronecker symbol (1 for $k = l$ and 0 otherwise). Every function $u_N(x) \in V_N$ therefore has an expansion (2.2.4) in terms of the basis functions ψ_k where the coefficients are local values of the unknown,

$$u_N(x) = \sum_{k=0}^N u_N(x_k)\psi_k(x). \tag{2.2.22}$$

We define the collocation approximation of $\hat{u}$ as the element $u_N \in V_N$ such that

$$\mathcal{L}u_N(x_i) = f(x_i) \quad \forall x_i \in \Omega, \tag{2.2.23}$$
$$\mathcal{B}u_N(x_i) = 0 \quad \forall x_i \in \partial\Omega. \tag{2.2.24}$$

The differential equation is satisfied at the interior nodes, while the boundary conditions are enforced at the end points. Putting the trial function (2.2.22) into (2.2.1) and imposing the relationships (2.2.23), (2.2.24) gives an algebraic system

$$L_C \underline{u} = \underline{f}, \tag{2.2.25}$$

where L_C is the collocation matrix and the vectors $\underline{u}$ and $\underline{f}$ are made of the nodal values of the unknown and of the right-hand side of (2.2.1).

In the case of Dirichlet boundary conditions at both end points, the matrix L_C has elements

$$L_{C,mn} = \mathcal{L}\psi_n(x_m), \quad 1 \leq m, n \leq N - 1. \tag{2.2.26}$$

At first glance, a comparison between (2.2.6) and (2.2.26) would suggest that the Galerkin and collocation approximations have little in common. In addition, although we do not yet have an expression of the matrix M in a Galerkin approximation, in the collocation case (2.2.26) this matrix, if any, must be equal to the identity matrix. Despite these apparent differences, we will show in Sections 2.4 and 2.5 that the two approximations have something in common.

Before leaving this section we remark that the right-hand side of (2.2.22) is a (Lagrangian) interpolation formula on the ordered set of nodes $\{x_i\}_{i=0}^N$. For simplicity, we henceforth will use the notation I_N to represent an *interpolation operator* from V onto V_N. Accordingly, Equation (2.2.22) might also be written $u_N(x) = I_N u(x)$, where $u(x) \in V$ in this case is the unknown.

2.3 Finite-Element Methods

So far, the approximation principles have not referred to any particular choice of trial and test functions $\{\psi_n(x)\}_{n=0}^N$. Finite elements are such a choice of considerable practical importance, because they enable the implementation of Galerkin approximations with great programming and computation efficiency. We describe their main features on the problem (2.1.9), emphasizing those aspects that will be used in later chapters. Further details may be found in [79, 71, 370]. Multidimensional problems will be treated in Chapter 4, essentially in a tensorized approach.

In any finite-element method (FEM) one starts with a partition of the integration domain into intervals (or elements). Let Δ_E denote a partition of (a, b) with E elements:

$$\Delta_E: \quad a = x_0 < x_1 < x_2 < \cdots < x_{E-1} < x_E = b. \tag{2.3.1}$$

In the sequel we will refer to these nodes as the *primary* nodes, or element boundary nodes. Here and throughout the text we use an index e to denote quantities related to the element Ω^e. In general, all elements $\Omega^e := \{x; x_{e-1} < x < x_e\}$, $1 \le e \le E$, might have different sizes $h_e = x_e - x_{e-1}$. For simplicity, however, we will assume in this section that the partition is homogeneous with a mesh size equal to $h = (b-a)/E$.

For the sake of convenience, we let $\hat{\Omega} := \{\xi | -1 \le \xi \le 1\}$ denote the parent (or reference) element, onto which each Ω^e may be mapped by using the affine transformation

$$\xi = \frac{2x - (x_{e-1} + x_e)}{h}, \quad 1 \le e \le E, \tag{2.3.2}$$

with inverse

$$x = \frac{1-\xi}{2} x_{e-1} + \frac{1+\xi}{2} x_e, \quad 1 \le e \le E. \tag{2.3.3}$$

We denote by $\mathbb{P}_p(\hat{\Omega})$ ($p = 1, \ldots$), the linear space of polynomials of degree p on the parent element $\hat{\Omega}$, with $p+1$ degrees of freedom.

The analysis of a FEM starts at the local level of the parent element, where a set of polynomial basis functions is introduced. These functions are the building blocks that, when pieced together on Δ_E, element by element, form the global basis $\{\psi_n(x)\}_{n=0}^N$ of V_N, the elements of which are used as trial functions in the expansion (2.2.4). The epithet "global," in this context, refers to the entire domain of integration, as opposed to "local," which applies to the element level. Thus, each member of the basis of V_N is a patchwork of local functions mapped from the parent element onto each interval Ω^e of Δ_E. The dimension of the basis (i.e., the number of degrees of freedom), $\mathcal{N}$, is related to E and to the polynomial degree p by $\mathcal{N} = Ep + 1$.

We examine two different finite-element approximations, known as the h- and p-versions of the FEM, as an example of implementation of Galerkin approximations. In each case we indicate the main steps that, starting from the parent element $\hat{\Omega}$, lead to the algebraic system (2.2.5).

2.3.1 The h-Version of Finite Elements

We start our analysis with a description of the h-version of the FEM.

Global Mesh and Set of Basis Functions

Every FEM of degree p is characterized by an expansion of the elements of $\mathbb{P}_p(\hat{\Omega})$, in terms of $p + 1$ basic polynomial functions $\{\pi_i(\xi)\}_{i=0}^p$ and their associated coefficients $\{\varrho_i\}_{i=0}^p$:

$$w_p(\xi) := \sum_{i=0}^p \varrho_i \pi_i(\xi), \qquad \xi \in \hat{\Omega}, \quad w_p \in \mathbb{P}_p. \tag{2.3.4}$$

The canonical form of w_p would correspond to $\{\pi_i(\xi) = \xi^i\}_{i=0}^p$. This turns out, in fact, to be a bad choice, for stability reasons. Most of the time, the expansion (2.3.4) is derived from Lagrangian or Hermitian interpolation theory, where the parameters $\{\varrho_i\}_{i=0}^p$ are nodal values or nodal derivatives of a function.

As a first example, consider an elemental grid on $\hat{\Omega}$ with $p + 1$ distinct nodes, $\Xi_{p+1} := \{\xi_0, \xi_1, \ldots, \xi_p\}$. The Lagrangian interpolation polynomial at Ξ_{p+1} of any smooth function $f(\xi)$ on $[-1, 1]$ is $I_p f \in \mathbb{P}_p$ with

$$I_p f(\xi) := \sum_{i=0}^p f(\xi_i) \pi_i(\xi), \qquad \xi \in \hat{\Omega}, \tag{2.3.5}$$

and

$$\pi_i(\xi) = \prod_{j \neq i} \frac{\xi - \xi_j}{\xi_i - \xi_j}, \qquad 0 \leq i, j \leq p. \tag{2.3.6}$$

The polynomials $\{\pi_i(\xi)\}_{i=0}^p$ form the Lagrangian interpolation basis on the parent element, satisfying the bi-orthonormality relationships (2.2.21) with the

2.3. Finite-Element Methods

nodal values on Ξ_{p+1}. The location of the nodes $\{\xi_i\}_{i=0}^p$ on $\hat{\Omega}$ is arbitrary. However, to comply with the C^0 continuity constraint on the trial functions derived in Section 2.1.3, the end points of $\hat{\Omega}$ must be included into Ξ_{p+1}. In this case, continuity is automatically ensured by the existence of a nodal value at each mesh point in Δ_E, and the resulting basis functions belong to the Sobolev space $H^1(a, b)$. This finite element is the simplest element belonging to the family of tensorized Lagrangian finite elements of degree p in d-dimensional space (here, $d = 1$), which in Ciarlet's taxonomy [79] are denoted by $\mathcal{Q}_p$. The $\mathcal{Q}_1$ element uses the polynomial basis

$$\pi_0(\xi) = \frac{1-\xi}{2} \quad \text{and} \quad \pi_1(\xi) = \frac{1+\xi}{2}, \tag{2.3.7}$$

while the $\mathcal{Q}_2$ element uses

$$\pi_0(\xi) = \frac{\xi(\xi-1)}{2}, \quad \pi_1(\xi) = 1-\xi^2, \quad \pi_2(\xi) = \frac{\xi(\xi+1)}{2}. \tag{2.3.8}$$

Another important example is the cubic Hermite finite element, where the functionals in the expansion are the nodal values and first-order derivatives of the function at the end points of $\hat{\Omega}$. Given a smooth function $f(\xi)$ on $[-1, +1]$, one writes its cubic Hermite interpolation $I_3^H f(\xi)$ as

$$I_3^H f(\xi) := \sum_{i=0,1} \sum_{j=\pm 1} f^{(i)}(\xi_j) \pi_{i,j}(\xi), \tag{2.3.9}$$

with $\xi_{\pm 1} := \pm 1$ and the basis functions

$$\pi_{0,j}(\xi) = \frac{1}{4}(1+j\xi)(-\xi^2 + j\xi + 2),$$

$$\pi_{1,j}(\xi) = \frac{(-1)^j}{4}(1+j\xi)^2(j\xi - 1). \tag{2.3.10}$$

The expansion (2.3.9) provides enhanced smoothness compared with (2.3.5), since continuity is ensured for $I_3^H f$ and its first-order derivative. This finite-element, denoted $\mathcal{P}_3$ in Ciarlet's nomenclature, belongs to the Sobolev space $H^2(a, b)$.

Having chosen a Lagrangian polynomial expansion (2.3.4) on $\hat{\Omega}$, one may construct the set of global basis functions $\{\psi_n(x)\}_{n=0}^N$. Using (2.3.3) to map the elemental grid Ξ_{p+1} onto each interval of Δ_E gives the global finite-element grid $\{x_n\}_{n=0}^N$, made of primary and secondary nodes. The primary nodes are the mesh points $\{x_e\}_{e=0}^E$ belonging to Δ_E, while the secondary nodes $\{x_{e+i/p}\}_{i=1}^{p-1}$ are the images of the inner nodes of Ξ_{p+1} onto each element Ω^e. In our example

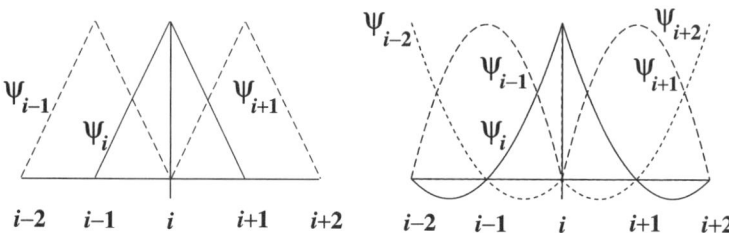

Figure 2.3.1. Piecewise linear and quadratic ($p = 1, 2$) finite elements.

the global grid contains a total of $\mathcal{N} = Ep + 1$ mesh points, with $E + 1$ primary nodes and $p - 1$ secondary nodes per element. Then, (2.3.2) is used to map the $p + 1$ Lagrangian interpolation polynomials (2.3.6) onto each finite element as well. Clearly, since every image of an interpolation polynomial is located in one element only, the global basis functions $\psi_n(x)$ have essentially *local support*: those corresponding to a secondary node vanish identically outside the element containing the node, whereas those corresponding to a primary node extend over the adjacent intervals and vanish identically elsewhere.

Figure 2.3.1 displays two sets of Lagrangian finite-element basis functions with piecewise linear and quadratic polynomials ($p = 1, 2$). They are typical examples of functions belonging to a Sobolev space H^1. The linear elements have no secondary nodes, while the quadratic elements have one secondary node per element.

In Chapter 4 we will introduce the concept of (multidimensional) *isoparametric* elements. At this stage we emphasize that such elements are characterized by polynomial expansions (2.3.5) where both the unknown and the element boundaries are represented by the same expansions. As an example, the polynomial functions (2.3.8) are the ingredients of the second-order isoparametric finite-element transformation where, in two space dimensions, part of an element boundary is described by

$$(x, y) = (x, y)_e \pi_0(\xi) + (x, y)_{e+1/2} \pi_1(\xi) + (x, y)_{e+1} \pi_2(\xi), \quad -1 \leq \xi \leq 1,$$

$(x, y)_{e+k/2}$ ($k = 0, 1, 2$) denoting points located on the boundary.

Stiffness and Mass Matrices

With the global finite-element basis $\{\psi_n(x)\}_{n=0}^N$ now unambiguously defined and

$$V_N := \text{span}\{\psi_0, \psi_1 \ldots, \psi_N\}, \quad N = Ep,$$

2.3. Finite-Element Methods

any trial function $u_N \in V_N$ may be decomposed as (2.2.4). We emphasize again that the parameters u_n in this case are nodal values on the finite-element grid because of the choice of Lagrangian bases (2.3.6). Hence, (2.2.4) may be used as a finite-element interpolation formula. Let $u(x)$ denote any function belonging to $H^1(a, b)$; its finite-element interpolation $I_N u$ into V_N is thus given by

$$I_N u(x) := \sum_{n=0}^{N} u(x_n) \psi_n(x). \qquad (2.3.11)$$

Going back to the Galerkin method, one can now evaluate the matrices H and M in (2.2.5) with more detail. The local character of the basis functions renders the task of evaluating these quantities easy and systematic. All calculations may be performed on $\hat{\Omega}$ by using the polynomial basis (2.3.6) and then may be translated onto all finite elements with a scaling factor to compensate for the affine transformation (2.3.2). To illustrate the method by an example, we perform the computations in the case of the linear Lagrangian element Q_1, assuming the coefficient $p(x)$ in (2.1.9) to be constant in the integration domain and equal to p. Furthermore, we set $q(x) = 0$.

The basic building block for the assembly of matrix H in this particular case is a 2×2 matrix originating from the differentiated term in $\mathcal{A}(u, v)$ on the parent element $\hat{\Omega}$:

$$\hat{K} := \left\{ \int_{-1}^{1} \pi'_m \pi'_n \, d\xi \right\}_{m,n=1,2} = \begin{pmatrix} 1/2 & -1/2 \\ -1/2 & 1/2 \end{pmatrix}, \qquad (2.3.12)$$

which is called the *element stiffness matrix* in the finite-element literature. Then, taking into account the appropriate scaling factor on each element Ω^e, one gets the *local stiffness matrix* K^e:

$$K^e := \frac{2}{h} \hat{K}, \qquad e = 1, \ldots, E.$$

Finally, the matrix operator H results from the assembly of *all* local stiffness matrices K^e. Using an element-by-element process of fetching and summing known as *direct stiffness summation*, one enters the individual contributions of all primary and secondary nodes into the appropriate components of H.

To complete the assembly process, one implements the boundary conditions. Following the previous discussion, this is straightforward in our particular example: a homogeneous Dirichlet condition is enforced on the numerical solution at the left, while at the right the variational principle provides the homogeneous Neumann condition naturally (i.e., the condition is automatically included).

In the particular case where $p = 1$ and $q = 0$ corresponding to the Laplace equation, we get the matrix

$$K = \frac{1}{h} \begin{pmatrix} 1 & 0 & & & & \\ -1 & 2 & -1 & & & \\ & \ddots & \ddots & \ddots & & \\ & & -1 & 2 & -1 & \\ & & & -1 & 1 & \end{pmatrix}. \qquad (2.3.13)$$

Other strategies for applying Dirichlet boundary conditions, which preserve the symmetry of K, are described in Sections 4.3.2 and 4.5.2.

In the finite-element literature the matrix K, related to the component $u'v'$ in the variational functional (2.1.12), is called the (global) *stiffness matrix* (see, for instance, [214]).

The right-hand side of (2.2.5) must now be evaluated taking into account the fact that $f(x)$ is an arbitrarily given function. Theoretically, all components $\mathcal{F}(\psi_m)$, $0 \le m \le N$, of (2.2.7) could be determined exactly (albeit numerically), at some computer expense. In practice, however, the computation is never done in this way. To evaluate $\mathcal{F}(\psi_m)$, one replaces the forcing term $f(x)$ by its finite-element interpolation in V_N given by (2.3.11). The components of the load vector are thus approximated by

$$\sum_{n=0}^{N} f(x_n) \int_a^b \psi_m(x) \psi_n(x) \, dx, \qquad 0 \le m \le N. \qquad (2.3.14)$$

The local character of the global basis functions $\{\psi_m(x)\}_{m=0}^{N}$ makes most of the integrals in (2.3.14) identical to zero. Those that do not vanish are computed on $\hat{\Omega}$.

One recognizes in (2.3.14) the components of a 2×2 matrix $\hat{M}$ analogous to $\hat{K}$. This matrix is called the *element mass matrix*. It is equal to

$$\hat{M} := \left\{ \int_{-1}^{1} \pi_m \pi_n \, d\xi \right\}_{m,n=1,2} = \begin{pmatrix} 2/3 & 1/3 \\ 1/3 & 2/3 \end{pmatrix}. \qquad (2.3.15)$$

Using the element mass matrix $\hat{M}$ and appropriate scaling factors (in this case $h/2$), one gets the *local mass matrices* M^e. Currently, they are all equal to

$$M^e := \frac{h}{2} \hat{M}, \qquad e = 1, \ldots, E.$$

The building process ends with the (global) *mass matrix* M obtained from the local mass matrices M^e by the same assembly process used to build K,

2.3. Finite-Element Methods

and with the r.h.s. of (2.2.5) resulting from the application of M to $\underline{f} = (f_0, f_1, \ldots, f_N)^T$, the vector of nodal values of $f(x)$ on the finite-element grid.

As a result, the algebraic system (2.2.5) produced by the finite-element approximation of the problem with $p = 1$ and $q = 0$ is

$$H\underline{u} = M\underline{f}, \tag{2.3.16}$$

with $H = K$ and with a mass matrix

$$M = \frac{h}{2} \begin{pmatrix} 0 & 0 & & & & \\ 1/3 & 4/3 & 1/3 & & & \\ & \ddots & \ddots & \ddots & & \\ & & & 1/3 & 4/3 & 1/3 \\ & & & & 1/3 & 2/3 \end{pmatrix}. \tag{2.3.17}$$

When $q(x) \neq 0$, the preceding procedure is still applicable. In this case, the mass matrix M also enters into the composition of the left-hand side (l.h.s.) matrix operator H. This is due to the uv component in the variational functional (2.1.12). For constant coefficients p and q, one has

$$H = pK + qM, \tag{2.3.18}$$

with K and M given by (2.3.13) and (2.3.17), respectively.

Finally, when the coefficients $p(x)$ and $q(x)$ in the problem (2.1.9) are non-constant, the evaluation of the stiffness and mass matrices K and M is somewhat different. In this case, $p(x)$ and $q(x)$ are replaced by their finite-element interpolation $I_N p(x)$ and $I_N q(x)$ in V_N, using (2.3.11). The local matrices K^e and M^e are then built from sets of elemental stiffness and mass matrices $\{\hat{K}_\ell\}_{\ell=1}^2$ and $\{\hat{M}_\ell\}_{\ell=1}^2$ by a local assembly process. When $\mathcal{Q}_1$ Lagrangian elements are used, each set has two members given by

$$\hat{K}_\ell := \left\{ \hat{K}_{\ell,mn} = \int_{-1}^{1} \pi_\ell \pi'_m \pi'_n \, d\xi, \right\}_{m,n=1,2},$$

$$\hat{M}_\ell := \left\{ \hat{M}_{\ell,mn} = \int_{-1}^{1} \pi_\ell \pi_m \pi_n \, d\xi, \right\}_{m,n=1,2},$$

with the index $\ell = 1, 2$. Once these elements have been computed, K and M are assembled element by element, with $\{\hat{K}_\ell\}_{\ell=1}^2$, $\{\hat{M}_\ell\}_{\ell=1}^2$, and the appropriate scaling factors and nodal values of $p(x)$ and $q(x)$ as weighting coefficients. The assembly process gives an algebraic system (2.3.16) with a symmetric matrix H.

The generation of the finite-element equations (2.3.16) is easily automated, irrespective of the polynomial degree. With the element $\mathcal{Q}_1$, the matrices, H and M, are tridiagonal. In the case of Lagrangian elements of degree p these matrices would have a band structure with a half bandwidth w equal to p. We recall that the half bandwidth $w(A)$ of a banded matrix A is defined as the smallest integer such that $a_{ij} = 0$ for all (i, j) satisfying $|i - j| > w(A)$. Since, most of the time, $w(H) \ll \mathcal{N}$, where $\mathcal{N}$ is the number of unknowns, the linear system (2.2.5) is sparse and may be solved by direct solution techniques. For example, LU decomposition would give the solution at a computational cost asymptotically proportional to $\mathcal{N}w^2(H)$ floating-point operations (flops) for the factorization stage, and proportional to $\mathcal{N}w(H)$ flops for the forward and backward substitutions. Other solution techniques may be used as well. We postpone further discussion of this topic to Section 2.7.

Having described the implementation technique of finite elements based on interpolation, we need only clarify the title of this section. Finite elements are said to be of *h-type* when the degree p of the interpolation formula is fixed and when any change of discretization to enhance accuracy implies a mesh refinement, that is, a reduction in h.

2.3.2 The p-Version of Finite Elements

In contrast with the preceding subsection, finite elements are said to be of *p-type* when the partition Δ_E with $E + 1$ mesh points (2.3.1) is fixed and any change of discretization is introduced through a modification in the degree of the polynomial space $\mathbb{P}_p$. The degree may depend on the element Ω^e in Δ_E, particularly if local refinements are imposed by the complexity of the problem.

FEMs of *p*-type are usually defined in terms of the Legendre polynomials $L_j(\xi)$ of degree j ($j = 2, \ldots, p$) (see, e.g., Szabó and Babuška [370]). The polynomial expansion of degree p on $\hat{\Omega}$ corresponding to (2.3.4), is

$$w_p(\xi) := w_p(-1)\frac{1-\xi}{2} + w_p(1)\frac{1+\xi}{2} + \sum_{j=2}^{p} \varrho_j \phi_j(\xi), \quad \xi \in \hat{\Omega}, \quad (2.3.19)$$

with

$$\phi_j(\xi) := \left(\frac{2j-1}{2}\right)^{1/2} \int_{-1}^{\xi} L_{j-1}(t)\,dt$$

$$= \frac{1}{\sqrt{2(2j-1)}}(L_j(\xi) - L_{j-2}(\xi)), \quad 2 \le j \le p. \quad (2.3.20)$$

The first two terms in the expansion (2.3.19) ensure C^0 continuity of the trial functions $\{\psi_n\}_{n=0}^{N}$ on the integration domain. The $p - 2$ remaining terms with

2.3. Finite-Element Methods

basis functions given by (2.3.20) are inner modes (also called *bubble functions*), which vanish identically on the boundaries of $\hat{\Omega}$ [see Equation (B.1.17)].

The p- and h-versions of FEMs differ in one main aspect: the unknowns ϱ_j in (2.3.19) are simply parameters (i.e., unspecified linear functionals), whereas all functionals in the h-version are nodal values or derivatives, leading to physical interpretation of the solution. In compensation, finite-element approximations of p-type possess the interesting property of forming a *hierarchical* structure in the sense that the expansion of polynomial degree p lies in the expansion of polynomial degree $p + 1$, allowing for adaptive computations to be made quite naturally.

The implementation procedure of the Galerkin method in the p-version of the FEM follows the same pattern as that indicated in the h-version. For instance, in the constant coefficient case, one ends up with an algebraic system

$$H\underline{u} = \underline{f}, \qquad (2.3.21)$$

where $H = pK + qM$, K and M being sparse stiffness and mass matrices. The r.h.s. of (2.3.21) results from a numerical integration of $\mathcal{F}(v)$ in (2.2.3), each component of $\underline{f}$ corresponding to setting v to a particular trial function $\psi_n(x)$ ($n = 0, \ldots, N$), a building block of the piecewise polynomial space V_N.

Figure 2.3.2 displays the (nonzero) structure of element stiffness matrix $\hat{K}$ and mass matrix $\hat{M}$ of a p-type FEM with $p = 11$. The ordering of the unknowns is $(w_p(-1), w_p(+1), \varrho_2, \ldots, \varrho_{11})^T$. The analytical form of the bubble functions (2.3.20) makes the stiffness matrix almost diagonal. Notice also that, because of the hierarchical property of the p-type FEM, the element stiffness and mass matrices corresponding to $p = 10$ can be obtained very simply from those with $p = 11$, by dropping the last rows and last columns of $\hat{K}$ and $\hat{M}$.

When the coefficients p and q in problem (2.1.9) are nonconstant, the procedure used for the h-type FEM to evaluate the components of the l.h.s. of (2.3.21) is no longer applicable. In this case, the matrix constituents of H must

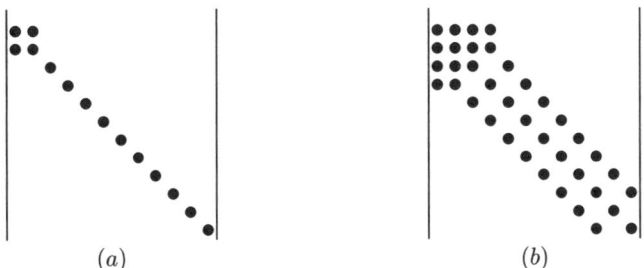

Figure 2.3.2. Structures of (*a*) element stiffness matrix $\hat{K}$ and (*b*) element mass matrix $\hat{M}$ of a p-type FEM with $p = 11$.

be computed element by element using a numerical quadrature

$$H_{mn} = \sum_{e=1}^{E} \left(\int_{\Omega^e} p(x) \frac{d\psi_m}{dx} \frac{d\psi_n}{dx} dx + \int_{\Omega^e} q(x) \psi_m(x) \psi_n(x) dx \right), \quad (2.3.22)$$

with $0 \le m, n \le N$. Once these coefficients are known, the l.h.s. matrix H is assembled as before. Most of the time, the element blocks are now full.

The vector of unknowns $\underline{u}$, which comprises the nodal values at the grid points of Δ_E and the coefficients of the bubble modes in each finite element, contains all the information needed to evaluate the approximate solution u_N in (a, b). Further details may be found in [370].

2.4 Spectral-Element Methods

Spectral-element methods (SEMs) form another family of approximation schemes based on the Galerkin method. These methods, which were introduced by Patera [297], have close connections with finite-element discretizations and share specific features with both the h- and p-versions of the FEM. Common to spectral elements and the h-version are a Lagrangian interpolation formula on the parent element $\hat{\Omega}$ and the fact that the basis functions have *local* support. With the p-version, spectral elements share the preferred use of high-degree polynomials on a fixed (coarse) geometric mesh, in order to take advantage of enhanced accuracy when solving smooth problems.

The specific character of SEMs consists in their close relationship with orthogonal polynomials and Gaussian quadrature. Orthogonality between basis functions in the h- and p-versions of the FEM is uniquely due to nonoverlapping local functions. This is where SEMs depart from the two other families; orthogonality is related to both the topological nature (local extension) and the *analytical* nature of the basis functions.

Two implementations have been proposed, based on Chebyshev [297] and on Legendre [336] polynomials. In both cases Lagrangian interpolation is performed on the related Gauss–Lobatto quadrature grid in order to ensure C^0 continuity of the solution and to benefit from the associated numerical quadrature schemes. Chebyshev polynomials were chosen because they offered the possibility of using fast transform techniques. However, the quadratures needed to evaluate the stiffness and mass matrices were performed analytically without introduction of the associated Chebyshev weighting factor $(\sqrt{1-x^2})^{-1}$. Although the results for smooth problems were showing the expected exponential behavior in accuracy, the implementation with Chebyshev polynomials was abandoned in favor of the more straightforward Legendre implementation. We will therefore concentrate on the latter.

2.4. Spectral-Element Methods

We consider again a partition Δ_E of (a, b) [see (2.3.1)]. However, as in the p-version of the FEM, this partition will be *frozen*. Let $\Xi_{N+1} = \{\xi_0, \xi_1, \ldots, \xi_N\}$ denote the ordered set of $N + 1$ Gauss–Lobatto–Legendre (GLL) quadrature nodes on $\hat{\Omega}$ [see Section B.2, Equations (B.2.5) and (B.2.9)]. These nodes are solutions to the equation

$$(1 - \xi^2)L'_N(\xi) = 0, \qquad \xi \in \hat{\Omega}, \qquad (2.4.1)$$

where L'_N is the derivative of the Legendre polynomial of degree N. The Lagrange interpolation polynomial, of degree N, of any regular function $u(\xi)$ at the GLL quadrature nodes Ξ_{N+1} is

$$I_N u(\xi) = \sum_{j=0}^{N} u(\xi_j)\pi_j(\xi), \qquad \xi \in \hat{\Omega}, \qquad (2.4.2)$$

with $\{\pi_j(\xi)\}_{j=0}^{N}$, the associated interpolation basis of degree N. One may show that the elements of this basis are given by

$$\pi_j(\xi) = \frac{-1}{N(N+1)} \frac{(1 - \xi^2)L'_N(\xi)}{(\xi - \xi_j)L_N(\xi_j)}, \qquad 0 \le j \le N, \qquad \xi \in \hat{\Omega}. \qquad (2.4.3)$$

Figure 2.4.1 shows an example of basis functions (2.4.3), corresponding to $N = 10$. Only half of the basis is displayed.

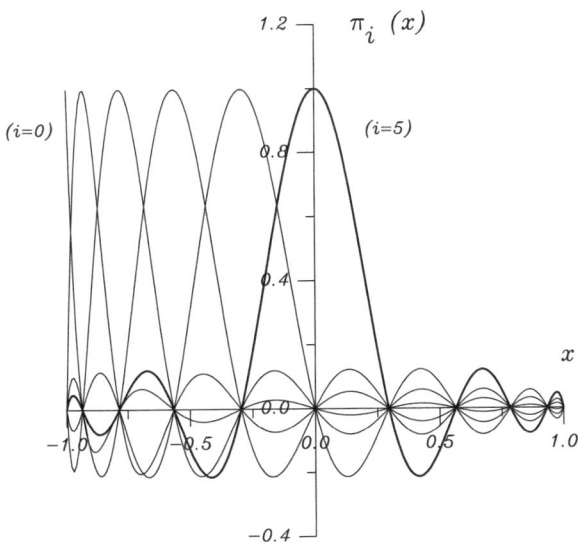

Figure 2.4.1. An example of Legendre spectral-element basis functions (2.4.3) corresponding to $N = 10$.

2. Approximation Methods for Elliptic Problems

The implementation of the Galerkin method (2.2.3) with spectral elements is now similar to the h-version of the FEM. We come back to the model problem (2.1.9) with coefficients $p(x), q(x)$, and with a partition Δ_E of $\Omega := (a, b)$. The spectral element approximation $u_N^e(x)$ of degree N in Ω^e ($e = 1, \ldots, E$), mapped onto $\hat{\Omega}$, is written

$$u_N^e(\xi) = \sum_{j=0}^{N} u_j^e \pi_j(\xi), \qquad \xi \in \hat{\Omega}, \tag{2.4.4}$$

where the coefficients $\{u_j^e\}_{j=0}^{N}$ are nodal values of the unknown in Ω^e. The local matrices H^e and M^e ($e = 1, \ldots, E$) have elements

$$H_{ij}^e := \frac{2}{h_e} \int_{-1}^{1} p^e \frac{d\pi_i}{d\xi} \frac{d\pi_j}{d\xi} d\xi + \frac{h_e}{2} \int_{-1}^{1} q^e \pi_i \pi_j d\xi, \tag{2.4.5}$$

$$M_{ij}^e := \frac{h_e}{2} \int_{-1}^{1} \pi_i \pi_j d\xi, \qquad 0 \leq i, j \leq N, \tag{2.4.6}$$

where p^e and q^e denote the (variable) coefficients of the differential equation mapped from Ω^e onto the parent element. With the particular selection of nodes Ξ_{N+1} in $\hat{\Omega}$, one is tempted to apply Gaussian quadrature for the evaluation of the matrix elements (2.4.5) and (2.4.6). This gives the local matrices with numerical quadrature resulting from the application of (2.2.18). The components of these matrices for the element Ω^e are

$$H_{N,ij}^e = \frac{2}{h_e} \sum_{m=0}^{N} \rho_m p_m^e D_{N,mi}^{(1)} D_{N,mj}^{(1)} + \frac{h_e}{2} \rho_i q_i^e \delta_{ij}, \tag{2.4.7}$$

$$M_{N,ij}^e = \frac{h_e}{2} \rho_i \delta_{ij}, \qquad 0 \leq i, j \leq N. \tag{2.4.8}$$

The coefficients $\{\rho_m\}_{m=0}^{N}$ denote the GLL quadrature weights (B.2.9). The r.h.s. of (2.4.7) uses the nodal values of p^e and q^e on the GLL grid, and the differentiation matrix (B.3.51). The matrix elements

$$D_{N,ij}^{(1)} := \left.\frac{d\pi_j}{d\xi}\right|_{\xi=\xi_i} = \begin{cases} \dfrac{L_N(\xi_i)}{L_N(\xi_j)} \dfrac{1}{\xi_i - \xi_j}, & i \neq j, \\ -\dfrac{(N+1)N}{4}, & i = j = 0, \\ \dfrac{(N+1)N}{4}, & i = j = N, \\ 0 & \text{otherwise} \end{cases} \tag{2.4.9}$$

are the nodal values of the first derivative of the GLL interpolation polynomials (2.4.3). Using the direct stiffness process mentioned in Section 2.3.1,

2.4. Spectral-Element Methods

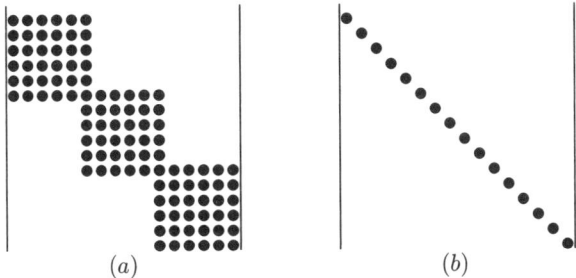

Figure 2.4.2. Global stiffness matrix (*a*) and mass matrix (*b*) structures of a spectral-element approximation with three elements and $N = 5$.

one obtains the algebraic system governing the spectral-element solution $u_N(x) := \{u_N^e(x)\}_{e=1}^E$ [$u_N(x) \in C^0(\Omega)$]:

$$H\underline{u} = M\underline{f}. \tag{2.4.10}$$

In this expression, the vectors $\underline{u}$ and $\underline{f}$ have $\mathcal{N}$ components ($\mathcal{N} = EN + 1$), with $\underline{u} := (u_0^1, u_1^1, \ldots, u_{N-1}^1, u_N^E)^T$, and the same structure for $\underline{f}$.

The stiffness and mass matrices of spectral elements and *p*-type finite elements are quite different in structure. While for *p*-type FEM $\hat{K}$ is almost diagonal (in the constant coefficient case), for spectral elements this matrix is full. On the other hand, due to the cardinality property of the Lagrangian basis (2.4.3) on the GLL grid, for spectral elements the matrix $\hat{M}$ is *diagonal* [see Equation (2.4.8)]. This property, resulting from Gaussian integration, is an important feature from the computational point of view. Figure 2.4.2 displays the structure of the *global* stiffness and mass matrices for a partition into three spectral elements (i.e. $E = 3$) with $N = 5$.

Finally, in order to improve numerical accuracy one usually keeps the partition Δ_E fixed and increases the polynomial degree inside the elements Ω^e, as is the case with *p*-type FEM. The polynomial degree may also be changed from element to element, depending on the nature of the problem and the accuracy requirements. However, in multidimensional problems, local solution refinements are more difficult to implement, as discussed in later chapters.

Remark 2.2 Here we make an important remark about spectral-element methods. Assume the problem (2.1.9) with *constant* coefficients p and q is solved on $\Omega = (-1, 1)$ using one Legendre spectral element of degree N. Let us denote by $\mathbb{P}_{N,0}(\hat{\Omega})$, the set of polynomials of degree N vanishing at $x = -1$. Then the Galerkin approximation with numerical quadrature (2.2.18) is: Find

$u_N \in \mathbb{P}_{N,0}(\hat{\Omega})$ such that

$$\mathcal{A}_N(u_N, v_N) = \mathcal{F}_N(v_N) \qquad \forall v_N \in \mathbb{P}_{N,0}(\hat{\Omega}), \qquad (2.4.11)$$

with

$$\mathcal{A}_N(u_N, v_N) := \sum_{j=0}^{N} \rho_j \left(p \frac{du_N}{dx} \frac{dv_N}{dx} + q u_N v_N \right)(\xi_j), \qquad (2.4.12)$$

$$\mathcal{F}_N(v_N) := \sum_{j=0}^{N} \rho_j f(\xi_j) v_N(\xi_j). \qquad (2.4.13)$$

In (2.4.12) and (2.4.13) we deliberately avoid the expansion of u_N and v_N in the basis functions π_j. Since $u_N, v_N \in \mathbb{P}_N(\hat{\Omega})$, the product $u'_N v'_N \in \mathbb{P}_{2N-2}(\hat{\Omega})$. Consequently, the GLL quadrature of the derivative term in (2.4.12) is *exact*, and one has

$$\sum_{j=0}^{N} \rho_j p \left(\frac{du_N}{dx} \frac{dv_N}{dx} \right)(\xi_j) = \int_{-1}^{1} p \frac{du_N}{dx} \frac{dv_N}{dx} dx$$

$$= -\int_{-1}^{1} p \frac{d^2 u_N}{dx^2} v_N \, dx + p u'_N v_N(1). \qquad (2.4.14)$$

As the product $u''_N v_N \in \mathbb{P}_{2N-2}(\hat{\Omega})$, the argument about the GLL quadrature may be repeated. It turns out that the Galerkin approximation (2.4.11) is equivalent to the following problem: Find $u_N \in \mathbb{P}_{N,0}(\hat{\Omega})$ such that

$$\sum_{j=0}^{N} \rho_j v_N(\xi_j)(\mathcal{L}u_N - f)(\xi_j) + p u'_N v_N(1) = 0 \qquad \forall v_N \in \mathbb{P}_{N,0}(\hat{\Omega}). \qquad (2.4.15)$$

One concludes that the spectral-element solution $\hat{u}_N$ to (2.4.11) satisfies the relationships

$$\mathcal{L}u_N(\xi_j) = f(\xi_j), \qquad 1 \le j \le N-1, \qquad (2.4.16)$$

$$u'_N(1) + \frac{1}{p} \rho_N (\mathcal{L}u_N - f)(1) = 0. \qquad (2.4.17)$$

According to (2.4.16) the differential equation is identically verified at all interior nodes of $\hat{\Omega}$, while the relationship (2.4.17) shows that the Neumann condition at the right boundary is enforced *only asymptotically*, since $\rho_0 = \rho_N \equiv 2/N(N+1)$ [see Equation (B.2.9)]. The equations (2.4.16)–(2.4.17) are similar to the collocation equations (2.2.23)–(2.2.24). Thus, with a slight reservation on the Neumann boundary condition, one may state that, in the case of constant coefficients, a Legendre spectral-element solution is a particular collocation solution of the problem.

The discussion has focused so far on Legendre spectral elements. Similar conclusions may be drawn for SEMs based on Chebyshev polynomials. We turn now to collocation calculations on Gaussian quadrature grids.

2.5 Orthogonal Collocation

We introduced the collocation approximation in Section 2.2.2 in very general terms. Nothing was said about the actual choice of the collocation nodes. It seems obvious from Remark 2.2 that a good choice of the collocation grid may improve the approximation error of the numerical solution of a BVP.

Orthogonal collocation in the engineering literature refers to the family of collocation methods with discretization grids associated to Gaussian quadrature methods (see, for instance, [400]). For the purpose of illustration, we consider again the problem (2.1.9) with space-dependent coefficients on the open reference interval $\Omega := (-1, +1)$. We first describe the orthogonal collocation procedure on a single domain. The approach is then extended to multidomain collocation, according to the technique introduced for the Navier–Stokes equations in higher space dimensions by Demaret and Deville [95].

2.5.1 Orthogonal Collocation in a Monodomain

In the single-domain case, the algorithm is simple. It starts with the selection of a collocation grid and the construction of the associated polynomial functions forming a bi-orthogonal Lagrangian interpolation basis. In order to get the greatest possible choice, we introduce a function related to the Jacobi orthogonal polynomial of degree N, $P_N^{\alpha,\beta}$ [see Equation (B.2.5)]:

$$\chi_N^{(\alpha,\beta)}(\xi) := (1 - \xi^2) \frac{d}{d\xi} P_N^{\alpha,\beta}(\xi), \qquad \xi \in \hat{\Omega}. \tag{2.5.1}$$

The Gauss–Lobatto–Jacobi (GLJ) quadrature nodes $\{\xi_i^J\}_{i=0}^N$ are the solutions to

$$\chi_N^{(\alpha,\beta)}(\xi) = 0. \tag{2.5.2}$$

Two particular cases are important: the Legendre polynomials $L_N(\xi)$ already met in the preceding section (corresponding to $\alpha = \beta = 0$), and the Chebyshev polynomials $T_N(\xi)$ (corresponding to $\alpha = \beta = -1/2$).

Most of the time the GLJ quadrature nodes $\{\xi_i^J\}_{i=0}^N$ must be evaluated numerically. Robust computation routines are quite common in the literature (see, for instance, Press et al. [316] or Villadsen and Michelsen [400]). Fortunately, in the Gauss–Lobatto–Chebyshev (GLC) case, the nodes $\{\xi_i^c\}_{i=0}^N$ are given explicitly by the expression (B.2.13). Whatever the choice of the Gaussian grid, a set of bi-orthonormal Lagrangian interpolation functions $\{\pi_i^J\}_{i=0}^N$ can be built.

Let $I_N u(\xi) \in \mathbb{P}_N$ denote the interpolation polynomial corresponding to the grid defined by (2.5.2):

$$I_N u(\xi) := \sum_{i=0}^{N} u(\xi_i^J)\pi_i^J(\xi), \qquad \xi \in \hat{\Omega}. \tag{2.5.3}$$

The basis functions $\pi_i^J(\xi)$ are given by

$$\pi_i^J(\xi) = C_i \frac{1-\xi^2}{\xi - \xi_i^J} \frac{d}{d\xi} P_N^{\alpha,\beta}(\xi), \tag{2.5.4}$$

where C_i is the normalization constant such that

$$\pi_i^J(\xi_k^J) = \delta_{ik}. \tag{2.5.5}$$

With a little algebra and with the use of the Jacobi polynomials properties, one gets

$$C_i = -\frac{1}{\bar{c}_i N(N + \alpha + \beta + 1)} \frac{1}{P_N^{\alpha,\beta}(\xi_i^J)} \tag{2.5.6}$$

with

$$\bar{c}_i = \begin{cases} 1/(\beta + 1), & i = 0, \\ 1, & i \in [1, N-1], \\ 1/(\alpha + 1), & i = N. \end{cases} \tag{2.5.7}$$

For the particular cases of Legendre and Chebyshev polynomials, the constant C_i is assigned to the classical values

$$C_i = -\frac{1}{N(N+1)L_N(\xi_i^J)} \tag{2.5.8}$$

and

$$C_i = \frac{(-1)^{i+1}}{\bar{c}_i N^2}, \qquad \bar{c}_0 = \bar{c}_N = 2, \quad \bar{c}_i = 1, \quad 1 \leq i \leq N-1, \tag{2.5.9}$$

2.5. Orthogonal Collocation

respectively. Notice that, in the Legendre case, the basis functions (2.5.4) are equal to (2.4.3).

The orthogonal collocation algorithm is now straightforward. Applying Equations (2.2.23)–(2.2.24) on the open domain $\Omega := (-1, +1)$ and its boundary $\partial\Omega$, one gets to the following problem: Find $u_N(x) \in \mathbb{P}_N$ in the form (2.5.3) such that

$$\mathcal{L}(u_N)(\xi_i^J) = f(\xi_i^J), \quad \forall \xi_i^J \in \Omega, \quad (2.5.10)$$

$$\mathcal{B}(u_N)(\xi_i^J) = 0, \quad \forall \xi_i^J \in \partial\Omega. \quad (2.5.11)$$

The ensuing algebraic system (2.2.25) is

$$L_C^J \underline{u} = \underline{f}, \quad (2.5.12)$$

where the vectors $\underline{u}$ and $\underline{f}$ contain the nodal values of the unknown and of the r.h.s. The Jacobi collocation matrix L_C^J has elements $L_{C,mn}^J = \mathcal{L}\pi_n^J(\xi_m^J)$, involving the first and second derivatives of the basis functions at the Gaussian nodes. Analytical expressions of these derivatives for the GLJ quadrature rule are given in Section B.3.3 of Appendix B (see also [100]).

We emphasize that the linear system related to collocation calculations has much less sparsity than do linear systems induced by low-order finite-element calculations. In the one-dimensional case, the matrix L_C^J is even full. In the multidimensional case it is banded, but the bandwidth is very large. We stress also that, although the differential equation (2.1.9) is self-adjoint, the linear system (2.5.12) is nonsymmetric. This is because for two different basis functions π_m^J and π_n^J, one usually has $\mathcal{L}\pi_m^J(\xi_n^J) \neq \mathcal{L}\pi_n^J(\xi_m^J)$. Finally, the matrix L_C becomes very *ill-conditioned* with increasing values of N, a concept that will be defined more precisely later in this chapter and that has to do with the sensitivity of the linear system solution to roundoff errors. These properties do not prevent orthogonal collocation methods from giving extremely accurate solutions to regular problems, however, as will be shown through some numerical examples.

2.5.2 Orthogonal Collocation in a Multidomain

The preceding algorithm can be extended to multidomains. For simplicity, we will limit ourselves to a two-domain approximation, and we will give the essential features of the algorithm in the Chebyshev case.

Let $\Omega^- := (-1, 0)$ and $\Omega^+ := (0, 1)$ denote the two subdomains of $\Omega := (-1, +1)$, where approximations of degree N, $u_N^-(x)$ and $u_N^+(x)$, to the solution of (2.1.9) are required. We denote by $\Omega^{\pm} := \Omega^{-} \cup \Omega^{+}$ and $\Gamma := \overline{\Omega^{-}} \cap \overline{\Omega^{+}}$, respectively, the interior and interface of the two subdomains. Both subdomains are mapped onto the reference element $\hat{\Omega}$ by using the mapping procedure described in Section 2.3. One is naturally led to the analytical expansions

$$u_N^{\pm}(\xi) = \sum_{j=0}^{N} u_j^{\pm} \pi_j^c(\xi), \qquad \xi \in \hat{\Omega}, \qquad (2.5.13)$$

where ξ is the coordinate inside the reference element $\hat{\Omega}$ and

$$\xi = \begin{cases} 2x + 1, & x \in \Omega^{-}, \\ 2x - 1, & x \in \Omega^{+}. \end{cases} \qquad (2.5.14)$$

Let Ξ_{N+1} denote the ordered set of secondary nodes on $\hat{\Omega}$, in this case the GLC quadrature nodes (B.2.13). A mapping of Ξ_{N+1} onto Ω^{-} and Ω^{+}, using the inverse of (2.5.14), gives the set of secondary nodes $\{x_j^{-}\}_{j=0}^{N}$ and $\{x_j^{+}\}_{j=0}^{N}$. The quantities $u_j^{\pm}$, $0 \leq j \leq N$, are the nodal values of the trial elements $u_N^{\pm}$ on the secondary nodes. The functions $\pi_j^c(\xi)$ are the Lagrangian interpolation polynomials on the Chebyshev grid (B.3.20):

$$\pi_j^c(\xi) = \frac{(-1)^{j+1}}{\bar{c}_j N^2} \frac{(1 - \xi^2) T_N'(\xi)}{\xi - \xi_j^c}, \qquad 0 \leq j \leq N, \qquad \xi \in \hat{\Omega}, \quad (2.5.15)$$

with the convention on $\bar{c}_j$ given by (2.5.9).

The orthogonal collocation algorithm is then straightforward. The key point is matching the partial solutions at the interface Γ. One requires maximum regularity of the approximation $u_N(x) = \{u_N^{-}(x), u_N^{+}(x)\}$, which, as explained earlier, belongs to $H^1(\Omega)$. To get $u_N \in C^1(\Omega)$, one must satisfy the constraints

$$\frac{d^k}{d\xi^k}(u_N^{-})(1) = \frac{d^k}{d\xi^k}(u_N^{+})(-1), \qquad k = 0, 1, \qquad (2.5.16)$$

called *patching* conditions.

This relationship assumes continuity of the diffusion coefficient $p(x)$. If $p(x)$ is discontinuous at the inner boundary Γ with left and right values p^{-} and p^{+}, then (2.5.16) should be replaced by a similar relationship for the quantity $p \, du_N/d\xi$ representing the local "flux" of the (physical) variable u:

$$p^{-} \frac{d}{d\xi}(u_N^{-})(1) = p^{+} \frac{d}{d\xi}(u_N^{+})(-1). \qquad (2.5.17)$$

Let us add the constraints on the interface (2.5.16) [or (2.5.17)] into the boundary conditions (2.2.24) of the problem. Orthogonal collocation on the multidomain now is

$$\mathcal{L}(u_N^\pm)(x_i^\pm) = f(x_i^\pm) \quad \forall x_i^\pm \in \Omega^\pm, \qquad (2.5.18)$$

$$\mathcal{B}(u_N)(x_i^\pm) = 0 \quad \forall x_i^\pm \in \partial\Omega \cup \Gamma. \qquad (2.5.19)$$

It gives an algebraic system (2.2.25) with sparser structure than (2.5.12).

2.6 Error Estimation

How far is the approximate solution u_N from $\hat{u}$, the exact solution to the problem? An error estimate may be obtained by using the inequality (2.2.10). By considering $\hat{u}_I := I_N \hat{u}$, the interpolant of $\hat{u}$ in V_N given by (2.3.11), one can get an upper bound of the approximation error in some V-norm in terms of the *interpolation* error in the same norm, since

$$\|u_N - \hat{u}\|_V \le C \min_{u \in V_N} \|u - \hat{u}\|_V \le C \|\hat{u}_I - \hat{u}\|_V.$$

We first look at the h-version of the FEM, assuming again a homogeneous partition of the domain with mesh size h and Lagrangian elements Q_1 for (2.1.9). Refining the mesh, one expects the error in the Sobolev norm $\|u_N - \hat{u}\|_{H^1}$ to go to zero as $h \to 0$. However, the *rate* at which this result is achieved depends essentially on the regularity properties of $\hat{u}$, as indicated below.

Using a Fourier expansion of the interpolation error $\varepsilon_I(x) := \hat{u}_I - \hat{u}$ on a particular mesh, one can show that if $\hat{u}(x) \in H^2(a, b)$, then the approximation error in the Sobolev norm $\|\cdot\|_{H^1}$ behaves like

$$\|u_N - \hat{u}\|_{H^1} \le Ch|\hat{u}|_{H^2}, \qquad (2.6.1)$$

where C is a generic constant, independent of the discretization parameter, and $|\cdot|_{H^2}$ denotes the Sobolev seminorm (2.1.36). Inequalities of the type (2.6.1) are easy to establish, because they are a direct consequence of the minimization property (2.2.14). One would like also to characterize the approximation in the more natural L^2 norm $\|\cdot\|$ defined by (A.4.8). No minimization property exists in this norm, however. To derive an error bound, one uses an argument due to Aubin and Nitsche, called *Nitsche's trick* [285]. Its application with the elements Q_1 yields

$$\|u_N - \hat{u}\| \le \tilde{C}h^2|\hat{u}|_{H^2}, \qquad (2.6.2)$$

where $\tilde{C}$ denotes another generic constant. The latter expression gives the correct convergence order in the L^2 error norm for linear finite elements. The difference between the results (2.6.1) and (2.6.2) is due to the presence of the first derivative in the Sobolev norm $\|\cdot\|_{H^1}$, which reduces the convergence order by one.

If the preceding results are generalized to the elements $\mathcal{Q}_k$, the Lagrange interpolation theory shows that for sufficiently smooth solutions (i.e. $\hat{u} \in H^{k+1}$), the L^2 error norm of the approximation behaves like

$$\|u_N - \hat{u}\| \le \hat{C} h^{k+1} |\hat{u}|_{H^{k+1}}. \qquad (2.6.3)$$

If $\hat{u}$ has only limited smoothness and belongs, for instance, to the Sobolev space H^σ ($\sigma > 0$), while the finite-element basis is made of the Lagrangian set $\mathcal{Q}_k$, a more general result applies in the L^2 norm, depending on the values of σ and k. One may show that (see [173, 174, 175])

$$\|u_N - \hat{u}\| = O(h^\beta), \quad \text{with} \quad \beta = \min\left(k+1, \sigma + \tfrac{1}{2}\right), \qquad (2.6.4)$$

illustrating the claim made earlier about the influence of the regularity of $\hat{u}$ upon the error decay rate.

Convergence is much faster in the p-version of the FEM and in the SEM, especially with smooth problems. We mention only the spectral-element results because of their importance in this book. Readers interested in the convergence results for the p-version should consult Szabó and Babuška [370].

Canuto and Quarteroni [68] have proven that for the case of a single domain, if one assumes that $\hat{u}$ belongs to a weighted Sobolev space H_w^σ with $\sigma > 0$, then the approximation error in H_w^0-norm is bounded by

$$\|u_N - \hat{u}\|_{H_w^0} \le C N^{-\sigma} \|\hat{u}\|_{H_w^\sigma}, \qquad (2.6.5)$$

where N is the degree of the polynomial approximation, and C a generic constant. The essential difference between (2.6.3) and (2.6.5) lies in the nature of the exponents in the r.h.s. of these inequalities. For very smooth solutions $\sigma \gg 1$, in the h-version the error decay rate is limited by the polynomial degree k. In the SEM, however, the decay rate is much higher, being essentially governed by the regularity of the solution. The h-version of FEM is characterized by an *algebraic* convergence of the numerical approximation, while spectral elements lead to *exponentially* convergent solutions. This is the key issue that makes spectral methods so attractive for regular problems.

Unfortunately, real-life problems are often nonregular. Their approximation by spectral elements becomes more challenging and poses some difficulties that will be faced in later chapters.

2.7 Solution Techniques

Whatever the choice of the approximation technique, as the preceding sections have shown, one ends up with an algebraic system to be solved. The choice of a solution algorithm is influenced by a number of factors, of which the spatial dimension is probably most important. Available options can be broadly partitioned into two categories: direct methods, such as Gaussian factorization with or without pivoting, that will yield the exact answer (assuming infinite-precision arithmetic) in a finite number of operations, and iterative methods, such as Jacobi or Gauss–Seidel relaxation, that generally do not exhibit a finite termination property. In between these are the semiiterative methods, such as the conjugate gradient (CG) and generalized minimum residual (GMRES) methods, which are based on projection. In principle, these methods also have algebraic work bounds, because they are guaranteed to converge in a finite number of iterations, assuming exact arithmetic. However, these bounds are generally too pessimistic, because the solution will usually converge to an acceptable precision long before the upper bound on the iteration count is reached, particularly if an appropriate preconditioner is used.

The choice of iterative versus direct solution techniques is based primarily on matrix sparsity and bandwidth considerations. Linear systems arising from unstructured discretizations of PDEs are generally quite sparse (i.e., have a bounded number of nonzeros per row, independent of the system size) and have bandwidth that increases with spatial dimension. For instance, the classical finite-difference five-point stencil for 2D second-order problems with lexicographical ordering leads to matrices with at most five nonzero coefficients per row. The LU factorization of this matrix generates $O(N^3)$ nonzeros in the resultant lower- and upper-triangular components, where N is the number of grid points per space direction. The cost of the factorization is $O(N^4)$, and the cost for solving several problems with the same system is $O(N^3)$ per r.h.s. In three dimensions, the situation is worse. For a problem having $\mathcal{N} = N^3$ degrees of freedom, the matrix bandwidth resulting from the lexicographical ordering is N^2. The number of nonzeros in the factors is thus $O(N^5)$, which is prohibitive for even moderate values of N. For sparse systems such as arise from low-order discretizations, nested dissection orderings improve these results in two and three dimensions to $O(N^2 \log N)$ and $O(N^4)$ storage, and $O(N^3)$ and $O(N^5)$

work, respectively [153]. Nevertheless, these are generally not applicable to high-order methods where formation of the operator is prohibitive (see, however, [272]). In these cases, or for 3D problems in general, iterative methods are usually faster.

Direct solution techniques, such as Gaussian elimination and its variants, are well known today and can be found in many textbooks. For details on these techniques we refer readers to the books by Golub and Van Loan [160], Strang [367], Trefethen and Bau [384], or Press et al. [316]. Texts specifically oriented to direct solution of sparse systems include those by George and Liu [153], Pissanetsky [313], and Duff et al. [117]. Recent texts on general iterative methods include works by Saad [339], and Axelsson [13].

We will start the discussion by introducing the concept of conditioning of a matrix, which is central in many developments. This is followed by a discussion of some iterative techniques that have gained great importance in recent times.

2.7.1 The Conditioning of a Matrix

Consider the algebraic system

$$A\underline{x} = \underline{b}, \quad (2.7.1)$$

where $A \in \mathbb{R}^{N \times N}$ is a nonsingular matrix with elements $\{a_{ij}\}_{i,j=1}^{N}$. The efficiency of several solution techniques depends crucially on the sensitivity of the system to small perturbations, either to the operator A or to the r.h.s. $\underline{b}$. If the sensitivity is high (in some sense to be defined below), the algebraic system will be called *ill-conditioned*, and several solution techniques will be quite inefficient. If, instead, the sensitivity is low, there will be far fewer problems in general. To evaluate the sensitivity, one has to look more closely at the spectral properties of the matrix A. First, we introduce the concept of matrix norm.

Matrix Norms and Spectral Radius

Section A.2 of Appendix A gives the basic axioms of a norm defined on a real vector space such as $\mathbb{R}^N$ [see (A.2.1)–(A.2.3)]. Here, we want to associate a norm to the operators acting on the elements of $\mathbb{R}^N$ (see also Section A.3.3).

Let $\|\cdot\|$ denote any vector norm on $\mathbb{R}^N$. To a matrix $M \in \mathbb{R}^{N \times N}$ we associate the induced *matrix norm*, defined as

$$\|M\| := \max_{\substack{\underline{x} \in \mathbb{R}^N \\ \underline{x} \neq 0}} \frac{\|M\underline{x}\|}{\|\underline{x}\|} = \max_{\substack{\underline{x} \in \mathbb{R}^N \\ \|\underline{x}\|=1}} \|M\underline{x}\|. \quad (2.7.2)$$

The ratio $\|M\underline{x}\|/\|\underline{x}\|$ being insensitive to a stretching of $\underline{x}$, the scalar $\|M\|$ is

2.7. Solution Techniques

simply the maximum amount by which a vector on the unit sphere in $\mathbb{R}^N$ (i.e. the set of vectors $\underline{x}$ such that $\|\underline{x}\| = 1$) can be stretched by the action of M. From (2.7.2) one deduces immediately the very useful inequality

$$\|M\underline{x}\| \leq \|M\|\|\underline{x}\|. \tag{2.7.3}$$

Furthermore, we define the *spectral radius* $\rho(M)$ as the radius of the disk centered at the origin of the complex plane containing all the eigenvalues of M. One has

$$\rho(M) := \max_i |\lambda_i|, \tag{2.7.4}$$

where λ_i is an eigenvalue (possibly complex) of M. Using the definitions (2.7.2) and (2.7.4), we list some useful identities for matrix norms:

(i) For $\|\underline{x}\|_1 := \sum_{i=1}^{N} |x_i|, \quad \|M\|_1 = \max_i \sum_{j=1}^{N} |m_{ij}|.$

(ii) For $\|\underline{x}\|_\infty := \max_i |x_i|, \quad \|M\|_\infty = \max_j \sum_{i=1}^{N} |m_{ij}|.$

(iii) For $\|\underline{x}\|_2 := \left(\sum_{i=1}^{N} |x_i|^2\right)^{1/2}, \quad \|M\|_2 = \sqrt{\rho(M^T M)}.$

(iv) $\rho(M) \leq \|M\|$ for all $\|\cdot\|$.

(v) For any $\epsilon > 0$, $\exists \|\cdot\|_*$ such that $\|M\|_* \leq \rho(M) + \epsilon$.

The first three identities give closed-form expressions for computing $\|M\|$ in the three most convenient vector norms. If M is symmetric, then, from the third identity, the Euclidean norm gives $\|M\|_2 = \rho(M)$.

The equivalence of vector norms in $\mathbb{R}^N$, stated in the following theorem, allows us to relate one matrix norm to another.

Theorem 2.4 (Equivalence of norms) *For any pair of vector norms $\|\cdot\|$ and $\|\cdot\|_*$ on $\mathbb{R}^N$, there exist constants c and C, possibly dependent on N, such that*

$$c\|\underline{x}\| \leq \|\underline{x}\|_* \leq C\|\underline{x}\| \qquad \forall \underline{x} \in \mathbb{R}^N.$$

This theorem, coupled with identities (iv), and (v) above, implies that $\|\cdot\|_* < 1$ is a sufficient condition, and $\rho(M) < 1$ is a necessary and sufficient condition for $\lim_{k \to \infty} \|M^k \underline{x}\| = 0$, for arbitrary $\underline{x}$ and $\|\cdot\|$.

We introduce an additional vector norm that we will need often. Recall that a matrix $A \in \mathbb{R}^{N \times N}$ is symmetric positive definite (SPD) if $A^T = A$, and

$$\forall \underline{u} \in \mathbb{R}^N, \quad (\underline{u}, A\underline{u}) := \underline{u}^T \cdot A\underline{u} \geq 0 \quad \text{with} \quad (\underline{u}, A\underline{u}) = 0 \Leftrightarrow \underline{u} = \underline{0}. \tag{2.7.5}$$

Consider the bilinear expression $(\underline{x}, \underline{y})_A$ defined by

$$(\underline{x}, \underline{y})_A := \underline{x}^T \cdot A\underline{y}, \quad \underline{x}, \underline{y} \in \mathbb{R}^N. \tag{2.7.6}$$

Using the definition (2.7.5) and the axioms (A.4.1)–(A.4.4), one can easily show that $(\underline{x}, \underline{y})_A$ is an inner product with associated norm $\|\underline{x}\|_A := (\underline{x}, \underline{x})_A^{1/2}$. The reader will recognize in $(\underline{x}, \underline{y})_A$ and $\|\underline{x}\|_A$, discrete analogs of the energy inner product and energy norm (2.1.14) and (2.1.15). In the sequel we will denote the quantities $(\underline{x}, \underline{y})_A$ and $\|\underline{x}\|_A$ as the A-inner product of $\underline{x}$ and $\underline{y}$ and A-norm of $\underline{x}$, respectively.

The Condition Number

Let $T \in \mathbb{R}^{N \times N}$ denote a (test) matrix such that $A + \varepsilon T$ is nonsingular for the range of a perturbation parameter ε (with $|\varepsilon| \ll 1$). Let $\underline{t} \in \mathbb{R}^N$ denote a test vector. The perturbed linear system (2.7.1) is

$$(A + \varepsilon T)\underline{x}(\varepsilon) = \underline{b} + \varepsilon \underline{t}, \tag{2.7.7}$$

where $\underline{x}(\varepsilon) = \underline{x} + \underline{\delta}(\varepsilon)$. We want to evaluate the magnitude of the perturbation effect, $\underline{\delta}(\varepsilon)$, on the solution of the unperturbed problem. Introducing (2.7.1) into (2.7.7), we can easily show that, because $A + \varepsilon T$ is nonsingular, to the first order in ε, $\underline{\delta}(\varepsilon)$ is given by

$$\underline{\delta}(\varepsilon) = \varepsilon A^{-1} \cdot (\underline{t} - T\underline{x}). \tag{2.7.8}$$

Using any vector norm $\|\cdot\|$ and the induced matrix norm (2.7.2), we get

$$\|\underline{\delta}(\varepsilon)\| \leq \varepsilon \|A^{-1}\| \cdot (\|\underline{t}\| + \|T\underline{x}\|), \tag{2.7.9}$$

or, also with (2.7.1) and (2.7.3),

$$\frac{\|\underline{\delta}(\varepsilon)\|}{\|\underline{x}\|} \leq \varepsilon \|A^{-1}\| \cdot \|A\| \left(\frac{\|\underline{t}\|}{\|\underline{b}\|} + \frac{\|T\|}{\|A\|} \right). \tag{2.7.10}$$

The quantity

$$\kappa(A) := \|A^{-1}\| \cdot \|A\| \tag{2.7.11}$$

2.7. Solution Techniques

in the r.h.s. of (2.7.10) is known as the *spectral condition number* of A. Large values of $\kappa(A)$ allow small perturbations of A or $\underline{b}$ to have large effects on the solution of the linear system, a mark of the ill-conditioning of A.

The definition (2.7.11) is quite inconvenient for evaluating $\kappa(A)$. Fortunately, as seen above, the Euclidean norm of a *symmetric* matrix A is $\|A\|_2 = \rho(A)$, with $\rho(A)$ the spectral radius of A. As a consequence, for symmetric matrices the spectral condition number in the Euclidean norm is also given by the relationship

$$\kappa_2(A) = |\lambda_M|/|\lambda_m|, \qquad (2.7.12)$$

where λ_M and λ_m are the algebraically largest and smallest eigenvalues of A, respectively. For the nonsymmetric matrices that we will consider, the r.h.s. of (2.7.12), though different from κ_2, gives a convenient estimate of their condition. Any mention of the spectral condition number in the sequel of the book will automatically refer to the relationship (2.7.12).

The Conditioning of Spectral Schemes

Let us look at the conditioning problems raised by high-order (spectral or collocation) schemes through an example, as this is a fundamental topic. For the purpose of illustration we consider the eigenvalue problem

$$-\frac{d^2 u}{dx^2} = \lambda u(x), \quad x \in (0, 1), \qquad u(0) = u(1) = 0. \qquad (2.7.13)$$

The eigensolutions $\{\lambda_k, \zeta_k(x)\}_{k=1}^{\infty}$ are

$$\lambda_k = k^2 \pi^2, \qquad \zeta_k(x) = \sin(k\pi x). \qquad (2.7.14)$$

Now consider the $\mathcal{Q}_1$ finite-element approximation of (2.7.13) on a uniformly spaced grid with $N+1$ nodes $\{x_j = jh\}_{j=0}^{N}$, and with $Nh = 1$. As shown in Section 2.3.1, the stiffness and mass matrices are equal to $K = \frac{1}{h}\text{tridiag}(-1, 2, -1)$ and $M = \frac{h}{6}\text{tridiag}(1, 4, 1)$. The generalized eigenvalue problem

$$K\underline{z} = \lambda^{\text{FE}} M\underline{z} \qquad (2.7.15)$$

has real positive solutions for λ^{FE}. Since the mass matrix M is symmetric positive definite (SPD) [see Equation (2.7.5)], we can use the Cholesky factorization $M = LL^T$ to rewrite (2.7.15) as

$$(L^{-1} K L^{-T})\underline{\tilde{z}} = \lambda^{\text{FE}} \underline{\tilde{z}}, \qquad (2.7.16)$$

where $\underline{\tilde{z}} := L^T \underline{z}$. Since L is invertible and K is SPD, we have

$$\underline{y}^T (L^{-1} K L^{-T}) \underline{y} = (L^{-T} \underline{y})^T K (L^{-T} \underline{y}) > 0 \quad \forall \underline{y} \neq \underline{0},$$

implying that $L^{-1} K L^{-T}$ is SPD, which establishes the property.

The problem (2.7.15) may be solved by recognizing that the associated eigenvectors are $\underline{\zeta}_k = \{(\underline{\zeta}_k)_j := \sin(jk\pi h)\}_{j=1}^{N-1}$. Inserting $\underline{\zeta}_k$ into the eigensystem (2.7.15) yields

$$-\frac{1}{h} \{\sin[(j-1)k\pi h] - 2\sin(jk\pi h) + \sin[(j+1)k\pi h]\}$$

$$= \frac{\lambda_k^{\text{FE}} h}{6} [\sin(k\pi(j-1)h) + 4\sin(k\pi jh) + \sin(k\pi(j+1)h)]. \quad (2.7.17)$$

With the help of trigonometric identities we deduce easily that, for $1 \leq k \leq N-1$,

$$\lambda_k^{\text{FE}} = \frac{6}{h^2} \frac{1 - \cos(k\pi h)}{2 + \cos(k\pi h)} = k^2 \pi^2 \left(1 + \frac{k^2 \pi^2 h^2}{12} - \cdots\right), \quad (2.7.18)$$

where the second expression is derived from a Taylor expansion in powers of h. From the result (2.7.18) we see that λ_k^{FE} approximates the smallest eigenvalues of the continuous problem (2.7.14) to second order in the mesh size $O(h^2)$. In particular,

$$\lambda_1^{\text{FE}} \sim \pi^2 \quad \text{and} \quad \lambda_{N-1}^{\text{FE}} \sim \frac{12}{h^2},$$

so that the $\mathcal{Q}_1$ finite-element matrix $L_{\text{FE}} := M^{-1} \cdot K$ has a spectral condition number proportional to the *square* of the number of degrees of freedom: $\kappa(M^{-1} \cdot K) \sim cN^2$, with $c = 12/\pi^2$.

At this point, we remark that when using Lagrangian bases in finite-element calculations, one often can use *mass lumping*, in which one replaces the (global) mass matrix $M := \{M_{ij} = \int_\Omega \psi_i \psi_j \, dx\}$ with the diagonal matrix $\tilde{M} := \text{diag}(\int_\Omega \psi_i \, dx)$. This is equivalent to replacing the product of the separately expanded functions $u(x) := \sum u_i \psi_i(x)$ and $v(x) := \sum v_j \psi_j(x)$ with the expansion of the product $u(x)v(x) := \sum u_i v_i \psi_i(x)$. From the algebraic standpoint, it is equivalent to replacing M with a diagonal matrix having the same row sum, which must be preserved if the mass matrix is to integrate the basis functions correctly and satisfy the condition

$$\int_{\hat{\Omega}} \psi_i \cdot 1 \, dx = \underline{\hat{e}}_i^T M \underline{\hat{e}} = \underline{\hat{e}}_i^T \tilde{M} \underline{\hat{e}}, \quad (2.7.19)$$

2.7. Solution Techniques

where $\hat{e}_i$ is the ith column of the identity and $\hat{e} = (1, 1, \ldots, 1)^T$. In the SEM, M is diagonal by construction, since high-order pointwise quadrature is used in evaluating all of the bilinear forms. Practical reasons for the use of mass lumping will appear naturally in the later chapters.

By replacing M in (2.7.15) with the lumped mass matrix $\tilde{M} = hI$ (with $I \in \mathbb{R}^{N-1 \times N-1}$, the identity matrix), we have a system that is equivalent to the standard second-order (centered) finite-difference approximation. The eigenvalue spectrum λ_k^{FD} is found by the same procedure to be

$$\lambda_k^{\text{FD}} = \frac{2}{h^2}[1 - \cos(k\pi h)] = k^2 \pi^2 \left(1 - \frac{k^2 \pi^2 h^2}{12} + \cdots \right), \quad (2.7.20)$$

and the same conclusions hold with regard to the dependence on N of $\kappa(\tilde{M}^{-1} \cdot K)$.

For the spectral case based on Legendre or Chebyshev polynomials, the problem is more complicated and no closed-form expression for the eigenvalues exists. Weideman and Trefethen [408] have shown that the first $2N/\pi$ eigenvalues closely match the continuous spectrum. They note that, heuristically, the fraction of converged eigenvalues corresponds to the number of full sinusoidal waves that can be accurately resolved on the grid. This is most readily seen for the GLC points (the Legendre points are similar):

$$\xi_j^c := \cos \frac{\pi j}{N}, \quad 0 \le j \le N$$

[see (B.2.13)]. As illustrated in Figure 2.7.1, these points correspond to the projection onto $[-1, 1]$ of the $N + 1$ points uniformly partitioning the unit semicircle. Asymptotically, the minimum and maximum spacings on $[0, 1]$ are

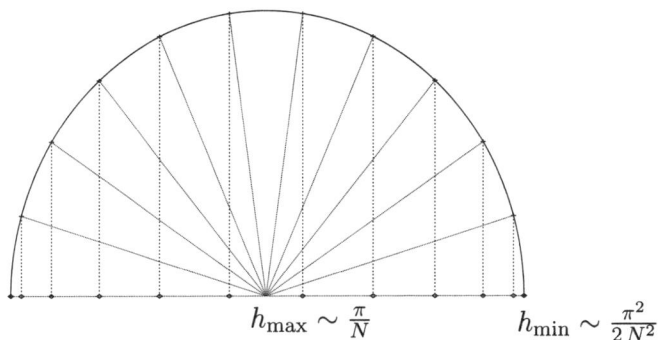

Figure 2.7.1. Distribution of Gauss–Lobatto–Chebyshev points on $[-1, 1]$.

2. Approximation Methods for Elliptic Problems

Table 2.7.1. *Eigenvalues of the* SEM ($E = 1$) *discretization matrix of a model diffusion problem, as a function of polynomial degree* N. *The condition number* $\kappa := \lambda_{N-1}/\lambda_1$ *increases as* $O(N^\gamma)$.

N	λ_1^{SE}	λ_{N-1}^{SE}	$\lambda_{N-1}^{SE}/\lambda_1^{SE}$	γ
4	9.862395e + 00	6.813760e + 01	6.908829e + 00	—
8	9.869604e + 00	5.872463e + 02	5.950049e + 01	3.11
16	9.869604e + 00	7.687161e + 03	7.788723e + 02	3.71
32	9.869604e + 00	1.136998e + 05	1.152020e + 04	3.89
64	9.869604e + 00	1.756206e + 06	1.779409e + 05	3.95
128	9.869604e + 00	2.763585e + 07	2.800097e + 06	3.98
256	9.869604e + 00	4.386215e + 08	4.444165e + 07	3.99

$h_{\min} \sim \pi^2/2N^2$ and $h_{\max} \sim \pi/N$. The minimum *fully resolved* wavelength is $l^* = 2h_{\max}$ (Nyquist criterion). Equating $\sin(k^*\pi x) = \sin(2\pi x/l^*)$, one sees that the maximum resolved wavenumber is $k^* = 2N/\pi$.

Vandeven [398] has further established that, for $1 \le k \le \alpha N$ with $0 \le \alpha < 2/\pi$, there exist constants C and ρ, with $0 < \rho < 1$, depending only on α, such that

$$|\lambda_k - k^2\pi^2| \le C\rho^{N^{2/3}}. \qquad (2.7.21)$$

Convergence for the lower $2N/\pi$ eigenvalues is thus nearly exponential. For the largest eigenvalues, he established the following asymptotic behavior for the Legendre–Galerkin method:

$$\lim_{N \to \infty} \lambda_{N-1}^{SE} = \frac{N^4}{\pi^2}, \qquad (2.7.22)$$

from which we conclude $\kappa(L_{SE}) = \kappa(M_{SE}^{-1} K_{SE}) \sim (N/\pi)^4$. Table 2.7.1 gives the minimum (λ_1^{SE}) and maximum (λ_{N-1}^{SE}) eigenvalues for the lumped-mass Legendre–Galerkin method (corresponding to the SEM with $E = 1$). The largest eigenvalue is seen to converge to N^4/π^2. Also illustrated is the rapid convergence of spectral methods for the lower eigenvalues. The first column reveals that the smallest eigenvalue is approximated to at least seven significant digits for $N \ge 8$.

We summarize these results with plots of the spectra for the model problem (2.7.13) resulting from several discretizations. The spectra for the two finite-element cases (M and $\tilde{M}$) are shown on Figure 2.7.2(a), along with the exact and fourth-degree spectral element spectra λ_k^{SE}. Notice that the eigenvalues of the continuous problem are framed by the numerical results obtained with M and $\tilde{M}$ as predicted by (2.7.18) and (2.7.20). Actually, the mean value $(\lambda_k^{FE} + \lambda_k^{FD})/2$

2.7. Solution Techniques

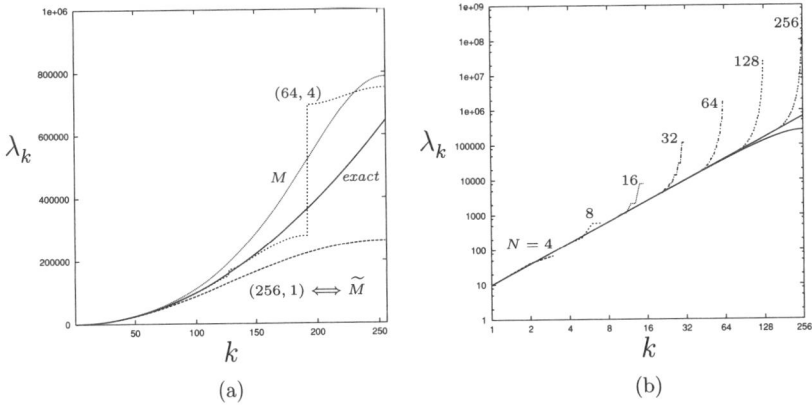

Figure 2.7.2. Comparison of eigenvalues for second-order diffusion operator, $-u_{xx} = \lambda u$, $u(0) = u(1) = 0$: (a) linear finite elements with tridiagonal (M) and lumped ($\tilde{M}$) mass matrices, spectral elements with $(E, N) = (64, 4)$, and analytical spectrum; (b) Legendre SEM with $E = 1$ and N ranging from 4 to 256. [Exact and FEM ($\tilde{M}$) curves are also shown.]

is equal to λ_k up to an error $O(h^4)$. Notice also the close agreement between the exact eigenvalues and the lower three-fifths of the λ_k^{SE} spectrum obtained with 64 spectral elements of degree 4. This is even more evident in Figure 2.7.2(b), comparing on a logarithmic scale the exact eigenvalues with the Legendre spectral-element ones evaluated on a single spectral element ($E = 1$) with a number of nodes, N, ranging from 4 to 256.

2.7.2 Basic Iterative Methods

A basic iterative method for solving $A\underline{x} = \underline{b}$ is given by the Richardson fixed-point iteration

$$\underline{x}^{k+1} = \underline{x}^k + \alpha P^{-1}(\underline{b} - A\underline{x}^k), \qquad k = 0, \ldots, \tag{2.7.23}$$

where P is an invertible matrix, α is an adjustable relaxation factor to ensure convergence of the scheme, and k is the iteration index. The initial vector $\underline{x}^0$ is given. The matrix P is usually called the *preconditioning matrix*, or more simply, the *preconditioner*. Eventually the solution to (2.7.23) has to be identical to the solution to (2.7.1), which raises the question of convergence of the iterative scheme and, more importantly, the *rate* of convergence, which is closely tied to the condition number $\kappa(P^{-1}A)$. The objective is to find matrices P as close as possible to A in some sense, for which it is easy (and, hopefully, cheap) to solve linear systems of the form $P\underline{z} = \underline{r}$, with $\underline{r}$ some given vector.

Classical examples of preconditioned Richardson iteration schemes are the Jacobi (J) and Gauss-Seidel (GS) iteration algorithms. The preconditioner in these cases is intimately related to A, since it is part of it. If $A = L + D + U$ is a matrix splitting in which L and U are strictly lower- and upper-triangular, respectively, and $D = \text{diag}(a_{ii})$, then the two choices $P := D$ and $P := L + D$ (with $\alpha = 1$, in both cases) yield

$$(\text{J}) \qquad \underline{x}^{k+1} = D^{-1}(\underline{b} - (L + U)\underline{x}^k),$$

$$(\text{GS}) \qquad \underline{x}^{k+1} = (L + D)^{-1}(\underline{b} - U\underline{x}^k).$$

The J and GS iteration schemes are guaranteed to converge for diagonally dominant systems. GS converges for SPD systems. J preconditioning has the advantage of being symmetric, which is important for preconditioned CG iteration introduced below. In addition, it vectorizes more readily, as discussed in Chapter 8. The general observation for systems arising from elliptic PDEs is that GS converges at about twice the rate of J iteration. For steady convection problems the improvement can be even more substantial if a "streamwise" ordering can be found.

Convergence Results

The convergence rate of the general Richardson iteration can be analyzed by subtracting the exact solution to (2.7.1) from both sides of (2.7.23). Since $\underline{b} = A\underline{x}$, we have

$$(\underline{x}^{k+1} - \underline{x}) = (\underline{x}^k - \underline{x}) - \alpha P^{-1}(A\underline{x}^k - A\underline{x}). \qquad (2.7.24)$$

Defining the error at the kth iteration as $\underline{e}^k := \underline{x}^k - \underline{x}$, we have

$$\underline{e}^{k+1} = (I - \alpha P^{-1}A)\underline{e}^k. \qquad (2.7.25)$$

Introducing any vector norm $\|\cdot\|$ and the induced matrix norm (2.7.2), we get

$$\|\underline{e}^{k+1}\| \leq \|I - \alpha P^{-1}A\| \|\underline{e}^k\|. \qquad (2.7.26)$$

If we assume $\underline{x}^0 := 0$, (always possible by redefining the problem as $A\underline{y} = \underline{b} - A\underline{x}_0$) we have $\|\underline{e}^0\| = \|\underline{x}\|$, and repeated application of the inequality (2.7.26) gives

$$\|\underline{e}^k\| \leq \|I - \alpha P^{-1}A\|^k \|\underline{x}\|. \qquad (2.7.27)$$

We conclude that if $\|I - \alpha P^{-1}A\| < 1$, then $\lim_{k \to \infty} \|\underline{e}^k\| = 0$. By the equivalence of vector norms in $\mathbb{R}^N$, if (2.7.27) converges for *any* vector norm,

2.7. Solution Techniques

then it converges in *every* norm. Specifically, the Richardson iteration scheme (2.7.23) will converge whenever the spectrum of $G := I - \alpha P^{-1}A$ lies within the unit disk centered in the complex plane and the rate of convergence will improve as $\rho(G)$ decreases.

It is easy to show that with λ_m and λ_M the algebraic smallest and largest eigenvalues of $P^{-1}A$, the optimal relaxation factor α and the associated spectral radius $\rho(G)$ are given by

$$\alpha = \frac{2}{\lambda_m + \lambda_M} \quad \text{and} \quad \rho(G) = \frac{\lambda_M - \lambda_m}{\lambda_M + \lambda_m}. \tag{2.7.28}$$

The convergence rate of the iterative scheme (2.7.23) depends on the spectral condition number κ. In view of the preceding results with the optimal relaxation factor taken into account, the inequality (2.7.26) becomes

$$\frac{\|\underline{e}^{k+1}\|}{\|\underline{e}^k\|} \leq \frac{\lambda_M - \lambda_m}{\lambda_M + \lambda_m} = \frac{\kappa - 1}{\kappa + 1}. \tag{2.7.29}$$

A level of machine accuracy equal to 10^{-14} (called *spectral* accuracy in this book) may then theoretically be reached in I_C iterations with

$$I_C \approx \frac{\log \zeta}{\log \frac{\kappa-1}{\kappa+1}}, \quad \text{where} \quad \zeta = \frac{10^{-14}}{\|\underline{e}^0\|}. \tag{2.7.30}$$

The reader has certainly noticed that, with regard to the preconditioning matrix P, no particular requirement was made, except for the following points: (a) algebraic systems $P\underline{z} = \underline{r}$ should be easy to solve, (b) the preconditioner P should be "close" to A, and (c) ideally, the eigenvalues of $P^{-1}A$ should be real. Point (b) is easily understood in view of the result (2.7.28). With P close to A, there will be some clustering of the eigenvalues of G in the vicinity of the origin in the complex plane, and quick convergence of the Richardson scheme (2.7.23) toward the solution of (2.7.1) can be expected.

2.7.3 Preconditioning Schemes of High-Order Methods

The ill-conditioning that arises from high-order discretizations is a major concern in their application. As noted in Section 2.7.1, $\kappa(M^{-1}K)$ scales as $O(N^4)$ for the SEM. The solution of linear systems in K is governed by $\kappa(P^{-1}K)$, where P is the preconditioning matrix. For a single element ($E = 1$), Bernardi and Maday [38] show that the unpreconditioned spectral-element stiffness matrix has $\kappa(K) = O(N^3)$. Maitre and Pourquier [265] prove that this property continues to hold in d dimensions when the discretization is a tensor product of the one-dimensional formulation. If P is taken to be the diagonal of K, Maday and Muñoz [257] find a closed-form solution for the eigenvalues and eigenvectors that establishes $\kappa(P^{-1}K) = O(N^2)$. However, this theory has not been extended to $d > 1$. Maitre and Pourquier [265] demonstrate numerically that the improvement with diagonal preconditioning is less dramatic in higher space dimensions. They also prove for the p-type FEM that $\kappa(K) = O(N^{4(d-1)})$ and $\kappa(P^{-1}K) = O(N^{2(d-1)})$ for the diagonally preconditioned case. Because of the rapid growth of κ with N, it is important to have more sophisticated preconditioning strategies. Here, we discuss some important advances in the one-dimensional case for a single domain. Preconditioners for general domains in d dimensions that build upon these ideas are presented in Chapter 7.

Preconditioned iterative solutions of collocation calculations were introduced by Morchoisne [280] and Orszag [288] at the end of the 1970s. Orszag advocated the use of preconditioning as a convenient way to avoid the memory storage and arithmetic operation count entailed by the collocation equations. He suggested that, for model problems like the Laplace equation, the preconditioning of Chebyshev collocation calculations with a standard finite difference (FD) scheme on the same grid would result in a rapidly converging algorithm. Indeed, he showed that, in the case of Fourier collocation of the one-dimensional Laplace equation, the spectral condition number of $L_{FD}^{-1}L_C$ is $\kappa = \pi^2/4$, yielding spectral accuracy in about 30 iterations. Here, L_{FD} denotes the matrix associated with the FD scheme on the collocation grid. From this result, he inferred an analogous property for Chebyshev collocation, the efficiency of the algorithm being based on the sparsity of L_{FD} (in contrast with the density of L_C) and on the existence of a fast Fourier transform (FFT) for the evaluation of the right-hand sides in the Richardson scheme (2.7.23). The conjecture for the Chebyshev polynomials was proven some time later by Haldenwang et al. [182].

Today's interest in preconditioned iterative solutions of high-order methods has another motivation. We have seen in Section 2.7.1 that the spectral condition number κ of high-order methods (collocation, spectral elements) grows as $O(N^4)$. As a consequence, high-order calculations become increasingly ill conditioned with mesh refinement, and machine accuracy may be completely lost. Orszag's conjecture on FD preconditioning of collocation calculations was the start of an intense research activity for the identification of preconditioners P of high-order methods with a spectral condition number $\kappa(P^{-1}A)$

2.7. Solution Techniques

as insensitive as possible to the value of N, as the easiest remedy to save spectral accuracy.

The numerical efficiency of Orszag's FD preconditioning scheme was assessed independently by Canuto and Quarteroni [69] and by Deville and Mund [99, 100]. These authors extended the study to low-order finite element (FE) preconditioning of collocation calculations on GLJ grids. They showed that FE preconditioning was even better than FD preconditioning. In particular, with FE preconditioning, the spectrum of the iteration operator $L_{\text{FE}}^{-1} L_C$ for a broad class of problems is located inside the unit circle in the complex plane. In that case, the relaxation factor may be set to the (nonoptimal) value $\alpha = 1$.

The $\mathcal{Q}_1$ FE preconditioning algorithm of Deville and Mund has been applied to diffusion-type equations and systems of equations using either collocation or spectral-element calculations (see, e.g., [104, 94, 95, 321, 73, 127, 282]). As an introduction, we give the various steps of the algorithm as they apply to the collocation solution of problem (2.1.9), for instance, on a GLJ grid with polynomials of degree N:

- Step 1. Compute the collocation grid on $\Omega = (a, b)$, using the solution to $\chi_N^{(\alpha,\beta)}(x) = 0$. See (2.5.1)–(2.5.2).
- Step 2. Assemble the collocation matrix L_C^J associated to the problem (2.5.10)–(2.5.11). We denote this problem symbolically as $\mathcal{A}_C u_N = f$, where $u_N \in \mathbb{P}_{N,0}(\Omega)$ is a polynomial of degree N satisfying the homogeneous Dirichlet condition at $x = a$. On the collocation grid, assemble also the $\mathcal{Q}_1$ FE matrix L_{FE} associated to the following problem: Find $w_h \in V_{N,0}$ such that

$$\int_\Omega (p w_h' v_h' + q w_h v_h - f v_h) d\Omega = 0 \quad \forall v_h \in V_{N,0}, \quad (2.7.31)$$

with the FE space $V_{N,0} \subset H^1(\Omega)$ such that

$$V_{N,0} := \{v_h \in C^0(\Omega) \quad v_h(a) = 0 \, \forall \Omega^e \in \Omega, \, v_{h|\Omega^e} \in \mathcal{Q}_1\}.$$

We denote the problem (2.7.31) symbolically as $\mathcal{A}_{\text{FE}} w_h = f$. The matrix L_{FE} is used as the preconditioner P in the iterative scheme (2.7.23).

- Step 3. Evaluate a first guess, u_h^0, such that $\mathcal{A}_{\text{FE}} u_h^0 = f$. Since u_h^0 is a $\mathcal{Q}_1$ FE, it is piecewise linear on the GLJ grid.
- Step 4. Evaluate the spectral residual r_N^0 associated to u_h^0. With I_N the spectral interpolation operator in $\mathbb{P}_N(\Omega)$, we have for $r_N^0 \in \mathbb{P}_N(\Omega)$

$$r_N^0 := \mathcal{A}_C u_N^0 - f, \quad \text{with} \quad u_N^0 := I_N u_h^0. \quad (2.7.32)$$

- Step 5. Iterate till convergence:

$$u_N^{k+1} = u_N^k - \alpha I_N \mathcal{A}_{FE}^{-1}[I_h^0(\mathcal{A}_C u_N^k - f)], \qquad k = 0, \ldots, \quad (2.7.33)$$

where $I_h^0 v$ is the $\mathcal{Q}_1$ FE interpolation of $v \in \mathbb{P}_N(\Omega)$.

The last step in the preconditioning algorithm is a functional formulation of the Richardson scheme (2.7.23). The presence of $I_N \mathcal{A}_{FE}^{-1} I_h^0 v$ (with $v := \mathcal{A}_C u_N^k - f$) in (2.7.33) shows that any correction to the approximate spectral solution u_N^k is provided by a FE calculation with the last residual as source term. The reader will find more detail in [100, 66], including the iterative scheme (2.7.33) for a problem with Neumann or Robin boundary conditions where extra terms due to the boundary conditions contribute also to the residual. Incidentally we note that the FD preconditioning of the GLJ collocation scheme may be cast in the same framework as outlined here, the quadratures in (2.7.31) being performed with the trapezoidal rule.

For advection-dominated diffusion problems, Funaro has shown that key in obtaining spectral accuracy is the coupling of a preconditioning algorithm (FD in this case) with collocation (or spectral elements) on a *staggered* grid [143, 144]. This strategy will be outlined in the next chapter.

2.7.4 Iterative Methods Based on Projection

Modern iterative methods such as conjugate gradients (CG) and the generalized method of residuals (GMRES) generate approximate solutions to the $N \times N$ linear system $A\underline{x} = \underline{b}$ by computing the projection of $\underline{x}$ onto a low-dimensional subspace $V_k \subset \mathbb{R}^N$. In this regard, these methods have much in common with the variational method introduced in the beginning of this chapter.

To illustrate the basic concepts, we begin with the case where A is SPD and make use of the associated norm $\|\cdot\|_A$ and inner product $(\cdot, \cdot)_A$ introduced in (2.7.6). Let V_k denote both the matrix of linearly independent vectors $(\underline{v}_1, \ldots, \underline{v}_k) \in \mathbb{R}^{N \times k}$ and the subset of $\mathbb{R}^N$ spanned by these vectors. Then the following problems are equivalent: Find $\underline{x}_k \in V_k$ such that

1. $\quad \|\underline{x} - \underline{x}_k\|_A \leq \|\underline{x} - \underline{v}\|_A \quad \forall \underline{v} \in V_k, \quad (2.7.34)$

2. $\quad \underline{v}^T \cdot A\underline{x}_k = \underline{v}^T \cdot \underline{b} \quad \forall \underline{v} \in V_k. \quad (2.7.35)$

3. $\quad \underline{x}_k = V_k(V_k^T A V_k)^{-1} V_k^T A\underline{x} \quad (2.7.36)$

4. $\underline{x}_k$ is the A-orthogonal projection of $\underline{x}$ onto V_k.

2.7. Solution Techniques

The first statement (2.7.34) corresponds to the minimization of the quadratic form $(\underline{x} - \underline{v})^T \cdot A(\underline{x} - \underline{v})$ with respect to $\underline{v} \in V_k$, which is the standard variational formulation of the problem. This is useful for estimating convergence rates as k, the dimension of the approximation space, is increased. The second statement (2.7.35) indicates that the residual $\underline{r}_k$ is orthogonal to V_k, or that the error vector $\underline{e}_k := \underline{x} - \underline{x}_k$ is orthogonal to V_k in the A-inner product, since this relationship is equivalent to $\underline{v}^T \cdot A(\underline{x} - \underline{x}_k) = 0$. This statement is most useful for constructing the solution. Despite appearances, for conditioning and efficiency reasons, the third form (2.7.36) is less useful for computing $\underline{x}_k$. However, we will encounter this form in later chapters, so it is important to recognize its equivalence to statement 4.

An interesting property of projection methods is that, in exact arithmetic, the solution will be reached in at most N iterations, since $V_N \equiv \mathbb{R}^N$. In practice, however, these methods typically generate an acceptable solution for $k \ll N$. The higher the value of k, the closer $\underline{x}_k$ will be to the solution of the problem $\underline{x}$. Again the connection may be established with the variational approach in the continuous case: the statement (2.7.34) is the discrete analog of the property (2.2.14), while the statement (2.7.35) corresponds to the Galerkin approximation (2.2.11).

To develop the projection technique (2.7.35), we postulate an A-orthogonal (or A-conjugate) basis $W_k := \{\underline{w}_1, \ldots, \underline{w}_k\}$ for V_k, satisfying $(\underline{w}_i, \underline{w}_j)_A = 0$, $i \neq j$, where $\underline{w}_i \in V_i, i = 1, \ldots, k$. Following the weighted-residual approach, we express the approximation in terms of this basis as $\underline{x}_k = \sum_{j=1}^{k} \alpha_j \underline{w}_j$ and require (2.7.35) to hold for $\underline{v} = \underline{w}_i$, $i = 1, \ldots, k$, leading to the $k \times k$ linear system

$$\tilde{A}\underline{\alpha} = \underline{\tilde{b}},$$

where $\underline{\alpha} \in \mathbb{R}^k$ is the vector of unknown basis coefficients, $\tilde{b}_i = \underline{w}_i^T \cdot \underline{b} = (\underline{w}_i, \underline{x})_A$, and $\tilde{a}_{ij} = \underline{w}_i^T \cdot A\underline{w}_j = (\underline{w}_i, \underline{w}_j)_A$. Because of the A-orthogonality of the basis vectors, $\tilde{A} \in \mathbb{R}^{k \times k}$ is diagonal. This implies that $\alpha_i = \tilde{b}_i/\tilde{a}_{ii}$ and leads to the following closed form for $\underline{x}_k$:

$$\underline{x}_k = \sum_{i=1}^{k} \alpha_i \underline{w}_i = \sum_{i=1}^{k} \frac{\underline{w}_i^T \cdot \underline{b}}{\underline{w}_i^T \cdot A\underline{w}_i} \underline{w}_i = \underline{x}_{k-1} + \frac{\underline{w}_k^T \cdot \underline{b}}{\underline{w}_k^T \cdot A\underline{w}_k} \underline{w}_k. \quad (2.7.37)$$

The significance of (2.7.37) is that the basis coefficients α_i are independent of the basis vectors $\underline{w}_j$, $j \neq i$, and $\underline{x}_k$ can therefore be constructed recursively as a sequence of approximations, as the final equality indicates. It is important in practice to replace $\underline{b}$ in (2.7.37) by the residual, $\underline{r}_{k-1} := \underline{b} - A\underline{x}_{k-1}$

($= \underline{r}_{k-2} - \alpha_{k-1} A \underline{w}_{k-1}$, $k > 1$), corresponding to a *modified Gram–Schmidt* (MGS) procedure [339]. Because $\underline{w}_k^T \cdot A \underline{x}_{k-1} = 0$, the two approaches are algebraically equivalent. However, loss of orthogonality in the $\underline{w}_i$'s due to round-off can compromise the stability of the classical Gram–Schmidt procedure (2.7.37).

To generate the A-orthogonal basis W_k from V_k, one could use the following Gram–Schmidt procedure (see Section A.4.7):

$\underline{w}_1 := \underline{v}_1$,

For $j = 2, \ldots, k$,

$$\underline{w}_j := \underline{v}_j - \sum_{i=1}^{j-1} \beta_{ij} \underline{w}_i, \qquad \beta_{ij} := \frac{(\underline{w}_i, \underline{v}_j)_A}{(\underline{w}_i, \underline{w}_i)_A}, \qquad (2.7.38)$$

End.

As a comparison with Equation (2.7.37) shows, the quantity $\sum_{i=1}^{j-1} \beta_{ij} \underline{w}_i$ is the A-projection of $\underline{v}_j$ onto W_{j-1}. Hence $\underline{w}_i$ given by (2.7.38) is A-orthogonal to the basis W_{j-1}, that is, $(\underline{w}_i, \underline{w}_j)_A = 0$, for $1 \leq i < j$. Again, from a stability standpoint, it may be necessary to replace (2.7.38) by an equivalent MGS scheme, or to employ *reorthogonalization*, where one makes a second pass through (2.7.38) with $\underline{v}_j := \underline{w}_j'$, where $\underline{w}_j'$ is the output of the first pass.

The convergence properties of $\underline{x}_k$ in (2.7.37) are determined by the choice of the approximation space V_k. Standard practice is to use the residual, setting $\underline{v}_j = \underline{r}_{j-1}$. In the preconditioned case, one uses $\underline{v}_j = P^{-1} \underline{r}_{j-1}$. Combining (2.7.37) and (2.7.38) with the latter choice leads to the following basic projection method.

Algorithm 2.1 (Basic projection scheme) Given an initial guess $\underline{x}_0$,

1. Compute $\underline{r}_0 := \underline{b} - A \underline{x}_0$,
2. For $j = 1, 2, \ldots$ until convergence Do:
3. $\quad \underline{v}_j := P^{-1} \underline{r}_{j-1}$,
4. $\quad \underline{w}_j := \underline{v}_j - \sum_{i=1}^{j-1} \beta_{ij} \underline{w}_i, \qquad \beta_{ij} := \frac{(\underline{w}_i, \underline{v}_j)_A}{(\underline{w}_i, \underline{w}_i)_A}$,
5. $\quad \underline{x}_j := \underline{x}_{j-1} + \alpha_j \underline{w}_j, \qquad \alpha_j := \frac{(\underline{w}_j, \underline{e}_{j-1})_A}{(\underline{w}_j, \underline{w}_j)_A}$,
6. $\quad \underline{r}_j := \underline{r}_{j-1} - \alpha_j A \underline{w}_j$,
7. EndDo.

Note that the inner product involving the error vector $\underline{e}_{j-1} := A^{-1} \underline{r}_{j-1}$ is readily computable as $\underline{w}_j^T \cdot \underline{r}_j$. This projection scheme calls for some remarks.

2.7. Solution Techniques

First, by construction, $\underline{x}_j$ is the best approximation to $\underline{x}$ in the A-norm. However, by changing the above inner products to $(\cdot, \cdot)_U$ for some SPD matrix U, the statement (2.7.34) would hold in the U-norm. In particular, setting $U = A^T A$ would yield the generalized conjugate residual (GCR) method suitable for nonsymmetric systems and algebraically equivalent to GMRES [339], provided it does not break down. In the nonsymmetric case, it is generally necessary to use a more stable procedure for computing $\underline{w}_j$ than the classical Gram–Schmidt algorithm in step 4. Moreover, breakdown can occur if $P^{-1}A$ is not positive definite, since it is possible to have $\alpha_j = 0$ before convergence is reached. The method will then stall because the linear independence of the subsequent $\underline{v}_j$'s is lost. GMRES avoids this by using a different choice for $\underline{v}_j$.

Second, we note that other choices for $\underline{v}_j$ can lead to different approximation spaces to be exploited within the same projection framework. For instance, when solving parameter- or time-dependent problems, it is often useful to set V_k to be the span of prior solutions [128]. We will consider this further in Section 4.7. Another possibility is to populate V_k with unit vectors, $\underline{\hat{e}}_j$, ordered in such a way that the orthogonalized basis W_k is as sparse as possible [391]. This is the basis for the parallel coarse grid solver considered in Section 8.5.4. An important case where one may need to modify the definition of $\underline{v}_j$ is if A has a nontrivial nullspace – e.g., the constant mode in the case of a pure Neumann problem, or the hydrostatic mode in the solution of a Stokes or Navier–Stokes problem [see (5.1.4)]. In this case, one can set

$$\underline{v}_j = (I - \Pi)P^{-1}\underline{r}_{j-1}, \tag{2.7.39}$$

where Π is the L^2 projector onto the nullspace of A. In the case where this is the constant mode, this amounts to subtracting $(\underline{\hat{e}}^T P^{-1}\underline{r}_{j-1})\underline{\hat{e}}$ from $P^{-1}\underline{r}_{j-1}$, where $\underline{\hat{e}} = (1, 1, \ldots, 1)^T$. Note that $\underline{r}_{j-1}$ will be in the range of A provided that $\underline{b}$ is. With the symmetry of Π, (2.7.39) is thus equivalent to applying the symmetric operator $(I - \Pi)^T P^{-1}(I - \Pi)$ and is therefore a suitable preconditioner for the CG method introduced below.

Finally, we note that for A and P SPD, with the choice $\underline{v}_j = P^{-1}\underline{r}_{j-1}$, one has

$$\underline{x}_j \in \text{span}\{P^{-1}\underline{b}, P^{-1}AP^{-1}\underline{b}, \ldots, (P^{-1}A)^{j-1}P^{-1}\underline{b}\} := \mathcal{K}_j(P^{-1}A; P^{-1}\underline{b}),$$

where $\mathcal{K}_j$ is referred to as the *Krylov subspace*. Therefore, for every polynomial $\mathbb{P}_{j-1}$ of degree $j-1$ or less,

$$\|\underline{x} - \underline{x}_j\|_A^2 \leq \|\underline{x} - \mathbb{P}_{j-1}(P^{-1}A)P^{-1}\underline{b}\|_A^2. \tag{2.7.40}$$

Because $\underline{x}_j$ is the (unique) minimizer of the expression on the right, any other Krylov method, such as Richardson iteration, will require at least as many iterations to reach the same error. By substituting a specific polynomial for $\mathbb{P}_{j-1}$ in (2.7.40) it is possible to obtain the error bound

$$\|\underline{x} - \underline{x}_j\|_A^2 \leq 2 \left(\frac{\sqrt{\kappa} - 1}{\sqrt{\kappa} + 1} \right)^j \|\underline{x}\|_A^2, \tag{2.7.41}$$

where κ is the condition number of $P^{-1}A$ [339]. This implies that $O(m\sqrt{\kappa})$ iterations are required to reduce the error by m digits. Moreover, it is straightforward to show that, for $i < j - 1$ in step 4, one has $\beta_{ij} = 0$, implying that only the most recent search vector $\underline{w}_j$ needs to be stored. One then obtains the preconditioned CG (PCG) algorithm, consisting of two recurrences of fixed length, independent of k. With some understanding of the genesis of projection methods, we now turn to the classical forms of the PCG and GMRES algorithms for linear systems [161, 339, 384].

Algorithm 2.2 (Preconditioned conjugate gradient) Starting with a guess $\underline{x}_0$, the standard PCG algorithm with P as preconditioner runs as follows:

1. Compute $\underline{r}_0 := \underline{b} - A\underline{x}_0$, $\underline{z}_0 = P^{-1}\underline{r}_0$, and $\underline{w}_0 := \underline{z}_0$,
2. For $j = 0, 1, \ldots$ until convergence Do:
3. $\alpha_j := (\underline{r}_j, \underline{z}_j)/(A\underline{w}_j, \underline{w}_j)$,
4. $\underline{x}_{j+1} := \underline{x}_j + \alpha_j \underline{w}_j$,
5. $\underline{r}_{j+1} := \underline{r}_j - \alpha_j A\underline{w}_j$,
6. $\underline{z}_{j+1} = P^{-1}\underline{r}_{j+1}$,
7. $\beta_j := (\underline{r}_{j+1}, \underline{z}_{j+1})/(\underline{r}_j, \underline{z}_j)$,
8. $\underline{w}_{j+1} := \underline{z}_{j+1} + \beta_j \underline{w}_j$,
9. EndDo.

The reader should notice a few differences between Algorithms 2.1 and 2.2. There is some reshuffling of the indices that is easily explained. The main differences lie in the calculation procedures for the coefficients α_j and β_j. The CG algorithm exploits a number of the polynomial identities associated with the elements of Krylov subspace to reduce the number of inner products to just two, namely, (r_j, z_j) and $(w_j, w_j)_A$. As we will see in Chapter 8, this is particularly important in parallel algorithms because inner products are a source of global communication. Finally, the recursion in step 5 allows the residual to be computed without an extra matrix–vector product.

For nonsymmetric matrices, the product $\underline{x}^T \cdot A\underline{x}$ no longer constitutes a basis for a norm. The error minimization process must be recast in terms of a different (computable) norm, with the $A^T A$-norm being the most common choice. Since $A\underline{e}_k = \underline{b} - A\underline{x}_k$, this is equivalent to the following minimization

2.7. Solution Techniques

problem: Find $\underline{x}_k \in V_k$ such that

$$\|\underline{b} - A\underline{x}_k\|_2 \leq \|\underline{b} - A\underline{v}\|_2 \quad \forall \underline{v} \in V_k, \tag{2.7.42}$$

where $\|\cdot\|_2$ denotes the Euclidean vector norm, and V_k is any approximation subspace in $\mathbb{R}^\mathcal{N}$.

Several approaches can be used to solve (2.7.42). As noted above, Algorithm 2.1 with all norms changed from A to $A^T A$ yields the GCR algorithm. However, efficient implementation of this scheme requires storage of both $\underline{w}_i$ and $A\underline{w}_i$, $i = 1, \ldots, k$ and, as noted earlier, GCR can break down. A significant advance in the solution of nonsymmetric systems was the development of the GMRES algorithm by Saad and Schultz [340]. To orthogonalize the Krylov subspace $\mathcal{K}_k$, GMRES uses an Arnoldi process that cannot prematurely break down. Breakdown (linear dependence of the Krylov vectors) at some step $j < k$ implies that the exact solution lies in $\mathcal{K}_j$, and this exact result will be computed by GMRES. In addition to being robust, GMRES exploits several identities associated with the Arnoldi process so that storage of only k vectors is required, roughly half as many as with GCR. The preconditioned variant of GMRES admits either left or right preconditioning, depending on how the basic linear system is multiplied by P^{-1}. The left-preconditioned GMRES algorithm applies to $P^{-1}A\underline{x} = P^{-1}\underline{b}$. It reads as follows:

Algorithm 2.3 (Left-preconditioned GMRES) Starting with a guess $\underline{x}_0$, perform the following set of operations:

1. Compute $\underline{r}_0 := P^{-1}(\underline{b} - A\underline{x}_0)$, $\beta := \|\underline{r}_0\|_2$, and $\underline{v}_1 := \underline{r}_0/\beta$,
2. For $j = 1, \ldots, k$ Do:
3.   Compute $\underline{w}_{j+1} := P^{-1}A\underline{v}_j$,
4.   For $i = 1, \ldots, j$ Do:
5. $h_{ij} := (\underline{w}_{j+1}, \underline{v}_i)$,
6. $\underline{w}_{j+1} := \underline{w}_{j+1} - h_{ij}\underline{v}_i$,
7.   EndDo.
8.   Compute $h_{j+1,j} = \|\underline{w}_{j+1}\|_2$ and $\underline{v}_{j+1} = \underline{w}_{j+1}/h_{j+1,j}$,
9. EndDo.
10. Define $V_k = [\underline{v}_1, \ldots, \underline{v}_k]$, and $\bar{H}_k = \{h_{ij}\}_{1 \leq i \leq k+1, 1 \leq j \leq k}$,
11. Compute $\underline{y}_k = \mathrm{argmin}_y \|\beta\underline{e}_1 - \bar{H}_k \underline{y}\|_2$,
12. Compute $\underline{x}_k = \underline{x}_0 + V_k \underline{y}_k$.
13. If satisfied Stop, else set $\underline{x}_0 := \underline{x}_k$ and go to 1.

As a brief comment on the GMRES algorithm, we mention that the Arnoldi iteration (steps 2–9) produces an $N \times k$ matrix V_k, whose column vectors have unit norm and are mutually orthogonal, and a $(k+1) \times k$ Hessenberg matrix $\bar{H}_k$, with the relationship

$$P^{-1}AV_k = V_{k+1}\bar{H}_k = V_k H_k + \underline{w}_{k+1}\underline{e}_k^T, \qquad (2.7.43)$$

where $\underline{e}_k$ denotes the last column of the $k \times k$ identity matrix. The matrix H_k is obtained by deleting the last row of $\bar{H}_k$. The nonzero elements of $\bar{H}_k$, h_{ij} ($i \le j$) are the quantities $(\underline{v}_i, A\underline{v}_j)$. If A is symmetric, one shows easily in the unpreconditioned case ($P \equiv I$) that H_k is tridiagonal. Further details about this algorithm can be found in the literature (see for instance [339, 384]).

2.8 A Numerical Example

To end this chapter, we discuss a numerical example illustrating the various approximation methods and solution techniques covered so far. We consider the model problem

$$-\frac{d^2u}{dx^2} + u(x) = f(x), \qquad x \in (-1, +1), \qquad (2.8.1)$$

with homogeneous Neumann and Dirichlet boundary conditions:

$$u'(-1) = u(1) = 0. \qquad (2.8.2)$$

The r.h.s. $f(x)$ is such that the solution to (2.8.1)–(2.8.2) is the C^∞ function

$$\hat{u}(x) = (1-x)^2(1+x)^2 e^{2x}. \qquad (2.8.3)$$

We solve the problem using (1) Lagrangian finite elements $\mathcal{Q}_1$ to $\mathcal{Q}_3$ with a series of uniformly spaced meshes with decreasing sizes, (2) orthogonal collocation on Legendre and Chebyshev grids with a set of increasing number of mesh points on a monodomain, and (3) Legendre spectral elements with increasing polynomial degrees, on one and two spectral elements. Approximation errors are collected in Tables 2.8.1 to 2.8.3.

Table 2.8.1 displays the discrete L^∞ norm and the Sobolev norm H^1 of the error $\varepsilon_N := \hat{u} - u_N$ for the Lagrangian finite-element solutions. Here, the integer N denotes the number of elements covering $(-1, +1)$. We define the discrete L^∞ error norm as

$$\|\varepsilon_N\|_{L^\infty} := \max_j |\hat{u}(x_j) - u_N(x_j)|,$$

2.8. A Numerical Example

Table 2.8.1. *Errors and convergence rates in discrete L^∞ and Sobolev H^1 norms for Lagrangian finite-element solutions of the model problem (2.8.1)–(2.8.3).*

	Q_1		Q_2		Q_3	
N	$\|\|\varepsilon\|\|_{L^\infty}$	τ	$\|\|\varepsilon\|\|_{L^\infty}$	τ	$\|\|\varepsilon\|\|_{L^\infty}$	τ
2	5.4206e+00		7.2612e−01		3.1590e−01	
4	1.4593e+00	1.89	1.4014e−01	2.37	4.1900e−02	2.91
8	3.9511e−01	1.88	3.4639e−02	2.02	3.9802e−03	3.40
16	9.5218e−02	2.05	5.9945e−03	2.53	3.0864e−04	3.69
32	2.4123e−02	1.98	8.8254e−04	2.76	2.1496e−05	3.84
64	6.0483e−03	2.00	1.1980e−04	2.88	1.4182e−06	3.92
128	1.6103e−03	1.91	1.5558e−05	2.94	9.0781e−08	3.97
256	4.2553e−04	1.92				
	$\|\|\varepsilon\|\|_{H^1}$	τ	$\|\|\varepsilon\|\|_{H^1}$	τ	$\|\|\varepsilon\|\|_{H^1}$	τ
2	2.7620e+00		5.3310e−01		3.3613e−01	
4	9.1175e−01	1.60	2.7263e−01	0.97	8.8407e−02	1.93
8	3.6444e−01	1.32	9.9528e−02	1.45	1.4163e−02	2.64
16	1.7829e−01	1.03	2.7604e−02	1.85	1.8824e−03	2.91
32	8.8750e−02	1.01	7.0841e−03	1.96	2.3602e−04	3.00
64	4.3663e−02	1.02	1.7798e−03	1.99	2.8473e−05	3.05
128	2.1121e−02	1.05	4.4259e−04	2.01	3.3173e−06	3.10
256	9.8408e−03	1.10				

on a set of two thousand points regularly spaced on $(-1, +1)$. The Sobolev norm H^1, defined earlier, is given by the expression (2.1.25). The coefficient τ in the table gives the (numerical) decay rate of the error norms. As the computation shows, with the Q_k Lagrangian elements, refinement of the mesh $h := 2/N$

Table 2.8.2. *Errors in discrete L^∞ and Sobolev H^1 norms for Legendre and Chebyshev collocation solutions of the model problem (2.8.1)–(2.8.3).*

	Legendre		Chebyshev	
N	$\|\|\varepsilon\|\|_{L^\infty}$	$\|\|\varepsilon\|\|_{H^1}$	$\|\|\varepsilon\|\|_{L^\infty}$	$\|\|\varepsilon\|\|_{H^1}$
3	1.1862e+00	9.2833e−01	1.1862e+00	9.2833e−01
5	5.3018e−01	5.8048e−01	3.2156e−01	4.8465e−01
7	3.1966e−01	1.4657e−01	1.5930e−01	1.2073e−01
9	3.2483e−02	1.2415e−02	1.3272e−02	8.1218e−03
11	1.1876e−03	4.3144e−04	4.2807e−04	2.4142e−04
13	2.2658e−05	8.0416e−06	7.4130e−06	4.0075e−06
15	2.6589e−07	9.3119e−08	8.0315e−08	4.2316e−08
17	2.1131e−09	7.3364e−10	5.9616e−10	3.0872e−10
19	1.2093e−11	4.1772e−12	3.2039e−12	1.6512e−12
21	3.2241e−14	1.7744e−13	2.7978e−14	1.6230e−13

Table 2.8.3. *Errors in discrete L^∞ and Sobolev H^1 norms for Legendre spectral-element solutions of the model problem (2.8.1)–(2.8.3).*

	$E = 1$		$E = 2$	
N	$\|\varepsilon\|_{L^\infty}$	$\|\varepsilon\|_{H^1}$	$\|\varepsilon\|_{L^\infty}$	$\|\varepsilon\|_{H^1}$
3	1.2875e + 00	8.7541e − 01	8.7804e − 01	6.7196e − 01
5	5.2201e − 01	5.7598e − 01	6.1434e − 02	1.3346e − 01
7	4.0851e − 02	1.0081e − 01	1.4025e − 03	5.1214e − 03
9	1.8256e − 03	6.1033e − 03	1.4042e − 05	6.8562e − 05
11	4.1146e − 05	1.7287e − 04	7.4085e − 08	4.4877e − 07
13	5.5352e − 07	2.7920e − 06	2.3767e − 10	1.7179e − 09
15	4.9252e − 09	2.8966e − 08	1.9130e − 12	4.6979e − 12
17	3.5640e − 11	2.0872e − 10	2.3826e − 12	3.6834e − 12
19	8.1912e − 12	6.4643e − 12	2.8688e − 12	5.2330e − 12
21	7.1029e − 12	7.9077e − 12	3.3997e − 12	6.6890e − 12

decreases the errors in such a way that

$$\|\varepsilon_N\|_{L^\infty} \propto h^{k+1} \quad \text{and} \quad \|\varepsilon_N\|_{H^1} \propto h^k. \tag{2.8.4}$$

The decay rate of the Sobolev H^1 error norm is one unit less than the rate associated to the L^∞ norm (or, also, to the L^2 norm), because of the presence of the first derivative in (2.1.25).

Table 2.8.2 gives the same error norms for Legendre and Chebyshev collocation calculations. Here, N denotes the polynomial degree of the approximation. Obviously, with orthogonal collocation the error decrease is much faster than with Lagrangian elements. In this particular case, spectral accuracy may be reached with approximately 20 degrees of freedom. The results displayed in the table have been obtained by using both a direct solution of the algebraic system (2.2.25) and the FE (or FD) preconditioning technique introduced in Section 2.7.3.

Table 2.8.3 gives the error norms obtained with two different Legendre spectral element calculations. Here, N denotes the number of GLL mesh points *per spectral element*, while E is the number of spectral elements covering the domain. The reader should be cautious when interpretating the numerical results, especially when making comparisons between the accuracies obtained with different numbers of spectral elements. The calculation shows that with a given number of degrees of freedom, the highest accuracy is reached with all degrees of freedom concentrated on the smallest number of spectral elements.

The results reported in Table 2.8.3 were also obtained by using both Gaussian elimination and the preconditioning techniques mentioned in this chapter. We

2.8. A Numerical Example

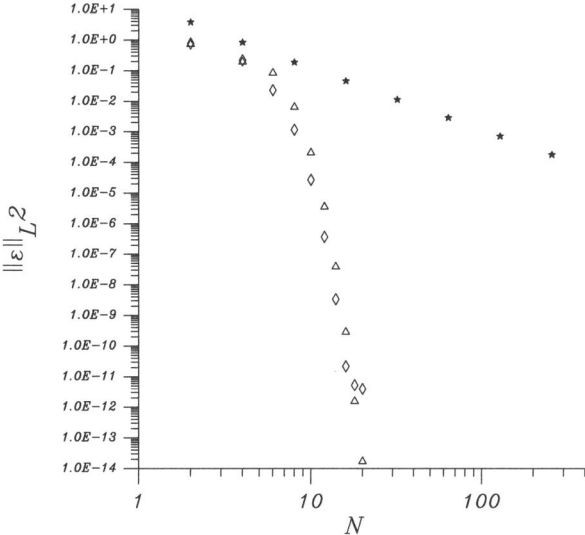

Figure 2.8.1. The L^2 error norms for the Q_1 FE ($\star$), Chebyshev collocation ($\triangle$), and Legendre spectral-element ($\diamond$) solutions of the model problem (2.8.1)–(2.8.3), as a function of the number of degrees of freedom, N.

emphasize that with only a few parameters, there is no ill-conditioning effect to be expected on the algebraic system, and both calculations should indeed give the same answer. In multidimensional calculations with large numbers of degrees of freedom per space variable, this may no longer be the case.

The large difference in accuracy between Lagrangian finite elements and high-order methods is also illustrated in Figure 2.8.1. This figure shows the decay of the L^2 error norms associated to the Q_1 FE solution ($\star$), the Chebyshev collocation ($\diamond$), and the Legendre spectral-element (monodomain) solutions ($\triangle$), as a function of N, the number of degrees of freedom. While the L^2 error convergence with Lagrangian finite elements is clearly *algebraic* [i.e. of type $O(N^{-\ell})$, with ℓ fixed and depending on the FE degree], the convergence rates of collocation and spectral-element solutions increase with increasing values of N, until roundoff error is reached (see Section 2.6).

Figures 2.8.2 and 2.8.3 illustrate the properties of the (low-order) preconditioning techniques introduced earlier in the chapter. Figure 2.8.2 displays the spectral condition number of three matrix operators involved in the Legendre collocation solution to (2.8.1)–(2.8.2) with Q_1 FE preconditioning. Calculations have been made with increasing polynomial degree N. The curve labeled (a) gives the spectral condition number of the collocation matrix $\kappa(L_C)$. The numerical computation illustrates the $O(N^4)$ growth of this quantity, a property

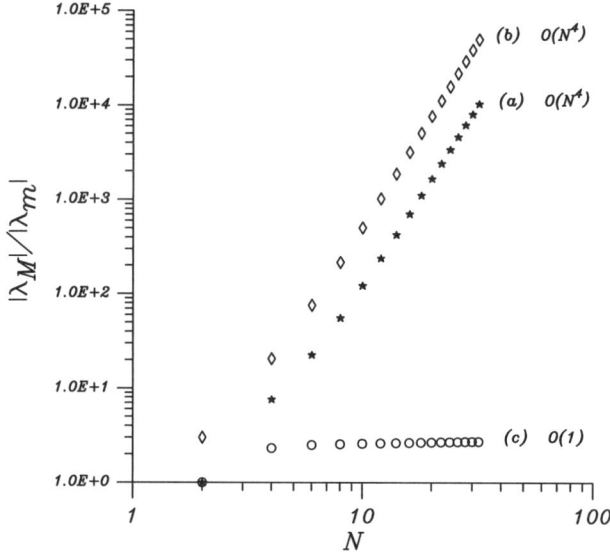

Figure 2.8.2. Spectral condition number of three matrix operators involved in a Legendre collocation scheme with $\mathcal{Q}_1$ FE preconditioning: (a) $\kappa(L_{FE})$, (b) $\kappa(L_C)$, and (c) $\kappa(L_{FE}^{-1} \cdot L_C)$. The parameter N is the polynomial degree of the approximation.

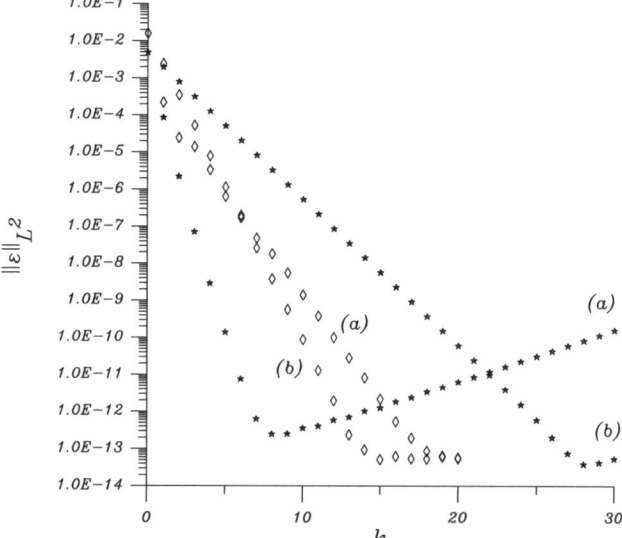

Figure 2.8.3. Decay of the L^2 error norm for FD ($\star$) and FE ($\diamond$) preconditioned solutions of a Chebyshev collocation calculation, as a function of the iteration index k. The problem solved is (2.8.1)–(2.8.3): (a) calculations with the relaxation factor $\alpha = 1$, and (b) with $\alpha = \alpha_{opt}$. The factors α_{opt} are equal to 0.590617 (FD) and 1.16086 (FE).

2.8. A Numerical Example

mentioned in Section 2.7 and resulting from the nonuniform spacing of the grid. The curve labeled (b), corresponding to $\kappa(L_{\text{FE}})$, where L_{FE} denotes the $\mathcal{Q}_1$ FE matrix built on the GLL collocation grid, exhibits similar asymptotic behavior, for the same reasons. The last curve, labeled (c), shows $\kappa(L_{\text{FE}}^{-1} \cdot L_C)$, the spectral condition number of the preconditioned Richardson iterative process (2.7.23). The main result here is that this quantity *does not depend on* N. Hence, with a suitable relaxation coefficient α, the preconditioned iterative process (2.7.23) will converge in almost the same number of iterations, whatever the value of N.

Figure 2.8.3 shows the evolution of the L^2 error norm during the FD (or FE) preconditioning iterations. The relaxation factor α has been set equal to 1 [see the curves labeled (a)] and to its optimal value α_{opt} [curves labeled (b)], which depends on the spectral properties of $P^{-1} \cdot L_C$ (with P, the preconditioner), according to (2.7.28). As the figure shows, with FD preconditioning and $\alpha = 1$, the error decays quite rapidly. However, since some eigenvalues of $P^{-1} \cdot L_C$ are located outside the unit circle in the complex plane, eventually the preconditioning iterations must diverge. One may argue that the accuracy reached before divergence is already quite good. This is purely related to the nature of the problem solved. If the differential problem is changed, the divergence of the iterative process may come much sooner. With $\alpha = \alpha_{\text{opt}}$, the error decay with FD preconditioning is slower, but the algorithm converges.

With the introduction of $\mathcal{Q}_1$ FE preconditioning, things are much simpler. Even with $\alpha = 1$, the eigenvalues of $L_{\text{FE}}^{-1} \cdot L_C$ are all located inside the unit circle, and the preconditioning process converges anyway. Francken et al. [140] have shown that this is related to the existence of the FE mass matrix that drives the eigenvalues of $P^{-1} \cdot L_C$ inside the unit circle of the complex plane. Setting $\alpha = \alpha_{\text{opt}}$ makes the convergence only a little faster, as shown in the figure.

3
Parabolic and Hyperbolic Problems

In the preceding chapter we introduced some of the basic principles and algorithms to solve second-order differential equations in the space variable. For convenience and in order to ease comprehension, the number of space dimensions was limited to one. In fact, most of the time real-life fluid flows are three-dimensional, which makes the numerical treatment of these problems much more sophisticated than explained in Chapter 2. Additionally, the flows may be time-dependent. The subject of multidimensional problems will be covered in Chapter 4. For the time being we stay within the sphere of one-dimensional problems in space as we deal with time dependence.

3.1 Introduction

The reader will remember from Chapter 1 that time-dependent fluid flows are governed by PDEs with the structure

$$\frac{\partial \bullet}{\partial t} = \mathcal{L} \bullet + \mathbf{f}, \tag{3.1.1}$$

where the bullet stands for the unknown velocity field, $\mathcal{L}$ is a (usually nonlinear) partial differential operator, and $\mathbf{f}$ denotes some forcing term [see for instance Equation (1.7.5)]. Time dependence in these problems obeys a first-order differential operator, which makes the time integration schemes different from those outlined in Chapter 2 for space dependence. Section 3.2 introduces the numerical schemes, as well as the basic concepts needed for their understanding. Section 3.3 describes the splitting technique combining integration schemes adapted to the underlying physics of the main components of the Navier–Stokes equations. Section 3.4 focuses on the parabolic case, where *dissipation* is the main characteristic. Section 3.5 deals with the hyperbolic case, in which, instead of dissipation, we have *advection* (i.e. transport) and *conservation*. In these two cases we will adopt different numerical schemes, for reasons that will be explained in detail. Sections 3.6 and 3.7 show how to deal with steady and unsteady problems combining diffusive and advection processes. Section 3.8 is devoted to the Burgers equation, which introduces nonlinearity. Finally, Sections 3.9 and 3.10 address advanced time discretizations that are tailored to CFD applications.

3.2 Time Discretization Schemes

For convenience we will consider in this section a generic set of equations

$$\frac{d\underline{u}}{dt} = \underline{f}(t, \underline{u}(t)), \quad \underline{u}(t^0) = \underline{u}^0, \quad t \in (t^0, T], \quad (3.2.1)$$

where $\underline{u}(t) \in \mathbb{R}^{\mathcal{N}}$ is a vector in the Euclidean space, T is some (finite) value of time, and $(t^0, T]$ denotes the interval open at t^0 and closed at T. In subsequent chapters we will meet situations in which $\underline{u}(t)$ is the set of coefficients resulting from a space discretization of the Navier–Stokes equations (1.6.3)–(1.6.4). In that case the r.h.s. of (3.2.1) is usually a nonlinear function of $\underline{u}(t)$ when convective terms are taken into account [see (1.7.5)]. For convenience we will also assume that the differential system (3.2.1) has been reduced to its *autonomous* form. This means that, on introducing an additional independent variable t^* satisfying the equation $t^* = t$ (or rather, $dt^*/dt = 1$), the r.h.s. of (3.2.1) no longer depends *explicitly* on t, and the differential problem becomes

$$\frac{d\underline{u}}{dt} = \underline{f}(\underline{u}(t)), \quad \underline{u}(t^0) = \underline{u}^0, \quad t \in (t^0, T], \quad (3.2.2)$$

where the new unknown vector $\underline{u}$, in addition to the old one, includes the variable t^*.

Before considering solution methods for (3.2.2) in detail, we make a few remarks about the underlying nature of $\underline{f}$ for problems arising in fluid flow applications. Significant insight can be gained by considering the model problem

$$\frac{d\underline{u}}{dt} = L\underline{u}, \quad (3.2.3)$$

where L is a linearized version of the differential operator in (3.1.1). The operator L corresponds to the Jacobian of $\underline{f}$, introduced below [see Equation (3.2.10)], and we assume that it may be diagonalized. If $Z = (\underline{z}_1, \ldots, \underline{z}_\mathcal{N})$ is the matrix of eigenvectors of L and $\Lambda = \text{diag}(\lambda_k)$ the corresponding set of eigenvalues, then (3.2.3) can be written as

$$\frac{d\underline{\hat{u}}}{dt} = \Lambda \underline{\hat{u}},$$

on making the substitutions $\underline{\hat{u}} = Z^{-1}\underline{u}$ and $\Lambda = Z^{-1}LZ$. Clearly, any numerical scheme used to advance (3.2.2) must be able to treat $u_t = \lambda u$ for all values of $\lambda = \lambda_k$ in practice. In fluid mechanics, the nature of these eigenvalues is fairly well understood. Diffusion contributes a negative real component to λ, while convection contributes a purely imaginary component. As we saw in Chapter 2, the eigenvalues for the second-order elliptic operator that governs diffusion scale as $O(N^4)$ for spectral discretizations of degree N, while the eigenvalues for the convection operator scale as $O(N^2)$. Because of these differences, and because the diffusive term is linear and symmetric while the convective term is nonlinear and nonsymmetric, these terms frequently are treated in different ways, giving rise to semiimplicit and operator-splitting time-stepping schemes.

In developing schemes for the solution of (3.2.2) we will momentarily ignore the physical meaning of $\underline{u}(t)$, keeping in mind that there is a distribution of eigenvalues in the left half of the complex plane, including the imaginary axis. We will use various approaches known as the *linear multistep*, the *predictor–corrector*, and the *Runge–Kutta methods*. The first and second approaches, developed in Sections 3.2.1 and 3.2.2, are based on a formal integration of (3.2.2), whereas the third approach, treated in Section 3.2.3, is based on the mean-value theorem and a systematic use of Taylor expansions.

3.2.1 Linear Multistep Methods

Formally, the solution of the differential system (3.2.2) at time t is

$$\underline{u}(t) - \underline{u}^0 = \int_{t^0}^{t} \underline{f}(\underline{u}(s))\,ds. \tag{3.2.4}$$

Many algorithms used in actual computations are based on an approximation of the r.h.s. of (3.2.4) such that if $\{t^i\}_{i=1}^{k}$ is a given set of discrete time values, one has

$$\int_{t^0}^{t} \underline{f}(\underline{u}(s))\,ds \approx \sum_{i=1}^{k} w_i \underline{f}(\underline{u}(t^i)), \quad t^i \leq t, \tag{3.2.5}$$

where the quantities w_i are quadrature weights. To distinguish the *numerical* solution to (3.2.2) from the exact solution $\underline{u}(t)$, we introduce the vector

3.2. Time Discretization Schemes

$\tilde{\underline{u}}(t) \in \mathbb{R}^{\mathcal{N}}$ resulting from the equation

$$\tilde{\underline{u}}(t) - \underline{u}^0 = \sum_{i=1}^{k} w_i \underline{f}(\tilde{\underline{u}}(t^i)), \quad t^i \leq t, \tag{3.2.6}$$

where we assume that all quantities except $\tilde{\underline{u}}(t)$ are known. Such a scheme used repetitively is called a *discrete-variable method* or a *forward step method*.

Suppose the time scale has been discretized with a constant step $\Delta t = h$.[1] The linear multistep methods are collectively characterized by the formula

$$\mathcal{L}_k[\tilde{\underline{u}}(t); h] := \tilde{\underline{u}}^{n+1} - \sum_{j=1}^{k} a_j \tilde{\underline{u}}^{n+1-j} - h \sum_{j=0}^{k} b_j \underline{f}^{n+1-j} = 0, \tag{3.2.7}$$

with the convention $\tilde{\underline{u}}^{n+1-j} := \tilde{\underline{u}}(t^{n+1-j})$, $\underline{f}^{n+1-j} := \underline{f}(\tilde{\underline{u}}^{n+1-j})$, and $k \geq 1$ the number of steps. The coefficients a_j and b_j depend on the choice of the numerical scheme, as we will see in various examples below. One easily recognizes in (3.2.7) a more general form of (3.2.6). In the simple case in which $k = 1$, the method is called a *single-step* method.

If the coefficient $b_0 = 0$, the numerical scheme (3.2.7) is called *explicit*. This means that, provided the set of values $\{\tilde{\underline{u}}^{n+1-j}\}_{j=1}^{k}$ is known at time t^n, the value of $\tilde{\underline{u}}^{n+1}$ may be readily computed by using (3.2.7) and one carries the time-marching process further to t^{n+1}. The computation cost is one evaluation of $\underline{f}$, the r.h.s. of (3.2.2), per time step. If the coefficient $b_0 \neq 0$, the numerical scheme is called *implicit*. To evaluate the vector $\tilde{\underline{u}}^{n+1}$, one has to solve a set of nonlinear equations at each time step:

$$\tilde{\underline{u}}^{n+1} - h b_0 \underline{f}(\tilde{\underline{u}}^{n+1}) = \sum_{j=1}^{k} a_j \tilde{\underline{u}}^{n+1-j} + h \sum_{j=1}^{k} b_j \underline{f}^{n+1-j}. \tag{3.2.8}$$

In practice, one normally uses a Taylor expansion of $\underline{f}(\tilde{\underline{u}}^{n+1})$ centered at $\tilde{\underline{u}}^n$ and limited to first order in $\tilde{\underline{u}}^{n+1} - \tilde{\underline{u}}^n$:

$$\underline{f}(\tilde{\underline{u}}^{n+1}) \approx \underline{f}(\tilde{\underline{u}}^n) + J(\tilde{\underline{u}}^n)(\tilde{\underline{u}}^{n+1} - \tilde{\underline{u}}^n), \tag{3.2.9}$$

where $J \in \mathbb{R}^{\mathcal{N} \times \mathcal{N}}$ is the Jacobian matrix associated with the transformation in the r.h.s. of (3.2.2), with elements

$$J_{ij} = \left.\frac{\partial f_i}{\partial u_j}\right|_{\tilde{\underline{u}}^n}, \quad i, j = 1, \ldots, \mathcal{N}. \tag{3.2.10}$$

---

[1] For the sake of convenience, we denote the time step in Sections 3.2 and 3.3 by h, despite the fact that this notation usually designates the spatial mesh.

Introducing the expansion (3.2.9) into (3.2.8), one concludes that the unknown at every time step results from the inversion of an $\mathcal{N} \times \mathcal{N}$ linear system

$$(I - hb_0 J)\underline{\tilde{u}}^{n+1} = \underline{q}^n, \quad (3.2.11)$$

$$\underline{q}^n = \sum_{j=1}^{k} a_j \underline{\tilde{u}}^{n+1-j} + h \sum_{j=1}^{k} b_j \underline{f}^{n+1-j} - hb_0 J \underline{\tilde{u}}^n, \quad (3.2.12)$$

with $I \in \mathbb{R}^{\mathcal{N} \times \mathcal{N}}$ the identity matrix.

Implicit multistep algorithms are therefore much more expensive in computing time than explicit multistep schemes. Quite generally, this is the price to pay for increased stability, a concept that we will define below. Similar conclusions will be reached later for Runge–Kutta integration schemes. In Section 3.2.2 we will indicate another technique to solve the nonlinear problem (3.2.8) that is cheaper than (3.2.11).

An important feature of multistep methods must be emphasized: these methods need starting procedures, since several numerical values are needed to advance the time-marching process, while only the initial condition is given. Usually, one uses a single-step or a Runge–Kutta method. Before we examine concrete examples of linear multistep methods, we give a short review of some essential theoretical concepts in the numerical analysis of ordinary differential equation (ODE) solvers.

Main Theoretical Concepts

For the sake of convenience, let us momentarily consider the problem of *one* nonlinear differential equation (3.2.2) with infinitely smooth solution $u(t) \in C^{\infty}(0, \infty)$. The introduction of $u(t)$ into the l.h.s. of (3.2.7) followed by a Taylor series expansion centered at t^{n+1} gives the local truncation error:

$$\mathcal{L}_k[u(t) - \tilde{u}(t); h] = \mathcal{L}_k[u(t); h] = \sum_{j=0}^{\infty} C_j \frac{h^j}{j!} u^{(j), n+1}, \quad (3.2.13)$$

where $u^{(j), n+1}$ denotes the jth derivative of u at t^{n+1}. The constants C_j depend on the parameters a_i and b_i, and with a little algebra one obtains

$$C_0 = 1 - \sum_{i=1}^{k} a_i, \quad C_j = \sum_{i=0}^{k} (-i)^{j-1}(ia_i - b_i), \quad j = 1, \ldots, \infty. \quad (3.2.14)$$

Three basic requirements for discrete-variable methods, which are linked together by a theorem due to Lax, are those of *consistency*, *convergence*, and *stability* (Theorem 3.1). Forward-step methods that satisfy any two of these three

3.2. Time Discretization Schemes

conditions automatically satisfy the third. Consistency compels the solution of the initial-value problem (3.2.2) to satisfy the difference scheme (3.2.7) apart from a small truncation error that should vanish with step size. In mathematical terms, we have the following.

Definition 3.1 *(Consistency)* A k-step multistep method (3.2.7) is consistent of order r if the local truncation error $\mathcal{L}_k[u(t); h] = O(h^{r+1})$, that is, if $C_0 = C_1 = \cdots = C_r = 0$ and $C_{r+1} \neq 0$.

In particular, a multistep method is said to be consistent if it is verified for all polynomials of degree ≤ 1.

Convergence is a key concern. It guarantees that with mesh refinement, the solution to the discrete-variable method (3.2.7) more and more closely follows the true solution of the problem.

Definition 3.2 *(Convergence)* The forward-step method (3.2.7) applied to any initial-value problem (3.2.2) with the r.h.s. satisfying a Lipschitz condition is convergent if

$$\lim_{h \to 0} \tilde{u}^n = u(t), \quad t = t^0 + nh \quad \forall t \in (t^0, T], \tag{3.2.15}$$

and the starting values tend to the initial condition as h tends to zero:

$$\lim_{h \to 0} \tilde{u}^i = u^0, \quad i = 0, \ldots, k-1.$$

Finally, it is important to prevent truncation errors or roundoff errors from destroying the solution by spurious amplification during the time-marching process. This requirement leads to the concept of *stability*, mathematically defined as follows.

Definition 3.3 *(Stability)* A multistep method (3.2.7) is stable iff for any differential problem (3.2.2) satisfying a Lipschitz condition, there exists a time step h_0 and a positive constant K such that for any two neighboring numerical solutions $\{\tilde{u}_1^n\}$ and $\{\tilde{u}_2^n\}$ of (3.2.7) with initial conditions u_1^0 and u_2^0, one has

$$\|\tilde{u}_1^n - \tilde{u}_2^n\| \leq K \|u_1^0 - u_2^0\| \quad \forall h, \quad 0 < h \leq h_0. \tag{3.2.16}$$

As this definition suggests, the stability property is strongly problem-dependent. To eliminate this constraint, one introduces a convenient test problem for $u \in \mathbb{R}$:

$$\frac{du}{dt} = \lambda u(t), \quad u(t^0) = 1, \quad t \in (t^0, T], \tag{3.2.17}$$

where λ is a complex constant. The relevance of this equation lies in the fact that (as already mentioned) λ must be regarded as an eigenvalue of the Jacobian J. To assess the stability properties of a forward-step method applied to a differential system (3.2.2), the analysis performed on (3.2.17) must be carried on the whole eigenspectrum of J, which might be spread over very large regions in the complex plane, especially if the problem is *stiff*.

The three concepts consistency, convergence, and stability are then combined in the *Lax equivalence theorem* (see [327, 244]).

Theorem 3.1 (Lax) *For a well-posed initial-value problem and a consistent discretization scheme, stability is the necessary and sufficient condition for convergence.*

Let us explore the consequences that are the most useful for practical purposes. We start with the multistep scheme (3.2.8). This gives a kth-order homogeneous difference equation with constant coefficients for $\tilde{u} \in \mathbb{R}$:

$$(1 - b_0 \lambda h)\tilde{u}^{n+1} = \sum_{j=1}^{k}(a_j + b_j \lambda h)\tilde{u}^{n+1-j}. \tag{3.2.18}$$

For a given value of $z := \lambda h$, this equation admits k polynomial solutions in the complex ζ-plane, $\tilde{u}_\ell^n = (\zeta_\ell)^n$ ($\ell = 1, \ldots, k$), such that

$$(1 - b_0 z)\zeta_\ell^k - \sum_{j=1}^{k}(a_j + b_j z)\zeta_\ell^{k-j} = 0, \quad \zeta_\ell = |\zeta_\ell| e^{i\phi_\ell}, \tag{3.2.19}$$

or

$$\rho(\zeta_\ell) - z\sigma(\zeta_\ell) = 0, \quad 1 \leq \ell \leq k. \tag{3.2.20}$$

The polynomials $\rho(\zeta) := \zeta^k - \sum_{j=1}^{k} a_j \zeta^{k-j}$ and $\sigma(\zeta) := \sum_{j=0}^{k} b_j \zeta^{k-j}$ in (3.2.20) are called the *characteristic* (or generating) polynomials of the multistep method.

Certainly, for eigenvalues λ in the left half complex plane [$\Re(\lambda) < 0$], the numerical solution of (3.2.17) using the linear multistep method (3.2.7) with time step h must eventually go to zero. This requires that the k solutions ζ_ℓ, $1 \leq \ell \leq k$, to Equation (3.2.20) satisfy the condition $|\zeta_\ell| < 1$. Ideally, this property

3.2. Time Discretization Schemes

should be verified for all $\Re(\lambda) < 0$. In practice, however, the property will be verified only in a subdomain of the left half complex plane, the *stability region* of the forward-step method, defined in somewhat broader terms.

Definition 3.4 *(Stability region)* The stability region of a forward step method is the set of values of h *(real, $h > 0$)* and of λ for which any perturbation to the solution introduced at any instant will not be amplified at later times.

In other words, the stability domain of a multistep scheme is defined as the (open) set of values in the complex z-plane such that the elements on the domain boundary correspond to $|\zeta| = 1$. Examples of stability regions that will be discussed below are given in Figures 3.2.1–3.2.3. Notice that if, for given values of λ and h, the product $z = \lambda h$ lies outside the stability region of the scheme, a homothetic transformation (for instance, a reduction of the mesh size) may be determined to enforce stability. This is not always needed however, as some forward-step methods have a stability region that *includes* the negative complex half plane $\Re(z) < 0$. Such forward-step methods are called *A-stable*, a concept due to Dahlquist [89, 151].

Definition 3.5 *(A-stability)* A forward-step method is A-stable if, applied to the test equation (3.2.17) with $\Re(\lambda) < 0$, $\forall h$ *(real, $h > 0$)*, one has $\lim_{n \to \infty} \tilde{y}(nh) = 0$.

A-stability is a convenient property in practice. When it is satisfied, one need not worry about nonphysical growth in the solutions. However, A-stability does not guarantee accuracy, especially for stiff problems, where the spread of eigenvalues (often over several decades) may require significant reduction in the time step in order to ensure $|\lambda h| \approx 1$.

We illustrate the concepts introduced so far with examples of linear multistep methods that have gained widespread popularity in numerical computation.

Adams–Bashforth and Adams–Moulton Schemes

Our first example deals with the k-step Adams–Bashforth (ABk) and Adams–Moulton (AMk) integration schemes, two families of explicit and implicit multistep methods built along the same principle with the use of Newton–Cotes quadrature schemes.

We approximate $\underline{f}(\underline{u})$ in (3.2.4) by a Lagrange interpolation polynomial of degree $k - 1$ at k consecutive mesh points. The ABk numerical scheme corresponds to the case in which these interpolation points are located in the interval

Table 3.2.1. *Coefficients for the k-step Adams–Bashforth* (ABk) *schemes with* $k \leq 6$. *For a given k*, $b_\ell = \gamma_\ell/\underline{\gamma}$ *with* $1 \leq \ell \leq k$ *[see (3.2.7)]. The* ABk *schemes have consistency order k.*

k	$\underline{\gamma}$	γ_1	γ_2	γ_3	γ_4	γ_5	γ_6
1	1	1					
2	2	3	−1				
3	12	23	−16	5			
4	24	55	−59	37	−9		
5	720	1901	−2774	2616	−1274	251	
6	1440	4277	−7923	9982	−7298	2877	−475

$[t^{n+1-k}, t^n]$. Then the approximation is

$$\underline{f}(\underline{u}) \approx \sum_{\ell=1}^{k} \underline{f}(\underline{u}^{n+1-\ell}) \pi_\ell(t), \qquad (3.2.21)$$

where the functions $\{\pi_\ell(t)\}_{\ell=1}^{k}$ form the cardinal Lagrangian basis satisfying the bi-orthogonality condition $\pi_\ell(t^{n+1-m}) = \delta_{\ell m}$, with $1 \leq \ell, m \leq k$. One verifies easily that the coefficients a_ℓ and b_ℓ ($1 \leq \ell \leq k$) in (3.2.7) are given by

$$a_\ell = \delta_{1\ell} \quad \text{and} \quad hb_\ell := \int_{t^n}^{t^{n+1}} \pi_\ell(t)\,dt. \qquad (3.2.22)$$

Numerical values of these coefficients for the ABk families with $k \leq 6$ are given in Table 3.2.1. Notice that, for a fixed k, the coefficients b_ℓ in (3.2.7) are given by $b_\ell = \gamma_\ell/\underline{\gamma}$. The single-step AB1 scheme is often referred to in the literature as the explicit (or "forward") Euler method (EM).

The AMk schemes are obtained by the same process on the interval $[t^{n+1-k}, t^{n+1}]$. The integer k is still the number of steps, but, since Adams–Moulton schemes are implicit, two single-step algorithms exist, corresponding to $k = 0$ and $k = 1$. The interpolation expansion for $\underline{f}(\underline{u})$ involves $k+1$ polynomials π_ℓ of degree k. One has an expansion (3.2.21) and associated coefficients a_ℓ and b_ℓ given by (3.2.22), where ℓ runs from 0 to k.

Table 3.2.2 displays the numerical values of the coefficients b_ℓ for the AMk schemes with $0 \leq k \leq 5$. Again the coefficients $b_\ell := \gamma_\ell/\underline{\gamma}$. The two single-step Adams–Moulton schemes are important in practice. They are usually known as the implicit (or "backward") Euler (IE) method, and as the trapezoidal rule, or Crank–Nicolson (CN) method. We give their characteristic equations (3.2.20),

3.2. Time Discretization Schemes

Table 3.2.2. *Coefficients for the k-step Adams–Moulton* (AMk) *schemes with* $k \leq 5$. *For a given k one has* $b_\ell = \gamma_\ell/\gamma$ *with* $0 \leq \ell \leq k$ *[see (3.2.7)]. The* AMk *schemes have consistency order* $k+1$.

k	γ	γ_0	γ_1	γ_2	γ_3	γ_4	γ_5
0	1	1					
1	2	1	1				
2	12	5	8	−1			
3	24	9	19	−5	1		
4	720	251	646	−264	106	−19	
5	1440	475	1427	−798	482	−173	27

together with the characteristic equation of the explicit Euler method:

$$\text{EM:} \quad \zeta = 1 + z, \tag{3.2.23}$$

$$\text{IE:} \quad \zeta = \frac{1}{1-z}, \tag{3.2.24}$$

$$\text{CN:} \quad \zeta = \frac{1+\frac{z}{2}}{1-\frac{z}{2}}. \tag{3.2.25}$$

These expressions are rational approximations of $\exp(z)$ with $z := \lambda h$, the exact solution of (3.2.17) at $t = h$.

The consistency property of the ABk and AMk schemes may be verified easily. Application of (3.2.13) with the coefficients borrowed from Tables 3.2.1 and 3.2.2 to infinitely smooth functions shows that the consistency orders are equal respectively to k and $k+1$. The stability regions for these methods (with $k \leq 4$) are shown in Figures 3.2.1 and 3.2.2.

The stability regions of the ABk schemes are located inside the closed curves labeled (k). Increasing the value of k certainly improves the consistency order but, at the same time, requires a refinement of h to deal with the eigenvalues of J in $\Re(\lambda) < 0$ having the largest module. Particularly difficult to solve are (pure) hyperbolic problems with eigenvalues on the imaginary axis. Among the four ABk schemes displayed in Figure 3.2.1, the only one including part of the imaginary axis in its stability region is AB3. The intersections of the stability curve and the y-axis are located at $y = \pm 0.72362$.

Similar comments apply to the AMk schemes with $k > 2$. These implicit schemes have larger stability regions than their Adams–Bashforth counterparts that are also located inside a closed curve. To ensure stability of pure hyperbolic problems, one has to take $k = 4$. The corresponding stability curve crosses the imaginary axis at $y = \pm 1.2118$.

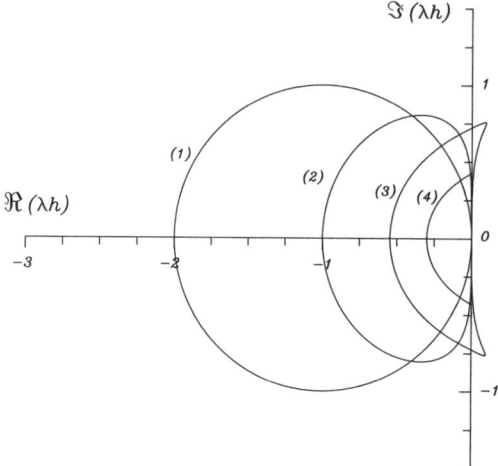

Figure 3.2.1. Stability regions for ABk schemes with $1 \leq k \leq 4$.

The IE and CN cases are somewhat different. The former scheme has its stability curve located in the right half plane $\Re(z) > 0$, the stability region lying *outside* the contour [and therefore including $\Re(z) < 0$], while the latter scheme has its stability curve superimposed on the imaginary axis. These two implicit schemes are thus A-stable. Although the IE scheme has a lower consistency order than does the CN scheme, it is better suited for the numerical solution of stiff problems, since $\lim_{z \to -\infty}(1-z)^{-1} = 0$, while for the CN

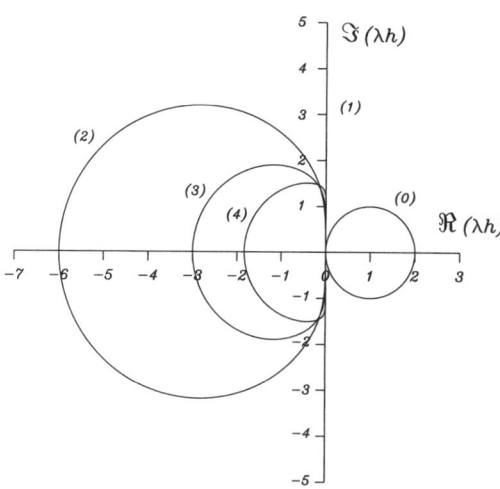

Figure 3.2.2. Stability regions for AMk schemes with $0 \leq k \leq 4$.

scheme $\lim_{z \to -\infty}(1 - z/2)^{-1}(1 + z/2) = -1$. A-stable numerical schemes approximating $\exp(z)$ and such that for $z \to -\infty$ the approximation tends to zero are called *L-stable*.

To conclude this example, we remark that, according to a property shown by Dahlquist [89], no A-stable linear multistep schemes exist with a consistency order higher than 2.

Backward Differencing Schemes

Our next example considers a family of *implicit* methods, the k-step backward differentiation formula (BDFk) based on a truncated Taylor expansion of the solution. Assuming once again a constant time step h, we introduce the backward difference operator ∇ such that

$$\nabla \underline{u}^n := \underline{u}^n - \underline{u}^{n-1}. \tag{3.2.26}$$

Let D denote the time differentiation operator $D := \frac{d}{dt}$. With the Taylor expansion of an infinitely regular function $f(t)$ around t^n, it is easy to show that ∇ and D are connected through the formal relationship

$$hD = -\log(1 - \nabla). \tag{3.2.27}$$

Multiplying the differential system (3.2.2) by h and using (3.2.27) to eliminate time differentiation, one gets, at time t^{n+1},

$$\sum_{\ell=1}^{\infty} \frac{1}{\ell} \nabla^\ell \underline{u}^{n+1} = h \underline{f}(\underline{u}^{n+1}), \tag{3.2.28}$$

where, starting from (3.2.26), the high-order backward difference may be obtained recursively by $\nabla^\ell \underline{u}^n = \nabla(\nabla^{\ell-1} \underline{u}^n)$. Truncation of the l.h.s. of (3.2.28) at order k and repeated application of (3.2.26) gives the BDFk scheme

$$\underline{\tilde{u}}^{n+1} = \sum_{j=1}^{k} a_j \underline{\tilde{u}}^{n+1-j} + hb_0 \underline{f}^{n+1}. \tag{3.2.29}$$

The coefficients b_0 and $\{a_\ell\}_{\ell=1}^k$ of (3.2.29) for values of $k \leq 6$ are listed in Table 3.2.3. The single-step BDF1 scheme is identical to the IE algorithm (3.2.24) derived in the preceding section.

Introduction of the coefficients of Table 3.2.3 into the truncation error (3.2.13) gives $\mathcal{L}_k[u(t); h] = O(h^{k+1})$, that is, a consistency order equal to k. The stability curves of the four lowest BDFk schemes are shown on Figure 3.2.3. The stability regions are located outside the contours labeled (k). Except for the

Table 3.2.3. *Coefficients for the k-step backward-difference* (BDFk) *schemes with $k \leq 6$. For a given k one has $b_0 = \gamma_0/\underline{\gamma}$ and $a_\ell = \gamma_\ell/\underline{\gamma}$ with $1 \leq \ell \leq k$ [see (3.2.29)]. The* BDFk *schemes have consistency order k.*

k	$\underline{\gamma}$	γ_0	γ_1	γ_2	γ_3	γ_4	γ_5	γ_6
1	1	1	1					
2	3	2	4	−1				
3	11	6	18	−9	2			
4	25	12	48	−36	16	−3		
5	137	60	300	−300	200	−75	12	
6	147	60	360	−450	400	−225	72	−10

two lowest schemes (with $k = 1, 2$), which are *A*-stable [and therefore have their contours entirely in $\Re(z) > 0$], the higher-order stability curves steadily penetrate into $\Re(z) < 0$ with increasing values of k. This might give unstable numerical solutions for problems having eigenvalues on or near the imaginary axis.

3.2.2 Predictor–Corrector Methods

We have seen that a standard approach of implicit linear multistep methods consists of linearizing $\underline{f}(\underline{u}(t))$, the r.h.s. in (3.2.2), and solving the algebraic system (3.2.11), where the matrix depends on the Jacobian *J* of the problem [see Equation (3.2.10)]. Obviously, this is exact for *linear* systems of ODEs. However,

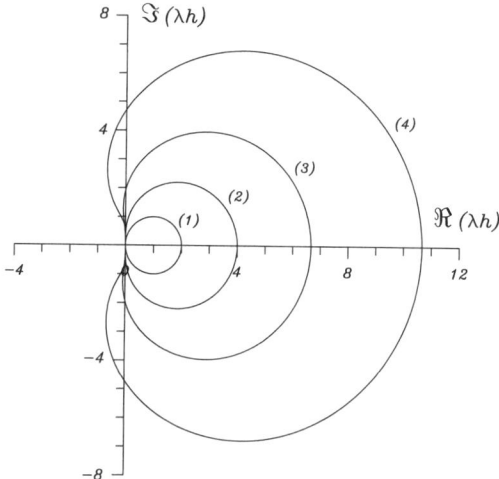

Figure 3.2.3. Stability regions for BDFk schemes with $1 \leq k \leq 4$.

3.2. Time Discretization Schemes

nonlinearity of the problem might be such that a more rigorous approach to (3.2.8) is needed. Predictor–corrector (PC) methods offer such an approach.

The main idea of PC methods consists in solving (3.2.8) iteratively to obtain $\underline{\tilde{u}}^{n+1}$. In what follows, the integer between square brackets in $\underline{\tilde{u}}_{[s]}^{n+1}$ refers to these iterations. The scheme is written as

$$\underline{\tilde{u}}_{[s+1]}^{n+1} = \sum_{j=1}^{k} a_j \underline{\tilde{u}}^{n+1-j} + hb_0 \underline{f}\left(\underline{\tilde{u}}_{[s]}^{n+1}\right) + h \sum_{j=1}^{k} b_j \underline{f}^{n+1-j}, \quad (3.2.30)$$

with the iteration index $s \geq 0$ and with $\underline{\tilde{u}}_{[0]}^{n+1}$ as a first guess. For the time being, we abstain from referring to the iterative process with regard to the quantities at instants before t^{n+1}, since this is discussed below in some detail.

One would like the difference between two iterates $\delta \underline{\tilde{u}}_{[m]}^{n+1} := \underline{\tilde{u}}_{[m+1]}^{n+1} - \underline{\tilde{u}}_{[m]}^{n+1}$ to go quickly to zero so that, for some integer $s = m - 1$, one has $\underline{\tilde{u}}_{[m]}^{n+1} \approx \underline{\tilde{u}}^{n+1}$, the solution to (3.2.8). In order to minimize the computational work, the first guess should be as close as possible to the solution of (3.2.8). A convenient procedure consists in selecting an explicit linear multistep method [with, say, k^* steps and coefficients $\{a_i^*\}_{i=1}^{k^*}$ and $\{b_i^*\}_{i=1}^{k^*}$], of order as high as possible, to *predict* a solution $\underline{\tilde{u}}_{[0]}^{n+1}$, leaving to the iterative scheme (3.2.30) the task of *correcting* this approximation to the value $\underline{\tilde{u}}_{[m]}^{n+1}$ (as close as possible to $\underline{\tilde{u}}^{n+1}$).

Let P denote an application of the predictor. The iterative scheme (3.2.30) shows that each correction iteration calls for an evaluation of the r.h.s. $\underline{f}(\underline{\tilde{u}}_{[s]}^{n+1})$. Let E denote such an evaluation, and let C denote an application of the corrector. Two PC algorithms can then be conceived on the basis of a fixed number of iterations m, according to whether the iterative process at some time step stops with the correction C or with the evaluation E. In the former case the algorithm symbolically is $P(EC)^m$:

$$\underline{\tilde{u}}_{[0]}^{n+1} = \sum_{j=1}^{k^*} a_j^* \underline{\tilde{u}}_{[m]}^{n+1-j} + h \sum_{j=1}^{k^*} b_j^* \underline{f}_{[m-1]}^{n+1-j},$$

$$\underline{\tilde{u}}_{[s+1]}^{n+1} = \sum_{j=1}^{k} a_j \underline{\tilde{u}}_{[m]}^{n+1-j} + hb_0 \underline{f}\left(\underline{\tilde{u}}_{[s]}^{n+1}\right) + h \sum_{j=1}^{k} b_j \underline{f}_{[m-1]}^{n+1-j}, \quad (3.2.31)$$

whereas in the latter case it is $P(EC)^m E$:

$$\underline{\tilde{u}}_{[0]}^{n+1} = \sum_{j=1}^{k^*} a_j^* \underline{\tilde{u}}_{[m]}^{n+1-j} + h \sum_{j=1}^{k^*} b_j^* \underline{f}_{[m]}^{n+1-j},$$

$$\underline{\tilde{u}}_{[s+1]}^{n+1} = \sum_{j=1}^{k} a_j \underline{\tilde{u}}_{[m]}^{n+1-j} + hb_0 \underline{f}\left(\underline{\tilde{u}}_{[s]}^{n+1}\right) + h \sum_{j=1}^{k} b_j \underline{f}_{[m]}^{n+1-j}. \quad (3.2.32)$$

In these two schemes we tacitly assume that $f_{[\ell]}^{n+1-j} := f(\tilde{u}_{[\ell]}^{n+1-j})$ and that $0 \leq s \leq m - 1$.

Two important questions are still pending: How does one select the value of m in practice? Among the schemes (3.2.31) and (3.2.32), which is to be preferred? Without entering into too much detail, let us mention that a convenient rule of thumb sets $m = p - p^* + 1$, where p and p^* denote the convergence orders of the corrector and predictor schemes, respectively. This is intuitively understood as follows: every correction iteration improves the (initial) consistency order of the numerical solution p^* by one unit up to p. An additional iteration is required, however, to reduce the error level to its minimum. With regard to the second question, we emphasize that $P(EC)^m E$ methods are usually preferred because of their better stability properties.

A simple example of the $P(EC)^m E$ algorithm combines the AB2 scheme used as predictor with the AM2 scheme used as corrector. The equations (3.2.32) thus are

$$\tilde{u}_{[0]}^{n+1} = \tilde{u}_{[m]}^n + \frac{h}{2}[3 f_{[m]}^n - f_{[m]}^{n-1}],$$
$$\tilde{u}_{[s+1]}^{n+1} = \tilde{u}_{[m]}^n + \frac{h}{12}[5 f(\tilde{u}_{[s]}^{n+1}) + 8 f_{[m]}^n - f_{[m]}^{n-1}],$$
(3.2.33)

with $0 \leq s \leq 1$, since the optimal value of m for this PC method is equal to 2. Figure 3.2.4 displays the stability regions of (3.2.33) with one and two correction

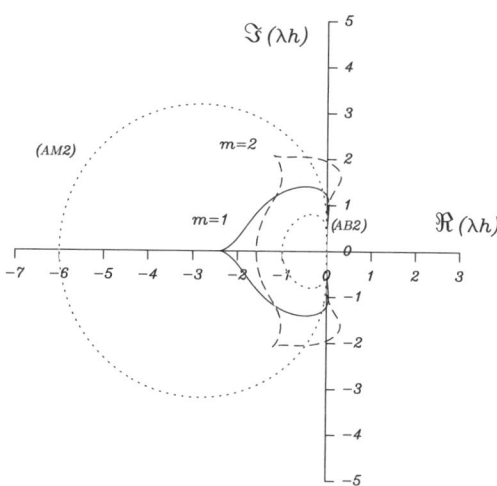

Figure 3.2.4. Stability regions for the $P(EC)^m E$ scheme (3.2.32) based on the AB2 and AM2 methods. The figure displays the stability contours of the PC scheme with $m = 1, 2$, as well as of the AB2 and AM2 schemes.

iterations. Also shown, for the purpose of comparison, are the stability regions of the basic AB2 and AM2 schemes. The $P(EC)^m E$ methods have larger stability regions than the predictor scheme, though not as large as that of the corrector scheme.

Our second example associates the AB4 and AM3 schemes both with order 4. Since in this case $p^* = p$, one correction iteration is enough ($m = 1$):

$$\tilde{\underline{u}}_{[0]}^{n+1} = \tilde{\underline{u}}_{[1]}^n + \frac{h}{24}[55\underline{f}_{[1]}^n - 59\underline{f}_{[1]}^{n-1} + 37\underline{f}_{[1]}^{n-2} - 9\underline{f}_{[1]}^{n-3}],$$

$$\tilde{\underline{u}}_{[1]}^{n+1} = \tilde{\underline{u}}_{[1]}^n + \frac{h}{24}[9\underline{f}(\tilde{\underline{u}}_{[0]}^{n+1}) + 19\underline{f}_{[1]}^n - 5\underline{f}_{[1]}^{n-1} + \underline{f}_{[1]}^{n-2}]. \quad (3.2.34)$$

Again, the stability region of (3.2.34) is located between those of the AB4 and AM3 algorithms. As is always the case with linear multistep methods, increasing the order goes in parallel with shrinking stability regions, hence with more drastic conditions on the time step.

3.2.3 Runge–Kutta Methods

Another approach for solving the differential system (3.2.2) uses the mean-value theorem, which states the existence of a time value ξ, *depending on the solution*, such that

$$\underline{u}(t) - \underline{u}^0 = (t - t^0)\underline{f}(\underline{u}(\xi)), \quad t^0 < \xi < t. \quad (3.2.35)$$

The r.h.s. of (3.2.35) is approximated by a linear combination of slopes at several points in the interval (t^0, t), in such a way that the Taylor expansion of the numerical solution $\tilde{\underline{u}}(t)$ coincides as much as possible with the Taylor expansion of $\underline{u}(t)$. This is the basic underlying principle of Runge–Kutta (RK) methods. Let us immediately emphasize that RK methods, unlike the previous integration schemes, are all one-step algorithms. Thus, they raise no problems with regard to initialization procedures and are often used as starting procedures for multistep integration methods. As before, we will distinguish between explicit and implicit RK methods according to whether the approximate solution $\tilde{\underline{u}}(t)$ obtains without or with the solution of a set of (nonlinear) auxiliary equations. Historically, explicit RK were introduced first and are probably the most widely used RK methods. They are easy to implement and have nice convergence and stability properties, as discussed hereafter. Implicit RK methods are more recent and, up to now, have received little attention in CFD applications. For more details on RK methods the reader should consult the books by Butcher [58] and by Gear [151], as well as [105] and a recent paper by Gottlieb et al. [164] that discusses strong stability-preserving high-order RK schemes.

Explicit Runge–Kutta Schemes

Let $\Delta t = h$ denote the time step. The canonical form of an m-stage RK integration scheme is given by

$$\underline{\tilde{u}}^{n+1} = \underline{\tilde{u}}^n + \sum_{j=1}^{m} w_j \underline{K}_j, \tag{3.2.36}$$

where the quantities $\underline{K}_j \in \mathbb{R}^{\mathcal{N}}$ are determined through the relationships

$$\underline{K}_j = h\underline{f}\left(\underline{\tilde{u}}^n + \sum_{\ell=1}^{m} a_{j\ell} \underline{K}_\ell\right), \quad 1 \leq j \leq m. \tag{3.2.37}$$

The set of $m(m+1)$ coefficients $\{w_j\}_{j=1}^{m}$ and $\{a_{j\ell}\}_{j,\ell=1}^{m}$ defines a particular RK scheme. A few examples that are important in practice will be given below. A convenient notation due to Butcher [58] for the m-stage RK method uses the coefficient matrix A and two vectors, $\underline{c}$ and $\underline{w}$, with the following pattern:

$$\frac{\underline{c} \mid A}{\mid \underline{w}^T} \equiv \begin{array}{c|cccc} c_1 & a_{11} & a_{12} & \cdots & a_{1m} \\ c_2 & a_{21} & a_{22} & \cdots & a_{2m} \\ \vdots & \vdots & \vdots & & \vdots \\ c_m & a_{m1} & a_{m2} & \cdots & a_{mm} \\ \hline & w_1 & w_2 & \cdots & w_m \end{array}, \tag{3.2.38}$$

where $\underline{w}^T$ denotes the transpose of $\underline{w}$. The components of $\underline{c}$ are related to the elements of A and are equal to the row sums: $c_i = \sum_j a_{ij}, 1 \leq i \leq m$.

An m-stage RK method is called *explicit* (ERK) when the coefficient matrix A is strictly lower triangular, that is, when $a_{ij} = 0, i \leq j$. In this case, the total number of coefficients is reduced to $m(m+1)/2$, and the relationship (3.2.37) becomes

$$\underline{K}_j = h\underline{f}\left(\underline{\tilde{u}}^n + \sum_{\ell=1}^{j-1} a_{j\ell} \underline{K}_\ell\right), \quad 1 \leq j \leq m, \tag{3.2.39}$$

with $a_{11} = 0$. One easily verifies that the auxiliary vectors $\underline{K}_j$ are evaluated one at a time, at the expense of one call of the r.h.s. of (3.2.2) per auxiliary vector and per time step (i.e., a total of m calls per time step). ERK integration schemes are therefore more expensive in computation time than the ABk algorithms introduced earlier. We will see below, through an example, that ERK schemes have close links with PC methods.

The nonlinearity of RK methods [see (3.2.37)] makes it much more difficult to derive general relationships between the degrees of freedom of an ERK

3.2. Time Discretization Schemes

Table 3.2.4. *Attainable order $r(m)$ of an explicit Runge–Kutta method as a function of the number of stages (see [58]).*

m	$r(m)$
1	1
2	2
3	3
4	4
5	4
6	5
7	6
≥ 8	$\leq m - 2$

scheme and its convergence order r. This topic has been thoroughly studied by Butcher, and we refer to [58] for further developments. Butcher's main results are summarized in Table 3.2.4, presenting the maximum attainable order of ERK methods as a function of the number of stages. A few comments are appropriate.

Consistency in the sense of Definition 3.1 requires that

$$\sum_{j=1}^{m} w_j = 1. \qquad (3.2.40)$$

The matching of higher terms in the Taylor series of the true solution induces further relationships between the elements of A and $\underline{w}$. A characteristic feature of ERK methods lies in the fact that, for a given convergence order, the number of relationships is usually less than the number of degrees of freedom. Consequently, one has for instance an infinity of two-stage ERK schemes of order 2, a double infinity of three-stage schemes with order 3, and so forth. The optimum is reached with the four-stage schemes of order 4. For $m > 4$, the complexity of the true solution's Taylor expansion becomes such that the number of degrees of freedom is insufficient to ensure $r = m$.

Let us now look at a few concrete examples. The arrays (3.2.41) define two classical two-stage ERK schemes of order two:

$$\begin{array}{c|cc} 0 & 0 & 0 \\ 1 & 1 & 0 \\ \hline & 1/2 & 1/2 \end{array} \, , \quad \begin{array}{c|cc} 0 & 0 & 0 \\ 1/2 & 1/2 & 0 \\ \hline & 0 & 1 \end{array} \quad [\text{ERK 2a,b}]. \qquad (3.2.41)$$

One can easily show that the ERK2a scheme (also known as "Euler–Cauchy") is equivalent to the $P(EC)E$ scheme (3.2.32) developed on the basis of the

EM scheme used as predictor, combined with CN as corrector. In practice, the preference goes to ERK2b (also known as "improved tangent") because of its slightly lower truncation error (though not optimal) and because there is only one call to f with the latter, instead of two with the former.

An important scheme in numerical computations is the four-stage ERK scheme of order 4, called the *classical* RK scheme and given by

$$
\begin{array}{c|cccc}
0 & 0 & 0 & 0 & 0 \\
1/2 & 1/2 & 0 & 0 & 0 \\
1/2 & 0 & 1/2 & 0 & 0 \\
1 & 0 & 0 & 1 & 0 \\
\hline
 & 1/6 & 1/3 & 1/3 & 1/6
\end{array}
\quad [\text{ERK4}]. \tag{3.2.42}
$$

A straightforward application of (3.2.41) and (3.2.42) to the test equation (3.2.17) shows that the approximate solutions at t^{n+1} are given by

$$\tilde{u}^{n+1} = R_p(z)\tilde{u}^n, \quad z := \lambda h, \tag{3.2.43}$$

where $R_p(z)$ ($p = 2, 4$) denotes polynomial approximations of the exponential function $\exp(z)$ such that

$$|\exp(z) - R_p(z)| = O(z^{p+1}).$$

The stability regions for the schemes (3.2.41) and (3.2.42) are displayed in Figure 3.2.5. The kidney shape of the ERK4 stability region is well known. The curve cuts the imaginary axis at $y = \pm 2.8283$.

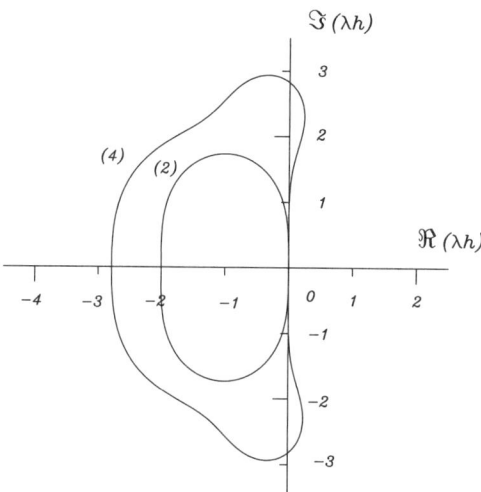

Figure 3.2.5. Stability regions for the Runge–Kutta schemes.

Implicit Runge–Kutta Schemes

When the matrix A is full, the m-stage RK scheme is called *implicit* (IRK), and, as was the case for implicit linear multistep methods, the solution of (3.2.2) obtains at the expense of an algebraic system to be solved at each time step. Let us deal with the test equation (3.2.17) first. With $z := \lambda h$ as before, and with the set of auxiliary vectors $\underline{K} := (K_1, K_2, \ldots, K_m)^T \in \mathbb{R}^m$, $\underline{w} := (w_1, w_2, \ldots, w_m)^T \in \mathbb{R}^m$, and $\underline{e} := (1, 1, \ldots, 1)^T \in \mathbb{R}^m$, the equations (3.2.37) are given by

$$(I - zA)\underline{K} = z\tilde{u}^n \underline{e}, \qquad (3.2.44)$$

with $I \in \mathbb{R}^{m \times m}$, the identity matrix. The solution (3.2.36) is given by

$$\tilde{u}^{n+1} = [1 + z\underline{w}^T \cdot (I - zA)^{-1}\underline{e}]\tilde{u}^n, \qquad (3.2.45)$$

where the quantity in brackets is now a *rational* approximation of $\exp(z)$.

For differential systems (3.2.2) where the unknown $\underline{u} \in \mathbb{R}^\mathcal{N}$, the algorithm gets more complicated, since each approximate slope $\underline{K}_i/h \in \mathbb{R}^\mathcal{N}$. This means that the algebraic system solved at each time step has dimension $m\mathcal{N}$. In anticipation of Chapter 4,[2] we introduce the tensor product of A and J, where J is the Jacobian operator (3.2.10). The tensor product $A \otimes J$ is an $m\mathcal{N} \times m\mathcal{N}$ matrix operator with components $a_{ij} J_{k\ell}$ such that

$$A \otimes J := \begin{pmatrix} a_{11}J & a_{12}J & \cdots & a_{1m}J \\ a_{21}J & a_{22}J & \cdots & a_{2m}J \\ \vdots & \vdots & & \vdots \\ a_{m1}J & a_{m2}J & \cdots & a_{mm}J \end{pmatrix}. \qquad (3.2.46)$$

Then, it is easy to show that with $\underline{K} := (\underline{K}_1, \underline{K}_2, \ldots, \underline{K}_m)^T \in \mathbb{R}^{m\mathcal{N}}$, and $\hat{\underline{f}} := (\underline{f}, \underline{f}, \ldots, \underline{f})^T \in \mathbb{R}^{m\mathcal{N}}$ the algebraic system (3.2.37) becomes

$$(I - hA \otimes J)\underline{K} = h\hat{\underline{f}}(\tilde{u}^n). \qquad (3.2.47)$$

With Gaussian elimination the solution of (3.2.47) requires asymptotically $m^3\mathcal{N}^3/3$ "add–multiplies," a considerable amount of work because it is repeated at each time step. This is about m^3 times as much as with any implicit linear multistep method. Nevertheless, IRK schemes have drawn considerable attention in recent years because their stability properties make them attractive for stiff ODE systems. Butcher [57] showed that, for every value of m, there

---

[2] The properties of such operators will be analyzed in Section 4.2.

exists an m-stage IRK scheme of order $2m$. Ehle [118] then demonstrated that all these IRK schemes are A-stable, a meaningful result in view of the limitation to order 2 for A-stable implicit linear multistep methods, as shown by Dahlquist [89].

More significant in practice is the existence of A-stable IRK schemes with a lower-triangular matrix A that has the same element along the main diagonal (i.e., $a_{ij} = 0, i < j$, and $a_{jj} = a \, \forall j$). In this case, the method is still implicit, but the vectors $\underline{K}_j$ may be obtained one at a time, and the size of the linear system to be solved at each step is reduced by a factor m. The computational work per time step becomes of the same order of magnitude as for an implicit multistep method [i.e., $O(\mathcal{N}^3)$] – actually a little higher, considering the fact that m linear $\mathcal{N} \times \mathcal{N}$ systems, which differ only by their right-hand sides, must be solved per time step. Such RK schemes are called *diagonally implicit* (DIRK). To this family belong the two-stage IRK schemes proposed by Rosenbrock [338] and Crouzeix [87], with the arrays

$$\begin{array}{c|cc} 1 - \sqrt{2}/2 & 1 - \sqrt{2}/2 & 0 \\ 1/2 & -1/2 + \sqrt{2}/2 & 1 - \sqrt{2}/2 \\ \hline & 0 & 1 \end{array} \quad \text{[Rosenbrock]} \quad (3.2.48)$$

and

$$\begin{array}{c|cc} 1/2 + \sqrt{3}/6 & 1/2 + \sqrt{3}/6 & 0 \\ 1/2 - \sqrt{3}/6 & -\sqrt{3}/3 & 1/2 + \sqrt{3}/6 \\ \hline & 1/2 & 1/2 \end{array} \quad \text{[Crouzeix]}. \quad (3.2.49)$$

Application of the linear system (3.2.44) to these schemes gives two rational approximations of the exponential function:

$$\begin{aligned} R_R(z) &= \frac{1 - (1 - \sqrt{2})z}{[1 - (1 - \sqrt{2}/2)z]^2}, \\ R_C(z) &= \frac{6 - 2\sqrt{3}z - (1 + \sqrt{3})z^2}{6 - 2(3 + \sqrt{3})z + (2 + \sqrt{3})z^2}. \end{aligned} \quad (3.2.50)$$

With a Maclaurin series, it is easy to show that

$$|e^z - R_R(z)| = O(z^3) \quad \text{and} \quad |e^z - R_C(z)| = O(z^4). \quad (3.2.51)$$

Both algorithms (3.2.48) and (3.2.49) are A-stable with stability contours in $\Re(\lambda h) > 0$. The consistency orders are 2 and 3 respectively; Rosenbrock's

3.3 Splitting Methods

lower order is compensated by the fact that this scheme is L-stable whereas Crouzeix's scheme is not, since $R_C(-\infty) \approx -0.732$.

3.3 Splitting Methods

Up to now, we have examined various numerical schemes to integrate the dynamical systems (3.2.1) arising from high-order space discretization of the time-dependent Navier–Stokes equations (1.6.3)–(1.6.4). Some of these schemes are well suited for the integration of diffusion problems but, being implicit, are rather costly for the treatment of (nonlinear) convection terms. Similarly, explicit schemes fitted for the integration of convection terms cannot be used for viscous terms without very small time steps to ensure stability control of the stiffness associated with the diffusion process.

In this section we examine splitting methods that allow implicit and explicit schemes to be used *simultaneously* (i.e., in the same equation) to handle the diffusion and convection terms separately. The procedure is analogous to the alternating-direction implicit (ADI) algorithm of Peaceman and Rachford [303, 401, 402], although here the splitting does not affect the space variables (or space and time variables), but rather the diffusion and advection contributions to the Navier–Stokes equations. We start with a very simple example showing the basic idea. This is followed by a more general setting of the problem that allows any combination of implicit and explicit algorithms described in the preceding section.

Assume the right-hand side of (3.2.2) is divided into two terms

$$\frac{d\underline{u}}{dt} = \underline{f}\left(\underline{u}(t)\right) + \nu \underline{g}\left(\underline{u}(t)\right), \quad \underline{u}(t^0) = \underline{u}^0, \quad t \in (t^0, T], \quad (3.3.1)$$

where $\underline{f}(\underline{u}(t)) \in \mathbb{R}^N$ and $\nu \underline{g}(\underline{u}(t)) \in \mathbb{R}^N$ represent the (space) discretized convection and diffusion terms, respectively, with $\nu = 1/Re$. Integrating both sides of (3.3.1) on (t^0, t) gives

$$\underline{u}(t) - \underline{u}^0 = \int_{t^0}^{t} \underline{f}\left(\underline{u}(s)\right) ds + \int_{t^0}^{t} \nu \underline{g}\left(\underline{u}(s)\right) ds. \quad (3.3.2)$$

Then, for the reasons already mentioned, we select explicit and implicit linear multistep methods with a constant time step h, to approximate the two quadratures in (3.3.2). For instance, the selection of AB2/AM1 gives the second-order Adams–Bashforth/Crank–Nicolson (ABCN) algorithm:

$$\underline{\tilde{u}}^{n+1} - \underline{\tilde{u}}^n = \frac{3}{2}h\underline{f}(\underline{\tilde{u}}^n) - \frac{1}{2}h\underline{f}(\underline{\tilde{u}}^{n-1}) + \frac{\nu}{2}[h\underline{g}(\underline{\tilde{u}}^{n+1}) + h\underline{g}(\underline{\tilde{u}}^n)], \quad (3.3.3)$$

which has been used by several authors [297, 22, 11]. In practice, Equation (3.3.3) is solved in two successive steps:

$$\begin{aligned}\underline{\tilde{u}}^{n+1/2} - \underline{\tilde{u}}^n &= \frac{3}{2} h \underline{f}^n - \frac{1}{2} h \underline{f}^{n-1}, \\ \underline{\tilde{u}}^{n+1} - \underline{\tilde{u}}^{n+1/2} &= \frac{\nu}{2} (h \underline{g}^{n+1} + h \underline{g}^n),\end{aligned} \quad (3.3.4)$$

hence the names "splitting," and "fractional step" for the method, which are common in the literature (see also [417]). In (3.3.4), we have used our earlier notation convention $[\underline{f}^{n+j} := \underline{f}(\underline{\tilde{u}}^{n+j})]$ for both $\underline{f}$ and $\underline{g}$.

The stability of (3.3.4) (though largely influenced by the convection term) must be studied per se, as shown below. Readers should convince themselves that there is more to splitting methods than a mere juxtaposition of explicit and implicit integration schemes. Splitting techniques are tailored to avoid the fully implicit solution of stiff systems resulting from the space discretization of the Navier–Stokes equations, with widespread complex eigenspectra comprising real negative eigenvalues due to the viscous contribution, and purely imaginary eigenvalues due to convection.

A systematic study of implicit–explicit algorithms for ODE systems in the context of spectral methods (and of their stability properties) can be found in Ascher et al. [11]. As an example, these authors show that, in the subclass of all second-order implicit–explicit schemes based on three successive time steps, the rule

$$\begin{aligned}\underline{\tilde{u}}^{n+1/2} - \underline{\tilde{u}}^n &= \tfrac{3}{2} h \underline{f}^n - \tfrac{1}{2} h \underline{f}^{n-1}, \\ \underline{\tilde{u}}^{n+1} - \underline{\tilde{u}}^{n+1/2} &= \nu \left(\tfrac{9}{16} h \underline{g}^{n+1} + \tfrac{3}{8} h \underline{g}^n + \tfrac{1}{16} h \underline{g}^{n-1} \right),\end{aligned} \quad (3.3.5)$$

which they call the *modified* Adams–Bashforth/Crank–Nicolson (MABCN) scheme, has the smallest truncation error. Strictly speaking, the implicit scheme (3.3.5) is not Crank–Nicolson, but, except for an additional vector storage $\underline{g}^{n-1}$, the numerical work is the same in (3.3.5) as in (3.3.4) with, in addition, a lower truncation error.

The stability of a splitting algorithm used to solve (3.3.1) depends on the spectral properties of the two components of the r.h.s. In other terms, the single test equation (3.2.17) introduced previously should now be replaced by two test equations, one for each component of the splitting. Thus, the stability of the ABCN scheme is governed by

$$\zeta^2 - \zeta = \frac{z_1}{2}(3\zeta - 1) + \frac{\nu z_2}{2}(\zeta^2 + \zeta), \quad (3.3.6)$$

3.3. Splitting Methods

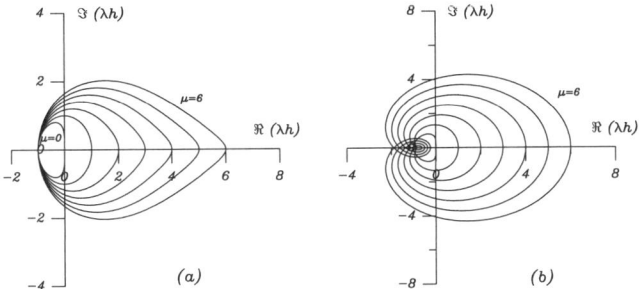

Figure 3.3.1. Stability regions for two second-order implicit–explicit schemes: (*a*) the ABCN scheme and (*b*) the MABCN scheme. Both set of curves depend on a parameter μ, with $0 \leq \mu \leq 6$ [see (3.3.7)].

with $z_i := \lambda_i h$ and $|\zeta| < 1$. For clarity we lump the diffusion contribution in (3.3.6) into a single parameter μ, and display in Figure 3.3.1(*a*) the stability region for

$$\zeta^2 - \zeta + \frac{\mu}{2}(\zeta^2 + \zeta) = \frac{z}{2}(3\zeta - 1) \tag{3.3.7}$$

as a function of μ ($0 \leq \mu \leq 6$). Notice that $\mu = 0$ corresponds to the AB2 scheme. Figure 3.3.1(*b*) displays the stability region for the MABCN algorithm, as a function of μ, with the same convention. Both calculations show that an adequate treatment of the diffusion term (in particular, of its fast-decaying components corresponding to large values of μ) may enlarge the stability region of the advection term.

Readers interested in other multistep implicit–explicit schemes or in other presentations of the stability conditions of these schemes are referred to [11].

3.3.1 The Operator-Integration-Factor Splitting Method

The derivation of splitting schemes based on BDF and/or RK algorithms is a little more complicated and must be done in a more general framework. For this purpose, Maday et al. [262] introduced a method called *operator-integration-factor-splitting* (OIFS). Our presentation closely follows their notation.

Assume the existence of an integration factor $\underline{\underline{Q}}_{\underline{f}}^{t^*}(t) \in \mathbb{R}^{\mathcal{N} \times \mathcal{N}}$ related to $\underline{f}(\underline{u})$ such that (3.3.1) may be cast into

$$\frac{d}{dt}\left[\underline{\underline{Q}}_{\underline{f}}^{t^*}(t) \cdot \underline{u}(t)\right] = \underline{\underline{Q}}_{\underline{f}}^{t^*}(t) \cdot v\underline{g}\left(\underline{u}(t)\right), \qquad t \in (t^0, T]. \tag{3.3.8}$$

The parameter t^* in the integrating factor is an arbitrary fixed time ($t^* \geq t$). We impose furthermore

$$\mathcal{Q}_f^{t^*}(t^*) = I, \tag{3.3.9}$$

where $I \in \mathbb{R}^{\mathcal{N} \times \mathcal{N}}$ is the identity matrix. From (3.3.1) and (3.3.8), one can easily show that the integration factor satisfies the differential system

$$\left(\frac{d}{dt}\mathcal{Q}_f^{t^*}(t)\right) \cdot \underline{u}(t) = -\mathcal{Q}_f^{t^*}(t) \cdot \underline{f}\left(\underline{u}(t)\right). \tag{3.3.10}$$

Let us now introduce an auxiliary variable $\underline{v}^{(t^*,t)}(s) \in \mathbb{R}^{\mathcal{N}}$, the solution to the following differential system on a shifted time axis s, with the origin set at time t:

$$\frac{d}{ds}\underline{v}^{(t^*,t)}(s) = \underline{f}\left(\underline{v}^{(t^*,t)}(s)\right), \quad 0 < s < t^* - t, \tag{3.3.11}$$

and with initial condition

$$\underline{v}^{(t^*,t)}(0) = \underline{u}(t). \tag{3.3.12}$$

Multiplying both sides of (3.3.11) by the integrating factor $\mathcal{Q}_f^{t^*}(t+s)$ (where t is fixed) and using (3.3.10), one gets

$$\frac{d}{ds}\left[\mathcal{Q}_f^{t^*}(t+s) \cdot \underline{v}^{(t^*,t)}(s)\right] = 0. \tag{3.3.13}$$

Then, with the initial condition (3.3.12) on the one hand, and the property (3.3.9) on the other hand, it is easy to show that

$$\mathcal{Q}_f^{t^*}(t) \cdot \underline{u}(t) = \underline{v}^{(t^*,t)}(t^* - t). \tag{3.3.14}$$

The importance of the integration factor lies in the fact that it allows a splitting of the basic problem (3.3.1) into two coupled differential problems (3.3.8) and (3.3.11)–(3.3.12) where the diffusion and convection contributions have been separated. Each of these problems may be solved with its own numerical scheme (say, S_g for diffusion and S_f for convection). As we will find out a little later, the time-step limitations to ensure stability are very different in the two cases, and with the decoupling provided by $\mathcal{Q}_f^{t^*}(t)$, the selection of different time steps h_t and h_s becomes straightforward. What is perhaps less transparent from the preceding analysis is that the integration factor never needs to be computed explicitly. It enters only as a splitting tool for the basic components of the problem, the evaluation of quantities such as $\mathcal{Q}_f^{t^*}(t) \cdot \underline{u}(t)$ being made with the relationship (3.3.14).

3.3.2 OIFS *Example: The* BDF3/RK4 *Scheme*

We illustrate the OIFS methodology with an example where for the advection component we choose the ERK4 (S_f) algorithm (see Section 3.2.3) while for the diffusion component we choose the implicit BDF3 (S_g) scheme (see Section 3.2.1).

We start with the integration of the diffusion problem (3.3.8) using a time step h_t, and we assume that the approximate solution is known up to time t^n. We first set $t^\star = t^{n+1}$. Then, with the appropriate coefficients borrowed from Table 3.2.3 and the property (3.3.9), the solution on the next time interval is written

$$\tilde{\underline{u}}^{n+1} = \sum_{j=1}^{3} a_j S_f\left(\mathcal{Q}_f^{t^{n+1}}(t^{n+1-j}) \cdot \tilde{\underline{u}}^{n+1-j}\right) + b_0 h_t \underline{v} g^{n+1}. \quad (3.3.15)$$

Recall that the quantities $\mathcal{Q}_f^{t^{n+1}}(t^{n+1-j}) \cdot \tilde{\underline{u}}^{n+1-j}$ entered into this equation are known as a result of the integration of the advection problem (3.3.11). The symbol S_f in the r.h.s. of (3.3.15) indicates this particular feature of the computation.

Let us now turn to the advection problem

$$\frac{d}{ds}\underline{v}^{(t^{n+1},t^{n+1-j})}(s) = \underline{f}\left(\underline{v}^{(t^{n+1},t^{n+1-j})}(s)\right), \quad (3.3.16)$$

solved by the ERK4 algorithm on the shifted time axis s, with the origin set at t^{n+1-j}. The initial condition is

$$\underline{v}^{(t^{n+1},t^{n+1-j})}(0) = \tilde{\underline{u}}^{n+1-j}. \quad (3.3.17)$$

The numerical integration uses a step size h_s such that $h_t = \gamma h_s$, where the integer $\gamma \geq 1$. Integration is carried over the interval $(0, \Delta s_j]$, with $\Delta s_j := t^{n+1} - t^{n+1-j}$. As shown earlier by (3.3.14), one has

$$S_f\left(\mathcal{Q}_f^{t^{n+1}}(t^{n+1-j}) \cdot \tilde{\underline{u}}^{n+1-j}\right) = \underline{v}^{(t^{n+1},t^{n+1-j})}(\Delta s_j). \quad (3.3.18)$$

Once the quantities (3.3.18) have been computed for all values of j, (3.3.15) is solved with respect to $\tilde{\underline{u}}^{n+1}$, and the integration process may be carried to the next time step (t^{n+1}, t^{n+2}).

Figure 3.3.2 shows an example of OIFS integration grids. The upper grid with time step h_t is associated to the implicit computation (BDF3), while the lower grids with time steps h_s correspond to the explicit computations (ERK4) carried to obtain the quantities (3.3.18). For each value of j, the figure shows the corresponding integration range Δs_j.

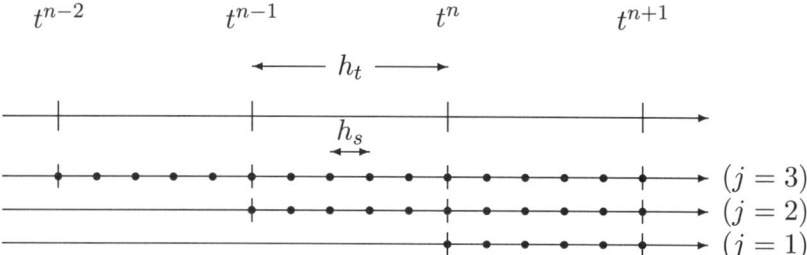

Figure 3.3.2. An example of OIFS integration grids with time steps h_t for the diffusion component (BDF3) and h_s for the advection component (ERK4). In this example $(h_t = 5 h_s)$.

With negative real eigenvalues for the diffusion problem (3.3.8), BDF3 induces no stability restriction on h_t. To minimize the computational work, one is therefore tempted to use values of h_t as large as possible. However, the time step h_t should not be so large that accuracy is lost by truncation errors. One should also give preference to L-stable implicit schemes to ensure correct handling of the fastest-decaying modes.

With regard to the convection problem (3.3.11)–(3.3.12) solved explicitly, the eigenvalues are purely imaginary and, for spectral elements, scale as $O(N^2)$. Most of the time, the associated time step h_s will be shorter and, as shown in Figure 3.3.2, will be a fraction of h_t.

Splitting methods based on integrating factors have been applied to incompressible fluid flows by Maday et al. [262] and by Couzy [83]. Other high-order splitting methods for incompressible Navier–Stokes calculations may be found in Karniadakis et al. [219] and in Heinrichs [191, 194].

3.4 The Parabolic Case: Unsteady Diffusion

At this point we reintroduce the physical meaning of the basic problem, and we consider the application of space *and* time discretization schemes to model problems of the form

$$\frac{\partial u}{\partial t} = \mathcal{L}u + f, \quad u(0, x) = u^0, \tag{3.4.1}$$

subject to boundary conditions appropriate for the operator $\mathcal{L}$. As noted in the preceding sections, the choice of a time-stepping scheme is primarily influenced by the nature of $\mathcal{L}$ along with the relative cost of applying it and/or its inverse. However, one must also pay attention to the nature of f if it is not smoothly varying in time, for example, in the case of stochastically forced turbulence

3.4. The Parabolic Case: Unsteady Diffusion

simulations [251, 253]. Here, we assume that f is smoothly varying in time, and we focus on the role of $\mathcal{L}$. If $-\mathcal{L}$ is linear and positive definite (i.e., has positive eigenvalues), the operator $\frac{\partial}{\partial t} - \mathcal{L}$ is parabolic, and we can expect solutions to the homogeneous problem associated to (3.4.1) to decay in time.

An example of a parabolic problem commonly encountered in fluid mechanics and heat transfer is the time-dependent diffusion (or heat) equation (1.9.4), given a more general form by

$$\frac{\partial u}{\partial t} = \frac{\partial}{\partial x}\left(\nu \frac{\partial u}{\partial x}\right) + f, \qquad u(t,0) = u(t,1) = 0, \quad u(0,x) = u^0(x). \tag{3.4.2}$$

We have encountered the steady-state version of this equation in (2.7.13). Equation (3.4.2) describes, for example, the time-varying temperature distribution in a rod with localized heat source $f(t, x)$, the temperature being fixed to zero at each end of the rod. It corresponds also to spatially developed unsteady viscous flow between walls having no-slip conditions. Neumann boundary conditions would correspond to a zero-heat-flux (perfectly insulated) condition in the thermal example and to a zero-stress (or symmetry) condition in the flow example. Unlike the steady-state case, this problem is well posed even if both ends of the domain have Neumann boundary conditions.

For ν constant and f time-invariant, the analytical solution to (3.4.2) is

$$u(t,x) = \sum_{k=1}^{\infty} \left[\left(\hat{u}_k^0 - \hat{u}_k^\infty\right) e^{\lambda_k t} + \hat{u}_k^\infty\right] \sin(\pi k x), \tag{3.4.3}$$

with $\lambda_k = -\nu k^2 \pi^2$ being the eigenvalues of the operator $\mathcal{L}$ with Dirichlet boundary conditions. The coefficients $\hat{u}_k^0$ and $\hat{u}_k^\infty$ in (3.4.3) are the Fourier coefficients

$$\hat{u}_0^k := 2 \int_0^1 u^0(x) \sin(\pi k x)\, dx, \qquad \hat{u}_k^\infty := \frac{2}{\nu k^2 \pi^2} \int_0^1 f(x) \sin(\pi k x)\, dx.$$

As the expansion (3.4.3) shows, the solution to (3.4.2) consists of two parts: a steady-state part, $u^\infty(x) := \sum_k \hat{u}_k^\infty \sin(\pi k x)$, determined by the inhomogeneity f, and a transient part having components $(\hat{u}_k^0 - \hat{u}_k^\infty) \sin(\pi k x)$ that decay as $e^{-\nu k^2 \pi^2 t}$. High-wavenumber (large-k) components decay quite rapidly, leaving after a short time only smooth low-wavenumber components active in the transient. Note that a change in f at some time $T > 0$ would correspond to establishing a new initial condition $u(T, x)$ and a new value of $u^\infty(x)$. In numerical computations, roundoff error plays the role of time-dependent forcing, and, as a result, high-wavenumber transients are always present. The chosen time-stepping scheme must be able to treat fast transients in a stable way, even when the solution is smooth and therefore evolving slowly.

3.4.1 Spatial Discretization

Our numerical treatment of (3.4.2) begins with the spatial discretization, based upon the equivalent weak form: Find $u(t, x) \in V$ such that

$$\int_\Omega \frac{\partial u}{\partial t} v \, dx = -\int_\Omega v \frac{\partial u}{\partial x} \frac{\partial v}{\partial x} \, dx + \int_\Omega v f \, dx \qquad \forall v(x) \in V,$$

where $V = H_0^1(\Omega)$ is the space of square-integrable functions satisfying the homogeneous boundary conditions on $\partial\Omega$. If Ω is invariant with respect to t, the time derivative can be moved outside the first integral and, using the notation introduced in (2.1.11)–(2.1.13), the problem becomes: Find $u(t, x) \in V$ such that

$$\frac{d}{dt}(u, v) = -\mathcal{A}(u, v) + \mathcal{F}(v) \qquad \forall v \in V.$$

Discretization proceeds by restricting the search space for u to a finite-dimensional subspace, $V_N \subset V$, having a basis $\{\psi_1, \ldots, \psi_N\}$. Letting

$$u_N(t, x) := \sum_j u_j(t) \psi_j(x) \qquad (3.4.4)$$

be the approximation to u, and testing for each basis element, $v = \psi_i$, one gets

$$M \frac{d\underline{u}}{dt} = -K\underline{u} + M\underline{f}, \qquad (3.4.5)$$

where $\underline{u} := \{u_j(t)\}_{j=1}^N$ is the vector of unknown coefficients, and $\underline{f}$ is the vector of coefficients representing the forcing term. We will assume the use of Lagrangian bases hereafter, such that the unknown and forcing terms represent nodal values. Here, K and M are the stiffness and mass matrices introduced in Chapter 2:

$$K_{ij} := \mathcal{A}(\psi_i, \psi_j), \quad M_{ij} := (\psi_i, \psi_j), \qquad 1 \le i, j \le N. \qquad (3.4.6)$$

As in the case of the elliptic problem, several possible spaces and associated bases may be considered in the formation of the semidiscrete system (3.4.5).

3.4.2 Time Advancement

To introduce the mechanics of the time-stepping procedures, we consider the two simple time advancement schemes introduced earlier: implicit Euler (IE) and explicit Euler (EM). For the homogeneous problem (3.4.5), the IE scheme is

$$M \frac{\underline{u}^{n+1} - \underline{u}^n}{\Delta t} = -K \underline{u}^{n+1},$$

or

$$H \underline{u}^{n+1} = M \underline{u}^n, \qquad (3.4.7)$$

where the Helmholtz system, $H := M + \Delta t K$, must be solved at each time step. In the one-dimensional case, the bandwidth of H is usually small enough (cf. Figures 2.3.2 and 2.4.2) that Cholesky factorization is the best approach, particularly since the factorization cost is amortized over many solves. In more than one space dimension, CG coupled with Jacobi preconditioning is usually preferable, because H is typically strongly diagonally dominant and the iteration converges quite rapidly. The iteration work can be further reduced by computing only the *change* in the solution, $\Delta \underline{u} := \underline{u}^{n+1} - \underline{u}^n$, by solving

$$H \Delta \underline{u} = M \underline{u}^n - H \underline{u}^n. \qquad (3.4.8)$$

This strategy avoids recomputation of components of $\underline{u}^{n+1}$ contained in $\underline{u}^n$ and yields an initial residual with norm scaling as $O(\Delta t)$, implying a substantial reduction in iteration effort. Additional savings may be obtained by computing the projection of $\underline{u}^{n+1}$ onto several previous solutions, as discussed in Sections 2.7.4 and 4.7.1.

We note that the overall accuracy of the solution will depend on the accuracy of both the temporal and spatial discretizations. Higher-order variations of (3.4.7) may be implemented with essentially the same computational complexity. For example, the BDF2 scheme, which approximates the time derivative $M d\underline{u}/dt$ by the second-order difference

$$M \frac{3\underline{u}^{n+1} - 4\underline{u}^n + \underline{u}^{n-1}}{2 \Delta t},$$

is implemented as

$$H \underline{u}^{n+1} = \frac{1}{2 \Delta t} M (4 \underline{u}^n - \underline{u}^{n-1}),$$

with $H := K + \frac{3}{2\Delta t} M$. Higher-order BDF schemes are similarly implemented with a single solve in H per step.

3. Parabolic and Hyperbolic Problems

As noted in Section 3.2, the stability of a time-stepping scheme is related to the eigenvalues of the Jacobian of the r.h.s. operator, which, in the case of the linear problem (3.4.5), is given by $J := -M^{-1}K$. We have shown earlier that, although J is nonsymmetric, its eigenvalues are real and negative [see Equation (2.7.16)]. Therefore, any time-stepping scheme whose stability region contains the entire negative real axis will be unconditionally stable. Crank–Nicolson and BDFk are examples of schemes suitable for implicit treatment of parabolic problems, as explained in Section 3.2.1.

The explicit counterpart (EM) to the implicit scheme (3.4.7) is

$$M\underline{u}^{n+1} = (M - \Delta t K)\underline{u}^n. \tag{3.4.9}$$

In contrast to finite differences, explicit methods for weighted residual techniques require a system solve because of the presence of the mass matrix on the l.h.s. of (3.4.9). While the same arguments regarding the solution of the implicit system (3.4.7) apply to (3.4.9), it is generally preferable to seek a formulation that yields a diagonal mass matrix, thus obviating the need for a system solve at every step. This explains the interest in the mass lumping scheme introduced in Section 2.7.1.

For explicit schemes, one must determine the minimum (negative) eigenvalue of J in order to select a time step Δt that is sufficiently small to ensure stability. For $\mathcal{Q}_1$ finite elements on a uniformly spaced grid, we determined in Chapter 2 closed forms for the spectrum λ_k^{FE} (without mass lumping) and λ_k^{FD} (with mass lumping). Using the relationships (2.7.18) and (2.7.20), one deduces easily that the minimum eigenvalues are equal to

$$\lambda_{N-1}^{\text{FE}} \approx -12\nu N^2 \quad \text{and} \quad \lambda_{N-1}^{\text{FD}} \approx -4\nu N^2. \tag{3.4.10}$$

Then, from the stability diagram for the EM scheme (Figure 3.2.1), we require $\lambda \Delta t \in (-2, 0)$. This implies that for the Lagrangian $\mathcal{Q}_1$ FE schemes without and with mass lumping, time steps are limited for stability reasons to

$$\Delta t \leq \frac{1}{6\nu N^2} \quad \text{and} \quad \Delta t \leq \frac{1}{2\nu N^2}, \tag{3.4.11}$$

respectively.

The analysis of the condition number of high-order schemes (collocation and spectral-element) in Section 2.7.1 leads to two conclusions. First, when coupling spectral methods with explicit schemes, stability dictates a very restrictive limitation:

$$\Delta t \leq C\frac{\pi^2}{\nu N^4}, \tag{3.4.12}$$

where C is an order-unity constant dependent on the particular time-stepping scheme. Because of this severe constraint on time-step size, implicit methods are generally preferred for time-dependent diffusion problems. This is more important for spectral methods than for low-order methods, owing to the different $[O(N^4)$ versus $O(N^2)]$ growth rates of $|\lambda_{\min}|$.

The second conclusion is that high-order methods are able to correctly capture roughly two-thirds of the spectrum, whereas linear finite elements only resolve about one-third, implying a two-to-one resolution saving *per space dimension* in the high-order case. From the spectral-element results with $(E, N) = (64, 4)$ shown in Figure 2.7.2(a), it is also clear that, with E spectral elements, one need not take an asymptotically large degree N to capture a significant portion of the resolvable spectrum.

3.5 The Hyperbolic Case: Linear Convection

We now consider hyperbolic problems of the form (1.9.5):

$$\frac{\partial u}{\partial t} + c\frac{\partial u}{\partial x} = 0 \quad \text{on } (a, b), \qquad u(0, x) = u^0(x), \tag{3.5.1}$$

subject to appropriate boundary conditions on $\partial \Omega$. This model problem corresponds to the transport of a quantity u by a velocity field $c(t, x)$. If c is constant, then the solution is $u(t, x) = u^0(x - ct)$, corresponding to a simple translation of the initial waveform $u^0(x)$ at speed c. The choice of boundary conditions depends on c. If $c > 0$, the transport is from left to right, and Dirichlet boundary conditions are needed on the left end of the domain. If $c < 0$, boundary conditions are required on the right, with nothing specified on the left. It is also often useful to consider periodic boundary conditions, $u(t, a) = u(t, b)$, which correspond to solving (3.5.1) on the infinite domain $(-\infty, \infty)$ with initial conditions and convective velocity that are $[a, b]$-periodic. This has the advantage of permitting long time integrations without having correspondingly long computational domains.

The nature of the solutions to (3.5.1) can be more clearly understood by considering the evolution of the energy, obtained by multiplying by u and integrating over Ω:

$$\int_\Omega \left(u\frac{\partial u}{\partial t} + cu\frac{\partial u}{\partial x} \right) dx = \frac{1}{2}\frac{d}{dt}\int_\Omega u^2\, dx + \frac{1}{2}\int_\Omega c\frac{\partial u^2}{\partial x}\, dx = 0.$$

The last term can be integrated by parts to obtain

$$\frac{d}{dt}\int_\Omega u^2\, dx = cu^2|_a - cu^2|_b + \int_\Omega u^2 \frac{\partial c}{\partial x}\, dx.$$

If c is constant, analogous to $\nabla \cdot \mathbf{c} = 0$ in the multidimensional case [cf. (1.6.4)], then only the boundary terms contribute to the growth or decay of energy, $\int_\Omega u^2 \, dx$. If $c > 0$ and $u(t, a) = 0$, the mean energy decays for any $u(t, b) \neq 0$, implying that energy leaves the domain through the right-hand end point. For periodic boundary conditions the mean energy is unchanged for all time. Unlike the parabolic case, where the signal quickly smooths out, the signal in the hyperbolic case persists for very long times (indefinitely if there is no diffusion), increasing the importance of accurate spatial and temporal discretizations for all resolved wavenumbers.

3.5.1 Spatial Discretization

As in the parabolic case, spatial discretization is based on the weak formulation: Find $u \in V_{N,B}$ such that

$$\frac{d}{dt} \int_\Omega u(t, x) v(x) \, dx = -\int_\Omega c(t, x) \frac{\partial u}{\partial x} v(x) \, dx \qquad \forall v \in W_{N,0},$$

where $V_{N,B}$ incudes functions satisfying inhomogeneous conditions on any boundary where the characteristics are incoming, and $W_{N,0}$ includes functions that vanish on those boundaries. Strictly speaking, if v is in H^1, u can be discontinuous, since we can transfer the x-derivative from u onto v through an integration by parts. Discontinuities may be present in the initial condition or may arise from smooth initial conditions if c is varying (inviscid Burgers equation). Here, we consider only solutions in H^1, in anticipation of adding viscosity when we consider the combined advection–diffusion equation or the full Navier–Stokes equations. For ease of notation, we also restrict our attention to the case of homogeneous or periodic boundary conditions, and drop the B and 0 subscripts on the subspaces.

We proceed with the Galerkin formulation, in which $W_N = V_N$. Letting $\{\psi_1, \ldots, \psi_N\}$ be a basis for V_N, we define

$$M_{ij} := \int_\Omega \psi_i \psi_j \, dx, \quad C_{ij} := \int_\Omega c \psi_i \frac{d\psi_j}{dx} \, dx, \quad 1 \leq i, j \leq N, \quad (3.5.2)$$

yielding the semidiscretization of (3.5.1) for $u_N(t, x)$ given by (3.4.4):

$$M \frac{d\underline{u}}{dt} = -C\underline{u}. \quad (3.5.3)$$

With the same partition Δ_E of $[a, b]$ as in Equation (2.3.1), the choice for ψ_j of the Legendre–Lagrange interpolation basis $\pi_j(x)$ of Equation (2.4.3) gives

3.5. The Hyperbolic Case: Linear Convection

the local spectral-element convective matrix

$$C_{ij}^e = \int_{-1}^{+1} c\pi_i \frac{d\pi_j}{d\xi} d\xi, \quad 1 \leq i, j \leq N. \tag{3.5.4}$$

With GLL quadrature, the matrix becomes

$$C_{N,ij}^e = \rho_i c_i^e D_{N,ij}^{(1)}, \quad 0 \leq i, j \leq N, \tag{3.5.5}$$

where the coefficients of the first Legendre collocation derivative $D_N^{(1)}$ are given by Equation (2.4.9), and c_i^e is the value of c evaluated at the point corresponding to ξ_i in Ω^e.

One has the choice of treating (3.5.3) either implicitly or explicitly. Most common is to use explicit methods, for the following reasons. First, the convection problem is nonsymmetric, implying that GMRES or some other solver appropriate for nonsymmetric problems must be employed. Second, since c is often a function of u, an implicit method would further require a nonlinear iteration such as Newton's method. Finally, the time-step restriction for stability is $\Delta t = O(\Delta x/c)$, generally less severe than the $O(\Delta x^2/\nu)$ restriction encountered in the parabolic case.

3.5.2 Eigenvalues of the Discrete Problem and CFL Number

To examine the question of stability, we consider the problem (3.5.1) on the interval $\Omega^+ = (0, 1)$ with periodic boundary conditions and a constant velocity c. As a preliminary remark, we note that the eigenvalue problem

$$c \frac{du}{dx} = \lambda u, \quad x \in \Omega^+, \quad u(0) = u(1), \tag{3.5.6}$$

has the simple solution

$$\lambda_k = \hat{\imath} c 2k\pi, \quad u_k(x) = \exp(\hat{\imath} 2k\pi x), \quad -\infty < k < \infty, \tag{3.5.7}$$

where $\hat{\imath} := \sqrt{-1}$.

The stability properties of (3.5.3) are governed by the eigenvalue problem

$$C\underline{z}_k = \lambda_k M \underline{z}_k. \tag{3.5.8}$$

Integrating (3.5.2) by parts, one readily verifies under our assumptions that C is skew-symmetric ($C = -C^T$), implying that its eigenvalues are purely imaginary. As in (2.7.15), the mass matrix can be split into $M = LL^T$, and (3.5.8) can be rewritten as a standard eigenvalue problem in the matrix $L^{-1}CL^{-T}$,

3. Parabolic and Hyperbolic Problems

which is also skew-symmetric. Hence, the eigenvalues all lie along the imaginary axis, implying that a scheme whose stability region encloses some portion of the imaginary axis, such as AB3 or RK4 (see Figures 3.2.1 and 3.2.5), is appropriate.

The maximum modulus of λ_k is of course dependent on the choice of basis. For uniform linear finite elements on the interval Ω^+, one has

$$C = \frac{c}{2}\begin{pmatrix} 0 & 1 & & & -1 \\ -1 & 0 & 1 & & \\ & -1 & \ddots & \ddots & \\ & & \ddots & \ddots & 1 \\ 1 & & & -1 & 0 \end{pmatrix}.$$

Using the mass matrix (2.3.17) and making the substitution $(\tilde{\underline{z}}_k)_j = e^{i2\pi k x_j}$ for the eigenmodes, we find

$$\lambda_k^{\text{FE}} = \hat{\imath} \frac{6c}{\Delta x} \frac{\sin(2\pi k \Delta x)}{4 + 2\cos(2\pi k \Delta x)}. \tag{3.5.9}$$

If instead we assume a lumped mass matrix $\tilde{M} := \Delta x I$, we get the eigenspectrum related to the classical FD discretization

$$\lambda_k^{\text{FD}} = \hat{\imath} \frac{c}{\Delta x} \sin(2\pi k \Delta x). \tag{3.5.10}$$

Figure 3.5.1(a) shows the eigenvalue distribution of the continuous operator (3.5.7), as well as the discrete (Q_1) FE and FD expressions (3.5.9) and (3.5.10).

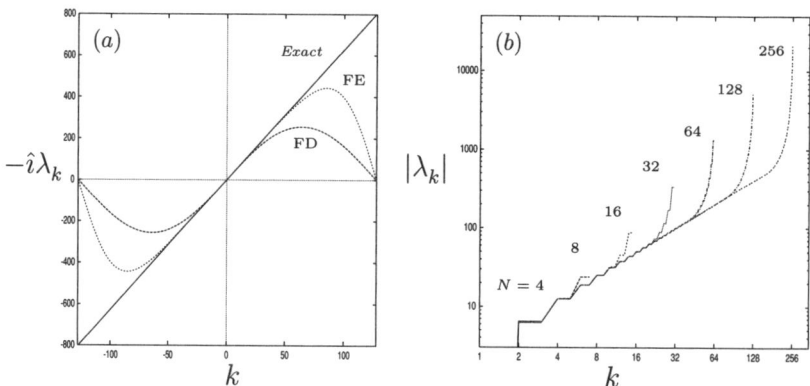

Figure 3.5.1. Comparison of eigenvalues for periodic convection operator, $u_x = \lambda u$, $u(0) = u(1)$: (a) exact, Q_1 linear finite elements with tridiagonal mass matix M (FE), and Q_1 linear finite elements with lumped mass matix $\tilde{M}$ (FD); (b) eigenvalues for Legendre spectral element method with $E = 1$ and N ranging from 4 to 256.

3.5. The Hyperbolic Case: Linear Convection

Computations have been made with $E = 256$. We note that the eigenvalues of the FE approximation in this case are closer to the continuous eigenvalues than are those of the FD approximation. This is in contrast to the parabolic case, where both errors were of the same order of magnitude [see Figure 2.7.2(a)]. For high-order discretizations, the spectral radius of $J := M^{-1}C$ scales as Δx_{min}^{-1}, as would be expected for a first-order operator. Figure 3.5.1(b) shows the eigenvalue distribution for Legendre spectral elements with $E = 1$ and N ranging from 4 to 256. As in the parabolic case, roughly two-thirds of the spectrum coincides with the analytic result, while the maximum moduli scale as $O(N^2) = O(\Delta x_{min}^{-1})$ (see Figure 2.7.1).

Because the eigenvalues for the convection operator (3.5.4) scale with $c/\Delta x$, the stability parameter can be expressed as the product

$$\max_k |\lambda_k \Delta t| = S \cdot \text{CFL}, \tag{3.5.11}$$

where

$$\text{CFL} := \max_{c, \Delta x} \left| \frac{c \Delta t}{\Delta x} \right| \tag{3.5.12}$$

is the convective Courant–Friedrichs–Lewy number corresponding to the ratio of two length scales (i.e., the distance a particle will travel in a given time step, and the local grid spacing), and S is an order-unity coefficient that depends on the discretization. For the FE and FD cases one finds from (3.5.9) and (3.5.10) $S = \sqrt{3}$ and $S = 1$, respectively. Computed numerical values of S for the SEM are plotted as a function of N in Figure 3.5.2(a). S ranges from unity for $N = 1$ to ≈ 1.52 for $N = 2$–3 and gradually settles in the range $1.2 \to 1.16$ for $N = 16 \to 256$.

One can use the CFL number and scaling parameter S in conjunction with the stability diagrams introduced in Section 3.2 to determine a stability criterion for a given space–time discretization pair. For example, from Figure 3.2.1, we see that the stability diagram for AB3 cuts the imaginary axis at ± 0.72362, implying that one must ensure that the CFL (3.5.12) is such that

$$S \cdot \text{CFL} \leq 0.72362 \tag{3.5.13}$$

to have a stable time integration.

The importance of satisfying (3.5.13) is illustrated in the example of Figure 3.5.2, in which unit-speed convection of a Gaussian pulse, $u^0 := e^{-\zeta^2}$, $\zeta := 15(x - 0.5)$, is simulated on the periodic domain [0, 1] using 256 linear finite elements with mass lumping [$(E, N) = (256, 1)$]. The solution is advanced by AB3 with (b) CFL $= 0.72$ and (c) CFL $= 0.770$. The initial pulse (bottom) moves to the right, leaves the domain at $x = 1$, and reenters at $x = 0$.

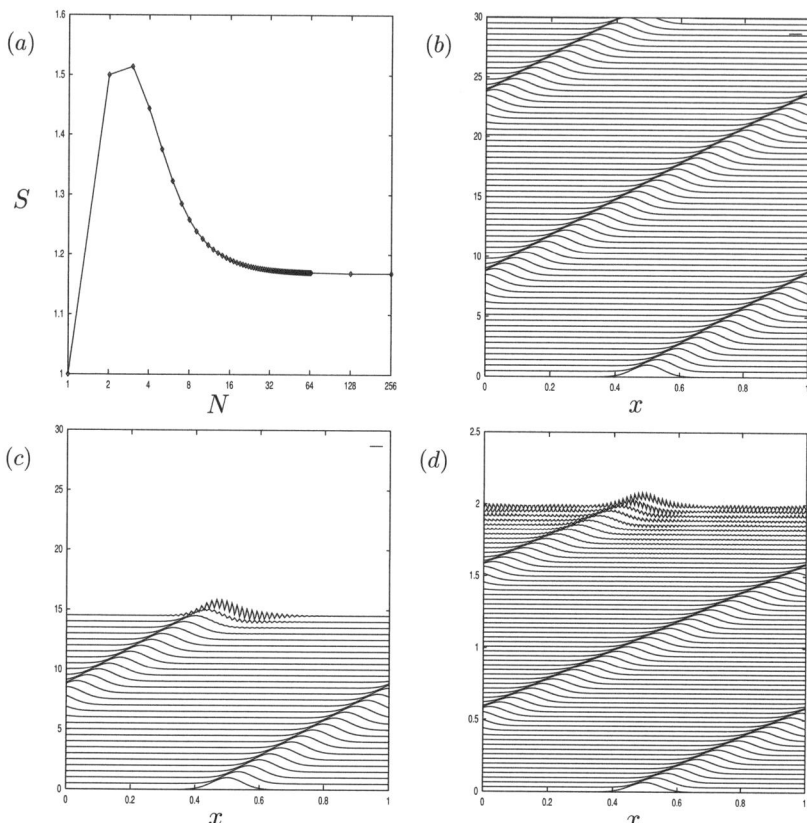

Figure 3.5.2. (a) Scale factors S for spectral-element discretizations. The other parts show SEM/AB3 solutions to $u_t + u_x = 0$, $u(0,t) = u(1,t)$ with (b) $(E, N) = (256, 1)$, CFL $= 0.72$; (c) $(E, N) = (256, 1)$, CFL $= 0.77$; (d) $(E, N) = (82, 3)$, CFL $= 0.72362/1.5$. The vertical axis in (b)–(d) is $t + u/15$.

For this case, $S = 1$, and the solution with CFL $= 0.720$ is consequently stable. The pulse convects indefinitely (with error increasing linearly with t). However, for CFL $= 0.77$, instabilities appear at time $t \approx 1$, shortly after which the solution exhibits explosive growth (not shown). The unstable mode consists of "$2 \Delta x$ waves" – the most oscillatory waves that can be represented on the grid, and those associated with the maximum (in modulus) eigenvalue. Similar behavior is shown in (d) for a spectral element calculation with $(E, N) = (82, 3)$ and CFL $= 0.72362/1.5$. Decreasing CFL to $0.72362/1.52$ yields a stable computation. Note that the time-step choice $\Delta t = 0.72 \Delta x$ for the $(E, N) = (256, 1)$ discretization is sufficient not only for stability, but also for accuracy. The temporal error is $O(\Delta t^3)$ and thus dominated by the spatial

3.5. The Hyperbolic Case: Linear Convection

error, which is $O(\Delta x^2)$ for this case. Reducing Δt does not further improve the accuracy.

3.5.3 Example of Temporal and Spatial Accuracy

We illustrate the convergence properties of combined AB3–spectral-element discretizations by considering periodic convection on $(0, 1)$. The initial condition, illustrated in Figure 3.5.3, is given by

$$u^0(x) = e^{-\zeta^2} + 0.2\cos(8\pi x),$$

where $\zeta := 15(x - 0.5)$. The maximum relative error

$$|\varepsilon|_{\max} := \frac{\max_{x_i, t^n} |u(x_i, t^n) - u_N(x_i, t^n)|}{\max_{x_i} |u(x_i, t^0)|} \quad (3.5.14)$$

is computed over the interval $t \in (0, 10]$, and the results are given in Table 3.5.1. The number of grid points, $\mathcal{N} := EN$, ranges from 16 to 256, with polynomial degree ranging from $N = 1$ to 32.

The dashed entries in Table 3.5.1 correspond to $|\varepsilon|_{\max} > 10^{10}$, implying that the discretization failed to satisfy the CFL condition (3.5.13). The first column shows that linear finite elements ($N = 1$) are inadequate for this problem, as they exhibit order-unity error, even at a resolution of $\mathcal{N} = 256$. From Figure 3.5.3, it is clear that this failure is a result of accumulated dispersion errors, *not* of an inability to resolve the signal. We note that dispersion errors prompt phase errors, which can be determined by Fourier analysis. In the case of linear finite

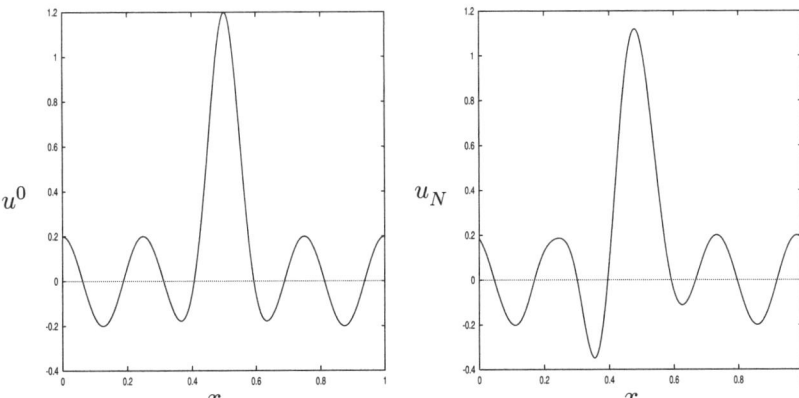

Figure 3.5.3. Initial condition (left) and final solution (right) for $(E, N) = (256, 1)$ results of Table 3.5.1.

Table 3.5.1. *Relative error (3.5.14) at time t = 10, hyperbolic problem.*

$\mathcal{N}$	Δt	\multicolumn{6}{c}{Relative error}					
		$N = 1$	$N = 2$	$N = 4$	$N = 8$	$N = 16$	$N = 32$
64	.0100000	6.87e − 1	—	—	—	—	—
	.0050000	7.88e − 1	1.56e − 1	1.43e − 1	—	—	—
	.0025000	8.09e − 1	1.29e − 1	3.19e − 2	2.91e − 2	—	—
	.0012500	8.13e − 1	1.41e − 1	1.27e − 2	4.58e − 3	4.11e − 3	—
	.0006250	8.13e − 1	1.42e − 1	1.22e − 2	1.08e − 3	5.43e − 4	5.22e − 4
	.0003125	8.14e − 1	1.43e − 1	1.23e − 2	8.27e − 4	8.90e − 5	6.55e − 5
128	.0100000	—	—	—	—	—	—
	.0050000	6.05e − 1	—	—	—	—	—
	.0025000	6.51e − 1	3.17e − 2	2.89e − 2	—	—	—
	.0012500	6.59e − 1	1.42e − 2	4.16e − 3	4.10e − 3	—	—
	.0006250	6.60e − 1	1.54e − 2	6.14e − 4	5.22e − 4	5.22e − 4	—
	.0003125	6.60e − 1	1.56e − 2	2.00e − 4	6.58e − 5	6.55e − 5	6.55e − 5
256	.0100000	—	—	—	—	—	—
	.0050000	—	—	—	—	—	—
	.0025000	2.88e − 1	—	—	—	—	—
	.0012500	3.01e − 1	4.66e − 3	4.10e − 3	—	—	—
	.0006250	3.03e − 1	1.51e − 3	5.26e − 4	5.22e − 4	—	—
	.0003125	3.03e − 1	1.71e − 3	7.02e − 5	6.55e − 5	6.55e − 5	—

elements, the dispersion errors are lagging phase errors. By contrast, higher-order schemes preserve the signal for much longer times. The error decreases by several orders of magnitude in the last row of each case ($\mathcal{N}$), indicating spectral convergence with increasing N, until a minimum error of 6.55e − 5 is reached, where $O(\Delta t^3)$ errors dominate. In the cases that are spatially well resolved (e.g., $N = 8, \mathcal{N} = 128$), each twofold reduction in Δt yields an eightfold reduction in error, reflecting the $O(\Delta t^3)$ behavior of AB3.

A potential drawback of high-order schemes is that, for a given resolution $\mathcal{N}$, they require smaller time steps for stability. However, this effect is mitigated by the accuracy gains with N that allow one to reduce $\mathcal{N}$. For example, for an error tolerance of 0.5 percent, attainable only with $\Delta t \leq 0.00125$, we see in Table 3.5.1 that the case ($\mathcal{N} = 256, N = 2$) is *less* than a factor of two below the critical CFL condition, as indicated by the dashed entry directly above. By the same reasoning, the corresponding ($\mathcal{N} = 128, N = 4$) and ($\mathcal{N} = 64, N = 8$) cases are *more* than a factor of two below their critical CFL values for the same Δt. Moreover, the fourfold reduction in grid points achieved by increasing N from 2 to 8 while decreasing E from 128 to 8 translates into a factor-of-64 reduction in grid points for 3D problems, with a concomitant reduction in work as discussed further in Chapter 4. For long-time integrations, we conclude that moderately high-order spatial discretizations can provide significant savings without undue temporal stability constraints.

3.5.4 Inflow–Outflow Boundary Conditions

The weighted residual implementation of inflow–outflow boundary conditions follows as in the Dirichlet–Neumann case for the elliptic problem. Wherever u is being convected *into* the domain [e.g., at a if $c(a) > 0$ and/or at b if $c(b) < 0$], the elements of the trial (test) space must be constrained to satisfy the essential (homogeneous) boundary condition; otherwise, they are unconstrained. It is also possible to impose inhomogeneous boundary conditions weakly (see, e.g., [319]). As noted earlier, if $c > 0$, energy leaves the system through the right boundary as the signal is convected out of the domain. We can expect therefore that at least some of the eigenvalues of $J = M^{-1}C$ will have a negative real part (C is no longer skew-symmetric). The continuous problem, however, has no eigenvalues. We discuss the consequences of this further in Section 3.7.

3.6 Steady Advection–Diffusion Problems

This chapter is devoted principally to time-dependent (hyperbolic and parabolic) problems in one space dimension. This section will be an exception, however, since, strictly speaking, the steady advection–diffusion problem

$$\mathcal{L}u := -\nu \frac{d^2u}{dx^2} + c\frac{du}{dx} = f \quad \text{in } \Omega := (-1, 1),$$
$$\mathcal{B}u := u(-1) = u(1) = 0 \tag{3.6.1}$$

is *elliptic*. Nevertheless, for high Reynolds numbers (i.e., vanishing diffusion), hyperbolic features of the problem, such as boundary layers, cannot be ignored. Equation (3.6.1) is said to be *singularly perturbed*, because the limiting case $\nu \to 0$ is not equivalent to the situation $\nu = 0$ [30, 368]. The latter case has only one boundary condition, while (3.6.1) has two.

In the numerical examples introduced in this section, we will systematically take $c = f = 1$. In that case one can show that the analytical solution to (3.6.1) is

$$u(x) = (x + 1) - 2\frac{e^{(x-1)/\nu} - e^{-2/\nu}}{1 - e^{-2/\nu}}. \tag{3.6.2}$$

With $\nu \to 0$, the solution corresponds to the hyperbolic result ($u = x + 1$) throughout most of the domain. The singular perturbation is manifested as a

138   3. Parabolic and Hyperbolic Problems

boundary layer of thickness $O(\nu)$ in which the steady-state solution changes by $O(1)$ in the interval $[1 - \nu, 1]$. This rapid change creates significant numerical difficulties and calls for special treatment, the subject of this section.

3.6.1 Spectral Elements and Bubble Stabilization

We start with the standard Galerkin formulation of the problem (3.6.1). Defining the functionals

$$\mathcal{A}(u, v) := \int_{-1}^{1} (\nu u'v' + cu'v)\, dx, \qquad u, v \in H_0^1(\Omega),$$
$$\mathcal{F}(v) := \int_{-1}^{1} fv\, dx, \qquad v \in H_0^1(\Omega),$$

(3.6.3)

where $H_0^1(\Omega)$ is the Sobolev space defined by (2.1.34), Equation (3.6.1) can be recast in the equivalent weak formulation: Find $u \in H_0^1(\Omega)$ such that

$$\mathcal{A}(u, v) = \mathcal{F}(v) \qquad \forall v \in H_0^1(\Omega). \tag{3.6.4}$$

The discretization procedure for (3.6.4) consists then in choosing a finite-dimensional trial and test space, $V_N \subset H_0^1(\Omega)$, with suitable basis functions $\{\psi_i\}_{i=1}^N$. In the Legendre spectral (or spectral-element) method, one takes

$$V_N := \mathbb{P}_N \cap H_0^1(\Omega),$$

where $\mathbb{P}_N$ is the space of polynomials of degree less than or equal to N. Normally, the basis is chosen to be the set of Lagrangian interpolants on the GLL grid $\pi_i(x) \in \mathbb{P}_N$ [see Equation (2.4.3)]. In addition, the integrals associated with the inner products in (3.6.3)–(3.6.4) are replaced by numerical quadratures, denoted as $(\cdot, \cdot)_N$, on the $N + 1$ GLL nodes.

Numerical results for the Legendre SEM ($E = 1$) applied to the model problem with $c = f = 1$ are assembled in Table 3.6.1, giving the maximum pointwise error,

$$|\varepsilon|_{\max, N} := \frac{\max_i |u_i - u(\xi_i)|}{\max_i |u(\xi_i)|}, \tag{3.6.5}$$

for several values of N and ν. Here, $\{u(\xi_i)\}$ is the exact solution (3.6.2) evaluated at the GLL points, while $\{u_i\}$ is the set of local values obtained from the spectral element. Exponential convergence is reached when the solution is smooth with respect to the grid spacing ($\nu = 1$ and diffusion is still important). However, the error becomes large for smaller values of ν, particularly for N even. Canuto

3.6. Steady Advection–Diffusion Problems

Table 3.6.1. *Spectral collocation, no stabilization.*

	$|\varepsilon|_{\max,N}$			
N	$\nu = 1$	$\nu = 0.1$	$\nu = 0.01$	$\nu = 0.001$
3	6.148e − 03	5.620e − 01	9.577e − 01	9.958e − 01
4	3.476e − 04	3.937e − 01	8.039e + 00	8.573e + 01
5	2.298e − 05	1.458e − 01	7.159e − 01	7.958e − 01
6	1.403e − 06	6.652e − 02	3.154e + 00	3.626e + 01
7	7.166e − 08	2.398e − 02	5.962e − 01	7.391e − 01
8	3.609e − 09	8.771e − 03	1.569e + 00	2.009e + 01
9	1.536e − 10	3.124e − 03	4.859e − 01	7.099e − 01
10	6.472e − 12	1.128e − 03	8.599e − 01	1.272e + 01
11	2.400e − 13	3.736e − 04	3.748e − 01	6.886e − 01

[62] noted that for N even, the spectral approximation to (3.6.4) is unbounded as $\nu \to 0$. We can see that this must be true for any Galerkin scheme. Since the basis functions vanish on the boundaries, a simple integration by parts in (3.5.2) reveals that $C_{ij} = -C_{ji}$, implying that C is skew-symmetric and has pure imaginary eigenvalues. However, if N is even, then the number of unknown coefficients ($N - 1$) is odd, and C must have at least one zero eigenvalue. In the limit $\nu \to 0$, $K + C$ is therefore singular.

This problem of instability (i.e., solution blowup with a vanishing parameter) is not unique to spectral methods. The same problem occurs for centered finite-difference (FD) approaches and classical h-type finite elements (FEs). A well-known cure in the case of FDs is to employ upwind differencing, which has the effect of adding artificial viscosity to the problem. For FEs, upwind stabilization methods have been developed by Brooks and Hughes [55], who introduced the streamline-upwind Petrov–Galerkin (SUPG) method, and by Johnson and Saranen [215], who introduced the streamline diffusion method. These are based on the Petrov–Galerkin method discussed in Chapter 2, in which the bases for the test space have an upwind bias. Somewhat later, Brezzi and coworkers [52, 19] developed a stabilization procedure in which the standard FE trial and test spaces are augmented with *bubble functions* that vanish on the boundary of each element. These local degrees of freedom are eliminated via static condensation so that the global system size remains unchanged. Under certain conditions it can be shown that the bubble stabilization technique is identical to a SUPG formulation [319].

Several stabilization techniques for advection-dominated diffusion problems and Navier–Stokes equations have been developed in the framework of high-order methods such as the use of bubble functions for spectral methods, introduced by Canuto and coworkers [63, 65, 67, 70, 317] and the use of (upwind) staggered grids for collocation methods. The latter will be examined in Section 3.6.2.

Bubble stabilization techniques look for solutions in an augmented space $W := V_N \oplus B_N$, where V_N is the standard spectral approximation space and B_N is a space of bubble functions spanned by $\{b_i\}_{i=1}^N$. The individual bubble functions satisfy $b_i \in H_0^1(\Omega_i)$, $b_i(x) = 0$ for $x \notin \Omega_i$, where $\Omega_i := (\xi_{i-1}, \xi_i)$. The generalized Galerkin formulation then reads: Find $u_N \in V_N, u_B \in B_N$ such that

$$\begin{aligned} \mathcal{A}(u_N, v_N) + \mathcal{A}(u_B, v_N) &= \mathcal{F}(v_N) & \forall v_N \in V_N, \\ \mathcal{A}(u_N, v_B) + \mathcal{A}(u_B, v_B) &= \mathcal{F}(v_B) & \forall v_B \in B_N. \end{aligned} \qquad (3.6.6)$$

Since V_N and $\mathcal{B}$ are discrete subspaces, we can cast this into matrix form as

$$\begin{pmatrix} A_{NN} & A_{NB} \\ A_{BN} & A_{BB} \end{pmatrix} \begin{pmatrix} \underline{u}_N \\ \underline{u}_B \end{pmatrix} = \begin{pmatrix} \underline{f}_N \\ \underline{f}_B \end{pmatrix}, \qquad (3.6.7)$$

where

$$\begin{aligned} (A_{NN})_{ij} &:= \mathcal{A}(\pi_i, \pi_j), & (A_{NB})_{ij} &:= \mathcal{A}(\pi_i, b_j), \\ (A_{BN})_{ij} &:= \mathcal{A}(b_i, \pi_j), & (A_{BB})_{ij} &:= \mathcal{A}(b_i, b_j), \\ (f_N)_i &:= (\pi_i, f_N), & (f_B)_i &:= (b_i, f_N), \end{aligned}$$

with $1 \leq i, j \leq N$. The element $f_N \in V_N$ is the spectral interpolant of $f(x)$. The complete solution is given by

$$u(x) = \sum_{i=0}^N u_{N,i} \pi_i(x) + \sum_{i=1}^N u_{B,i} b_i(x).$$

In general, one is interested only in the component in V_N, represented by the nodal values $u_{N,i}$, and the coefficients $u_{B,i}$ are not computed.

The precise definition of the bubble function b_i varies from one implementation to the next and can be regarded as a free parameter in the formulation. In the residual-free bubble (RFB) formulation presented by Canuto et al. [70], it is suggested to let b_i satisfy the differential equation

$$\begin{aligned} -\nu b_i'' + \bar{c}_i b_i' &= \bar{f}_i & \text{in } \Omega_i, \\ b_i(\xi_{i-1}) = b_i(\xi_i) &= 0, \end{aligned} \qquad (3.6.8)$$

where the coefficients and data are approximated on Ω_i by piecewise constants, $\bar{c}_i$ and $\bar{f}_i$, respectively. The potential of such an approach is realized in the case

3.6. Steady Advection–Diffusion Problems

Table 3.6.2. *Spectral collocation,* RFB *with exact integrals.*

	$|\varepsilon|_{\max,N}$			
N	$\nu = 1$	$\nu = 0.1$	$\nu = 0.01$	$\nu = 0.001$
3	3.814e − 04	1.260e − 01	3.164e − 01	3.406e − 01
4	1.092e − 05	4.289e − 02	1.418e − 01	1.643e − 01
5	9.922e − 07	2.082e − 02	2.325e − 01	2.811e − 01
6	8.027e − 08	9.852e − 03	1.239e − 01	1.698e − 01
7	5.254e − 09	4.030e − 03	1.770e − 01	2.576e − 01
8	3.094e − 10	1.525e − 03	9.725e − 02	1.711e − 01
9	1.467e − 11	5.533e − 04	1.272e − 01	2.425e − 01
10	6.598e − 13	2.119e − 04	6.866e − 02	1.699e − 01
11	4.051e − 14	7.140e − 05	8.461e − 02	2.300e − 01

where V_N is taken to be the standard FE space of piecewise linear functions on each Ω_i, and $c = f = 1$. In this case, the analytical solution is contained in W, and the nodal values resulting from solving (3.6.7) are exact (to machine precision) regardless of the boundary-layer resolution.

In the spectral-element formulation the bubble-stabilized result is no longer exact, since the analytic solution cannot be represented as a combination of the polynomial interpolant and the bubble functions defined by (3.6.8). However, as Table 3.6.2 shows, the solution is improved over the unstabilized spectral collocation scheme of Table 3.6.1, particularly for small values of ν, and exponential convergence is still attained in the well-resolved cases. This latter point results from the fact that the bubble formulation will always be superior to the standard spectral method because it is derived from an enlarged approximation space. The results of Table 3.6.2 were computed by replacing the upper left block in (3.6.7) with

$$(A_{NN})_{ij} := \mathcal{A}_N(\pi_i, \pi_j)$$

and retaining exact integrals for the other terms. For practical computations, Canuto and coworkers use static condensation to explicitly eliminate the bubble coefficients from (3.6.7), taking advantage of the fact that the block A_{BB} is diagonal. The key to the efficiency of their approach in higher space dimensions is the use of approximations that reduce the off-diagonal blocks to banded matrices without compromising spectral accuracy.

3.6.2 Collocation and Staggered Grids

Let us now examine orthogonal collocation solutions of the problem (3.6.1). As outlined in Section 2.5, we must first select a collocation grid with the help of the

Gauss–Lobatto–Jacobi (GLJ) grid-generating function $\chi_N^{(\alpha,\beta)}(x)$ (B.2.5). Here, we focus on Gauss–Legendre and Gauss–Chebyshev collocation calculations.

In the Legendre case, any trial element $\hat{u}_N(x)$ with $N+1$ degrees of freedom is written as

$$\hat{u}_N(x) := \sum_{j=0}^{N} u(\xi_j)\pi_j(x), \qquad x \in \hat{\Omega} := [-1, +1], \qquad (3.6.9)$$

where the basis functions $\pi_j(x)$ are given by (2.4.3). In the Chebyshev case the basis functions are denoted $\pi_j^c(x)$ and are given by (2.5.15).

According to (2.5.10)–(2.5.11), Legendre collocation produces the equations

$$\sum_{j=0}^{N}[-\nu\pi_j''(\xi_k) + c\pi_j'(\xi_k)] = 1 \qquad \forall \xi_k \in \Omega,$$
$$u(-1) = u(+1) = 0, \qquad (3.6.10)$$

with $\{\xi_j\}_{j=0}^{N}$, the set of collocation nodes. The equations (3.6.10) lead to an algebraic system (2.5.12). As mentioned earlier, the collocation derivatives that appear in (3.6.10) are explicitly given in Section B.3.3 of Appendix B. Similar equations apply, of course, in the Chebyshev case.

Figure 3.6.1 displays the numerical results obtained on the model problem with $c = f = 1$. The left part of the figure (a) shows the Legendre collocation solution (3.6.10) for $\nu = 0.005$, with $N = 9$ ($\star$), while the right part (b) shows the Chebyshev collocation results when $\nu = 0.001$ and $N = 25$ ($\star$). The reader should notice that, although the last mesh space in the GLL partition has roughly the size of the boundary layer (i.e., ≈ 0.224), the approximate solution

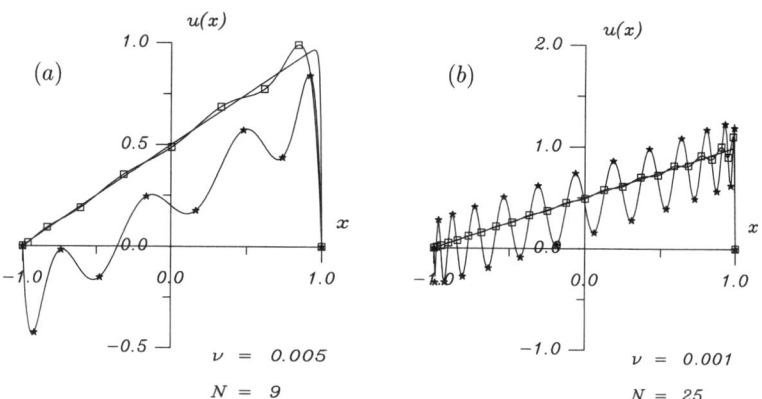

Figure 3.6.1. Collocation solutions of advection-dominated diffusion problems on (a) Legendre and (b) Chebyshev grids.

3.6. Steady Advection–Diffusion Problems

Table 3.6.3. *Orthogonal collocation on Gaussian and staggered grids.*

N	$\nu = 0.1$		$\nu = 0.001$	
	GLL	SL	GLC	SC
3	2.728e − 01	4.928e − 02	1.321e + 00	2.983e − 01
4	9.207e − 02	1.692e − 02	3.603e + 01	1.830e − 01
5	2.322e − 02	6.506e − 03	1.077e + 00	1.944e − 01
6	5.677e − 03	2.891e − 03	1.436e + 01	1.672e − 01
7	1.376e − 03	7.258e − 04	9.989e − 01	1.739e − 01
8	3.173e − 04	1.725e − 04	7.614e + 00	1.609e − 01
9	6.544e − 05	4.157e − 05	9.436e − 01	1.645e − 01
10	1.298e − 05	8.457e − 06	4.633e + 00	1.567e − 01
11	2.451e − 06	1.668e − 06	8.891e − 01	1.585e − 01
31	4.976e − 15	1.394e − 14	2.253e − 01	8.207e − 02

is very poor, with a maximum pointwise error $|\varepsilon|_{\max,N} = 0.4859$. With very small values of ν, collocation on the GLC grid gives a solution exhibiting large oscillations ($|\varepsilon|_{\max,N} = 0.3813$).

Table 3.6.3 gives $|\varepsilon|_{\max,N}$ for $\nu = 0.1$ and $\nu = 0.001$ as a function of N, for Legendre and Chebyshev collocation, respectively. In the first case, diffusion is still important ($\nu = 0.1$) and the errors on the GLL grid decrease toward round-off, while in the second case, advection dominates ($\nu = 0.001$) and one recognizes the same problem of instability as those identified above with spectral-element calculations, such as larger errors with even values of N and an error blowup with $\nu \to 0$.

This problem has led to a stream of efforts to develop stabilization techniques based on staggered grids. Funaro [143, 144] has shown that the use of a suitable staggered grid for collocation, combined with a shift operator from the staggered to the GLL grid, restores the efficiency of FD preconditioning, even in the presence of strong advection terms. For the time being, we deal with the generation of the staggered grid; the Funaro FD preconditioning technique will be described in Chapter 4, since it is a truly multidimensional algorithm.

The generation of the staggered grid starts from the recognition that, for a simple diffusion operator such as $\mathcal{L}_D := -d^2/dx^2$, the GLL collocation (or spectral element) grid points defined by (2.4.1) are also solutions to the polynomial equation

$$\mathcal{L}_D \chi_N^{(0,0)}(x) = 0, \qquad (3.6.11)$$

where $\chi_N^{(0,0)}$ is defined by (2.5.1). Adapting this characterization of the grid points to the advection–diffusion operator $\mathcal{L}_{AD} := -\nu d^2/dx^2 + c d/dx$ and using the fundamental property (B.1.11) of the Legendre polynomial $L_N(x)$,

$$-\frac{d}{dx}[(1-x^2)L'_N(x)] = N(N+1)L_N(x),$$

one finds that the inner collocation nodes (or inner spectral-element nodes) in the reference domain $\hat{\Omega}$, solutions to $\mathcal{L}_{AD}\chi_N^{(0,0)}(x) = 0$, satisfy the polynomial equation

$$\nu L'_N(x) - cL_N(x) = 0. \qquad (3.6.12)$$

Obviously, for a purely diffusive problem one recovers the GLL grid with the inner nodes $\{\xi_\ell\}_{\ell=1}^{N-1}$, the solutions to $L'_N(x) = 0$. For an advection-dominated problem with $c/\nu \gg 1$, the boundary nodes remain unchanged, while the inner collocation nodes tend toward $N-1$ (out of the N) of the Gauss–Legendre quadrature nodes $\{\zeta_i\}_{i=1}^{N}$ [the roots of $L_N(x)$], with the ordering relationship $\cdots < \zeta_i < \xi_i < \zeta_{i+1} < \cdots$. A first-order expansion of τ_i, the solution to (3.6.12) in the vicinity of ξ_i, gives

$$\tau_i \approx \xi_i - \frac{c}{\nu}\frac{1-\xi_i^2}{N(N+1)}, \qquad (3.6.13)$$

showing that the collocation (or spectral-element) nodes are moved in the *opposite* direction to the local advection motion. This explains why the resulting collocation grid is usually called the "upwind" grid. Notice that the displacement increases with c/ν and is more important at the center of the domain than close to the boundary. In practice, of course, c/ν is space-dependent, and (3.6.12) has to be solved in the vicinity of every GLL node using, for instance, a few iterations of the Newton–Raphson algorithm starting from, say, $(\xi_i + \zeta_i)/2$ (when $c > 0$) or $(\xi_i + \zeta_{i+1})/2$ (when $c < 0$).

An example is given in Figure 3.6.2, where the upper part shows the GLL and the staggered grid (the bullets) for the operator $\mathcal{L}_{AD} := -\nu d^2/dx^2 + x d/dx$ (with $\nu = 10^{-2}$ and $N = 10$), while the lower part displays the Gauss–Legendre

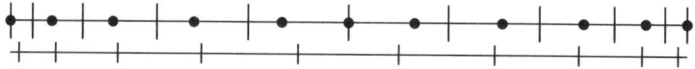

Figure 3.6.2. An example of a staggered grid (bullets) for the advection–diffusion problem $-\nu d^2/dx^2 + x d/dx$ on $\hat{\Omega}$ (with $\nu = 10^{-2}$ and $N = 10$). The upper and lower parts of the figure display the GLL and Gauss–Legendre grids, respectively.

grid. Notice that for $x > 0$ the nodes are moved to the left, while for $x < 0$, they are moved to the right, as emphasized earlier.

A similar analysis for the collocation and preconditioning of advection-dominated diffusion problems on Chebyshev grids has been performed by Pinelli et al. [308, 311]. Although in this case the GLC mesh points [the solution to $\chi_N^{(-1/2,-1/2)}(x) = 0$] do *not* satisfy the polynomial equation $\mathcal{L}_D \chi_N^{(-1/2,-1/2)}(x) = 0$, the authors use some heuristic arguments to show that the optimal choice for the inner staggered mesh points $\{\tau_i\}_{i=1}^{N-1}$ coincides with the roots of the polynomial equation

$$\nu T_N''(x) - c T_N'(x) = 0. \quad (3.6.14)$$

It comes as no surprise that the conclusions with regard to the displacement of the mesh points are the same as in the Legendre case. FD preconditioning on the (upwind) staggered grid is then straightforward; it will be explained later.

Figure 3.6.1 displays also the numerical results obtained with collocation on the staggered grids. On the left, for $\nu = 0.005$ and $N = 9$, with collocation on the Legendre staggered grid ($\square$) one readily captures the essential features of the analytical solution ($|\varepsilon|_{\max,N} = 0.07751$). The improvement is more significant on the right (*b*). Instead of the large oscillations resulting from collocation on the GLC grid, the staggered Chebyshev grid gives a very accurate solution in the convection region. However, in the very sharp transition from the convection region to the boundary layer (due to the very small ν), there is an "overshoot" in the numerical solution, and $|\varepsilon|_{\max,N} = 0.1144$, not very much lower than with orthogonal collocation ($|\varepsilon|_{\max,N} = 0.3813$). An L^2 error in this case would give significantly different results.

Similar conclusions may be drawn from the data in Table 3.6.3 for the staggered Legendre (SL) and staggered Chebyshev (SC) grids. The improvement increases with decreasing values of ν. However, for $\nu \to 0$ the convergence of the errors with N is far from being exponential.

3.7 Unsteady Advection–Diffusion Problems

We now consider the unsteady advection–diffusion problem

$$\frac{\partial u}{\partial t} + c \frac{\partial u}{\partial x} = \frac{\partial}{\partial x}\left(\nu \frac{\partial u}{\partial x}\right) + f, \quad u(0, x) = u^0, \quad (3.7.1)$$

subject to appropriate boundary conditions. If f, ν, and c are time-independent, u will evolve to the solution of the steady-state problem considered in the preceding section, and all the issues regarding (spatial) stability in that case must also apply to (3.7.1). Generally, if one is interested only in the steady-state solution, then it is more efficient to attack (3.6.1) directly rather than marching (3.7.1) forward in time. However, when time-dependent solutions are sought, the linear systems to be solved can be simplified through the use of various operator splitting techniques discussed earlier in this chapter.

3.7.1 Spatial Discretization

Spatial discretization proceeds as in the previous unsteady cases. Following the Galerkin projection procedure, we multiply (3.7.1) by a test function $v \in V \subset H_0^1$, integrate over Ω, and require that u be such that equality holds for all $v \in V$. Substituting (3.4.4) and testing with $v = \psi_i(x)$ for $i = 1, \ldots, N$, we get the evolution equation for the basis coefficients $\underline{u} := \{u_j\}_{j=1}^N$:

$$M \frac{d\underline{u}}{dt} = -K\underline{u} - C\underline{u} + M\underline{f}, \qquad (3.7.2)$$

where M and K are defined as in (3.4.6) and C is given by (3.5.2).

The temporal stability properties of (3.7.2) are determined by the corresponding eigenvalue problem

$$(K + C)\underline{z}_k = -\lambda_k M \underline{z}_k. \qquad (3.7.3)$$

The eigenvalues will be either complex or negative real, depending on the boundary conditions and discretization. We consider three possible boundary conditions for (3.7.1)–(3.7.3):

(i) periodic: $\quad u^{(j)}(t, 0) = u^{(j)}(t, 1), \quad j = 0, 1,$
(ii) Dirichlet: $\quad u(t, 0) = u(t, 1) = 0$
(iii) Dirichlet–Neumann: $\quad u(t, 0) = 0, \; \frac{\partial u}{\partial x}\big|_{x=1} = 0$

The eigenvalues and eigenfunctions for the corresponding continuous problems are

Case	λ_k	$\phi_k(x)$	$u(x)$
(i)	$2\pi \hat{\imath} k - 4v\pi^2 k^2$	$e^{2\pi \hat{\imath} k x}$	
(ii)	$-\dfrac{1 + 4(v\pi k)^2}{4v}$	$e^{x/2v} \sin(k\pi x)$	$x - \dfrac{e^{(x-1)/v} - e^{-1/v}}{1 - e^{-1/v}}$
(iii)	$-\dfrac{1 + (v(2k+1)\pi)^2}{4v}$	$e^{x/2v} \sin(\tfrac{2k+1}{2}\pi x)$	$x - v e^{(x-1)/v}$

where $\hat{\imath} := \sqrt{-1}$. Also shown is $u(x)$, the solution to (3.6.1) when $f = 1$. Here we have assumed that the velocity $c = 1$.

3.7. Unsteady Advection–Diffusion Problems

We make several observations about the analytical solutions. In cases (ii) and (iii), the eigenvalues are negative real, with $\lim_{\nu \to 0} \lambda_k = -\infty$. These cases are singularly perturbed, and the eigenfunctions are far from orthogonal. The singular perturbation manifests itself as a boundary layer of thickness $O(\nu)$ in which the steady-state solution changes by $O(1)$ in the interval $[1 - \nu, 1]$ in case (ii), and (mildly) by $O(\nu)$ in case (iii). By contrast, in the periodic case the eigenvalues are complex and bounded, and the eigenfunctions orthogonal, for all ν. The difficulty in the advection–diffusion problem arises from the boundary conditions, not from small values of ν per se. However, in situations where c is varying, such as the Burgers equation (Section 3.8), one may encounter internal layers that exhibit features similar to case (ii).

Figure 3.7.1 shows the eigenvalues for (3.7.3) discretized by spectral elements with $(E, N) = (1, 64)$. A few extreme negative values have been excluded from (b)–(f) to permit a reasonable scaling of the real axis. Figure 3.7.1(a) shows the spectrum for the pure convection problem with inflow–outflow (implemented as Dirichlet–Neumann) boundary conditions to be essentially imaginary, with a negative real component about one-fourth the magnitude. This spectrum has no continuous analogue. The eigenvalues for the periodic case are shown in (b), along with the values for the continuous problem. As in the isolated convection and diffusion problems, roughly two-thirds of the spectrum is accurately captured by the discretization.

Figure 3.7.1(c)–(f) show the Neumann and Dirichlet cases for $\nu = 0.1$ and $\nu = 0.001$. The $\nu = 0.1$ spectra are essentially negative real, with a small imaginary part arising for the Dirichlet case. By contrast, the $\nu = 0.001$ spectra have strong imaginary components, apparently having little in common with their continuous counterparts. Real spectra can be obtained by increasing the spatial resolution, but this is typically too costly to be practical.

Reddy and Trefethen [323] have studied the continuous and analytical spectra and note that spectral analysis in the cases of small diffusion is of limited practical value, because of the nonorthogonality of the eigenfunctions. They note that it would require very large coefficients of alternating signs to expand an order-unity function in terms of the eigenbasis and that the result would be lost because of roundoff in the exponentially large terms. For example, if $\nu = 0.001$, one would need over 200 digits of precision to use the eigenfunctions effectively as a basis. Reddy and Trefethen [323] suggest that the discrete systems give a reasonable representation of the *pseudospectra*, defined roughly as the surface in the complex λ-plane where $\|(\lambda I - \mathcal{L})^{-1}\|$ is large. In the normal case, this surface is limited to small regions near the eigenvalues. In the advection–diffusion case, it is more extensive and incorporates the part of the complex plane mapped by the discrete eigenvalues in Figure 3.7.1.

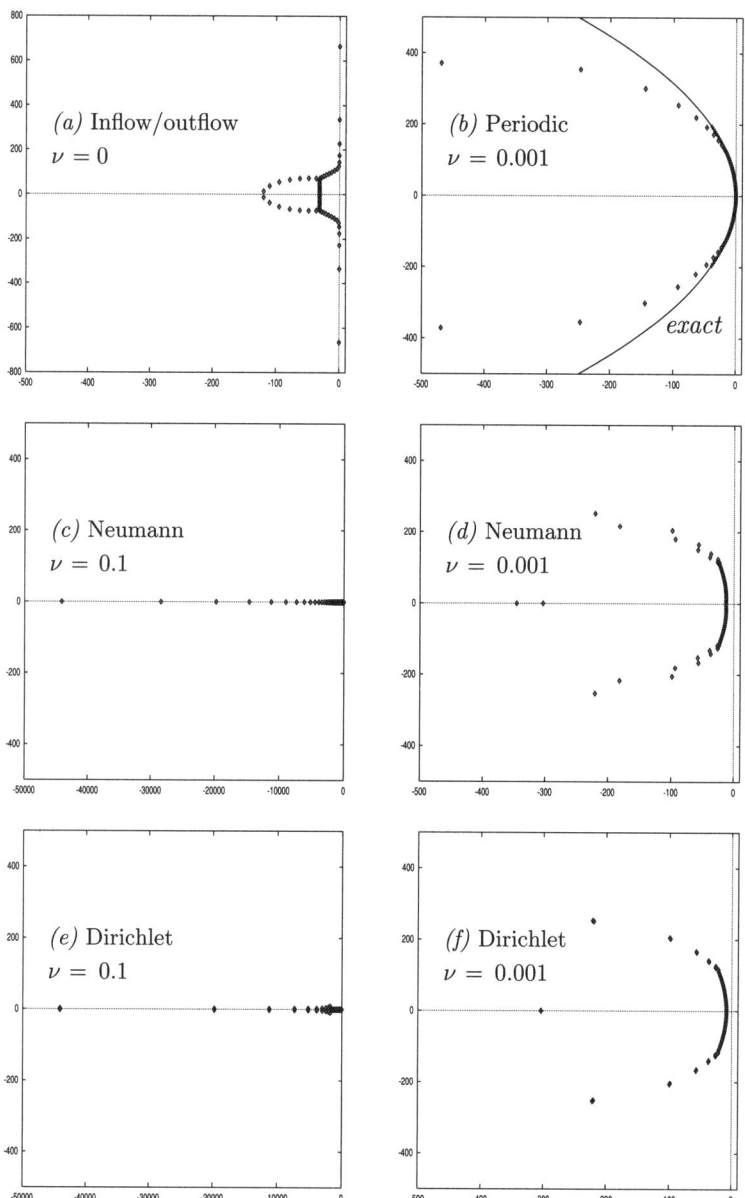

Figure 3.7.1. Spectra for spectral-element discretization of the advection–diffusion problem with $(E, N) = (1, 64)$. A few isolated negative eigenvalues are off the scale in (b)–(f).

3.7.2 Temporal Discretization

The choice of time-stepping scheme is decidedly influenced by the eigenvalue analysis of the preceding section. The presence of the large negative eigenvalues induces one to use a semiimplicit, or split, formulation based on implicit treatment of the diffusion term and explicit treatment of the convective term. This has the advantage of removing the stiffness while retaining a symmetric positive definite system to be solved. A typical formulation is the MABCN scheme (3.3.5)

$$\frac{1}{\Delta t} M(\underline{u}^{n+1} - \underline{u}^n) = -K(\alpha_0 \underline{u}^{n+1} + \alpha_1 \underline{u}^n + \alpha_2 \underline{u}^{n-1})$$
$$- C\left(\tfrac{3}{2}\underline{u}^n - \tfrac{1}{2}\underline{u}^{n-1}\right) - M\left(\tfrac{3}{2}\underline{f}^n - \tfrac{1}{2}\underline{f}^{n-1}\right). \quad (3.7.4)$$

For $\alpha_i = (\tfrac{1}{2}, \tfrac{1}{2}, 0)$ one obtains the standard ABCN scheme, while for $\alpha_i = (\tfrac{9}{16}, \tfrac{6}{16}, \tfrac{1}{16})$ one obtains the optimally accurate modified scheme. At each step, one must solve the system

$$\left(\frac{1}{\Delta t} M + \alpha_0 K\right) \underline{u}^{n+1} = \underline{g}^{n+1}, \quad (3.7.5)$$

where $\underline{g}^{n+1}$ accounts for all terms in (3.7.4) known at time levels t^n and t^{n-1}. The system (3.7.5) is SPD and diagonally dominant, so Jacobi preconditioned CG is an appropriate solution approach, as noted in (3.4.7)–(3.4.8).

3.7.3 Outflow Conditions and Filter-Based Stabilization

To illustrate the effects of boundary layers on the solution, we consider the example of a convecting Gaussian pulse, $u^0 = e^{-\zeta^2}$, $\zeta := 15(x - 0.5)$, on the unit interval $[0, 1]$, with $c = 1$, $\nu = 0.001$, $f = 0$, and $u(t, 0) = 0$. Figure 3.7.2 shows the results of using a Neumann condition, $u_x(t, 1) = 0$, in (a) versus the disastrous consequences of using the Dirichlet condition, $u(t, 1) = 0$. In the former case, the signal simply convects out of the domain, while in the latter, the solution breaks down as the pulse tries to match the Dirichlet condition at outflow. The breakdown is more contained in the linear FEM case (b) than in the spectral-element case (c), which suffers from oscillations several elements upstream. However, the FEM result suffers from dispersion errors at earlier times, as noted by undershoots in the signal. Part (d) shows SEM results when a filtering procedure (described below) is applied at each step. The boundary-layer breakdown is confined to the last element, and the solution does not suffer from early dispersion errors.

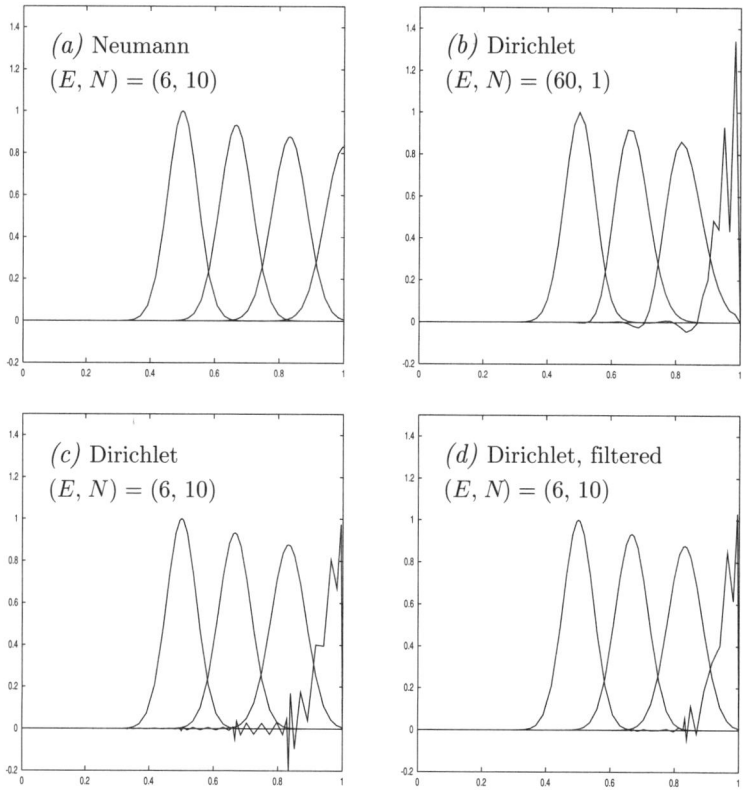

Figure 3.7.2. Advection–diffusion results.

Filtering is an alternative means of stabilizing spectral methods that has found widespread use in hyperbolic applications [106, 245, 254]. The results in Figure 3.7.2(d) were obtained by solving (3.7.5) at each step, followed by a filtering procedure

$$\underline{u}^{n+1} \longleftarrow F_\alpha \underline{u}^{n+1},$$

developed in [130]. The filter F_α is defined as follows. Let I_m be the operator that interpolates u onto the $m+1$ GLL points ξ_i^m, and let $\Pi_{N-1} := I_N I_{N-1}$ be a projector from $\mathbb{P}_N$ to $\mathbb{P}_{N-1}$. Then

$$F_\alpha := \alpha \Pi_{N-1} + (1-\alpha)I.$$

Because the nodal basis points ξ_i^N interlace ξ_i^{N-1}, this filter will dampen any oscillations arising from the most oscillatory mode. Note that $\alpha = 1$ corresponds to a full projection onto $\mathbb{P}_{N-1}$, effectively yielding a sharp cutoff in modal space, whereas $0 < \alpha < 1$ yields a smoother decay. Navier–Stokes applications of the filter are discussed in Chapter 6.

3.8 The Burgers Equation

As a last example in this chapter on parabolic and hyperbolic equations we deal with the one-dimensional Burgers equation (1.9.7), an unsteady advection–diffusion equation in which the convective term is *nonlinear*. This equation often serves as a benchmark for numerical methods, for two reasons: it has much in common with the Navier–Stokes equations, as a quick comparison with (1.7.5) reveals, and exact solutions (in space *and* time) can be obtained with the Cole–Hopf transformation for a variety of initial and boundary conditions (see [81]). The Burgers equation with homogeneous Dirichlet conditions on the reference domain is

$$\frac{\partial u}{\partial t} + u\frac{\partial u}{\partial x} = \nu\frac{\partial^2 u}{\partial x^2} \quad \text{in } \Omega := (-1, 1),$$
$$u(t, -1) = u(t, +1) = 0. \tag{3.8.1}$$

With the initial condition

$$u(0, x) = u^0(x) := -\sin(\pi x), \tag{3.8.2}$$

the analytical solution of (3.8.1) obtained by Cole and compiled by Benton and Platzmann is (see [31])

$$u(t, x) = 4\pi \nu \frac{\sum_{n=1}^{\infty} n a_n e^{-\nu n^2 \pi^2 t} \sin(n\pi x)}{a_0 + 2\sum_{n=1}^{\infty} n a_n e^{-\nu n^2 \pi^2 t} \cos(n\pi x)}. \tag{3.8.3}$$

In this expression, the coefficients a_n are given by

$$a_n = (-1)^n I_n(1/2\pi \nu),$$

where $I_n(z)$ denotes the modified Bessel function of first kind and order n. Unfortunately, the numerical evaluation of (3.8.3) is quite hard, especially at small values of t and of ν. The reason is that, for large values of z, $I_n(z)$ behaves asymptotically as $e^z(2\pi z)^{-1/2}$, independently of the value of n.

3.8.1 Space and Time Discretization

As the Burgers equation is the first nonlinear case we have confronted, it deserves some closer analysis about space and time discretization. Here, we will consider only two cases, both of which have been extensively treated in the literature: the orthogonal collocation method and the spectral-element method.

Orthogonal Collocation

Orthogonal collocation requires that the residuals vanish at the collocation points. Resorting to the methodology described in Section 2.5.1, we first define a polynomial approximation of the solution based on the GLJ points $\{\xi_i^J\}_{i=0}^N$:

$$u_N(t, x) := \sum_{i=0}^{N} u_i(t)\pi_i^J(x), \qquad x \in \hat{\Omega}. \tag{3.8.4}$$

In this expansion, the basis functions $\pi_i^J(x)$ are given by (2.5.4); these functions are such that $u_N(t, \xi_k^J) = u_k(t)$. Then, applying Equation (2.5.10), we get a set of semidiscrete equations

$$\frac{du_i(t)}{dt} + u_i(t)\frac{\partial u_N}{\partial x}(t, \xi_i^J) = \nu \frac{\partial^2 u_N}{\partial x^2}(t, \xi_i^J), \quad 1 \leq i \leq N-1, \tag{3.8.5}$$

with the boundary conditions $u_0(t) = u_N(t) = 0$. Typically, we end up with a set of (coupled) nonlinear ODEs, and the system has the form (3.3.1).

Splitting techniques are most appropriate. The standard choice is to use an explicit scheme for the nonlinear term, avoiding a Newton linearization and the inversion of a large nonsymmetric system, and an implicit scheme for the viscous term, avoiding the severe stability restriction $\Delta t \propto N^{-4}$. A common practice for about a decade has been to use AB2 for the advection part and CN for the viscous part.

The evaluation of the nonlinearity is straightforward. Suppose we know all the u_i^n. We apply the matrix derivative to compute the first-order term. The computation of the nonlinear term is then simply the local multiplication of the nodal value times its first-order derivative.

Spectral-Element Method

Let us now focus on the spectral-element discretization of the Burgers equation, especially with regard to the nonlinear term. We deal with a Legendre spectral-element approximation on one element ($E = 1$). Within the generalized Galerkin framework introduced in Section 2.2.1, the weak form of the problem is as follows: Find an approximation $u_N(t, x) \in \mathbb{P}_N \cap H_0^1(\hat{\Omega})$,

$$u_N(t, x) := \sum_{i=0}^{N} u_i(t)\pi_i(x), \tag{3.8.6}$$

such that

$$\frac{d}{dt}(u_N, v_N)_N + \left(u_N \frac{\partial u_N}{\partial x}, v_N\right)_N = \left(\nu \frac{\partial^2 u_N}{\partial x^2}, v_N\right)_N \tag{3.8.7}$$

3.8. The Burgers Equation

$\forall v_N(x) \in \mathbb{P}_N \cap H_0^1(\hat{\Omega})$. The GLL Lagrangian basis $\{\pi_i(x)\}_{i=0}^N$ is given by (2.4.3).

Again, the temporal term involves the mass matrix M, while the viscous term involves the stiffness matrix K. Let us look at the nonlinear term in the l.h.s. of (3.8.7). Unlike the methods for compressible fluid flows, where integration by parts is carried out to generate (conservation of) fluxes, the incompressible approach leaves this term unchanged. In the particular case where $v_N(x) = \pi_i(x)$, the inner product $C_i := (u_N u'_{N,x}, \pi_i)_N$, with $u'_{N,x}$ denoting the partial derivative of u_N with respect to x, becomes

$$C_i = \sum_{m=0}^N \rho_m \left(\sum_{k=0}^N \sum_{\ell=0}^N u_k u_\ell \pi_k(\xi_m) \frac{d\pi_\ell}{dx}(\xi_m) \right) \pi_i(\xi_m). \qquad (3.8.8)$$

Using the bi-orthogonality properties of the basis functions, we can reduce this expression to

$$C_i = \sum_{\ell=0}^N C_{i\ell} u_\ell := \sum_{\ell=0}^N \rho_i u_i D_{N,i\ell}^{(1)} u_\ell. \qquad (3.8.9)$$

where $D_N^{(1)}$ is the first-order Legendre derivative (2.4.9). In matrix form, Equation (3.8.7) is

$$M \frac{d\underline{u}}{dt} + K\underline{u} + C(\underline{u})\underline{u} = 0, \qquad (3.8.10)$$

where $C(\underline{u})$ is the symbolic notation to express the nonlinear discrete operator with components $C_{i\ell} := \rho_i u_i D_{N,i\ell}^{(1)}$ acting on the vector of unknowns $\underline{u} \in \mathbb{R}^\mathcal{N}$. We observe that the spectral-element calculation of the nonlinear term is of collocation type, since the matrix element $C_{i\ell}$ is the local value of the unknown times the derivative matrix. The nonlinearity generates a quadratic expression. If one wants to avoid explicit time integration schemes for stability reasons, a Newton linearization may be carried out on the discrete equations. In spectral methods, we prefer to compute the Fréchet derivative of the operator (see Section A.3.5) and discretize in space and time afterwards.

3.8.2 Numerical Results

Here, we are interested in the advection-dominated case with very small values of the viscosity. Such low values of ν lead to a rapid variation of the solution that goes from the initial trigonometric profile to a (time-decaying) sawtooth profile with a thin internal layer located at the origin that must be resolved if

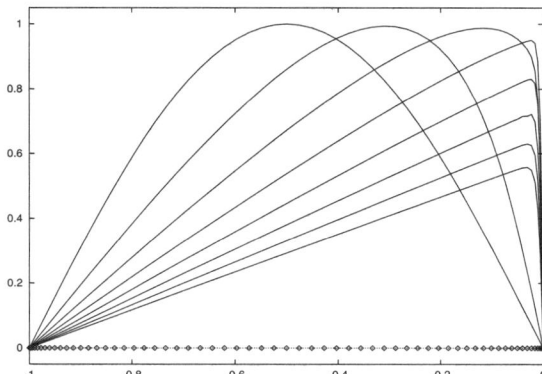

Figure 3.8.1. Solution to Burgers equation on $[-1, 0]$ using the Legendre spectral method, $(E, N) = (1, 64)$, with AB3CN.

accurate numerical solutions are to be obtained. The (exact) solution exhibits two particular features:

- A sudden increase of the slope $|\frac{\partial u}{\partial x}(t, 0)|$ versus time until this parameter reaches a maximum, and then decreases.
- A shift of $x_{\max}$, the abscissa of $u_{\max}$, versus time from $x = \pm 0.5$ initially, toward $x = 0$, and then, after some time, back again (see Figure 3.8.1).

Basdevant et al. [22] have studied numerical solutions to (3.8.1)–(3.8.2) for the case $\nu = 10^{-2}/\pi$ by a variety of combinations of discretizations in space (e.g., Fourier–Galerkin, Fourier pseudospectral, Chebyshev tau, Chebyshev collocation, spectral element, finite difference) and time (e.g., IE, ABCN, RK3). The results obtained by these authors are summarized in Table 3.8.1. The table gives the numerical values of the maximum slope at the origin, $|\frac{\partial u}{\partial x}(t, 0)|_{\max}$, and the time $t_{\max}$ at which this maximum occurs for the various algorithms used in the study, as well as the reference value taken from [31].

The authors conclude that, if one treats the full domain $(-1, +1)$, the sharp sawtooth profile at the center of the domain requires many grid points (or expansion functions) to resolve the internal layer. However, a well-chosen coordinate transformation may alleviate this difficulty. Also, the spectral schemes perform poorly with a time step $\Delta t = 10^{-2}/\pi$, even with large values of N/M, where N is the number of degrees of freedom and M is the length of the fast Fourier transform. Shortening the time step improves the numerical results, however.

If the equation is solved on half the domain, the accumulation of mesh points near the boundary (in this case, $x = 0$), characteristic of Chebyshev or Legendre spectral methods, improves the attainable precision. For instance,

Table 3.8.1. *Maximum absolute value of the slope at the origin and time t_{max} at which this maximum value occurs for all the numerical algorithms and for the analytical solution. N is the number of degrees of freedom, and M the length of the fast Fourier transform (see [22]).*

Method	Ω	$\left|\frac{\partial u}{\partial x}\right|_{max}$	πt_{max}	N/M	$\Delta t \cdot \pi$
Fourier–Galerkin	$(-1, 1)$	151.942	1.6035	682/1024	$5.0e-4$
		142.665	1.60	682/1024	$1.0e-2$
		148.975	1.603	170/256	$5.0e-4$
		142.313	1.60	170/256	$1.0e-2$
Fourier	$(-1, 1)$	142.606	1.60	256/256	$1.0e-2$
pseudospectral		144.237	1.60	128/128	$1.0e-2$
ABCN collocation +	$(-1, 1)$	145.877	1.60	512	$5.0e-3$
coordinate transf.	$(-1, 1)$	152.123	1.60	64	$1.0e-2$
Spectral element	$(-1, 1)$	152.04	1.6033	16×4	$(1.0e-2)/6$
FD	$(-1, 1)$	152.1	1.63	81	$1.0e-2$
Chebyshev:					
ABCN spectral	$(0, 1)$	152.05	1.60	64	$1/300$
Rosenbrock spectral	$(0, 1)$	151.998	1.60	64	$1.0e-2$
	$(0, 1)$	150.144	1.60	32	$1.0e-2$
ABCN collocation	$(0, 1)$	152.126	1.60	64	$1.0e-2$
Analytical		152.00516	1.6037		

the Rosenbrock (Chebyshev) spectral method with 64 degrees of freedom and $\Delta t \cdot \pi = 10^{-2}$ gives a relative error of 4.71×10^{-5} on the maximum slope, while the Fourier–Galerkin method with ten times as many mesh points and a time step twenty times shorter gives a relative error of 4.16×10^{-4} only. Figure 3.8.1 shows numerical solutions to (3.8.1)–(3.8.2) on half the $\hat{\Omega}$ domain, evaluated at various times before and after the maximum slope, with the spectral-element ABCN method. The curves were obtained with 64 degrees of freedom and a time step $\Delta t \approx 10^{-3}$.

3.9 The OIFS Method and Subcycling

The OIFS method introduced in Section 3.3.1 organizes the combination of any two time integration methods for the linear and nonlinear terms. Here, we will again treat the Burgers equation in the Legendre spectral-element framework, and we will interpret the OIFS method as a subcycling technique.

Proceeding with the space discretization, we obtain the set of nonlinear ODEs (3.8.10)

$$M\frac{d\underline{u}}{dt} + K\underline{u} + C(\underline{u})\underline{u} = 0.$$

For the Burgers equation (3.8.1), the corresponding Reynolds number is $Re = 1/\nu$ if one chooses as reference velocity and length scale $U = L = 1$. The Reynolds number may be viewed as the ratio of two time scales relevant to the physics, namely, the diffusive time scale L^2/ν over the inertial time scale L/U. Therefore, at moderate and high Reynolds numbers, the dynamics of the physics is driven essentially by inertia, but viscous effects play their role over longer time intervals. From this interpretation, we may conclude that it would be wise to integrate more often the part of the operator – namely, advection – that produces the changes over the shorter scales. This is the idea of the so-called method of subcycling. This method will implement the time integration in such a way that γ subcycles of time step Δs will be carried out on the explicit, cheap part of the convection term, while the implicit, more computation-intensive, viscous part is integrated less often, at every time step $\Delta t = \gamma \, \Delta s$. The stability condition will gain from being applied to Δs and not to Δt. The OIFS method allows the user to set up the subcycling technique in an efficient manner.

We begin by writing the basic equation (3.8.10) in terms of an integration factor $Q_C^{t^*}(t)$ related to the nonlinear operator $C(\underline{u})$:

$$\frac{d}{dt}\left[Q_C^{t^*}(t) \cdot M\underline{u}(t)\right] = -Q_C^{t^*}(t) \cdot K\underline{u}(t). \tag{3.9.1}$$

A combination of (3.8.10) and (3.9.1) shows that the integrating factor obey the equation

$$\frac{d}{dt}Q_C^{t^*}(t) \cdot M = Q_C^{t^*}(t) \cdot C(\underline{u}(t)), \qquad Q_C^{t^*}(t^*) = I, \tag{3.9.2}$$

where I is the identity matrix. As mentioned earlier, this relationship is not used to evaluate the integrating factor $Q_C^{t^*}(t)$ explicitly. Only quantities such as $Q_C^{t^*}(t) \cdot M\underline{u}(t)$ are evaluated. Specifically, we solve a set of differential equations involving an auxiliary vector $\underline{v}^{(t^*,t)}(s) \in \mathbb{R}^N$, on a shifted time axis s with the origin set at time t:

$$M\frac{d}{ds}\underline{v}^{(t^*,t)}(s) = -C\left(\underline{v}^{(t^*,t)}(s)\right)\underline{v}^{(t^*,t)}(s), \tag{3.9.3}$$

$0 < s < t^* - t$. Initial conditions for (3.9.3) are given by (3.3.12):

$$\underline{v}^{(t^*,t)}(0) = \underline{u}(t).$$

Following the analysis made in Section 3.3.1, one can show that

$$Q_C^{t^*}(t) \cdot M\underline{u}(t) = M\underline{v}^{(t^*,t)}(t^* - t). \tag{3.9.4}$$

3.9. The OIFS Method and Subcycling

Let us now apply the OIFS method based on BDF3/RK4 to solve the nonlinear time-dependent problem (3.8.10). The computation involves two time scales: Δt, related to the viscous component, which is treated implicitly (BDF3), and Δs, related to the advective component, which needs to be handled explicitly. The time scales are selected so that $\Delta t = \gamma \Delta s$, where γ is an integer. Obviously, the values of these parameters are strongly influenced by stability conditions.

Assume that the approximate solution is known up to time t^n on the time scale of the (implicit) calculation (see Figure 3.3.2). Applying the BDF3 integration scheme to (3.9.1) gives

$$M\underline{u}^{n+1} = \sum_{j=1}^{3} a_j S_C\left(\mathcal{Q}_C^{t^{n+1}}(t^{n+1-j}) \cdot M\underline{u}^{n+1-j}\right) - b_0 \Delta t K\underline{u}^{n+1}, \quad (3.9.5)$$

where the coefficients a_j and b_0 are taken from Table 3.2.3. We emphasize that the first contribution to the r.h.s. of (3.9.5) results from the integration of a set of nonlinear problems related to the advection term; hence the presence of the symbol S_C, referring to the explicit scheme (ERK4) used to solve these equations.

The set of auxiliary advection equations and corresponding initial conditions to evaluate $\mathcal{Q}_C^{t^{n+1}}(t^{n+1-j}) \cdot M\underline{u}^{n+1-j}$ for some value of j is

$$M\frac{d}{ds}\underline{v}^{(n+1,n+1-j)}(s) = -C\left(\underline{v}^{(n+1,n+1-j)}(s)\right)\underline{v}^{(n+1,n+1-j)}(s) \quad (3.9.6)$$

and

$$\underline{v}^{(n+1,n+1-j)}(0) = \underline{u}^{n+1-j}. \quad (3.9.7)$$

Integration of (3.9.6) on the interval $(0, \Delta s_j]$ along the shifted time axis with origin set at t^{n+1-j} and $\Delta s_j := t^{n+1} - t^{n+1-j} = j\gamma \Delta s$ gives the parameters

$$M\underline{\tilde{u}}_j \equiv S_C\left(\mathcal{Q}_C^{t^{n+1}}(t^{n+1-j}) \cdot M\underline{u}^{n+1-j}\right) = M\underline{v}^{(n+1,n+1-j)}(\Delta s_j) \quad (3.9.8)$$

needed to complete the implicit calculation (3.9.5). The expression on the left in the last relationship, $M\underline{\tilde{u}}_j$, is defined only to simplify the notation. The most important member is the last one, which is the final result in the (explicit) integration process for the current value of j.

Once the auxiliary advection calculations have been performed for all values of the integer j, and the results (3.9.8) have been put into (3.9.5) with the appropriate BDF3 coefficients, the implicit calculation may be completed to determine $\underline{u}^{n+1}$:

$$M\left(\underline{u}^{n+1} - \frac{1}{11}[18\underline{\tilde{u}}_1 - 9\underline{\tilde{u}}_2 + 2\underline{\tilde{u}}_3]\right) = -\frac{6\Delta t}{11} K \underline{u}^{n+1}. \quad (3.9.9)$$

The computation process then is carried to the next time interval (t^{n+1}, t^{n+2}).

We emphasize that, in ease of implementation, $\gamma = 1$ is the best solution. We will prove in Chapter 4 that the subcycling method ($\gamma > 1$) is in effect equivalent to the method of characteristics. Subcycling is expensive, however, in function evaluations.

3.10 Taylor–Galerkin Time Integration

The Taylor–Galerkin method was introduced by Donea [107] for pure advection problems. Instead of going the classical way, namely, first carrying out the space discretization and then applying a time integration scheme to the set of ODEs, Donea reverses the process. The Taylor–Galerkin approach may be considered as a time integration technique where the solution is first expanded in a Taylor series in time up to the third-order term and then all time derivatives (if possible) are replaced by space derivatives through the repeated use of the hyperbolic equation. Eventually, the resulting equation is approximated by a standard Galerkin discretization method.

In this section, we consider the Burgers equation (3.8.1), and the Galerkin approximation will be based on Legendre spectral elements. The Taylor–Galerkin method was first proposed in this setting by Timmermans et al. [380]. We expand the solution at (t^n, x) in a Taylor series including the third-order term:

$$u(t^{n+1}, x) = u(t^n, x) + \Delta t \frac{\partial u}{\partial t}(t^n, x) + \frac{(\Delta t)^2}{2} \frac{\partial^2 u}{\partial t^2}(t^n, x)$$

$$+ \frac{(\Delta t)^3}{6} \frac{\partial^3 u}{\partial t^3}(t^n, x) + O((\Delta t)^4). \quad (3.10.1)$$

However, for the sake of simplicity, we will restrict ourselves in the following development to the second-order term; the reader is referred to Donea and Quartapelle [109] for the full treatment. From the Taylor series, we obtain

$$\frac{u^{n+1} - u^n}{\Delta t} = \frac{\partial u}{\partial t}(t^n, x) + \frac{\Delta t}{2} \frac{\partial^2 u}{\partial t^2}(t^n, x). \quad (3.10.2)$$

Let us replace in (3.10.2) the first-order time derivative by the expression coming

3.10. Taylor–Galerkin Time Integration

from the Burgers equation (3.8.1), while the second-order time derivative is evaluated as

$$\frac{\partial^2 u}{\partial t^2}(t^n, x) = \frac{\partial}{\partial t}\left(-u\frac{\partial u}{\partial x} + \nu\frac{\partial^2 u}{\partial x^2}\right)(t^n, x). \quad (3.10.3)$$

Expanding the r.h.s. of (3.10.3) and approximating each of the time derivatives by $(u^{n+1} - u^n)/\Delta t$, in order to avoid the presence of third-order space derivatives (that would have required C^2 continuity in the Galerkin method), gives

$$\frac{\partial^2 u}{\partial t^2}(t^n, x) = -\frac{1}{\Delta t}\frac{\partial}{\partial x}[u(u^{n+1} - u^n)] + \nu\frac{\partial^2}{\partial x^2}\frac{u^{n+1} - u^n}{\Delta t}, \quad (3.10.4)$$

with u in the r.h.s. of (3.10.4) taken as $u(t^n, x)$. With all the terms collected, the relation (3.10.2) becomes

$$\left[1 - \frac{1}{2}\Delta t\left(-\frac{\partial u}{\partial x} - u\frac{\partial}{\partial x} + \nu\frac{\partial^2}{\partial x^2}\right)\right]\frac{u^{n+1} - u^n}{\Delta t} = -u\frac{\partial u}{\partial x} + \nu\frac{\partial^2 u}{\partial x^2}. \quad (3.10.5)$$

If the time derivative in Equation (3.10.3) is computed backward as $\partial u/\partial t = (u^n - u^{n-1})/\Delta t$, the method generated by the combination of (3.10.2) and (3.10.3) is completely explicit in time. Applying the space discretization to Equation (3.10.5), we are led to the Taylor–Galerkin SEM.

3.10.1 Nonlinear Pure Advection

Going back to the continuous problem (3.8.1) with a vanishing viscosity, and more specifically to Equation (3.10.2), we can now express the second-order time derivative as

$$\frac{\partial^2 u}{\partial t^2}(t^n, x) = -\frac{\partial u}{\partial t}\frac{\partial u}{\partial x} - u\frac{\partial}{\partial x}\left(\frac{\partial u}{\partial t}\right)$$

$$= -\frac{\partial u}{\partial t}\frac{\partial u}{\partial x} + u\frac{\partial}{\partial x}\left(u\frac{\partial u}{\partial x}\right). \quad (3.10.6)$$

Introducing (3.10.6) in (3.10.2), we obtain

$$\frac{u^{n+1} - u^n}{\Delta t} = -\left(u + \frac{\Delta t}{2}\frac{\partial u}{\partial t}\right)\frac{\partial u}{\partial x} + \frac{\Delta t}{2}u\frac{\partial}{\partial x}\left(u\frac{\partial u}{\partial x}\right). \quad (3.10.7)$$

The time derivative in the r.h.s. can be approximated by a forward formula, leading to an implicit scheme

$$u + \frac{\Delta t}{2}\frac{\partial u}{\partial t} = u^n + \frac{\Delta t}{2}\left(\frac{u^{n+1} - u^n}{\Delta t} + O(\Delta t)\right)$$

$$= \frac{u^{n+1} + u^n}{2} + O(\Delta t^2)$$

$$= u^{n+\frac{1}{2}} + O(\Delta t^2), \qquad (3.10.8)$$

or by a backward formula, leading to an explicit scheme

$$u + \frac{\Delta t}{2}\frac{\partial u}{\partial t} = u^n + \frac{\Delta t}{2}\left(\frac{u^n - u^{n-1}}{\Delta t} + O(\Delta t)\right)$$

$$= \frac{3u^n - u^{n-1}}{2} + O(\Delta t^2). \qquad (3.10.9)$$

The second approach is favored inasmuch as it keeps the convection step fully explicit. We recognize that with this last relation, the approximation of the first term on the r.h.s. of Equation (3.10.7) would yield an AB2 scheme. By applying the space discretization, we derive the second-order accurate explicit Euler Taylor–Galerkin method.

The appearance of the second-order term in (3.10.7) slows the evaluation of the nonlinearity. To avoid this computational burden, we resort to a two-step Taylor–Galerkin scheme. The Taylor series (3.10.1) with a time-centered approximation is

$$u(t^{n+1}, x) = u(t^n, x) + \Delta t \frac{\partial u}{\partial t}(t^{n+\frac{1}{2}}, x) + O((\Delta t)^3), \qquad (3.10.10)$$

where the time derivative is now

$$\frac{\partial u}{\partial t}(t^{n+\frac{1}{2}}, x) = \frac{\partial u}{\partial t}(t^n, x) + \frac{\Delta t}{2}\frac{\partial^2 u}{\partial t^2}(t^n, x). \qquad (3.10.11)$$

Using the two last relations, one easily gets the following two-step scheme:

$$u^{n+\frac{1}{2}} = u^n - \frac{\Delta t}{2}\left(u^n \frac{\partial u^n}{\partial x}\right),$$

$$u^{n+1} = u^n - \Delta t\left(u^{n+\frac{1}{2}} \frac{\partial u^{n+\frac{1}{2}}}{\partial x}\right). \qquad (3.10.12)$$

3.10.2 Taylor–Galerkin and OIFS Methods

Because of viscous stability restrictions, one must resort to implicit time integration for the viscous term in the set of nonlinear differential equations (3.8.10). In this framework, the OIFS technique helps to design an elegant and efficient scheme within which an implicit BDFk scheme handles the linear term while the explicit Taylor–Galerkin scheme (3.10.12) deals with the nonlinear term. Both schemes must be of the same order of accuracy. Since we have chosen to truncate the Taylor series after the second-order term, we apply BDF2 to Equation (3.9.1), giving

$$M\underline{u}^{n+1} = \sum_{j=1}^{2} a_j S_C\big(\mathcal{Q}_C^{t^{n+1}}(t^{n+1-j}) \cdot M\underline{u}^{n+1-j}\big) - b_0 \,\Delta t\, K\underline{u}^{n+1}. \quad (3.10.13)$$

As before, the terms involving the integrating factor in the r.h.s. of (3.10.13) are obtained from the solution of a set of nonlinear initial-value problems coping with the advection term only:

$$M\frac{d}{ds}\underline{v}^{(n+1,n+1-j)}(s) = -C\big(\underline{v}^{(n+1,n+1-j)}(s)\big)\underline{v}^{(n+1,n+1-j)}(s), \quad (3.10.14)$$

with $j = 1, 2$ and an initial condition

$$\underline{v}^{(n+1,n+1-j)}(0) = \underline{u}^{n+1-j}. \quad (3.10.15)$$

Integration of the problems (3.10.14)–(3.10.15) on the interval $(0, \Delta s_j]$ along the shifted time axis using the two-step scheme (3.10.12) gives the quantities

$$S_C\big(\mathcal{Q}_C^{t^{n+1}}(t^{n+1-j}) \cdot M\underline{u}^{n+1-j}\big) = M\underline{v}^{(n+1,n+1-j)}(\Delta s_j), \quad (3.10.16)$$

to be entered in (3.10.13), for $j = 1, 2$. Here again one has an additional degree of freedom in the choice of γ, the ratio of time steps in the implicit and explicit parts of the computation. A value of γ should be selected that not only ensures stability of the algorithm but also minimizes the computer work.

The two-step Taylor–Galerkin time integration technique has been applied in the OIFS framework to the Burgers equation by Timmermans et al. [380]. In their calculations, a one-dimensional Gaussian hill is traveling with a unit constant velocity and spreads isotropically with a viscosity $\nu = 1/200$. The authors show that the Taylor–Galerkin scheme achieves second-order accuracy in time and is faster than fully implicit time integration. Taylor–Galerkin schemes therefore seem very well suited for nonlinear and multidimensional problems.

4
Multidimensional Problems

Chapters 2 and 3 introduced several variational methods for one-dimensional problems. In this chapter, these techniques are extended to d space dimensions, where $d = 2$ or 3. We initially consider the case where the domain can be expressed as a tensor product of one-dimensional intervals; that is, it is a rectangle in $\mathbb{R}^2$ or a box in $\mathbb{R}^3$. We then consider the more general case where the domain is an image of the parent element, $\hat{\Omega} = [-1, 1]^d$, obtained via a smooth invertible mapping. Fully complex geometries can be handled via mapping coupled with domain decomposition, which is introduced here and considered further in Chapter 7.

4.1 Introduction

We begin in Section 4.2 with a review of tensor products, which are central to the efficiency of high-order methods. Tensor-product forms are used to develop spectral operators for elliptic problems in regular parallelepipeds in Section 4.3, and extended to deformed hexahedra in Section 4.4. Multidomain SEMs are introduced in Section 4.5, and collocation methods are treated in Section 4.6. Parabolic and hyperbolic problems are considered in Sections 4.7 and 4.8, respectively, and the combination of these is treated in the discussion of advection–diffusion problems in Section 4.9. Further reading and advanced topics are addressed in Section 4.10.

4.2 Tensor Products

In this section we introduce a number of basic properties of matrices based on tensor-product forms. Such matrices frequently arise in the numerical solution of PDEs. Their importance for the development of fast Poisson solvers was

4.2. Tensor Products

recognized in an early paper by Lynch et al. [250]. In 1980, Orszag [288] pointed out that tensor-product forms were the foundation for efficient implementation of spectral methods.

The relevance of tensor-product matrices to multidimensional discretizations is illustrated by considering a tensor-product polynomial on the reference domain, $(x, y) \in \hat{\Omega} := [-1, 1]^2$:

$$u(x, y) = \sum_{i=0}^{M} \sum_{j=0}^{N} u_{ij} \pi_{M,i}(x) \pi_{N,j}(y). \quad (4.2.1)$$

Here, $\pi_{M,i}$ ($\pi_{N,j}$) is the Lagrangian interpolant of degree M (N) based on Legendre polynomials introduced in (2.4.3). Consequently, the basis coefficients, u_{ij}, are also nodal values of u on the tensor product of Gauss–Lobatto–Legendre (GLL) quadrature points, as shown in Figure 4.2.1. Useful vector representations of the coefficients are denoted by

$$\underline{u} := (u_1, u_2, \ldots, u_l, \ldots, u_{\mathcal{N}})^T := (u_{00}, u_{10}, \ldots, u_{ij}, \ldots, u_{MN})^T, \quad (4.2.2)$$

where $\mathcal{N} = (M + 1)(N + 1)$ is the number of basis coefficients, and the mapping $l = 1 + i + (M + 1)j$ translates to standard vector form the two-index

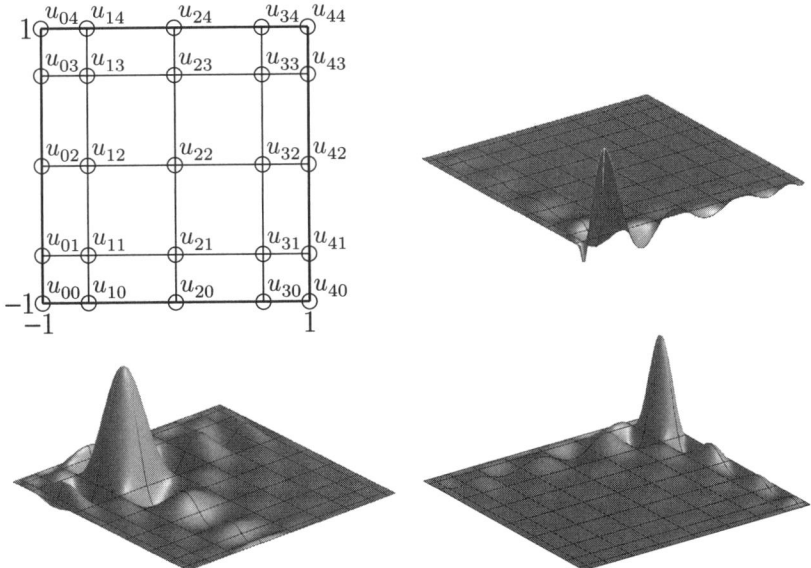

Figure 4.2.1. Clockwise from upper left: GLL nodal point distribution on $\hat{\Omega}$ for $M = N = 4$, and Lagrangian basis functions for $M = N = 10$: $\pi_{10}\pi_2$, $\pi_2\pi_9$, and $\pi_3\pi_4$.

coefficient representation, with the leading index advancing more rapidly. This is referred to as the natural, or lexicographical, ordering and is used throughout this text.

The tensor-product form (4.2.1) allows significant simplifications in the application of linear operators to u. For example, suppose w_{pq} is to represent the x-derivative of u at the GLL points, $(\xi_{M,p}, \xi_{N,q})$, $p, q \in \{0, \ldots, M\} \times \{0, \ldots, N\}$:

$$w_{pq} := \frac{\partial u}{\partial x}(\xi_{M,p}, \xi_{N,q}) = \sum_{i=0}^{M} \sum_{j=0}^{N} u_{ij} \pi'_{M,i}(\xi_{M,p}) \pi_{N,j}(\xi_{N,q})$$

$$= \sum_{i=0}^{M} u_{iq} \pi'_{M,i}(\xi_{M,p}). \qquad (4.2.3)$$

Expressed as a matrix–vector product, (4.2.3) reads

$$\underline{w} = D_x \underline{u} := \begin{bmatrix} \hat{D}_x & & & \\ & \hat{D}_x & & \\ & & \ddots & \\ & & & \hat{D}_x \end{bmatrix} \begin{pmatrix} u_{00} \\ u_{10} \\ \vdots \\ u_{MN} \end{pmatrix}. \qquad (4.2.4)$$

Here, $\hat{D}_x$ is the one-dimensional derivative matrix (2.4.9) associated with the $M + 1$ GLL points on the reference interval $[-1, 1]$ and is applied to *each row*, $(u_{0j}, u_{1j}, \ldots, u_{Mj})^T$, $j = 0, \ldots, N$, in the computational grid of Figure 4.2.1. To compute derivatives with respect to y, one applies the corresponding matrix $\hat{D}_y$ to *each column*. This latter case does not yield a simple block-diagonal matrix, because of the ordering of the coefficients in $\underline{u}$. Nonetheless, D_x and D_y are conveniently expressed as $D_x = I \otimes \hat{D}_x$ and $D_y = \hat{D}_y \otimes I$, where $\otimes$ indicates the tensor (or Kronecker) product, which we now introduce.

Let A and B be $k \times l$ and $m \times n$ matrices, respectively, and consider the $km \times ln$ matrix C, given in block form as

$$C := \begin{pmatrix} a_{11}B & a_{12}B & \ldots & a_{1l}B \\ a_{21}B & a_{22}B & \ldots & a_{2l}B \\ \vdots & \vdots & & \vdots \\ a_{k1}B & a_{k2}B & \ldots & a_{kl}B \end{pmatrix}. \qquad (4.2.5)$$

C is said to be the tensor (or Kronecker) product of A and B, denoted as

$$C = A \otimes B. \qquad (4.2.6)$$

4.2. Tensor Products

From the definition (4.2.5) one can easily verify that an entry c_{ij} ($1 \le i \le km$, $1 \le j \le ln$) in C is equal to

$$c_{ij} = a_{pq}b_{rs},$$

where the index pairs (pq) and (rs) satisfy the relationships

$$i = r + (p-1)m, \qquad j = s + (q-1)n.$$

We will occasionally also refer to c_{ij} as $c_{rp,sq}$.

Using the standard definition of matrix multiplication, one may show that

$$(A \otimes B)(F \otimes G) = (AF \otimes BG) \qquad (4.2.7)$$

for rectangular matrices A, B, F, and G appropriately dimensioned so that the products AF and BG are well defined. It follows that if A and B are square invertible matrices of orders m and n, respectively, then

$$(A^{-1} \otimes B^{-1})(A \otimes B) = I \otimes I, \qquad (4.2.8)$$

where I denotes the identity matrix of appropriate order. Thus, the inverse of $A \otimes B$ is $A^{-1} \otimes B^{-1}$.

Suppose A and B are diagonalizable square matrices of order m and n, respectively, with similarity transformations

$$A = S\Lambda S^{-1}, \qquad B = T\mathcal{M}T^{-1}, \qquad (4.2.9)$$

where the columns of the matrices S and T are the eigenvectors of A and B, and the entries in the diagonal matrices Λ and $\mathcal{M}$ are the respective eigenvalues. Then

$$(S^{-1} \otimes T^{-1})(A \otimes B)(S \otimes T) = \Lambda \otimes \mathcal{M}.$$

Note that the tensor product of two diagonal matrices is also diagonal. Thus the diagonalization of C in (4.2.6) may be obtained from the diagonalization of the two much smaller systems, A and B.

The following form frequently arises in FD discretization of 2D BVPs:

$$C = A \otimes I + I \otimes B$$

(e.g., when A and B represent one-dimensional difference operators). Here the inverse is not so easily obtained as in (4.2.8). However, if A and B are diagonalizable as in (4.2.9), then the diagonalization of C is

$$C = (S \otimes T)(\Lambda \otimes I + I \otimes \mathcal{M})(S^{-1} \otimes T^{-1}), \qquad (4.2.10)$$

from which it follows that

$$C^{-1} = (S \otimes T)(\Lambda \otimes I + I \otimes M)^{-1}(S^{-1} \otimes T^{-1}). \quad (4.2.11)$$

Note that $\Lambda \otimes I + I \otimes M$ is diagonal and therefore trivially inverted.

Weighted residual techniques frequently give rise to systems of a more general form, for example,

$$C = M_y \otimes K_x + K_y \otimes M_x, \quad (4.2.12)$$

where K_* and M_* are the one-dimensional stiffness and mass matrices associated with their respective spatial directions. Here the appropriate similarity transformation comes from a generalized eigenvalue problem of the form

$$K_* \underline{s}_i = \lambda_i M_* \underline{s}_i.$$

If K_* is symmetric and M_* is symmetric positive definite, then there exists a complete set of eigenvectors that are orthonormal with respect to the M_* inner product, that is, $\underline{s}_i^T M_* \underline{s}_j = \delta_{ij}$. If $\Lambda = \text{diag}(\lambda_i)$ is the diagonal matrix of eigenvalues and $S = (\underline{s}_1, \ldots, \underline{s}_N)$ the corresponding matrix of eigenvectors, then the following similarity transformation holds:

$$S^T K_* S = \Lambda, \qquad S^T M_* S = I. \quad (4.2.13)$$

Suppose there exist matrices Λ_x and S_x such that (4.2.13) is satisfied for K_x and M_x, and similarly for the y-operators. Then the inverse of C (4.2.12) is given by

$$C^{-1} = (S_y \otimes S_x)(I \otimes \Lambda_x + \Lambda_y \otimes I)^{-1}(S_y^T \otimes S_x^T). \quad (4.2.14)$$

The preceding tensor-product factorizations readily extend to more dimensions. Rather than presenting the general form, we merely illustrate the procedure by giving the result for the 3D case. As we will demonstrate in the next section, weighted residual techniques for 3D BVPs frequently give rise to system matrices of the form

$$C = M_z \otimes M_y \otimes K_x + M_z \otimes K_y \otimes M_x + K_z \otimes M_y \otimes M_x.$$

If one assumes that similarity transforms of the form (4.2.13) exist for each (K_*, M_*) pair, then C^{-1} is given by

$$C^{-1} = (S_z \otimes S_y \otimes S_x)D^{-1}(S_z^T \otimes S_y^T \otimes S_x^T), \quad (4.2.15)$$

4.2. Tensor Products

where

$$D = I \otimes I \otimes \Lambda_x + I \otimes \Lambda_y \otimes I + \Lambda_z \otimes I \otimes I$$

is the diagonal matrix containing the eigenvalues of the one-dimensional operators.

The forms (4.2.11) and (4.2.15) are referred to as *fast* diagonalization methods, because they permit inversion of an $N^d \times N^d$ matrix in only $O(N^{d+1})$ operations (as the next section will show). They were proposed for the solution of FD problems in [250] and are often used in the solution of global spectral methods. [For Fourier and Chebyshev methods, the work complexity can be reduced to $O(N^d \log N)$ through the use of fast-transform methods that allow the one-dimensional operators S_* and S_*^T to be applied in $O(N \log N)$ operations.] Although strictly applicable only to constant-coefficient problems, the fast diagonalization method has also been shown to be effective as an approximate local solver for domain-decomposition-based preconditioners [83, 129] (see Section 7.2).

Operator Evaluation

The preceding section illustrates several useful properties of matrices based on tensor-product forms. Essential for operator evaluation in numerical solution of PDEs is the fact that matrix–vector products of the form $\underline{v} = (A \otimes B)\underline{u}$ can be evaluated in an order of magnitude fewer operations than would be possible in the general case.

Suppose for ease of exposition that A and B are square matrices of order n and m, respectively, and that $\underline{u}$ is an mn-component vector with entries denoted by a double subscript, u_{ij}, $i = 1, \ldots, m$, $j = 1, \ldots, n$. We associate a natural ordering of these entries with the $\hat{\imath}$th component of $\underline{u}$: $u_{\hat{\imath}} = u_{ij}$ iff $\hat{\imath} = i + m(j-1)$. With this convention, the product is

$$v_{ij} = \sum_{l=1}^{n} \sum_{k=1}^{m} a_{jl} b_{ik} u_{kl} \qquad i = 1, \ldots, m, \quad j = 1, \ldots, n. \quad (4.2.16)$$

If A and B are full matrices, then $C = A \otimes B$ is full, and computation of $C\underline{u}$ by straightforward matrix–vector multiplication nominally requires $nm(2nm - 1)$ operations, or $O(n^4)$ if $m = n$. (We define an *operation* as either floating-point addition or multiplication.) In practice, however, the matrix C is never explicitly formed; one evaluates only the *action* of C on a vector. The usual approach is to exploit the relationship (4.2.7) and the associativity of matrix multiplication to rewrite $C\underline{u}$ as

$$C\underline{u} = (A \otimes I)(I \otimes B)\underline{u}. \quad (4.2.17)$$

One then computes $\underline{w} = (I \otimes B)\underline{u}$:

$$w_{ij} = \sum_{k=1}^{m} b_{ik} u_{kj}, \quad i = 1, \ldots, m, \quad j = 1, \ldots, n, \quad (4.2.18)$$

followed by $\underline{v} = (A \otimes I)\underline{w}$:

$$v_{ij} = \sum_{l=1}^{n} a_{jl} w_{il}, \quad i = 1, \ldots, m, \quad j = 1, \ldots, n,$$

$$= \sum_{l=1}^{n} w_{il} a_{lj}^T, \quad i = 1, \ldots, m, \quad j = 1, \ldots, n. \quad (4.2.19)$$

The total work is only $2nm(n + m - 1)$, or $O(n^3)$ if $n = m$.

The 3D case follows similarly and yields even greater reduction in the number of operations. Consider the tensor product of three full $n \times n$ matrices, $C = A_z \otimes A_y \otimes A_x$. If formed explicitly, C will contain n^6 nonzeros, and the matrix–vector product $\underline{v} = C\underline{u}$ will require $2n^6$ operations. However, if C is factored as $C = (A_z \otimes I \otimes I)(I \otimes A_y \otimes I)(I \otimes I \otimes A_x)$ and the matrix–vector product is evaluated as above for the 2D case, the total cost is only $3 \times 2n^4$ operations. In general, the rule for a matrix operator resulting from a discretization with n mesh points per space dimension, in d space dimensions (full coupling in *each* spatial direction), is that it should be possible to evaluate the action of an operator in only $O(n^{d+1})$ operations.

We are now in a position to establish an important result for multidimensional problems: *Tensor-product-based high-order methods become increasingly effective as the spatial dimension d increases.* Consider an example where the number of points required to resolve a function in d space dimensions to a given accuracy is N^d for a high-order method and $\hat{N}^d$ for a low-order scheme with $\sigma := N/\hat{N} < 1$. For explicit methods and for implicit schemes with iterative solvers, the leading-order cost is associated with operator evaluation, which scales as $c_h N^{d+1}$ in the high-order case and $c_l \hat{N}^d$ in the low-order case. The work ratio is thus

$$\frac{\text{work(high-order method)}}{\text{work(low-order method)}} = \frac{c_h}{c_l} N \sigma^d.$$

Clearly, any high-order advantage for $d = 1$ is further increased as d is increased from 1 to 3. We will see in Section 4.4 that $c_h/c_l \approx 1$; we have already seen in Table 3.5.1 that σ can be significantly less than unity.

It is important to note that, if we view the entries of $\underline{u}$ as a matrix U with $\{U\}_{ij} := u_{ij}$, then the forms (4.2.18)–(4.2.19) can be recast as *matrix–matrix* products

$$(A \otimes B)\underline{u} = BUA^T. \quad (4.2.20)$$

A similar form exists for the 3D case. Because of their low memory–bandwidth requirements and high degree of parallelism, optimized matrix–matrix product routines are available that are extremely fast on modern cache-based and pipelined architectures. This feature is particularly significant for 3D applications, where typically more than 90% of the operation count is expended on evaluating forms similar to (4.2.20). We consider these performance and implementation issues in detail in Chapter 8.

4.3 Elliptic Problems

In this section, we consider the extension of the one-dimensional elliptic problem (2.1.9) to d dimensions. The most general form of the problem is given by

$$-\sum_{i=1}^{d}\sum_{j=1}^{d} \frac{\partial}{\partial x_i}\left(p_{ij}(\mathbf{x})\frac{\partial u}{\partial x_j}\right) + q(\mathbf{x})u(\mathbf{x}) = f(\mathbf{x}) \quad \text{in } \Omega, \quad (4.3.1)$$

subject to Neumann, Dirichlet, or mixed (Robin) boundary conditions on the boundary $\partial\Omega$. In analogy to the one-dimensional case, this problem will be elliptic if the $d \times d$ coefficient matrix $p_{ij}(\mathbf{x})$ is symmetric positive definite (SPD) and $q(\mathbf{x}) \geq 0$ for all $\mathbf{x} := (x_1, \ldots, x_d) \in \Omega$. Presently, we will consider a more restricted class of problems where the coefficient matrix is simply block-diagonal with $p_{ij} = \delta_{ij} p(\mathbf{x})$, which corresponds to an isotropic medium. In this case, the elliptic problem can be expressed in compact vector notation as

$$-\nabla \cdot [p(\mathbf{x})\nabla u(\mathbf{x})] + q(\mathbf{x})u(\mathbf{x}) = f(\mathbf{x}) \quad \text{in } \Omega. \quad (4.3.2)$$

The more general formulation (4.3.1) will resurface in the study of (4.3.2) in transformed coordinates. [Between these two cases, one has also the case of an anisotropic medium, with $p_{ij} = \delta_{ij} p_i(\mathbf{x})$.] Unless otherwise specified, we will assume that homogeneous boundary conditions of the Neumann or Dirichlet type are imposed:

$$u = 0 \quad \text{on } \partial\Omega_D, \qquad \nabla u \cdot \mathbf{n} = 0 \quad \text{on } \partial\Omega_N,$$

where $\mathbf{n}$ is the outward-pointing normal on the boundary $\partial\Omega := \partial\Omega_D \cup \partial\Omega_N$.

4.3.1 Weak Formulation and Sobolev Spaces

The weak formulation of (4.3.2) is given by the following: Find $u \in V_0$ such that

$$\mathcal{A}(u, v) = \mathcal{F}(v) \quad \forall v \in V_0, \quad (4.3.3)$$

where the space of admissible functions is given by

$$V_0 := \{v \in H^1(\Omega); v = 0 \text{ on } \partial\Omega_D\}.$$

Here, the definitions of $\mathcal{A}$ and $\mathcal{F}$ introduced in (2.1.11)–(2.1.13) must be modified to reflect the fact that Ω is no longer one-dimensional and that we need to differentiate with respect to more than one spatial variable. Let

$$(u, v) := \int_\Omega u(\mathbf{x})v(\mathbf{x})\,d\mathbf{x}, \quad (4.3.4)$$

$$\mathcal{A}(u, v) := \int_\Omega [p(\mathbf{x})\nabla u \cdot \nabla v + q(\mathbf{x})uv]\,d\mathbf{x}, \quad (4.3.5)$$

$$\mathcal{F}(v) := \int_\Omega f(\mathbf{x})v(\mathbf{x})\,d\mathbf{x}. \quad (4.3.6)$$

Similarly, $H^1(\Omega)$ is the d-dimensional extension of the Sobolev space $H^1(a, b)$ defined by (2.1.33).

Discretization of (4.3.3) proceeds much as in the one-dimensional case: suitable finite-dimensional test and trial spaces are chosen along with appropriate bases, and the bilinear forms (4.3.4)–(4.3.6) are evaluated for each test- and trial-basis pair. We will generally consider two finite-dimensional approximation spaces, $V_N \subset H^1(\Omega)$ and $V_{N,0} \subset V_N$, with the latter simply being the subset that satisfies the homogeneous boundary conditions on $\partial\Omega_D$. In the d-dimensional case there is a broad range of options in selecting V_N. For example, the linear basis span $\{1, x\}$ can be extended to two dimensions by considering the strictly linear basis span $\{1, x, y\}$ or the tensor product of the linear basis given by span $\{1, x, y, xy\}$. We will primarily consider the tensor-product case, because of the computational efficiencies alluded to in the preceding section. In addition, as is common in practice, we will restrict ourselves to the case where the polynomial degree is the same in each spatial dimension. Thus, if $\{\psi_n(x_i)\}_{n=0}^N$ is the chosen one-dimensional basis for $\mathbb{P}_N(x)$, we will assume that $V_N = \text{span } \otimes_{i=1}^d \{\psi_n(x_i)\}_{n=0}^N$.

With this choice of approximation space, the discrete counterpart to (4.3.3) is as follows: Find $u \in V_{N,0}$ such that

$$\mathcal{A}(u, v) := \mathcal{F}(v) \quad \forall v \in V_{N,0}. \quad (4.3.7)$$

4.3. Elliptic Problems

Before dealing with specific examples of multidimensional elliptic problems, we briefly revisit the definition of Sobolev spaces given in Section 2.1.3, in order to stress the changes brought by higher space dimensions.

Let $\alpha := (\alpha_1, \ldots, \alpha_d)$ denote a multiindex of nonnegative integers with $|\alpha| := \alpha_1 + \cdots + \alpha_d$, $w(\mathbf{x})$ denote a weight function, and

$$D^\alpha u := \frac{\partial^{|\alpha|} u}{\partial x_1^{\alpha_1} \ldots \partial x_d^{\alpha_d}}$$

represent a distributional derivative of $u(\mathbf{x}) \in L^2_w(\Omega)$. By definition, the d-dimensional extension of $H^k_w(a, b)$ given by (2.1.35) is

$$H^k_w(\Omega) := \left\{ u \in L^2_w(\Omega); \text{ for } |\alpha| \leq k, D^\alpha u \in L^2_w(\Omega) \right\}.$$

$H^k_w(\Omega)$ is a Hilbert space with respect to the Sobolev inner product

$$(u, v)_k := \sum_{|\alpha| \leq k} \int_\Omega w(\mathbf{x}) D^\alpha u(\mathbf{x}) D^\alpha v(\mathbf{x}) \, d\mathbf{x} \tag{4.3.8}$$

and induced norm

$$\|u\|_{H^k_w(\Omega)} := \left(\sum_{\alpha \leq k} \|D^\alpha u\|^2_{L^2_w(\Omega)} \right)^{1/2}. \tag{4.3.9}$$

The closure in $H^k_w(\Omega)$ of functions belonging to $C^k(\Omega)$ with compact support in Ω is the Sobolev space

$$H^k_{0,w}(\Omega) := \left\{ u \in H^k_w(\Omega); \text{ for } 0 \leq \ell \leq k-1, \frac{\partial^\ell u}{\partial n^\ell} = 0 \text{ on } \partial\Omega \right\},$$

where the normal derivatives $\partial^\ell u / \partial n^\ell$ must be interpreted in the distributional sense.

$H^k_w(\Omega)$ and $H^k_{0,w}(\Omega)$ are the relevant Hilbert spaces for multidimensional problems into which finite-dimensional subspaces provide trial and test functions for FE approximations. Embedding properties such as those introduced in Section 2.1.3 also exist for Sobolev spaces in multidimensional problems. In particular, one can show that for $\Omega \subset \mathbb{R}^d$ ($d = 2, 3$), one has $H^k(\Omega) \subset C^{k-2}(\bar{\Omega})$ with $k \geq 2$ [14, 64]. Whereas in one dimension all elements of H^1 are continuous functions, this is not the case for $d > 1$.

All differential problems analyzed in this chapter involve a single (unknown) field to be approximated in one of the Sobolev spaces defined so far. However, in forthcoming chapters, the differential problems to be tackled will involve *vector* fields for which still other spaces of functions should be introduced.

Let again $\Omega \in \mathbb{R}^d$ denote a bounded open domain in d-dimensional space and consider the vector $\mathbf{v}$ with components v_ℓ ($\ell = 1, \ldots, d$) depending on all space variables. With the same spirit as before, we introduce the Sobolev space $H_w^k(\Omega)^d$ of vector functions,

$$H_w^k(\Omega)^d := \{\mathbf{v};\ v_\ell \in H_w^k(\Omega),\ \ell = 1, \ldots, d\}. \qquad (4.3.10)$$

Quite naturally, in order to belong to $H_w^k(\Omega)^d$, a vector $\mathbf{v}$ should have all its components inside the Sobolev space $H_w^k(\Omega)$. It is then easy to show that $H_w^k(\Omega)^d$ is a Hilbert space with respect to the inner product

$$(\mathbf{u}, \mathbf{v})_k := \sum_{\ell=1}^d \sum_{|\alpha| \le k} \int_\Omega w(\mathbf{x}) D^\alpha u_\ell(\mathbf{x}) D^\alpha v_\ell(\mathbf{x})\, d\mathbf{x} \qquad (4.3.11)$$

and induced norm

$$\|\mathbf{u}\|_{H_w^k(\Omega)^d} := \left(\sum_{\ell=1}^d \sum_{|\alpha| \le k} \|D^\alpha u_\ell\|_{L_w^2(\Omega)}^2 \right)^{1/2}, \qquad (4.3.12)$$

which are straightforward extensions to the vector case of the same quantities (4.3.8) and (4.3.9) for scalar functions.

The definition of the Sobolev space $H_{0,w}^k(\Omega)^d$ that generalizes to vectors the Sobolev space of functions with compact support for scalars follows exactly the same pattern, and we have quite simply

$$H_{0,w}^k(\Omega)^d := \{\mathbf{v};\ v_\ell \in H_{0,w}^k(\Omega),\ \ell = 1, \ldots, d\}. \qquad (4.3.13)$$

Having specified all the spaces of functions needed in this book, we proceed with the treatment of some examples.

4.3.2 A Constant-Coefficient Case

To illustrate the use of the tensor-product bases, we consider an example on the rectangular geometry $\Omega := [0, L_1] \times [0, L_2]$ in $\mathbb{R}^2$. We map each physical coordinate $\mathbf{x} := (x, y) := (x_1, x_2)$ onto respective reference coordinates $\mathbf{r} := (r, s) = (r_1, r_2) \in [-1, 1]^2$ via the affine transformation $x_i = (L_i/2)(r_i + 1)$. A typical function $u(\mathbf{x}) \in V_N$ has the representation

$$u(\mathbf{x}) = \sum_{i=0}^N \sum_{j=0}^N u_{ij} \psi_i(r) \psi_j(s), \qquad (4.3.14)$$

4.3. Elliptic Problems

with $\mathbf{x}(\mathbf{r})$ implied by the coordinate transformation. Note that differentiation with respect to, say, x, is expressed in the reference coordinates as

$$\frac{\partial u}{\partial x} = \sum_{i=0}^{N}\sum_{j=0}^{N} u_{ij} \frac{2}{L_1} \psi_i'(r)\psi_j(s), \qquad (4.3.15)$$

where the prime denotes differentiation with respect to the argument.

The 2D mass matrix is derived by simply evaluating (u, v) for all $u, v \in V_N$:

$$(u, v) = \int_\Omega vu \, dV = \sum_{\hat{i}\hat{j}}\sum_{ij} v_{\hat{i}\hat{j}} \left(\int_\Omega \psi_{\hat{i}}\psi_{\hat{j}}\psi_i\psi_j \, d\mathbf{x}\right) u_{ij} \qquad (4.3.16)$$
$$= \underline{v}^T M \underline{u},$$

where $\underline{v} = v_{\hat{i}\hat{j}}$ and $\underline{u} = u_{ij}$ are the vectors of basis coefficients. Associating a natural ordering of the basis coefficients, $\hat{k} = 1 + \hat{i} + (N+1)\hat{j}$, $k = 1 + i + (N+1)j$, we give the entries of the mass matrix by

$$M_{\hat{k}k} = \int_{-1}^{1}\int_{-1}^{1} \psi_{\hat{i}}(r)\psi_i(r)\psi_{\hat{j}}(s)\psi_j(s) \frac{L_1}{2} dr \frac{L_2}{2} ds$$
$$= \frac{L_1 L_2}{4} \left(\int_{-1}^{1} \psi_{\hat{i}}(r)\psi_i(r)\,dr\right)\left(\int_{-1}^{1} \psi_{\hat{j}}(s)\psi_j(s)\,ds\right)$$
$$= \frac{L_1 L_2}{4} \hat{M}_{\hat{i}i}\hat{M}_{\hat{j}j},$$

or, equivalently,

$$M = \frac{L_1 L_2}{4} \hat{M} \otimes \hat{M}. \qquad (4.3.17)$$

Here, $\hat{M}$ is the mass matrix associated with the one-dimensional reference domain $\hat{\Omega} = [-1, 1]$. From the above considerations, it is quite apparent that the tensor-product-based mass matrix in $\mathbb{R}^3$ will be

$$M = \frac{L_1 L_2 L_3}{8} \hat{M} \otimes \hat{M} \otimes \hat{M}$$

for the regular parallelepiped $\Omega = [0, L_1] \times [0, L_2] \times [0, L_3]$.

Notice that if $\hat{M}$ is full, then M is full, comprising $(N+1)^{2d}$ nonzero entries in the d-dimensional case. Nonetheless, because of the tensor-product form, it is possible to compute the action of M (or M^{-1}) on a vector in only $O(N^{d+1})$ operations, as shown in the preceding section. If the one-dimensional mass matrix is diagonal, however, its d-dimensional counterpart will also be diagonal, and application of M (or M^{-1}) to a vector requires only $O(N^d)$ operations.

The availability of a diagonal mass matrix is a particularly useful feature in time-stepping schemes that require frequent application of M^{-1} or in solution of more complex problems, such as the unsteady Stokes problem, in which the mass matrix plays an integral role. Clearly, $\hat{M}$ will be diagonal if the basis functions are orthogonal with respect to the inner product (4.3.4). One possibility is to choose $\{\psi_i\}_{i=0}^N$ to be a set of orthogonal functions such as Legendre polynomials. However, these will not automatically satisfy the essential boundary conditions associated with V, and an additional constraint would be required in order to discretize the BVP (4.3.3). As an alternative to the fully orthogonal basis, one can use localized Lagrangian interpolants coupled with *mass lumping*, in which the mass matrix is replaced by a diagonal matrix with an identical row sum. This is achieved by the SEM in a more formal setting.

The SEM is defined not only by its choice of Lagrangian basis functions, $\psi_i := \pi_i$ (2.4.3), but also by the associated quadrature rule or inner-product definition. In this case, the inner product $(\cdot,\cdot)$ of Equation (4.3.4) is approximated by the discrete inner product $(\cdot,\cdot)_N$ given by Gauss–Lobatto quadrature (see Appendix B) in each spatial direction. For a single coordinate direction and with $\{\xi_0, \xi_1, \ldots, \xi_N\}$ and $\{\rho_0, \rho_1, \ldots, \rho_N\}$ denoting respectively the quadrature nodes and weights, one has

$$\int_{-1}^{1} g(r)\,dr \approx \sum_{k=0}^{N} \rho_k g(\xi_k). \tag{4.3.18}$$

Each inner product in the SEM is computed by first evaluating the integrand, then substituting quadrature (4.3.18) for integration. Thus, entries in $\hat{M}$ become

$$\hat{M}_{ij} := \sum_{k=0}^{N} \rho_k \pi_i(\xi_k)\pi_j(\xi_k).$$

However, because the basis is Lagrangian [i.e., $\pi_i(\xi_j) = \delta_{ij}$, where δ_{ij} is the Kronecker delta], it is clear that $\hat{M} = \mathrm{diag}(\rho_i)$.

The stiffness matrix K is derived in a similar manner. We begin with the case where $p(\mathbf{x})$ and $q(\mathbf{x})$ are constant, leading to a particularly simple form that is amenable to both fast evaluation and fast inversion.

In $\mathbb{R}^2$, the energy inner product is

$$\mathcal{A}(u,v) = \int_{\Omega} \left(p\frac{\partial v}{\partial x}\frac{\partial u}{\partial x} + p\frac{\partial v}{\partial y}\frac{\partial u}{\partial y} + quv \right) d\mathbf{x}.$$

4.3. Elliptic Problems

Using the expansions (4.3.14)–(4.3.15) for u and v, we calculate the first term on the right:

$$\int_\Omega p \frac{\partial v}{\partial x} \frac{\partial u}{\partial x} d\mathbf{x} = \sum_{\hat{i}\hat{j}} \sum_{ij} v_{\hat{i}j} p \frac{L_2}{L_1} \left(\int_{-1}^1 \psi_{\hat{j}} \psi_j \, ds \right) \left(\int_{-1}^1 \psi'_{\hat{i}} \psi'_i \, dr \right) u_{ij}$$

$$= \sum_{\hat{i}\hat{j}} \sum_{ij} v_{\hat{i}\hat{j}} p \frac{L_2}{L_1} \hat{M}_{\hat{j}j} \hat{K}_{\hat{i}i} u_{ij},$$

where $\hat{K}$ is the one-dimensional stiffness matrix on $[-1, 1]$. Using tensor notation as in (4.3.17) gives

$$\int_\Omega p \frac{\partial v}{\partial x} \frac{\partial u}{\partial x} d\mathbf{x} = p \frac{L_2}{L_1} \underline{v}^T (\hat{M} \otimes \hat{K}) \underline{u}. \tag{4.3.19}$$

Numerical quadrature can be applied also for the evaluation of $\hat{K}$:

$$\hat{K}_{ij} = \sum_{k=0}^N \rho_k \psi'_i(\xi_k) \psi'_j(\xi_k). \tag{4.3.20}$$

Contrary to the mass matrix, in this case the result is exact, because the $(N+1)$-point Gauss–Lobatto quadrature rule is exact for all polynomials of degree $2N - 1$ or less, and $\psi'_i \psi'_j$ is of degree $2N - 2$.

Combining (4.3.19) with a similar expression for the y-derivatives and with the mass matrix derived above yields

$$\mathcal{A}(u, v) = \underline{v}^T \left(p \frac{L_2}{L_1} (\hat{M} \otimes \hat{K}) + p \frac{L_1}{L_2} (\hat{K} \otimes \hat{M}) + q \frac{L_1 L_2}{4} (\hat{M} \otimes \hat{M}) \right) \underline{u} \tag{4.3.21}$$

and

$$\mathcal{F}(v) = \frac{L_1 L_2}{4} \underline{v}^T (\hat{M} \otimes \hat{M}) \underline{f}. \tag{4.3.22}$$

The second expression results from the insertion of the interpolant of $f(\mathbf{x})$ into (4.3.6).

Equations (4.3.17)–(4.3.22) describe the essential mechanics for evaluating the bilinear forms (4.3.4)–(4.3.6) for any element pair $u, v \in V_N$. The discrete Helmholtz operator

$$\bar{H} := p \left[\frac{L_2}{L_1} (\hat{M} \otimes \hat{K}) + \frac{L_1}{L_2} (\hat{K} \otimes \hat{M}) \right] + q \frac{L_1 L_2}{4} (\hat{M} \otimes \hat{M}) \tag{4.3.23}$$

is referred to as the *Neumann operator*, because it is the system governing the homogeneous Neumann problem. It is symmetric positive definite unless $q = 0$, in which case it has a one-dimensional nullspace corresponding to the constant mode.

If u and v are to satisfy the essential boundary condition, that is, $u(\mathbf{x}) = v(\mathbf{x}) = 0$ on $\partial\Omega_D$, then the additional boundary constraints must be applied before the weighted residual problem (4.3.3) can be properly formulated. For Lagrangian bases, this simply entails formulating the problem in terms of restricted index ranges on $\hat{\imath}$, $\hat{\jmath}$, i, and j when formulating the problem in terms of v_{ij} and u_{ij}. The final solution is then obtained by extending the basis coefficient set by zero for those coefficients corresponding to homogeneous Dirichlet boundary values. Formally, we can write the restricted coefficient set as $\underline{\tilde{u}} = R\underline{u}$ and the extended, or prolongated, solution as $\underline{u} = R^T \underline{\tilde{u}}$. Here R is an $\mathcal{N} \times (N+1)^d$ restriction operator comprising $\mathcal{N} \leq (N+1)^d$ columns of the $\mathcal{N} \times \mathcal{N}$ identity matrix with columns of zeros interspersed at locations corresponding to nodal points on $\partial\Omega_D$. Generally, R will also have a tensor-product form, $R = R_1 \otimes \cdots \otimes R_d$. Note that R is never explicitly formed, since only the *action* of R is required. This is readily implemented through the use of an index set pointing to the Dirichlet nodes on $\partial\Omega_D$.

Applying (4.3.3) leads then to the linear system for $\underline{u}$:

$$H\underline{\tilde{u}} = RM\underline{f},$$
$$\underline{u} = R^T\underline{\tilde{u}}, \qquad (4.3.24)$$

with

$$H := R\bar{H}R^T \qquad (4.3.25)$$

and M the mass matrix defined by (4.3.17). The careful implementation of the boundary conditions outlined above implies that the restricted Helmholtz operator H is symmetric positive definite, and (4.3.24) can consequently be solved using PCG (see Section 2.8).

If u is to satisfy inhomogeneous Dirichlet boundary conditions, the solution can be split into $u := u_h + u_b$, where u_h satisfies homogeneous boundary conditions on $\partial\Omega$ and u_b is any function in V_N that satisfies $u_b = u$ on $\partial\Omega$. In this case, a symmetric system analogous to (4.3.24) results:

$$H\underline{\tilde{u}}_h = RM\underline{f} - R\bar{H}\underline{u}_b,$$
$$\underline{u} = R^T\underline{\tilde{u}}_h + \underline{u}_b. \qquad (4.3.26)$$

4.3. Elliptic Problems

Since H has a form similar to (4.2.12), its forward and inverse action can be evaluated in $O(N^3)$ operations. The inversion can be effected using the fast diagonalization method (4.2.15) once the generalized eigenvalue problem associated with each one-dimensional operator pair, $(R_i \hat{K} R_i^T, R_i \hat{M} R_i^T)$, $i = 1, \ldots, d$, has been solved in a preprocessing step. We emphasize, however, that the fast diagonalization method breaks down under all but the simplest of geometry deformations or if either $p(\mathbf{x})$ or $q(\mathbf{x})$ cannot be cast in tensor-product form [see (4.4.8)]. In these more complex cases, it is nonetheless possible to exploit the fast diagonalization method as a preconditioner, as demonstrated in [83, 129].

4.3.3 The Variable-Coefficient Case

To develop the system matrices for the case of variable $p(\mathbf{x})$, we require evaluation of integrals of the form

$$\mathcal{A}_x(u, v) := \int_{-1}^{1} \int_{-1}^{1} p(r, s) \frac{\partial v}{\partial r} \frac{\partial u}{\partial r} \, dr \, ds, \qquad (4.3.27)$$

which constitutes a single term in the energy inner product $\mathcal{A}(u, v)$. To generate the discrete operators, we insert the expansions (4.3.14)–(4.3.15) for u, v, and p into (4.3.27):

$$\mathcal{A}_x(u, v) = \sum_{ij} \sum_{\hat{i}\hat{j}} v_{ij} \sum_{mn} p_{mn} \left(\int_{-1}^{1} \psi_i' \psi_{\hat{i}}' \psi_m \, dr \right) \left(\int_{-1}^{1} \psi_j \psi_{\hat{j}} \psi_n \, ds \right) u_{\hat{i}\hat{j}}. \qquad (4.3.28)$$

If left in this form, the cross term p_{mn} destroys the tensor-product form and leads to unacceptable fill in the stiffness matrix.

The SEM avoids this difficulty through the use of a high-order quadrature rule coupled with Lagrangian basis functions based on the GLL points ξ_i. Employing the basis functions $\psi_i = \pi_i(r)$ described in Chapter 2, we approximate the first integral on the right in (4.3.28) by

$$\sum_{k=0}^{N} \pi_i'(\xi_k) \pi_{\hat{i}}'(\xi_k) \pi_m(\xi_k) \rho_k = \pi_i'(\xi_m) \pi_{\hat{i}}'(\xi_m) \rho_m = \hat{D}_{mi} \hat{D}_{m\hat{i}} \rho_m, \qquad (4.3.29)$$

where $\hat{D}_{mi} := \pi_i'(\xi_m)$. The second integral in (4.3.28) is approximated by

$$\sum_{k=0}^{N} \pi_j(\xi_k) \pi_{\hat{j}}(\xi_k) \pi_n(\xi_k) \rho_k = \delta_{jn} \delta_{\hat{j}n} \rho_n. \qquad (4.3.30)$$

Let $P := \text{diag}(p_{\hat{m}})$ and $W := \text{diag}(w_{\hat{m}})$ be the diagonal matrices having entries $p_{\hat{m}} := p_{mn}$ and $w_{\hat{m}} := p_{mn}\rho_m\rho_n$, respectively, where $\hat{m} = 1 + m + (N+1)n$ corresponds to the natural ordering of the nodes. W can be expressed in terms of tensor-product forms in $\hat{M} = \text{diag}(\rho_i)$,

$$W = P(\hat{M} \otimes \hat{M}), \tag{4.3.31}$$

and the integral expression (4.3.28) recast in terms of W,

$$\mathcal{A}_x(u, v) = \underline{v}^T (I \otimes \hat{D}^T) W (I \otimes \hat{D}) \underline{u}.$$

From this we can conclude that the spectral-element stiffness matrix for (4.3.2) with variable $p(\mathbf{x})$ and $q = 0$ is of the form

$$\bar{K} = (I \otimes \hat{D}^T) W (I \otimes \hat{D}) + (\hat{D}^T \otimes I) W (\hat{D} \otimes I). \tag{4.3.32}$$

As in (4.3.25), Dirichlet boundary conditions are imposed by restricting the index set of the trial (u) and test (v) functions to arrive at an invertible system $K := R\bar{K}R^T$.

The presence of the matrix P (which is generally not in tensor form) in (4.3.31) spoils the overall tensor-product form of (4.3.32), and therefore the fast diagonalization method cannot be used to invert K. However, because the cost of applying the diagonal matrix W to a vector is only $O(N^d)$, the leading-order complexity of forward application of K is governed by the differentiation associated with the matrices $\hat{D}$ and $\hat{D}^T$ and is only $O(N^{d+1})$. Note that if W were full rather than diagonal, the cost of applying K would be $O(N^{2d})$. It is precisely the use of the diagonal mass matrix $\hat{M}$ [i.e., the approximations (4.3.29)–(4.3.30)] that leads to a favorable complexity estimate in the variable-coefficient case and that is central to the utility of high-order methods in complex geometries.

4.4 Deformed Geometries

We now consider the case where Ω may be deformed, as illustrated in Figure 4.4.1. After suitable transformations to the computational domain $\hat{\Omega}$, the preceding methodology developed for the variable-coefficient case can be readily extended to develop a compact formulation of the stiffness matrix in the case of deformed geometries. To highlight the many symmetries in the problem, we derive the result for the case $d = 3$.

4.4. Deformed Geometries

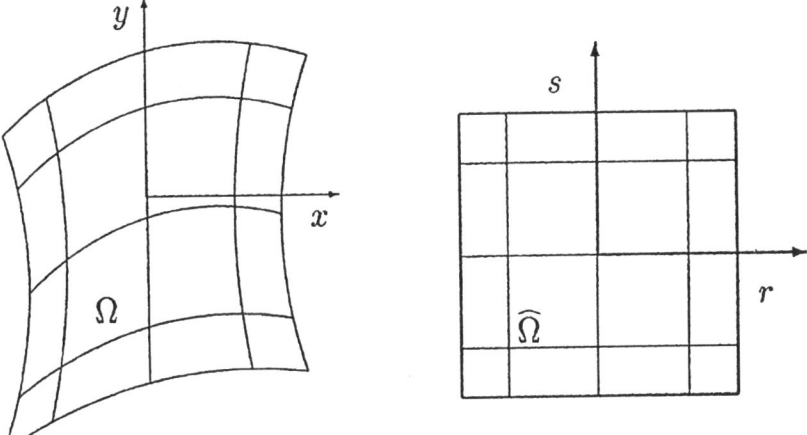

Figure 4.4.1. Sketch of coordinate transformation from physical domain Ω to computational domain $\hat{\Omega}$.

Assume that there exists an invertible map $\mathbf{x}(\mathbf{r})$ from the physical deformed domain $\Omega(\mathbf{x})$ to the reference domain $\hat{\Omega}(\mathbf{r}) := [-1, 1]^d$ for which the Jacobian

$$J(\mathbf{r}) = \det \begin{pmatrix} \dfrac{\partial x_1}{\partial r_1} & \cdots & \dfrac{\partial x_1}{\partial r_d} \\ \vdots & & \vdots \\ \dfrac{\partial x_d}{\partial r_1} & \cdots & \dfrac{\partial x_d}{\partial r_d} \end{pmatrix}$$

is nonvanishing and therefore of the same sign everywhere on $\hat{\Omega}$. (Specifically, all vertex angles should be bounded away from 0 and π.) Without loss of generality, assume also that the Jacobian is positive, implying that an element volume in the transformed coordinates is positive.

Assuming $q = 0$ for simplicity, the energy inner product (4.3.5) is

$$\mathcal{A}(u, v) = \int_\Omega p(\mathbf{x}) \nabla v \cdot \nabla u \, d\mathbf{x}, = \sum_{k=1}^d \int_\Omega p \frac{\partial u}{\partial x_k} \frac{\partial v}{\partial x_k} \, d\mathbf{x}. \qquad (4.4.1)$$

Going from Ω to $\hat{\Omega}$, partial derivatives of u are evaluated according to the chain rule:

$$\frac{\partial u}{\partial x_k} = \sum_{i=1}^d \frac{\partial u}{\partial r_i} \frac{\partial r_i}{\partial x_k}, \quad k = 1, \ldots, d.$$

Combining with a similar expression for v leads to

$$\mathcal{A}(u, v) = \sum_{k=1}^{d} \int_{\hat{\Omega}} p \left(\sum_{i=1}^{d} \frac{\partial v}{\partial r_i} \frac{\partial r_i}{\partial x_k} \right) \left(\sum_{j=1}^{d} \frac{\partial u}{\partial r_j} \frac{\partial r_j}{\partial x_k} \right) J(\mathbf{r}) \, d\mathbf{r}$$

$$= \sum_{i=1}^{d} \sum_{j=1}^{d} \int_{\hat{\Omega}} p \frac{\partial v}{\partial r_i} \left(\sum_{k=1}^{d} \frac{\partial r_i}{\partial x_k} \frac{\partial r_j}{\partial x_k} J(\mathbf{r}) \right) \frac{\partial u}{\partial r_j} \, d\mathbf{r}, \qquad (4.4.2)$$

where, in the last equation, the geometric factors associated with the metrics, $\partial r_i / \partial x_k$, and the Jacobian have been assembled within a set of functions

$$\mathcal{G}_{ij}(\mathbf{r}) := \sum_{k=1}^{d} \frac{\partial r_i}{\partial x_k} \frac{\partial r_j}{\partial x_k} J(\mathbf{r}), \qquad 1 \leq i, j \leq d. \qquad (4.4.3)$$

We see that (4.4.2) has a form similar to (4.3.27). As in that case, it is evaluated by using numerical quadrature on the tensor product of the Gauss–Lobatto grid points, $\otimes_{i=1}^{d} \Xi_{N+1}^{i}$, yielding

$$\mathcal{A}(u, v) = \sum_{i=1}^{d} \sum_{j=1}^{d} \sum_{klm} \left[\frac{\partial v}{\partial r_i} p \mathcal{G}_{ij} \frac{\partial u}{\partial r_j} \right]_{(\xi_k, \xi_l, \xi_m)} \rho_k \rho_l \rho_m. \qquad (4.4.4)$$

The coefficient p, geometric terms $\mathcal{G}_{ij}$, and quadrature weights ρ_* can all be conveniently combined into a set of d^2 diagonal matrices, G_{ij}, $i, j \in \{1, \ldots, d\}^2$. Let

$$(G_{ij})_{\hat{k}\hat{k}} := [p\mathcal{G}_{ij}]_{(\xi_k, \xi_l, \xi_m)} \rho_k \rho_l \rho_m, \qquad (4.4.5)$$

with $\hat{k} := 1 + k + (N+1)l + (N+1)^2 m$, $k, l, m \in \{0, \ldots, N\}^3$, defining a natural ordering of the quadrature points. Note that multiplication of $\underline{u}$ by each G_{ij} simply involves pointwise multiplication (collocation) of the nodal values u_{klm} with the terms on the right of (4.4.5), while the derivatives in (4.4.4) are evaluated, for example, as

$$\left. \frac{\partial u}{\partial r_1} \right|_{klm} = \sum_{p=0}^{N} \hat{D}_{kp} u_{plm}, \qquad k, l, m \in \{0, \ldots, N\}^3$$

$$= (I \otimes I \otimes \hat{D}) \underline{u}.$$

Defining

$$D_1 := (I \otimes I \otimes \hat{D}), \quad D_2 := (I \otimes \hat{D} \otimes I), \quad D_3 := (\hat{D} \otimes I \otimes I), \qquad (4.4.6)$$

4.4. Deformed Geometries

we can combine the derivative operators with the geometric factors (4.4.5) to yield a final compact form for the energy inner product (4.4.2):

$$\mathcal{A}(u,v) = \underline{v}^T \begin{pmatrix} D_1 \\ D_2 \\ D_3 \end{pmatrix}^T \begin{pmatrix} G_{11} & G_{12} & G_{13} \\ G_{21} & G_{22} & G_{23} \\ G_{31} & G_{32} & G_{33} \end{pmatrix} \begin{pmatrix} D_1 \\ D_2 \\ D_3 \end{pmatrix} \underline{u}$$

$$= \underline{v}^T D^T G D \underline{u}. \tag{4.4.7}$$

Since $G_{ij} = G_{ji}$, only six of the geometric factors need to be computed. The leading-order storage requirement for the factored stiffness matrix is thus only $6(N+1)^3$, compared with $(N+1)^6$ for the full matrix.

The work required for a matrix–vector product is similarly reduced. In applying $\bar{K} := D^T G D$ to a vector $\underline{u}$, one begins with d tensor-product-based derivative evaluations, $\hat{\underline{u}}_j := D_j \underline{u}$, followed by multiplication with the geometric factors, $\tilde{\underline{u}}_i := \sum_j G_{ij} \hat{\underline{u}}_j$, followed finally with a sum across the transposed derivative operators, $\bar{K}\underline{u} = \sum_i D_i^T \tilde{\underline{u}}_i$, for a total operation count of $12(N+1)^4 + 15(N+1)^3$. This is a significant improvement over the $2(N+1)^6$ cost incurred if the stiffness matrix is computed and stored explicitly.

If the physical domain is an $L_1 \times L_2 \times L_3$ rectangular parallelepiped of *arbitrary* orientation, the metrics and Jacobian simply become constants with $\mathcal{G}_{ij} \equiv 0$, $i \neq j$. If, in addition, p is constant or has tensor-product form, $p(\xi_i, \xi_j, \xi_k) = p_1(\xi_i) p_2(\xi_j) p_3(\xi_k)$, then $\bar{K}$ simplifies to

$$\bar{K} = W(I \otimes I \otimes \tilde{K}_1 + I \otimes \tilde{K}_2 \otimes I + \tilde{K}_3 \otimes I \otimes I), \tag{4.4.8}$$

with $W := (L_1 L_2 L_3 / 8)(P_3 \hat{M} \otimes P_2 \hat{M} \otimes P_1 \hat{M})$, $\tilde{K}_i := \frac{4}{L_i^2}(P_i \hat{M})^{-1} \hat{D}^T P_i \hat{M} \hat{D}$, and $P_i := \text{diag}(p_i(\xi_j))_{j=0}^N$. Matrix–vector product evaluation using (4.4.8) requires half as many operations as using (4.4.7).

The extension of the mass matrix to the deformed geometry case is a straightforward application of quadrature and leads (in $\mathbb{R}^3$) to the diagonal form

$$M_{\hat{\imath}\hat{\imath}} = J(\xi_i, \xi_j, \xi_k) \rho_i \rho_j \rho_k, \qquad \hat{\imath} := 1 + i + (N+1)j + (N+1)^2 k. \tag{4.4.9}$$

If $q(\mathbf{x})$ is nonzero in (4.3.1), the corresponding Helmholtz operator is created by augmenting the stiffness matrix $\bar{K}$ as

$$\bar{H} := \bar{K} + QM, \tag{4.4.10}$$

where Q is the diagonal matrix corresponding to nodal values $q(\mathbf{x}(\xi_i, \xi_j, \xi_k))$. Although the presence of the variable q and Jacobian degrades the accuracy of the quadrature somewhat, the fact that it is high-order tends to diminish the severity of this "variational crime" [369] (see, however, Figure 4.4.5). The possibility of using higher-order integration rules to overcome this difficulty was investigated by Maday and Rønquist [264], who concluded that, for the added cost, it was better to recover the accuracy by simply increasing N.

It is often convenient for preconditioning purposes to have a closed form for the diagonal of $\bar{K}$. To illustrate the basic steps for its derivation, we introduce the unit vector $\hat{\underline{e}}_{ijk}$, which is 1 at the point (ξ_i, ξ_j, ξ_k) and 0 at all other points in $\hat{\Omega}$. The desired diagonal elements of $\bar{K}$ are given by

$$d_{ijk} = \hat{\underline{e}}_{ijk}^T \bar{K} \hat{\underline{e}}_{ijk} = \sum_{l=1}^{d}\sum_{m=1}^{d}(D_l\hat{\underline{e}}_{ijk})^T G_{lm} D_m \hat{\underline{e}}_{ijk} \qquad (4.4.11)$$

for $i, j, k \in \{0, \ldots, N\}^3$. For $m = 1$, a single element of the rightmost term in (4.4.11) is

$$\hat{\underline{e}}_{pqr}^T D_1 \hat{\underline{e}}_{ijk} := \hat{\underline{e}}_{pqr}^T (I \otimes I \otimes \hat{D}) \hat{\underline{e}}_{ijk} = \hat{D}_{pi}\delta_{qj}\delta_{rk},$$

where $\hat{D}_{pi}$ is an element of the one-dimensional derivative matrix $\hat{D}$, and δ_{qj} and δ_{rk} are the usual Kronecker delta functions. From (2.4.9), the diagonal of $\hat{D}$ is zero in all but the first and last columns; that is, for the interior points $i = 1, \ldots, N-1$ the derivative of each $\pi_i(\xi)$ vanishes at ξ_i. Therefore, for $i = 1, \ldots, N-1$,

$$\hat{\underline{e}}_{pqr}^T D_1 \hat{\underline{e}}_{ijk} = (1 - \delta_{pi})\hat{D}_{pi}\delta_{qj}\delta_{rk}. \qquad (4.4.12)$$

Recalling that the (diagonal) matrices in G represent a simple pointwise scaling, we see from (4.4.12) that the off-diagonal blocks G_{lm}, $l \neq m$, will make no contribution to (4.4.11) for points in the domain interior. For instance, with $l = 2$ and $m = 1$, we have for $i = 1, \ldots, N-1$

$$(D_2\hat{\underline{e}}_{ijk})^T G_{21} D_1 \hat{\underline{e}}_{ijk} = \sum_{\substack{p,q,r \\ =(0,0,0)}}^{(N,N,N)} (\delta_{pi}\hat{D}_{qj}\delta_{rk})(G_{21})_{pqr}[(1-\delta_{pi})\hat{D}_{pi}\delta_{qj}\delta_{rk}],$$

which vanishes because $\delta_{pi}(1 - \delta_{pi}) \equiv 0$. Continuing in this fashion, one concludes that

$$d_{ijk} := \sum_{p=0}^{N}\left[\hat{D}_{pi}^2(G_{11})_{pjk} + \hat{D}_{pj}^2(G_{22})_{ipk} + \hat{D}_{pk}^2(G_{33})_{ijp}\right] \qquad (4.4.13)$$

correctly determines the diagonal of $\bar{K}$ (4.4.7) for all points in the interior of $\hat{\Omega}$, that is, for $i, j, k \in \{1, \ldots, N-1\}^3$.

4.4. Deformed Geometries

Additional terms are required for the surface values (where the diagonal of $\hat{D}$ is nonvanishing). Following a similar approach, and exploiting the symmetry of G, one finds that the correct updates will be generated by augmenting the surface values in (4.4.13) according to

$$d_{ijk} := d_{ijk} + \hat{D}_{ii}\hat{D}_{jj}(G_{12})_{ijk} + \hat{D}_{ii}\hat{D}_{kk}(G_{13})_{ijk},$$
$$i = 0, N, \quad j = 0, \ldots, N, \quad k = 0, \ldots, N, \quad (4.4.14)$$

for the $r = \pm 1$ surfaces, with analogous corrections on the $s = \pm 1$ and $t = \pm 1$ surfaces.

4.4.1 Generation of Geometric Deformation

The functions $\mathcal{G}_{ij}(\mathbf{r})$ defined by (4.4.3) collect all the information about the deformed geometry and must be analyzed case by case. Common practice regarding domain mappings relies usually on the use of isoparametric mappings in which the geometry, $\mathbf{x}$, is represented in terms of the same polynomial basis used for the solution and data:

$$\mathbf{x}(r, s, t) = \sum_{i=0}^{N}\sum_{j=0}^{N}\sum_{k=0}^{N} \mathbf{x}_{ijk}\pi_i(r)\pi_j(s)\pi_k(t). \quad (4.4.15)$$

To accurately evaluate the integrand in (4.4.1), one must find a set of values $\mathbf{x}_{ijk}$ that will yield a smooth distribution of $\mathbf{x}(\mathbf{r})$ in the interior of $\hat{\Omega}$ and that will satisfy boundary conditions $\mathbf{x}_{ijk}|_{\partial\hat{\Omega}} = \tilde{\mathbf{x}}(\xi_i, \xi_j, \xi_k)|_{\partial\hat{\Omega}}$, where $\tilde{\mathbf{x}}$ is the prescribed surface geometry.

A useful mapping from $\hat{\Omega}$ to Ω is provided by a blending technique derived by Gordon and Hall [162], which is based on the Boolean sum of interpolation operators. Let I_r, I_s, and I_t denote the one-dimensional Lagrange interpolation operators such that $I_{r_*} f(\mathbf{r})$ is linear in the r_*-direction and coincides with f at $r_* = \pm 1$. Introducing the Boolean sum $I_{r_i} \oplus I_{r_j}$ of any pair of operators I_{r_i}, I_{r_j} ($i \neq j$),

$$I_{r_i} \oplus I_{r_j} := I_{r_i} + I_{r_j} - I_{r_i} I_{r_j},$$

one can show that the relationships

$$\mathbf{x}(\mathbf{r}) = (I_r \oplus I_s)\tilde{\mathbf{x}}(\mathbf{r}) \quad \text{(in } \mathbb{R}^2\text{)}, \quad (4.4.16)$$
$$\mathbf{x}(\mathbf{r}) = (I_r \oplus I_s \oplus I_t)\tilde{\mathbf{x}}(\mathbf{r}) \quad \text{(in } \mathbb{R}^3\text{)} \quad (4.4.17)$$

provide an approximation of $\tilde{\mathbf{x}}(\mathbf{r})$ that coincides with this function along the continuous set of boundary points.

The most striking property of (4.4.16)–(4.4.17) lies in the fact that no other information is needed to construct **x** than the boundary values, which are reproduced exactly. This fulfills precisely the task set earlier: extend, inside the domains Ω and $\hat{\Omega}$, the correspondence between these domains set by their boundaries.

It is convenient to implement the Gordon–Hall procedure as a sequence of approximations, each incorporating boundary information of successively higher dimension, $0, \ldots, d-1$. Using the identity $I_{r_*} I_{r_*} = I_{r_*}$, we can express this sequence in the 2D case as

$$(I_r \oplus I_s)\tilde{\mathbf{x}}(r,s) = I_r I_s \tilde{\mathbf{x}} + I_r(\tilde{\mathbf{x}} - I_r I_s \tilde{\mathbf{x}}) + I_s(\tilde{\mathbf{x}} - I_r I_s \tilde{\mathbf{x}}).$$

The first term on the right is the bilinear interpolant of the vertex values, $\tilde{\mathbf{x}}(\pm 1, \pm 1)$. Noting that $\tilde{\mathbf{x}} - I_r I_s \tilde{\mathbf{x}}$ vanishes at the vertices, we see that the second term incorporates the s-edge variation as a linear interpolant in r, and the third incorporates the r-edge variation as a linear interpolant in s.

A 3D application of the Gordon–Hall procedure is illustrated in Figure 4.4.2. We assume that $\tilde{\mathbf{x}}(\mathbf{r})$ is expressed in the usual form (4.4.15), with interior

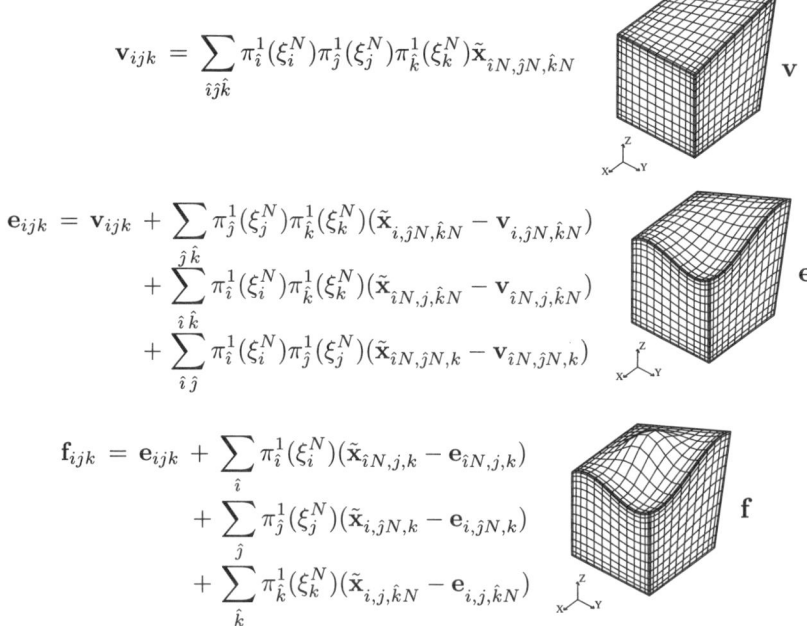

Figure 4.4.2. Application of the Gordon–Hall algorithm in $\mathbb{R}^3$. Subscripts i, j, and k range from 0 to N. Summations involving $\hat{\imath}$, $\hat{\jmath}$, and $\hat{k}$ range from 0 to 1, accounting for contributions from opposing faces in each of the three coordinate directions.

4.4. Deformed Geometries

coefficients to be determined. The first transformation (**v**) defines a mapping on $\hat{\Omega}$ that incorporates the vertex information only–the function $\mathbf{v} = I_r I_s I_t \tilde{\mathbf{x}}$ coincides with the prescribed data at the element vertices and uses trilinear interpolation elsewhere. The second transformation (**e**) adds to the vertex function any edge deformation induced by the perturbation $(\tilde{\mathbf{x}} - \mathbf{v})$. The final transformation (**f**) extends the edge–vertex map by the face perturbation $(\tilde{\mathbf{x}} - \mathbf{e})$. One can stop at any point in the sequence and set **x** equal to **v**, **e**, or **f**, with each approximation incorporating more information than the preceding one. For example, if one has only wire-frame information about the mesh, then **e** provides an interpolant onto the faces and into the interior of Ω. Or (as is frequently the case), if data are specified on only one face, one can employ all the available vertex and edge data and limit the sums in the **f**-computation to range only over the specified face(s).

Other interpolation schemes are possible. For example, one may solve Laplace's equation on $\hat{\Omega}$ for each component x_i, $i = 1, \ldots, d$:

$$\frac{\partial^2 x_i}{\partial r^2} + \frac{\partial^2 x_i}{\partial s^2} = 0 \quad \text{in } [-1, 1]^2, \qquad x_i|_{\partial \hat{\Omega}} = \tilde{x}_i|_{\partial \hat{\Omega}}, \qquad (4.4.18)$$

which can easily be done by using (4.3.26) in conjunction with (4.2.14).

Whichever interpolation procedure is employed, it is necessary first to determine a distribution of nodal points on each of the $2d$ faces constituting $\partial \Omega$. Common practice in two dimensions is to use an arclength-preserving distribution or a projection onto the chord that joins the vertices of the given edge. In $\mathbb{R}^3$ the edge distribution alone may be sufficient to bootstrap the mesh generation in the absence of face data. For the Gordon–Hall method, the approximation **e** provides a reasonable interpolation of edge data onto the faces and into the interior. For the Laplacian method, one can simply use the 2D algorithm (4.4.18) to determine the surface node distribution on each of the six faces, followed by an extension of the data into the volume by solving the 3D Laplace equation.

Once **x** is obtained, the values of the metrics, $\partial x_i / \partial r_j$, are readily computed with the differentiation matrices D_j (4.4.6). The Jacobian and the inverse metrics are computed pointwise, with the latter satisfying (in $\mathbb{R}^2$)

$$\begin{pmatrix} \frac{\partial x_1}{\partial r_1} & \frac{\partial x_1}{\partial r_2} \\ \frac{\partial x_2}{\partial r_1} & \frac{\partial x_2}{\partial r_2} \end{pmatrix}_{\xi_p, \xi_q} \begin{pmatrix} \frac{\partial r_1}{\partial x_1} & \frac{\partial r_1}{\partial x_2} \\ \frac{\partial r_2}{\partial x_1} & \frac{\partial r_2}{\partial x_2} \end{pmatrix}_{\xi_p, \xi_q} \equiv \begin{pmatrix} 1 & 0 \\ 0 & 1 \end{pmatrix}$$

for each $p, q \in \{0, \ldots, N\}^2$. A similar identity holds in three dimensions.

4.4.2 Surface Integrals and Robin Boundary Conditions

Robin boundary conditions commonly arise in heat transfer problems where the energy flux on the domain boundary is proportional to the temperature difference between the surface and a given external value [see (1.6.17)]. In the multidimensional case, the pointwise boundary terms applied in the one-dimensional example (2.1.23) must be extended to boundary integrals. Let u satisfy

$$-\nabla \cdot [p(\mathbf{x})\nabla u(\mathbf{x})] = f(\mathbf{x}), \qquad \mathbf{x} \in \Omega, \qquad (4.4.19)$$

$$\nabla u \cdot \hat{\mathbf{n}} = 0 \qquad \text{on } \partial\Omega_N, \qquad (4.4.20)$$

$$\nabla u \cdot \hat{\mathbf{n}} + \alpha u = \beta \qquad \text{on } \partial\Omega_R. \qquad (4.4.21)$$

Following the standard approach, we multiply (4.4.19) by a test function v, and integrate over Ω to obtain

$$\int_\Omega p\nabla v \cdot \nabla u \, dV - \int_{\partial\Omega} vp\nabla u \cdot \hat{\mathbf{n}} \, dS = \int_\Omega vf \, dV. \qquad (4.4.22)$$

Using the boundary conditions (4.4.20)–(4.4.21) and making the substitution $\nabla u \cdot \hat{\mathbf{n}} = \beta - \alpha u$ on $\partial\Omega_R$, we obtain the weak formulation of this problem: Find $u \in V$ such that

$$\int_\Omega p\nabla v \cdot \nabla u \, dV + \int_{\partial\Omega_R} vp\alpha u \, dS = \int_\Omega vf \, dV + \int_{\partial\Omega_R} vp\beta \, dS \qquad \forall v \in V. \qquad (4.4.23)$$

Spectral-element discretization of (4.4.23) proceeds as before, with Lagrangian bases representing u, v, and f. The surface integrals in (4.4.23) are treated via numerical quadrature. Since only a small number of the test functions v are nonzero on $\partial\Omega_R$, the resultant system will be largely unchanged from the homogeneous Neumann system (4.4.10).

The procedure is illustrated for the domain shown in Figure 4.4.3. Here, $\partial\Omega_R$ is the image in physical coordinates of $\partial\hat{\Omega}_R := (r, s, 1)$. For the $(N+1)^2$ equations associated with nodes $\mathbf{x}_{ij}^R$ (where the superscript denotes the restriction to $\partial\Omega_R$), one adds to the corresponding diagonal entries of $\bar{K}$ the coefficients $\tilde{J}_{ij}^R \rho_i \rho_j \alpha p_{ij}^R$ and to the r.h.s. the terms $\tilde{J}_{ij}^R \rho_i \rho_j \beta p_{ij}^R$, where $p_{ij}^R = p(\mathbf{x}_{ij}^R)$ and $\tilde{J}_{ij}^R$ is the Jacobian associated with the surface integral over $\partial\Omega_R$.

The surface Jacobian is determined by noting that an infinitesimal displacement dr on $\partial\hat{\Omega}_R$ gives rise to a corresponding displacement

$$\epsilon_r := \frac{\partial \mathbf{x}}{\partial r} dr = \left(\frac{\partial x}{\partial r}, \frac{\partial y}{\partial r}, \frac{\partial z}{\partial r}\right)^T dr$$

4.4. Deformed Geometries

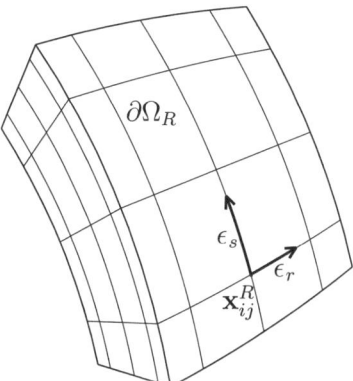

Figure 4.4.3. Description of surface geometry in $\mathbb{R}^3$.

on $\partial \Omega_R$, while an orthogonal displacement ds yields $\epsilon_s = \frac{\partial \mathbf{x}}{\partial s} ds$. It is clear from Figure 4.4.3 that an infinitesimal area on the physical surface is given by

$$dA = \|\epsilon_r \times \epsilon_s\| = \left\| \frac{\partial \mathbf{x}}{\partial r} \times \frac{\partial \mathbf{x}}{\partial s} \right\| dr\, ds,$$

where $\|\cdot\|$ denotes the standard Euclidean norm. Thus, the surface Jacobian at a point $\mathbf{x}_{ij}^R$ is

$$\tilde{J}_{ij}^R := \left\| \left.\frac{\partial \mathbf{x}}{\partial r}\right|_{ij}^R \times \left.\frac{\partial \mathbf{x}}{\partial s}\right|_{ij}^R \right\|.$$

The associated unit normal in physical space is

$$\hat{\mathbf{n}}_{ij} := \frac{1}{\tilde{J}_{ij}^R} \left(\left.\frac{\partial \mathbf{x}}{\partial r}\right|_{ij}^R \times \left.\frac{\partial \mathbf{x}}{\partial s}\right|_{ij}^R \right).$$

Deformed-Geometry Example: Hamel Flow

To illustrate the effects of mesh distortion, we consider an exact solution to the steady Stokes equations presented in Chapter 5. For a 2D divergent channel, illustrated in Figure 4.4.4, Hamel [183] gives the exact solution

$$u = \frac{Qx}{x^2+y^2}\left(\frac{x^2-y^2}{x^2+y^2} - \cos(2\alpha)\right) \times \frac{1}{\sin(2\alpha) - 2\alpha\cos(2\alpha)},$$

$$v = \frac{Qy}{x^2+y^2}\left(\frac{x^2-y^2}{x^2+y^2} - \cos(2\alpha)\right) \times \frac{1}{\sin(2\alpha) - 2\alpha\cos(2\alpha)},$$

$$p = p_0 - \frac{2\mu Q}{x^2+y^2}\frac{x^2-y^2}{x^2+y^2}\frac{1}{\sin(2\alpha) - 2\alpha\cos(2\alpha)},$$

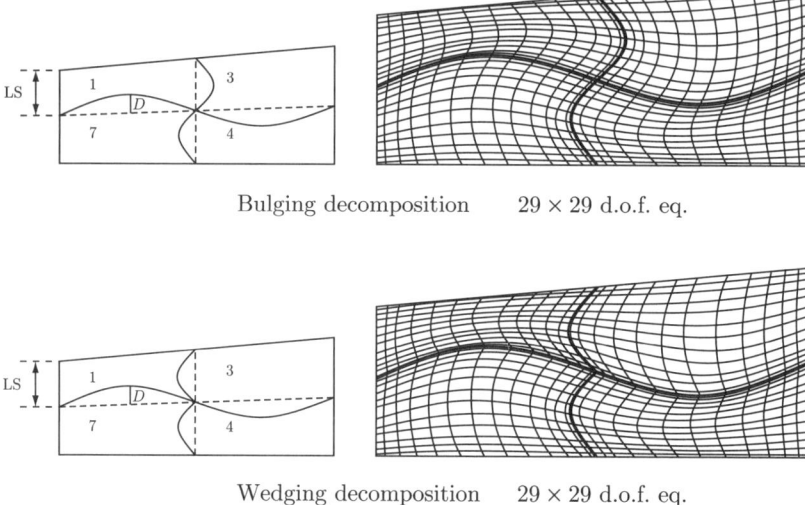

Figure 4.4.4. Computational domains for Hamel flow.

where (u, v) is the velocity vector, p is the pressure, Q is the flow rate, 2α is the total included angle of the aperture, and p_0 is a constant reference pressure.

Schneidesch and Deville [345] consider the computational domains shown in Figure 4.4.4. The bottom of the domain coincides with the x-axis, where a symmetry condition ($\frac{\partial u}{\partial y} = 0, v = 0$) is applied. The exact solution is applied on the left (inlet) and right (outlet) boundaries. The velocity is zero along the top wall, which is inclined at an angle of $\alpha = 5°$ with respect to the horizontal. The domains are partitioned into four subdomains by a sinusoidal interface in either *bulging* or *wedging* configurations, as illustrated in Figure 4.4.4. The parameter D specifies the ratio of the amplitude of the sinusoid to the subdomain height at the inlet. The maximum pointwise velocity error in each of the subdomains is shown in Figure 4.4.5 as D is varied from 0.0 to 0.4. The most striking feature of these results is the four-orders-of-magnitude reduction in accuracy as D is increased from 0 to 0.05, which is attributed to the error incurred in evaluating expressions of the form (4.4.2) with quadrature. Despite the significant difference in errors, all the cases exhibit exponential convergence as the polynomial degree is increased from $N = 6$ (13 × 13 case) to $N = 14$ (29 × 29 case).

4.5 Spectral-Element Discretizations

The methodology presented in the preceding section suffices for moderate geometric deformation. However, more complex geometries or widely disparate

4.5. Spectral-Element Discretizations

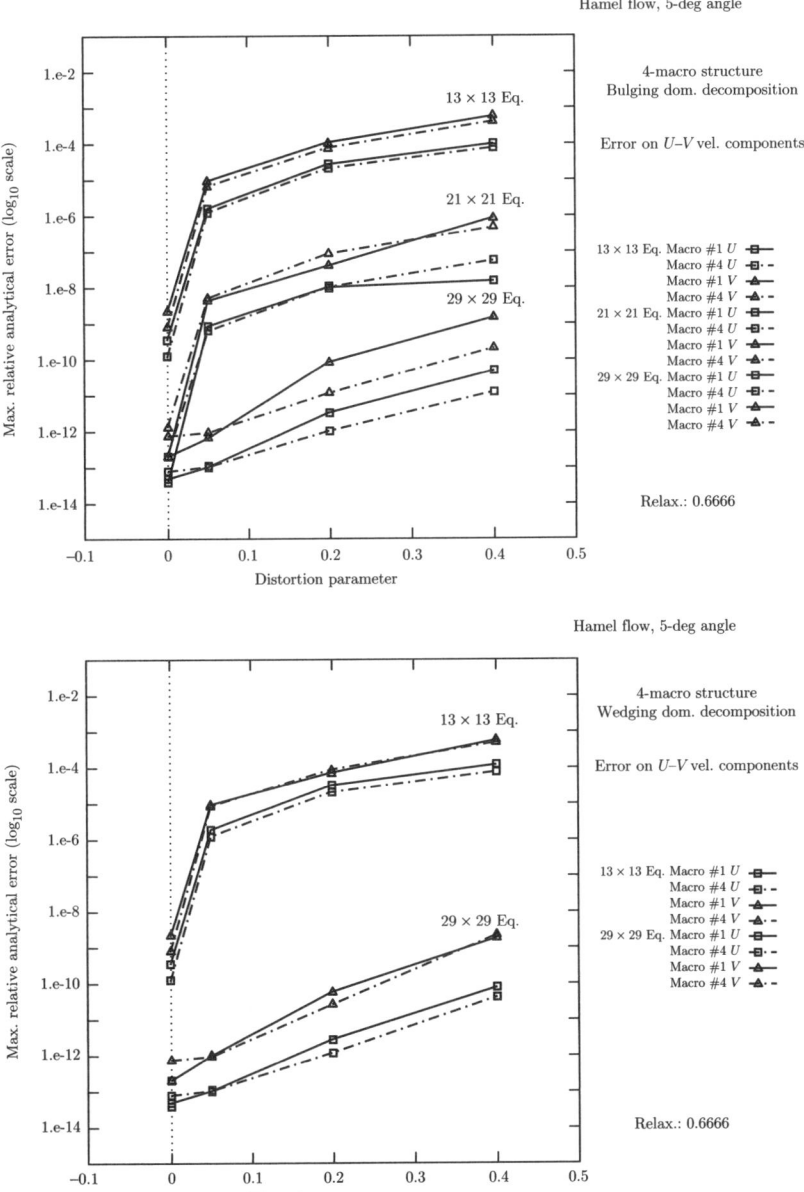

Figure 4.4.5. Influence of mesh distortion D on convergence for Hamel flow example (from [345]).

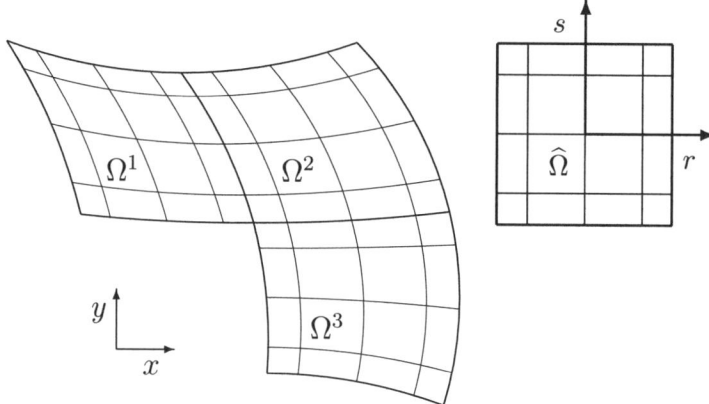

Figure 4.5.1. Example of spectral-element discretization in $\mathbb{R}^2$, showing GLL nodal lines for $(E, N) = (3, 4)$.

resolution requirements call for greater flexibility in the nodal point distribution. Such flexibility has been the hallmark of FEMs from their inception. To combine the high-order convergence rates and fast operator evaluation properties of spectral methods with the geometric flexibility of FEs, Patera [297] developed the SEM for simulation of incompressible flows. The original implementation was based on Chebyshev collocation within each element. Subsequently, Maday and Patera [261] developed a Galerkin formulation based on GLL quadrature, which is described here.

The SEM is based on a decomposition of the global domain, $\bar{\Omega} = \Omega \cup \partial \Omega$, into E nonoverlapping subdomains (or *elements*), Ω^e, $e = 1, \ldots, E$. As illustrated in Figure 4.5.1, the intersection of the closure of any two subdomains is void, a vertex, an entire edge, or an entire face (in $\mathbb{R}^3$). (Nonconforming discretizations that relax this condition are treated in Chapter 7.) Each subdomain is a deformed quadrilateral in $\mathbb{R}^2$ or deformed parallelepiped in $\mathbb{R}^3$ under a mapping $\mathbf{x}^e(\mathbf{r}) \in \Omega^e \Longrightarrow \mathbf{r} \in \hat{\Omega}$, with well-defined inverse $\mathbf{r}^e(\mathbf{x}) \in \hat{\Omega} \Longrightarrow \mathbf{x} \in \Omega^e$, and thereby admits the use of the tensor-product bases introduced in Section 4.3. Generally, the subdomain geometry will be represented by an isoparametric mapping similar to Equation (4.4.15).

The underlying approximation space in the SEM is $V_N := \mathbb{P}_{N,E} \cap H^1$, where $\mathbb{P}_{N,E}$ is the space of piecewise continuous functions that map to polynomials in the reference domain,

$$\mathbb{P}_{N,E}(\Omega) := \{v(\mathbf{x}^e(\mathbf{r}))|_{\Omega^e} \in \mathbb{P}_N(r_1) \otimes \cdots \otimes \mathbb{P}_N(r_d), e = 1, \ldots, E\}. \tag{4.5.1}$$

4.5. Spectral-Element Discretizations

Here, $\mathbb{P}_N(r)$ is the space of all polynomials of degree less than or equal to N in the argument. For efficiency reasons noted previously, and to permit trivial enforcement of interelement continuity, the basis within each element is taken to be a tensor product of Lagrangian interpolants. Thus, for 2D problems, function representation takes the form

$$u(\mathbf{x})|_{\Omega^e} = \sum_{i=0}^{N}\sum_{j=0}^{N} u_{ij}^e \pi_i(r)\pi_j(s), \qquad r,s \in [-1,1]^2. \quad (4.5.2)$$

Here, u_{ij}^e represents nodal values for element e; $\pi_i(r)$ and $\pi_j(s)$ are the Lagrangian interpolant basis functions based on the GLL points, as introduced in (2.4.3); and $\mathbf{x}$ is an isoparametric mapping of the local geometry,

$$\mathbf{x}(r,s)|_{\Omega^e} := \sum_{i=0}^{N}\sum_{j=0}^{N} \mathbf{x}_{ij}^e \pi_i(r)\pi_j(s), \qquad r,s \in [-1,1]^2.$$

Within the framework of multielement discretizations, the preceding discrete operators for the scalar product, the bilinear form, and the source term are defined by

$$(u,v)_N := \sum_{e=1}^{E} \int_{\Omega^e} u(\mathbf{x})v(\mathbf{x})\,d\mathbf{x},$$

$$\mathcal{A}_N(u,v) := \sum_{e=1}^{E} \int_{\Omega^e} (p(\mathbf{x})\nabla u \cdot \nabla v + q(\mathbf{x})uv)\,d\mathbf{x}, \quad (4.5.3)$$

$$\mathcal{F}_N(v) := \sum_{e=1}^{E} \int_{\Omega^e} f(\mathbf{x})v(\mathbf{x})\,d\mathbf{x}.$$

In the SEM, quadrature is substituted for the integrals, resulting in local forms similar to the single-domain case introduced in Sections 4.3–4.4. The Dirichlet problem (4.3.3) becomes the following: Find $u \in V_{N,0}$ such that

$$\mathcal{A}_N(u,v) := \mathcal{F}_N(v) \qquad \forall v \in V_{N,0}, \quad (4.5.4)$$

where $V_{N,0} := \{v \in V_N : v|_{\partial\Omega_D} = 0\}$, and $\partial\Omega_D$ is the subset of $\partial\Omega$ on which homogeneous Dirichlet boundary conditions are enforced.

4.5.1 Continuity and Direct Stiffness Summation

To construct $V_N := \mathbb{P}_{N,E} \cap H^1$, we restrict the space $\mathbb{P}_{N,E}$ (4.5.1) to admit only functions that are continuous across element interfaces. One of the major

192   4. Multidimensional Problems

advantages of the Lagrangian basis (4.5.2) is that function continuity is enforced by simply equating coincident nodal values, that is,

$$\mathbf{x}_{ij}^e = \mathbf{x}_{\hat{i}\hat{j}}^{\hat{e}} \implies u_{ij}^e = u_{\hat{i}\hat{j}}^{\hat{e}}. \qquad (4.5.5)$$

If $\bar{\mathcal{N}}$ is the number of distinct nodes in Ω, then (4.5.5) represents $(N+1)^d E - \bar{\mathcal{N}}$ constraints on the choice of local nodal values, u_{ij}^e.

It is convenient for notational purposes to cast the constraint (4.5.5) in matrix form. Let $\underline{u} \in \mathbb{R}^{\bar{\mathcal{N}}}$ denote the vector of nodal values associated with a global numbering of the distinct nodes in all of Ω, as illustrated in Figure 4.5.2. Let $\underline{u}^e \in \mathbb{R}^{(N+1)^d}$ denote the vector of local basis coefficients associated with Ω^e:

$$\underline{u}^e := \left(u_{00}^e, u_{10}^e, \ldots, u_{NN}^e\right)^T, \quad e = 1, \ldots, E,$$

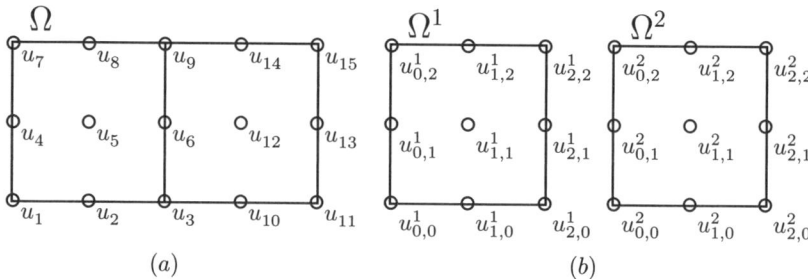

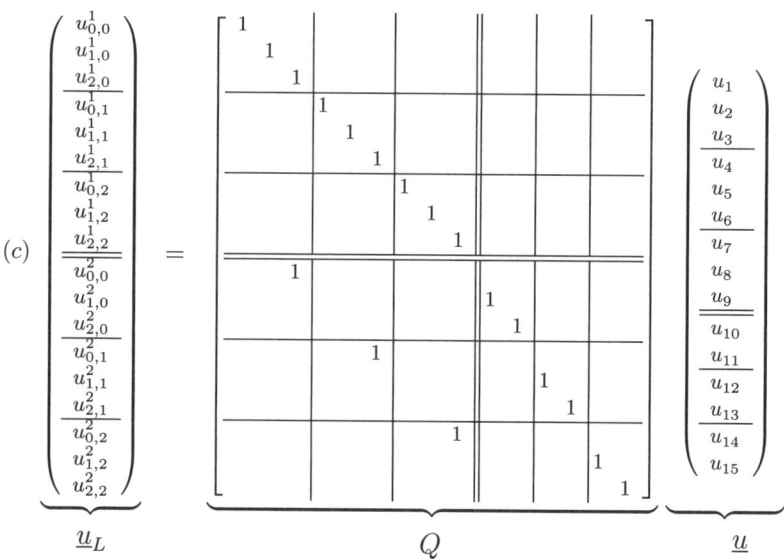

Figure 4.5.2. Example of spectral-element mesh for $(E, N) = (2, 2)$: (a) global values are mapped to (b) local values by (c) the operation $\underline{u}_L = Q\underline{u}$.

4.5. Spectral-Element Discretizations

and let $\underline{u}_L$ be the collection of these local vectors:

$$\underline{u}_L := \begin{pmatrix} \underline{u}^1 \\ \underline{u}^2 \\ \vdots \\ \underline{u}^e \\ \vdots \\ \underline{u}^E \end{pmatrix} = \begin{pmatrix} u_{0,0}^1 \\ u_{1,0}^1 \\ \vdots \\ u_{N,N}^1 \\ u_{0,0}^2 \\ \vdots \\ u_{N,N}^E \end{pmatrix}.$$

If u is to be continuous, then there exists a Boolean connectivity matrix Q that maps $\underline{u}$ to $\underline{u}_L$, ensuring that (4.5.5) is satisfied, as illustrated in Figure 4.5.2. The operation

$$\underline{u}_L = Q\underline{u} \tag{4.5.6}$$

is referred to as a *scatter* from the global ($\underline{u}$) to the local ($\underline{u}_L$) vector. For example, in Figure 4.5.2 the global value u_3 is copied to local coefficients $u_{2,0}^1$ and $u_{0,0}^2$ as a result of the two unit entries in the third column of Q.

Note that for every global vector $\underline{u}$ there is a corresponding local vector, $\underline{u}_L$, given by (4.5.6). The converse is not true, because Q is not invertible. However, we will frequently employ the closely related *gather* operation

$$\underline{v} = Q^T \underline{u}_L \tag{4.5.7}$$

and denote the ouput ($\underline{v}$) with a different notation from the input ($\underline{u}_L$). Whereas the action of Q is to *copy* entries of $\underline{u}$ to $\underline{u}_L$, the action of Q^T is to *sum* entries from corresponding nodes. This is clear from Figure 4.5.2, where it can be seen that Q^T will have multiple 1s on several rows, implying, for example, that $v_3 = u_{2,0}^1 + u_{0,0}^2$.

In practice, the matrix Q is never constructed. Rather, the *actions* of Q and Q^T are implemented via indirect addressing, as illustrated by the pseudocode in Figure 4.5.3. For the example of Figure 4.5.2, the local-to-global mapping array, *global_index()*, would contain {1 2 3 4 5 6 7 8 9} for element 1, and {3 10 11 6 12 13 9 14 15} for element 2.

For high-order methods, it is natural to store data in terms of local coefficients u_{ij}^e, as this readily allows evaluation of derivatives, integrals, and so forth without having to first map data from global to local coordinates. This is particularly true in parallel, where such mappings imply interprocessor communication. Consequently, a gather typically is followed immediately by a scatter back to local form. We denote the combined gather–scatter operation by

$$\Sigma' := QQ^T \tag{4.5.8}$$

```
Procedure u_L = Qu          Procedure v = Q^T u_L
do e = 1, ..., E:           v = 0
 do j = 0, ..., N:          do e = 1, ..., E:
  do i = 0, ..., N:          do j = 0, ..., N:
   î = global_index(i, j, e)  do i = 0, ..., N:
   u^e_ij = u_î                î = global_index(i, j, e)
enddo                          v_î = v_î + u^e_ij
                            enddo
```

Figure 4.5.3. Pseudocode for serial implementation of Q and Q^T in $\mathbb{R}^2$.

and refer to application of Σ' as *direct stiffness summation*. Direct stiffness summation is a noninvertible, local-to-local transformation that amounts to summing shared interface variables and redistributing them to their original locations, leaving interior nodes unchanged. Parallel implementations of Σ' are discussed in Section 8.5.

4.5.2 Spectral–Element Operators

To develop the spectral-element stiffness and mass matrices, we extend the notion of local and global vector representations introduced in the preceding section to system matrices.

For a single domain Ω^e, we identify with the bilinear form $\mathcal{A}_N$ the integral

$$\int_{\Omega^e} p \nabla v \cdot \nabla u \, d\mathbf{x} = (\underline{v}^e)^T K^e \underline{u}^e, \qquad (4.5.9)$$

where K^e is the *local* stiffness matrix, evaluated by using quadrature to yield a form similar to (4.4.7) or (4.4.8) according to whether Ω^e is deformed or not. We further define the *unassembled* stiffness matrix as the block-diagonal collection of local stiffness matrices

$$K_L := \begin{bmatrix} K^1 & & & \\ & K^2 & & \\ & & \ddots & \\ & & & K^E \end{bmatrix}.$$

Clearly we can extend the bilinear form (4.5.9) to the whole domain via

$$\int_{\Omega} p \nabla v \cdot \nabla u \, d\mathbf{x} = \sum_{e=1}^{E} (\underline{v}^e)^T K^e \underline{u}^e$$

$$= \underline{v}_L^T K_L \underline{u}_L.$$

4.5. Spectral-Element Discretizations

However, this holds only for continuous u and v, implying that there exist vectors $\underline{u}$ and $\underline{v}$ such that $\underline{u}_L = Q\underline{u}$ and $\underline{v}_L = Q\underline{v}$. Consequently,

$$\forall v, u \in V_N, \quad \int_\Omega p \nabla v \cdot \nabla u \, d\mathbf{x} = \underline{v}^T Q^T K_L Q \underline{u}. \quad (4.5.10)$$

The $\bar{\mathcal{N}} \times \bar{\mathcal{N}}$ matrix $\bar{K} := Q^T K_L Q$ is referred to as the *assembled* Neumann operator, since the approximation space has yet to be restricted for enforcement of essential, or Dirichlet, boundary conditions on $\partial \Omega_D$. $\bar{K}$ has a one-dimensional nullspace associated with the constant vector $\underline{\hat{e}} := (1, 1, \ldots, 1)^T$.

The derivation of the spectral-element mass matrix is similar to (4.5.9)–(4.5.10), resulting in the following expression for the bilinear form (v, u):

$$\forall v, u \in V_N, \quad \int_\Omega vu \, d\mathbf{x} = \underline{v}^T Q^T M_L Q \underline{u}, \quad (4.5.11)$$

where M_L is the block-diagonal matrix comprising local mass matrices M^e, $e = 1, \ldots, E$. The local mass matrices are evaluated by using quadrature, yielding diagonal forms similar to (4.3.17) or (4.4.9). The assembled mass matrix $\bar{M} := Q^T M_L Q$ is thus also diagonal. Note that, because $f \in L^2$, the inner product associated with the linear functional $\mathcal{F}_N(v)$ differs slightly from (4.5.11):

$$\forall v \in V_N, f \in \mathbb{P}_{N,E}, \quad \int_\Omega vf \, d\mathbf{x} = \underline{v}^T Q^T M_L \underline{f}_L.$$

Here, f has been replaced by its local interpolant, which may be discontinuous across element boundaries.

Dirichlet boundary conditions can be enforced as in (4.3.24)–(4.3.26) by introducing a Boolean restriction matrix R, which enumerates all but the Dirichlet nodes, to yield an invertible stiffness matrix $K := R^T Q^T K_L Q R$. Alternatively, one can use a mask array to restrict nodal values on $\partial \Omega_D$ to zero. The mask can be applied in either global or local form, as illustrated in Figure 4.5.4. The global form amounts to pre- and postmultiplying $\bar{K}$ by a diagonal matrix, to yield $K := \mathcal{M} Q^T K_L Q \mathcal{M}$. The local form amounts to restricting $u \in V_N$ to $V_{N,0} := V_N \cap H_0^1$ by multiplying $Q\underline{u}$ by a mask array $\mathcal{M}_L$, and yields the stiffness matrix $K = Q^T \mathcal{M}_L K_L \mathcal{M}_L Q$. Both masks are essentially the identity matrix, save that diagonal entries corresponding to nodes on $\partial \Omega_D$ are set to zero, and both forms yield the same stiffness matrix. Although the local form is more useful in practice (one simply sets appropriate boundary values to zero in each element), the global form is conceptually convenient because the details of the boundary conditions are decoupled from the structure of the assembled Neumann operator, which can be quite complex, particularly in the nonconforming case (see Section 7.3).

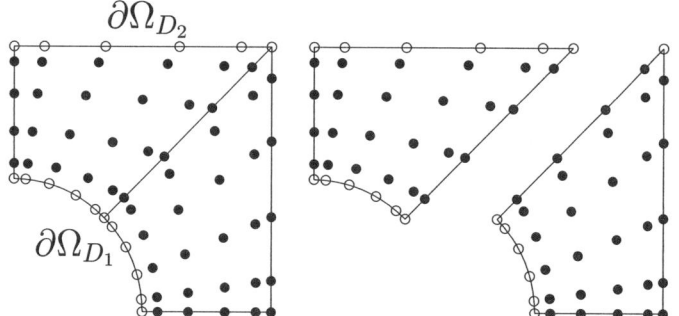

Figure 4.5.4. Global (left) and local (right) mask arrays. Open circles imply $(\mathcal{M})_i = (\mathcal{M}_L)_{ij}^e = 0$. Solid circles imply $(\mathcal{M})_i = (\mathcal{M}_L)_{ij}^e = 1$.

Strictly speaking, the masked form of K is still not invertible, since it has a nontrivial nullspace associated with values on $\partial \Omega_D$. However, for iterative solution techniques, in which solutions are constructed as linear combinations of other vectors, the nullspace is not a problem provided that all of the underlying basis vectors (search directions) vanish on $\partial \Omega_D$, that is, that all vectors are masked. Note that the masks are projectors (i.e., $\mathcal{M} = \mathcal{M}^2$) and that the local mask commutes with direct stiffness summation, that is, $\mathcal{M}_L Q Q^T = Q Q^T \mathcal{M}_L$.

In practice, $\mathcal{M}_L$ is constructed by initializing a local array to unity, $\underline{\tilde{m}}_L = \underline{\hat{e}}$; zeroing each entry in $\underline{\tilde{m}}^e$, $e = 1, \ldots, E$, on any face where Dirichlet conditions are applied; performing "direct-stiffness multiply," $\underline{\tilde{m}}_L := \Sigma^* \underline{\tilde{m}}_L$; and setting $\mathcal{M}_L = \text{diag}(\tilde{m}_{ij}^e)$. The application of Σ^*, which is defined by replacing summation in *Procedure* Q^T with multiplication, ensures that *all* vertices common to a Dirichlet surface are masked, even if an element is otherwise interior to Ω or has only Neumann boundaries. The upper-right-corner vertex in Figure 4.5.4 is one such example.

4.5.3 Inhomogeneous Dirichlet Problems

We illustrate implementation of inhomogeneous boundary conditions by considering the Helmholtz problem

$$-\Delta u + q u = f \quad \text{in } \Omega,$$
$$u = g \quad \text{on } \partial \Omega_D, \quad \nabla u \cdot \hat{\mathbf{n}} = 0 \quad \text{on } \partial \Omega_N, \quad (4.5.12)$$

with $q > 0$ a constant. The weighted residual formulation of (4.5.12) reads as follows: Find u in H_b^1 such that

$$\int_\Omega (\nabla v \cdot \nabla u + v q u) \, dV = \int_\Omega v f \, dV \quad \forall v \text{ in } H_0^1, \quad (4.5.13)$$

4.5. Spectral-Element Discretizations

where H_b^1 is the space of functions in H^1 satisfying the inhomogeneous boundary conditions on $\partial\Omega_D$, and H_0^1 is the space satisfying homogeneous boundary conditions.

Using superposition, we first symmetrize (4.5.13) by decomposing the solution into $u = u_0 + u_b$, where u_b is any known function in H_b^1, and moving the inhomogeneous boundary term to the r.h.s. The new unknown, u_0, is therefore in H_0^1, implying that the trial and test spaces coincide. Application of the spectral-element operators introduced in the preceding section leads to the symmetric positive-semidefinite linear system

$$\mathcal{M}Q^T H_L Q \mathcal{M} \underline{u}_0 = \mathcal{M}Q^T (M_L \underline{f}_L - H_L \underline{u}_{b,L}),$$
$$H_L := K_L + q M_L.$$
(4.5.14)

This system is solvable because the r.h.s. lies in the range of the symmetric matrix on the left. Note that the matrix $\mathcal{M}Q^T$ appearing on the right arises from the requirement $v \in H_0^1$. However, $Q\mathcal{M}$ does not appear on the right, because neither u_b nor f vanishes on $\partial\Omega_D$, and continuity is assumed for u_b and not required for f. We stress that u_b *must* be continuous in order for the final result $u := u_0 + u_b$ to be so. Any inexpensive projection step can be employed to guarantee this. For example, $\underline{u}_b := M^{-1} \Sigma' M_L \underline{u}_{b,L}$ yields a weighted average of the interface values.

The form (4.5.14) is convenient for analysis, because it clearly reveals the symmetric form of H. However, the following local form is often used in practice: Find $\underline{u}_{0,L} \in \mathcal{R}(M_L Q)$ such that

$$M_L \Sigma' H_L \underline{u}_{0,L} = M_L \Sigma' (M_L \underline{f}_L - H_L \underline{u}_{b,L}).$$
(4.5.15)

Equation (4.5.15) is obtained by multiplying (4.5.14) by Q and using the equivalence properties of $\mathcal{M}$ and $\mathcal{M}_L$. The advantage of the local form is that data are immediately ready for application of tensor-product operators without the need for additional scatter operations, which require data communication on parallel computers. Also, combining the gather and scatter (QQ^T) into a single operation (Σ') reduces communication overhead (see Section 8.5).

4.5.4 Iterative Solution Techniques

Because the underlying system is symmetric positive (semi)definite, PCG can be used to solve (4.5.14). To work with the local form (4.5.15), we modify Algorithm 2.2 as follows.

Algorithm 4.1 (Preconditioned conjugate gradient) Starting with $\underline{x}_L = 0$, $\underline{r}_L = M_L \Sigma' (M_L \underline{f} - H_L \underline{u}_{b,L})$,

1. Compute $\underline{z}_L = P_L^{-1}\underline{r}_L$, $\underline{w}_L = \underline{z}_L$, $\rho_1 = \underline{r}_L^T W \underline{z}_L$.
2. For $j = 0, 1, \ldots$ until convergence Do:
3. $\quad \underline{q}_L = \mathcal{M}_L \Sigma' H_L \underline{w}_L$
4. $\quad \alpha = \rho_1 / \underline{w}_L^T W \underline{q}_L$,
5. $\quad \underline{x}_L = \underline{x}_L + \alpha \underline{w}_L$,
6. $\quad \underline{r}_L = \underline{r}_L - \alpha \underline{q}_L$,
7. $\quad \underline{z}_L = P_L^{-1} \underline{r}_L$,
8. $\quad \rho_0 = \rho_1$, $\rho_1 = \underline{r}_L^T W \underline{z}_L$, $\beta = \rho_1 / \rho_0$,
9. $\quad \underline{w}_L = \underline{z}_L + \beta \underline{w}_L$,
10. EndDo.
11. $\underline{u}_L = \underline{x}_L + \underline{u}_{b,L}$.

Here, the matrix–vector product $\underline{q} = H\underline{w}$ has been cast into its constituent components of local Neumann operator (H_L), direct-stiffness summation (Σ'), and mask ($\mathcal{M}_L$). In addition, the inner products have been weighted by a diagonal scaling matrix W to account for multiply represented nodal values arising from the local storage format. We define as the multiplicity

$$\underline{\hat{m}}_L := \Sigma' \underline{\hat{e}}_L, \tag{4.5.16}$$

where $\underline{\hat{e}}_L$ is the unit vector equal to one at each (local) gridpoint. W is the inverse of the multiplicity,

$$W = \text{diag}\left(\frac{1}{\hat{m}_{L,i}}\right). \tag{4.5.17}$$

Note that, with unpreconditioned iterative methods, no special effort is required to enforce the restriction of $\underline{u}_{0,L}$ ($:= \underline{x}_L$) to $\mathcal{R}(\mathcal{M}_L Q)$ when solving (4.5.15), as it is a natural consequence of the Krylov subspace approximation, $\underline{x}_L \in \mathcal{K}_k(\mathcal{M}_L \Sigma' H_L, \underline{b}_L)$, where $\underline{b}_L := \mathcal{M}_L \Sigma'(M_L \underline{f}_L - H_L \underline{u}_{0,L})$. However, in the preconditioned case, one must ensure that the solution satisfies the homogeneous boundary conditions and is continuous. This is straightforward in the case of diagonal preconditioning. If $\underline{\tilde{d}}_L = \text{diag}(d_{ij}^e)$ and $\underline{m}_L = \text{diag}(m_{ij}^e)$ are the diagonals of the unassembled stiffness and mass matrices, K_L (4.4.13)–(4.4.14) and M_L, respectively, then $\underline{w}_L = \Sigma'(\underline{\tilde{d}}_L + q\underline{m}_L)$ contains the diagonal of the assembled Neumann operator $\bar{H}$, and $P_L^{-1} = \text{diag}(1/w_{ij}^e)$ is the desired diagonal preconditioner in local form.

4.5.5 Two-Dimensional Examples

As in the single-domain case, the matrices in (4.5.15) are not explicitly constructed. Instead, the *action* of the operators is effected via procedural calls. One

4.5. Spectral-Element Discretizations

simply needs to construct the necessary pointwise or tensor-product operators that are to be applied to nodal entries in each subdomain. Implementation of (4.5.15) thus requires four basic steps:

- identification and development of local tensor-product-based operators in K_L and M_L (as in Section 4.3),
- identification of interelement continuity requirements, and development of the operators Q, Q^T, and Σ' (Figure 4.5.3),
- application of boundary conditions via $\mathcal{M}_L$, and
- development of a preconditioner for the linear system (4.5.15).

We illustrate these points through the following examples.

Example 1

Consider solving (4.5.12) in the square domain, $[0, 1]^2$, having a unit-diameter disk cut out of the lower left corner, as depicted in Figure 4.5.5. Let $\partial \Omega_D := \partial \Omega_{D_1} \cup \partial \Omega_{D_2}$, $\partial \Omega_N := \partial \Omega \setminus \partial \Omega_D$,

$$f = \begin{cases} 0 & x \geq y, \\ 1 & x < y, \end{cases} \quad \text{and} \quad g = \begin{cases} 1 & \text{on } \partial \Omega_{D_1}, \\ 2 & \text{on } \partial \Omega_{D_2}. \end{cases}$$

Using the numbering in Figure 4.5.5, we define the following arrays for $N = 2$:

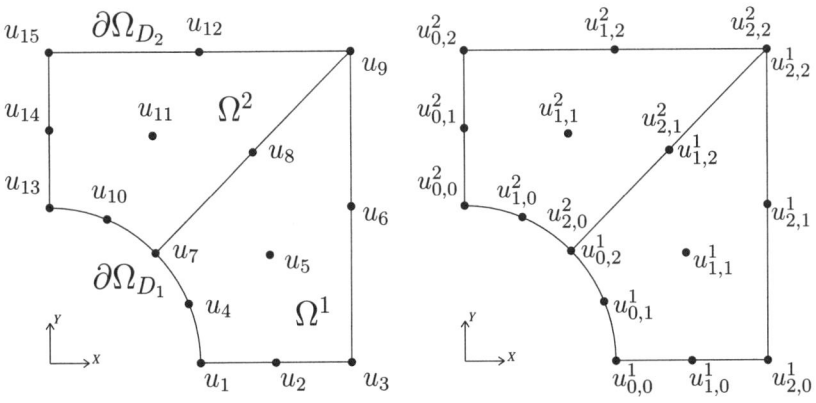

Figure 4.5.5. Global (left) and local (right) node numbering for spectral-element example with $(E, N) = (2, 2)$. Dirichlet boundary segments are $\partial \Omega_{D_1}$ at $r = 0.5$ and $\partial \Omega_{D_2}$ at $y = 1$.

ij	glob	$\mathcal{M}_{ij}^1$	$\hat{m}_{ij}^1$	$u_{b,ij}^e$	f_{ij}^e	glob	$\mathcal{M}_{ij}^2$	$\hat{m}_{ij}^2$	$u_{b,ij}^e$	f_{ij}^e
00	1	0	1	1	0	13	0	1	1	1
10	2	1	1	0	0	10	0	1	1	1
20	3	1	1	0	0	7	0	2	1	1
01	4	0	1	1	0	14	1	1	0	1
11	5	1	1	0	0	11	1	1	0	1
21	6	1	1	0	0	8	1	2	0	1
02	7	0	2	1	0	15	0	1	2	1
12	8	1	2	0	0	12	0	1	2	1
22	9	0	2	2	0	9	0	2	2	1

Here, glob corresponds to the *global_index* arrays of Figure 4.5.3, $\mathcal{M}_{ij}^e$ corresponds to the mask, and $\hat{m}_{ij}^e$ is the multiplicity (4.5.16). Note that f is allowed to be multivalued along $y = x$ in order to obtain a good approximation to (v, f).

Example 2

The preceding example can be readily extended to allow $q(\mathbf{x})$ to vary by replacing q in (4.5.14) with q_{ij}^e at each nodal point, collocated with the entries in the diagonal mass matrix, M_L. Robin boundary conditions can be implemented in a similar manner. For example, suppose we have

$$\nabla u \cdot \mathbf{n} + \alpha u = \beta(\mathbf{x}) \qquad \text{on } \partial\Omega_R \qquad (4.5.18)$$

and homogeneous conditions elsewhere. To derive the weighted residual formulation for this new problem, consider the effect of integration by parts on $-v\Delta u$:

$$\int_\Omega -v\Delta u \, dV = \int_\Omega \nabla v \cdot \nabla u \, dV - \int_{\partial\Omega} v \nabla u \cdot \mathbf{n} \, dS$$

$$= \int_\Omega \nabla v \cdot \nabla u \, dV - \int_{\partial\Omega_R} v(\beta - \alpha u) \, dS.$$

The second expression follows from (4.5.18), coupled with the constraints $v|_{\partial\Omega_D} = 0$, $\nabla u \cdot \mathbf{n}|_{\partial\Omega_N} = 0$. With the nonvanishing boundary term, the weighted residual formulation of modified system thus reads as follows: Find u in H_0^1 such that

$$\int_\Omega (\nabla v \cdot \nabla u + vqu) \, dV + \int_{\partial\Omega_R} v\alpha u \, dS = \int_\Omega vf \, dV + \int_{\partial\Omega_R} v\beta \, dS \quad \forall v \text{ in } H_0^1.$$

Note that inhomogeneous Neumann conditions are included in this framework by simply taking $\alpha = 0$.

4.5. Spectral-Element Discretizations

Since equation generation results from successively choosing Lagrangian basis elements for v, most of the boundary integral terms vanish. Only those equations corresponding to nodes on $\partial \Omega_R$ differ from the homogeneous case. If $\partial \Omega_R$ corresponds, say, to the right edge of the domain in Figure 4.5.5 ($x = 1, 0 \leq y \leq 1$), one would update the diagonal entries of $K_L + qM_L$ and the r.h.s. entries corresponding to $e = 1, i = N$, and $j = 0, \ldots, N$. For example, for $N = 2$, the terms to be added to each are

e, i, j	diag_{H_L}	R.h.s.
1, 2, 0	$\frac{1}{2}\rho_0 \alpha$	$\frac{1}{2}\rho_0 \beta_0$
1, 2, 1	$\frac{1}{2}\rho_1 \alpha$	$\frac{1}{2}\rho_1 \beta_1$
1, 2, 2	$\frac{1}{2}\rho_2 \alpha$	$\frac{1}{2}\rho_2 \beta_2$

where ρ_j denotes a GLL quadrature weight, β_j denotes the value of β at the nodal point $\mathbf{x}_{2j}^1$, and the factor of $\frac{1}{2}$ corresponds to the surface Jacobian associated with the transformation of $\partial \Omega_R$ to the reference interval $[-1, 1]$.

Example 3
The results of diagonal preconditioning for the Helmholtz problem (4.5.12) are illustrated in Figure 4.5.6, which shows the CG iteration counts for a 10^{-10} residual reduction when a random forcing f is applied to the the domain depicted on the right. The mesh parameter γ is a measure of the range of scales in the solution. The left value in each table entry corresponds to the unpreconditioned case, while the right value is the Jacobi preconditioned result. We see that for $q = 0$ (Poisson's equation), the diagonal scaling is useful in eliminating the extreme mesh dependencies but, as we might anticipate, does little to improve the balanced result ($\gamma \approx 0.5$). As q increases, the relative improvement increases but does not become significant until $q > 10^3$.

	Iteration count							
	$q = 0$		$q = 10^1$		$q = 10^3$		$q = 10^5$	
γ	U.	P.	U.	P.	U.	P.	U.	P.
.95	142	89	138	85	96	63	77	16
.75	89	83	86	75	47	36	52	7
.50	74	71	72	68	38	30	48	7
.25	89	79	87	77	54	39	69	9
.05	117	88	116	87	84	58	131	20

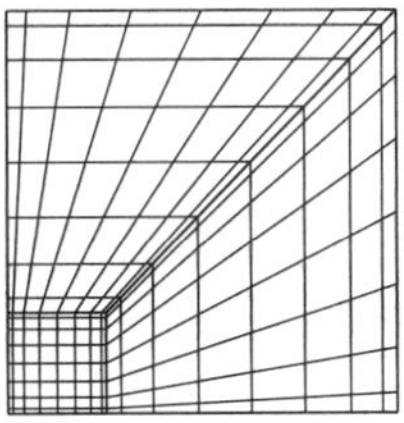

Figure 4.5.6. Diagonal preconditioning results for $(E, N) = (3, 8)$.

4.6 Collocation Discretizations

We now turn to collocation discretizations to solve multidimensional elliptic problems. Again, our purpose is to explain the different steps leading to the implementation of the algorithm. Systematic use of tensor products makes it easy to write the equations in compact form for 3D problems. For the sake of simplicity we will, however, restrict ourselves to the case $d = 2$. We start with the collocation solution of diffusion problems on mono- and multidomains. Next, we deal with advection-dominated diffusion problems.

4.6.1 The Diffusion Case

Consider the BVP on the rectangular monodomain $\Omega := (0, L_1) \times (0, L_2) \subset \mathbb{R}^2$:

$$\begin{aligned} \mathcal{L}u &:= -\nabla \cdot (p\nabla u) + qu = f \quad \text{in } \Omega, \\ \mathcal{B}u &= 0 \quad \text{on } \partial\Omega, \end{aligned} \tag{4.6.1}$$

where the coefficients $p > 0$ and $q \geq 0$ may be space-dependent. However, for the time being we assume these quantities are constants. Also, we deal with homogeneous Dirichlet boundary conditions. Neumann boundary conditions are treated later on.

We first map the physical domain Ω onto the reference domain $\hat{\Omega} := [-1, +1]^2$ using an affine transformation between the physical coordinates of the problem, $(x, y) \in \Omega$, and the reference coordinates $(r, s) \in \hat{\Omega}$. Then, according to the orthogonal collocation methodology described in Section 2.5, we select a collocation grid related to a Gauss–Lobatto–Jacobi (GLJ) quadrature rule with $M + 1$ nodes in r, and $N + 1$ nodes in s. Let $\mathcal{G}_{MN}^{(\alpha,\beta)}$ denote the tensor product of these one-dimensional GLJ quadrature grids,

$$\mathcal{G}_{MN}^{(\alpha,\beta)} := G_{r,M}^{(\alpha,\beta)} \otimes G_{s,N}^{(\alpha,\beta)}, \tag{4.6.2}$$

where $G_{r,M}^{(\alpha,\beta)}$ and $G_{s,N}^{(\alpha,\beta)}$ are the ordered sets of solutions to Equation (2.5.2) in r and s. As already mentioned, the two most important choices in practice are the GLC (Chebyshev) and GLL (Legendre) grids with $\alpha = \beta = -1/2$ and $\alpha = \beta = 0$, respectively.

Any approximate solution to (4.6.1) then is

$$\hat{u}(\mathbf{x}) = \sum_{i=0}^{M} \sum_{j=0}^{N} u_{ij} \pi_{M,i}^J(r) \pi_{N,j}^J(s), \quad (r, s) \in \hat{\Omega}, \tag{4.6.3}$$

4.6. Collocation Discretizations

where the interpolation basis functions $\pi_{M,i}^J$ ($\pi_{N,j}^J$) of degree M (N) are defined by Equations (2.5.4)–(2.5.9). The collocation solution $\hat{u}$ results from (2.5.10)–(2.5.11):

$$\begin{aligned}\mathcal{L}\hat{u}(\xi_{ij}^J) &= f(\xi_{ij}^J) & \forall \xi_{ij}^J \in \mathcal{G}_{MN}^{(\alpha,\beta)} \cap \hat{\Omega}, \\ \hat{u}(\xi_{ij}^J) &= 0 & \forall \xi_{ij}^J \in \mathcal{G}_{MN}^{(\alpha,\beta)} \cap \partial\hat{\Omega},\end{aligned} \quad (4.6.4)$$

with $\xi_{ij}^J := (\xi_{M,i}^J, \xi_{N,j}^J)$.

One verifies easily that, because of the normalization property (2.5.5) of the basis functions $\pi_{M,i}^J$ and $\pi_{N,j}^J$, the second-order derivatives of the approximation (4.6.3) at the collocation nodes are equal to

$$\frac{\partial^2 \hat{u}}{\partial x^2}(\xi_{k\ell}^J) = \left(\frac{2}{L_1}\right)^2 \sum_{i=0}^{M} u_{i\ell} \frac{d^2 \pi_{M,i}^J}{dr^2}(\xi_{M,k}^J), \quad (4.6.5)$$

$$\frac{\partial^2 \hat{u}}{\partial y^2}(\xi_{k\ell}^J) = \left(\frac{2}{L_2}\right)^2 \sum_{j=0}^{N} u_{kj} \frac{d^2 \pi_{N,j}^J}{ds^2}(\xi_{N,\ell}^J), \quad (4.6.6)$$

with $0 \leq k \leq M$ and $0 \leq \ell \leq N$. For simplicity we will now assume that the polynomial degree is equal to N in both space directions, denoted (r_1, r_2). We also drop the reference to N in the basis functions and the GLJ quadrature nodes.

With homogeneous Dirichlet boundary conditions everywhere along $\partial \Omega$, one can easily show, using the tensor-product notation, that the collocation equations (4.6.4) lead to the algebraic system $L_C \underline{u} = \underline{f}$, with a matrix

$$L_C = -p\left[\left(\frac{2}{L_1}\right)^2 (I_2 \otimes D_1^{(2)}) + \left(\frac{2}{L_2}\right)^2 (D_2^{(2)} \otimes I_1)\right] + q(I_2 \otimes I_1). \quad (4.6.7)$$

Here, I_s, $D_s^{(2)} \in \mathbb{R}^{N \times N}$, $s = 1, 2$, denote the identity matrices in r_1 $(= r)$ and r_2 $(= s)$, and the one-dimensional second-order derivative matrices whose elements

$$[D_s^{(2)}]_{k\ell} := \frac{d^2 \pi_\ell^J}{dr_s^2}(\xi_k^J) \quad (4.6.8)$$

may be evaluated by using the relationships (B.3.45)–(B.3.50) The components of the vectors of unknowns $\underline{u}$ and forcing terms $\underline{f}$ are set in lexicographical ordering.

Multiplying both sides of the collocation system $L_C \underline{u} = \underline{f}$ by $L_1 L_2/4$ gives a new system $L_C \underline{u} = L_1 L_2 \underline{f}/4$, where we incorporate the constant factor into the collocation matrix:

$$L_C = -p\left[\frac{L_2}{L_1}(I_2 \otimes D_1^{(2)}) + \frac{L_1}{L_2}(D_2^{(2)} \otimes I_1)\right] + q\frac{L_1 L_2}{4}(I_2 \otimes I_1). \quad (4.6.9)$$

The reader should notice the similarity between the operators (4.6.9) and (4.3.23) (except for a minus sign), although the latter corresponds to the weak formulation of a problem with homogeneous Neumann boundary conditions. The weak formulation introduces local mass and stiffness matrices ($\hat{M}$ and $\hat{K}$) where collocation imposes identity and second-order derivative matrices.

In Chapter 2, using integration by parts, we concluded that Legendre spectral elements for a Dirichlet problem were equivalent to the Legendre collocation (see Remark 2.2). This is exactly what we have here, provided each row in the operator (4.6.9) (and the corresponding component of the r.h.s.) is multiplied by the appropriate scaling factor equal to the GLJ quadrature weight associated to the collocation node $\xi_{k\ell}^J$. Remembering that the mass matrix for Legendre spectral elements is diagonal, that integration by parts leading to the stiffness matrix elements may be run backwards (thus introducing a minus sign), and finally that the quadratures are evaluated exactly using GLL quadrature, we conclude that the two methods are indeed the same for Dirichlet boundary conditions.

Figure 4.6.1(a) shows the topological structure of the collocation matrix L_C for the problem (4.6.1) with homogeneous Dirichlet conditions on the whole

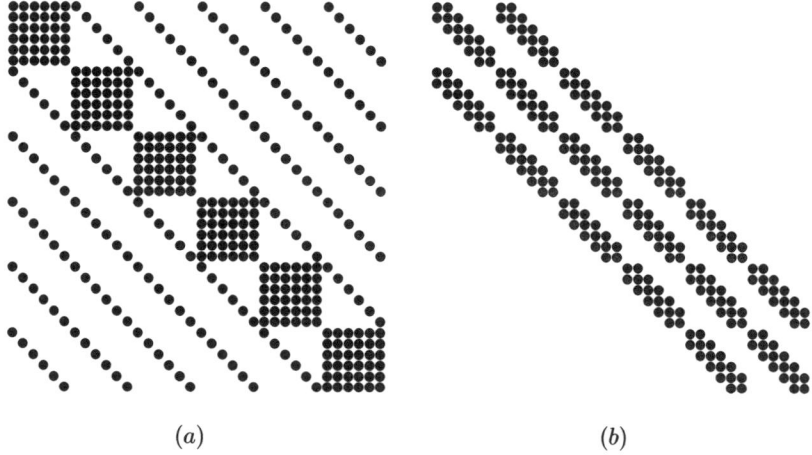

(a)   (b)

Figure 4.6.1. (a) Topological structure of the collocation matrix L_C for the elliptic problem (4.6.1). In this example $N = 7$. (b) Topological structure of the $\mathcal{Q}_1$ Lagrangian FE matrix L_P for the same problem on the same grid.

4.6. Collocation Discretizations

boundary. In this example, $N = 7$. The diagonal blocks correspond to the operator $I_2 \otimes D_1^{(2)}$, while off-diagonals are the elements of $D_2^{(2)} \otimes I_1$. The matrix has a sparse structure, but long-range coupling between the unknowns gives a large bandwidth. Obviously, other unknown ordering schemes might be chosen. But in that case the structure shown in the figure is lost, and the bandwidth is almost the same.

At this stage we notice that for diffusion problems with a factorizable space-dependent diffusion coefficient [i.e., $p = p_1(r_1) p_2(r_2)$], the collocation matrix L_C has a compact algebraic structure, similar to (4.6.9). We leave to the reader (as an exercise) the task of showing that, with the definitions of the matrix operators $\hat{P}_s$, $\hat{P}'_s$, and $\hat{D}_s$ ($s = 1, 2$) given by

$$\hat{P}_s := \mathrm{diag}(p_s(\xi_0^J), \ldots, p_s(\xi_N^J)),$$
$$\hat{P}'_s := \mathrm{diag}(p'_s(\xi_0^J), \ldots, p'_s(\xi_N^J)), \qquad (4.6.10)$$
$$\hat{D}_s := \hat{P}'_s \cdot D_s^{(1)} + \hat{P}_s \cdot D_s^{(2)},$$

the collocation matrix L_C in this case is

$$L_C = -\left[\frac{L_2}{L_1}(\hat{P}_2 \otimes \hat{D}_1) + \frac{L_1}{L_2}(\hat{D}_2 \otimes \hat{P}_1)\right] + q\frac{L_1 L_2}{4}(I_2 \otimes I_1). \quad (4.6.11)$$

The operator $D_s^{(1)}$ in (4.6.10) is, with simplified notation, the first-order derivative matrix of the GLJ basis functions on the Gaussian grids $D_N^{J,(1)}$, given in Appendix B [see (B.3.41)–(B.3.44)]. We remark also that $D_s^{(2)} = [D_s^{(1)}]^2$.

With factorizable p, the topological structure of the collocation matrix (4.6.11) is the same as before. This is true also for more general functions $p(r_1, r_2)$. In this case, however, no compact algebraic relationship such as (4.6.9) or (4.6.11), with the associated potential savings in matrix construction, will exist.

Iterative techniques are fundamental in solving the algebraic system $L_C \underline{u} = \underline{f}$ related to collocation discretizations. We have seen in preceding chapters that the spectral condition number κ of algebraic systems related to high-order methods grows like $O(N^4)$, where N is the number of unknowns in a single space dimension. To avoid any loss of accuracy resulting from the ill-conditioning of L_C, one evaluates $\underline{u}$ iteratively using a low-order (FD or FE) preconditioning scheme as outlined in Section 2.7.3.

We denote by L_P the matrix related to a low-order approximation (FD or FE) of the basic problem (4.6.1) that is built on the same GLJ grid. With lexicographical ordering of the unknowns L_P may be cast in compact form. The associated low-order FE problem is

$$\tilde{L}_P \underline{u} = \frac{L_1 L_2}{4} M_1 \otimes M_2 \underline{f}, \qquad (4.6.12)$$

with

$$\tilde{L}_P = p\left[\frac{L_2}{L_1}(M_2 \otimes K_1) + \frac{L_1}{L_2}(K_2 \otimes M_1)\right] + q\frac{L_1 L_2}{4}(M_2 \otimes M_1). \quad (4.6.13)$$

It turns out that

$$L_P = \left(\frac{L_1 L_2}{4} M_1 \otimes M_2\right)^{-1} \cdot \tilde{L}_P, \qquad (4.6.14)$$

where, in the FE case, the sets (K_1, M_1) and (K_2, M_2) refer to the one-dimensional tridiagonal stiffness and mass matrices of Q_1 Lagrangian FEs with N_1 and N_2 elements in the corresponding directions. The topological structure of L_P for the example is shown in Figure 4.6.1(b). In the FD case the mass matrices M_s are diagonal, and L_P has the familiar five-point stencil structure of the 2D Laplacian operator.

The matrix L_P can be used as a preconditioner of L_C. For instance, the preconditioned Richardson iteration scheme (2.7.23) is

$$L_P \underline{u}^{(k+1)} = L_P \underline{u}^{(k)} - \alpha_k \left(L_C \underline{u}^{(k)} - \underline{f}\right), \quad k = 0, 1, \ldots, \qquad (4.6.15)$$

where α_k is a relaxation factor needed to ensure convergence of the iterative algorithm, which may depend on the iteration index. As outlined in Section 2.7.3 using a functional formulation, the calculation starts with the determination of an initial (FE or FD) approximation $\underline{u}^{(0)}$ such that $L_P \underline{u}^{(0)} = \underline{f}$. Most of the time the subsequent iterative process (4.6.15) runs on a correction $\delta \underline{u}^{(k)}$ such that $\underline{u}^{(k+1)} = \underline{u}^{(k)} + \delta \underline{u}^{(k)}$, with

$$L_P \delta \underline{u}^{(k)} = -\alpha_k \left(L_C \underline{u}^{(k)} - \underline{f}\right), \qquad k = 0, 1, \ldots. \qquad (4.6.16)$$

This iterative process runs till a convergence criterion is satisfied. The convergence rate depends on the spectral radius of the operator $I - \alpha_k L_P^{-1} \cdot L_C$, as outlined in Section 2.7.2. We emphasize that the algebraic system (4.6.16) itself may be solved by using direct or iterative solution techniques. Whereas Deville and Mund [100] essentially used direct solution techniques, many authors have implemented standard iterative techniques. For instance, Canuto and Pietra [66] applied ADI techniques to solve two- and three-dimensional elliptic problems by Chebyshev collocation in mono- and multidomains, and

4.6. Collocation Discretizations

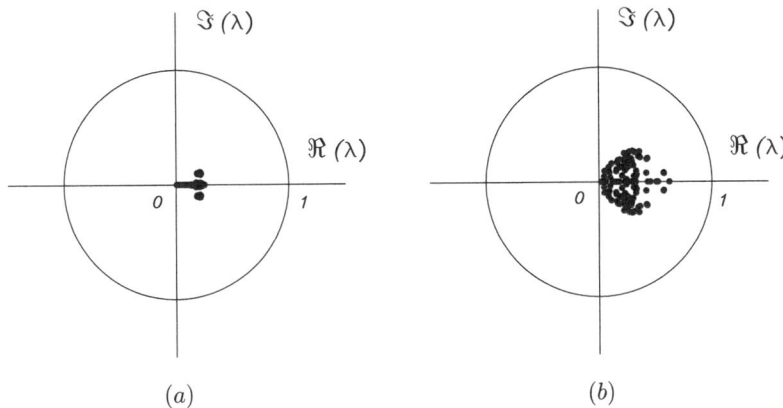

Figure 4.6.2. Spectrum of $I - L_P^{-1} \cdot L_C$ for $\mathcal{Q}_1$ FE preconditioning of two diffusion operators $\mathcal{L}u := -\nabla \cdot (p\nabla u) + u$, with Dirichlet conditions on the reference square $\hat{\Omega}$, and collocation on the Chebyshev grid with $N = 16$: (a) $p = 1$, and (b) $p = 1 + 10 x^2 y^2$.

Quarteroni and Zampieri [321] used the minimum residual (MR) and bi-CGSTAB iterative methods.

Figures 4.6.2, 4.6.3, and 4.6.4 display a few examples of eigenspectra of the operator $I - L_P^{-1} \cdot L_C$ for 2D diffusion problems with Dirichlet conditions, including problems with variable coefficients p. Figures 4.6.2 and 4.6.3 are related to FE preconditioning, whereas Figure 4.6.4 is related to FD preconditioning. In Figure 4.6.2, the preconditioner is the $\mathcal{Q}_1$ FE operator associated to

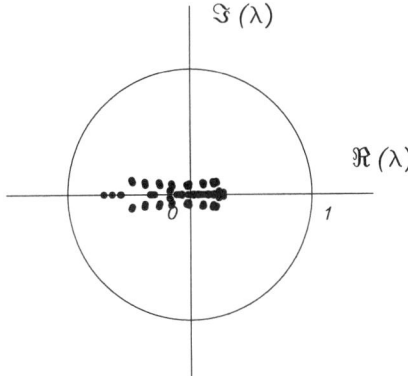

Figure 4.6.3. Spectrum of $I - L_P^{-1} \cdot L_C$ for $\mathcal{Q}_1$ FE preconditioning of $\mathcal{L}u := -\nabla \cdot [(1 + x^2 y^2)\nabla u] + u$, with Dirichlet conditions on the reference square $\hat{\Omega}$, and collocation on the Chebyshev grid with $N = 16$. In this case, the preconditioner L_P results from the FE approximation of $\mathcal{L}u := -\Delta u + u$.

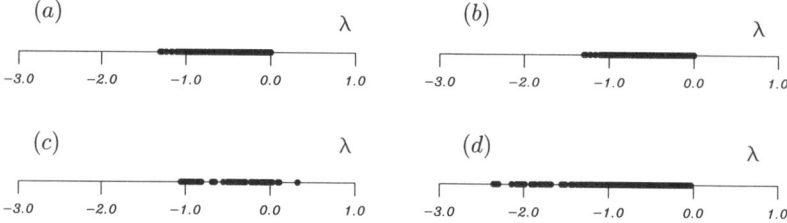

Figure 4.6.4. Spectrum of the operator $I - L_P^{-1} \cdot L_C$ for FD preconditioning of $\mathcal{L}u := -\nabla \cdot (p\nabla u) + u$, with Dirichlet conditions on the reference square $\hat{\Omega}$, and collocation on the Chebyshev grid with $N = 16$: (a) $p = 1$, (b) $p = 1 + x^2 y^2$, (c) $p = 1 + 10 x^2 y^2$, and (d) $p = 1 + x^2 y^2$, the preconditioner L_P being such that $L_P u$ is the FD approximation of $\mathcal{L}u := -\Delta u + u$.

the basic problem on the same Gaussian grid. The complex eigenvalues are located well inside the unit circle, an indication that fast convergence of (4.6.16) should be obtained with the relaxation factor $\alpha_k = 1$, even for the (nonfactorizable) diffusion coefficient $p = 1 + x^2 y^2$ – an indication of the robustness of the algorithm.

This robustness is even clearer in Figure 4.6.3, where L_P is the Q_1 FE operator associated to an auxiliary problem that differs from the basic one. Again, the eigenvalues are located inside the unit circle, and the algorithm should converge, even with $\alpha_k = 1$, though at a slower pace. Finally, Figure 4.6.4 shows that with FD preconditioning, the (real) spectrum occupies a domain larger than the unit interval and that the algorithm cannot converge without a relaxation factor $\alpha_k \neq 1$.

These numerical results on the eigenvalues of $L_P^{-1} \cdot L_C$, as well as those published by Quarteroni and Zampieri [321] and Zampieri [421], for a long time received no theoretical proof. However, in recent years Parter and Rothman [294] and Kim and Parter [224, 225] have obtained some first results for the FE and FD preconditioning of Legendre and Chebyshev collocation approximations to diffusion–advection problems. Their work gives some bounds to the eigenspectrum of $L_P^{-1} \cdot L_C$, depending on the properties of the preconditioner. Nevertheless, these encouraging results, the theoretical foundation of the (FE and FD) preconditioning of spectral calculations in the general case remains largely an open question.

Figure 4.6.5 shows the convergence histories in L^∞ norm for the Chebyshev collocation solution of a 2D diffusion problem on the reference square $\hat{\Omega} := [-1, +1]^2$, with Dirichlet boundary conditions:

$$-\nabla \cdot [(1 + x^2 y^2)\nabla u(x, y)] + u(x, y) = f, \qquad (x, y) \in \hat{\Omega}, \qquad (4.6.17)$$

4.6. Collocation Discretizations

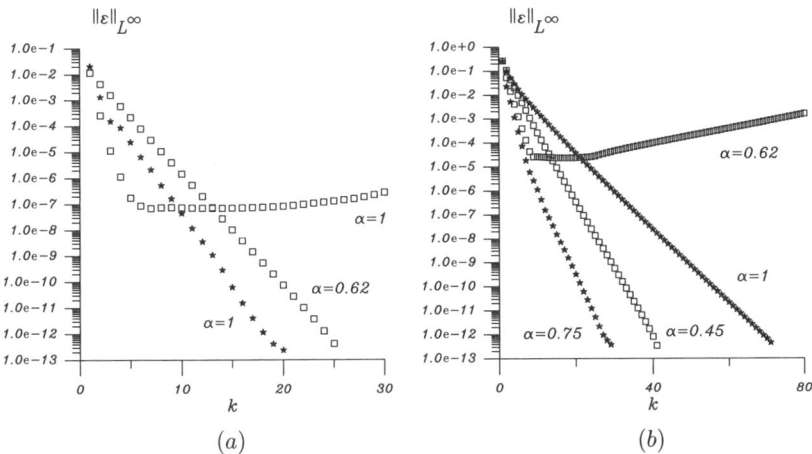

Figure 4.6.5. Convergence histories for the Chebyshev collocation solution of (4.6.17) with $N = 17$: (a) the preconditioner results from the FE ($\star$) or FD ($\square$) approximation of (4.6.17), (b) the preconditioner results from the FE or FD approximations of $\mathcal{L}u := -\Delta u + u$.

where the r.h.s. $f(x, y)$ is such that the solution to the problem is $u(x, y) = (1 - x^2)(1 - y^2)\exp(x + y)$. The Chebyshev expansion uses $N_x = N_y = 17$, and both preconditioning techniques ($\star$ for FE and $\square$ for FD) are applied. Figure 4.6.5(a) corresponds to the case where the preconditioner is the FE (or FD) approximation of the problem on the same grid; Figure 4.6.5(b) shows the results when the preconditioner results from the FE (or FD) approximation of the auxiliary expression $(-\Delta u + u)$ on the same grid.

One should notice that, in the regular case (a), although there is a sharp error drop with FD preconditioning and $\alpha = 1$, the computation ultimately diverges. The choice of α with FD preconditioning is even more important in case (b). In comparison, FE preconditioning is more robust, although in case (b) an adequate choice of α may decrease the iteration number till convergence by almost a factor of two. We note that all of these results can be further accelerated by using a projection-based iteration, such as the GMRES method discused in Section 2.8.

Inhomogeneous Neumann Boundary Conditions

Let us now look at the case of Neumann boundary conditions and consider the problem (4.6.1) with inhomogeneous conditions $\mathcal{B}u(\mathbf{x})$, which split into

$$u(\mathbf{x}) = g(\mathbf{x}) \quad \forall \mathbf{x} \in \partial\Omega_D,$$
$$-p\frac{\partial u}{\partial n}(\mathbf{x}) = h(\mathbf{x}) \quad \forall \mathbf{x} \in \partial\Omega_N, \quad (4.6.18)$$

with $\partial\Omega = \partial\Omega_D \cup \partial\Omega_N$. We assume that boundary conditions of the same type apply on whole sides of Ω. The symbol $\partial/\partial n$ denotes the outer normal derivative at the boundary. Furthermore, we suppose that Dirichlet conditions are enforced on those vertices of Ω that belong to the closure of $\partial\Omega_D$.

With regard to the collocation equations on the $(N+1) \times (N+1)$ GLJ grid, we have as before

$$\mathcal{L}\hat{u}(\xi_{ij}^J) = f(\xi_{ij}^J) \quad \forall \xi_{ij}^J \in \mathcal{G}_N^{(\alpha,\beta)} \cap \hat{\Omega},$$

$$\hat{u}(\xi_{ij}^J) = g(\xi_{ij}^J) \quad \forall \xi_{ij}^J \in \mathcal{G}_N^{(\alpha,\beta)} \cap \partial\hat{\Omega}_D, \qquad (4.6.19)$$

$$-p\frac{\partial \hat{u}}{\partial n}(\xi_{ij}^J) = h(\xi_{ij}^J) \quad \forall \xi_{ij}^J \in \mathcal{G}_N^{(\alpha,\beta)} \cap \partial\hat{\Omega}_N.$$

A difficulty arises at corners that are common to Neumann sides, because enforcing the values of the two normal derivatives at these points would lead to an overdetermined linear system. As we will see below, this difficulty may be overcome by using the FE–FD preconditioning scheme.

We have learned in Section 2.1 that, in the variational formulation of BVPs, Neumann conditions are satisfied "naturally," whereas Dirichlet conditions have to be included in the space of test functions. The variational formulation of Equation (4.6.1) with boundary conditions (4.6.18) is as follows: Find $u(\mathbf{x}) \in H_g^1 := \{v \in H^1(\Omega) | v(\mathbf{x}) = g(\mathbf{x}) \forall \mathbf{x} \in \partial\Omega_D\}$ such that

$$\int_\Omega (p\nabla u \cdot \nabla v + quv)\,d\mathbf{x} = \int_\Omega f v\,d\mathbf{x} - \int_{\partial\Omega_N} hv\,dS \qquad (4.6.20)$$

for all $v(\mathbf{x}) \in H^1(\Omega)$, where the second term in the r.h.s. includes the inhomogeneous condition on the (outer) normal derivative of the solution. After discretization, the problem (4.6.20) becomes: Find $u(\mathbf{x}) \in V_{N,g}$ such that, for all v in $V_N \subset H^1(\Omega)$,

$$\int_\Omega (p\nabla u \cdot \nabla v + quv)\,d\mathbf{x} = \int_\Omega f v\,d\mathbf{x} - \int_{\partial\Omega_N} hv\,dS, \qquad (4.6.21)$$

where V_N and $V_{N,g}$ are piecewise polynomial spaces, and the elements of $V_{N,g} \subset H_g^1$ satisfy the inhomogeneous Dirichlet boundary conditions on $\partial\Omega_D$.

In the variational formulation (4.6.21), the boundary condition on the normal derivative at $\partial\Omega_N$ is reached asymptotically with increasing dimension of the space of trial functions $V_{N,g}$. This means that, during the preconditioning process, the constraint (4.6.19) on $\partial\Omega_N$ is not satisfied exactly. One has therefore to include an additional residual for the boundary condition on $\partial\Omega_N$.

4.6. Collocation Discretizations

The functional formulation of the Q_1 FE preconditioning scheme introduced in Section 2.7.3 must therefore be adapted. The main steps of the algorithm (i.e., steps 3 to 5) become:

- Step 3. Evaluate a first guess u_h^0 such that $\mathcal{A}_{\text{FE}} u_h^0 = f$, where the operator $\mathcal{A}_{\text{FE}}$ relates to the variational functional (4.6.21); that is: Find $u_h^0(\mathbf{x}) \in V_{N,g}$ such that, for all v in $V_N \subset H^1(\Omega)$

$$\int_\Omega \left(p \nabla u_h^0 \cdot \nabla v + q u_h^0 v \right) d\mathbf{x} = \int_\Omega f v \, d\mathbf{x} - \int_{\partial \Omega_N} h v \, dS,$$

- Step 4. Evaluate the spectral residuals r_N^0 and $r_{N,B}^0$ associated with the interior nodes and the boundary nodes on $\partial \Omega_N$, respectively:

$$r_N^0 := \mathcal{L} u_N^0 - f, \qquad r_{N,B}^0 := p \frac{\partial u_N^0}{\partial n} + h. \qquad (4.6.22)$$

The element $u_N^0 := I_N u_h^0$ is the spectral interpolation of the FE solution u_h^0 in $\mathbb{P}_N(\Omega)$.

- Step 5. Iterate till convergence:

$$u_N^{k+1} = u_N^k - \alpha I_N \mathcal{A}_{\text{FE}}^{-1} \left[I_h^0 \left(\mathcal{L} u_N^k - f \right), J_h^0 \left(p \frac{\partial u_N^k}{\partial n} + h \right) \right]. \qquad (4.6.23)$$

In this relationship, I_h^0 and J_h^0 are two FE interpolation operators, inside Ω and along $\partial \Omega_N$ respectively. The operator I_h^0 is the two-dimensional Q_1 FE interpolation operator, obtained by tensor product of the one-dimensional FE interpolation operator defined in Chapter 2 (see Section 2.2.2). The operator J_h^0 interpolates linearly the values of the residual $r_{N,B}^0$ on the boundary nodes along $\partial \Omega_N$. On a vertex belonging to $\partial \Omega_D$, the nodal value of $r_{N,B}^0$ is set equal to zero. The two entries inside the brackets are the source terms in Equation (4.6.21) during the iteration process. Both terms should become vanishingly small as k increases to infinity.

An important result due to Canuto and Pietra [66] is that the Neumann boundary conditions are enforced asymptotically (i.e., with $k \to \infty$) at the boundary nodes along a Neumann side when its vertices belong to $\partial \Omega_D$. When $\partial \Omega_N$ contains several sides having a common vertex, the asymptotic error on the Neumann boundary condition is globally small. This is because the boundary residual is orthogonal to the trace of all Q_1 test functions, with respect to numerical quadrature along $\partial \Omega_N$.

Table 4.6.1. *Relative errors in the maximum norm of a problem with the exact solution* $u(x, y) = \sin 2\pi x \sin \pi y$, *for several combinations of boundary conditions (see [66]).*

	Relative error				
$N_x \times N_y$	DDDD ($q=0$)	NDDD ($q=0$)	NDND ($q=0$)	NNDD ($q=0$)	NNNN ($q=1$)
8×8	0.30e − 01	0.10e + 00	0.97e − 01	0.30e + 00	0.26e + 00
16×16	0.23e − 06	0.54e − 05	0.53e − 05	0.84e − 05	0.76e − 05
32×32	0.27e − 14	0.54e − 12	0.15e − 11	0.23e − 11	0.12e − 11

These conclusions are illustrated in Table 4.6.1, taken from [66]. The table gives some numerical results for the problem (4.6.1) with boundary conditions (4.6.18), on the reference domain $\hat{\Omega}$. Computations have been made with Q_1 FE and Chebyshev collocation, on the Laplace and Helmholtz equations (i.e., $p = 1$ and $q = 0, 1$) with exact solution $u(x, y) = \sin 2\pi x \sin \pi y$. The table gives the relative errors in maximum norm for some combinations of boundary conditions. The convergence is clearly exponential for every combination of boundary conditions. The errors show slightly larger values, however, in the cases where the Neumann conditions apply on sides having a common vertex (see, for instance, NNDD).

The Multidomain Approach

The handling of Neumann boundary conditions introduced in the previous section may be used to implement a multidomain collocation solution of the problem. We consider again Equation (4.6.1) with variable coefficients, but for simplicity we will assume that $u(\mathbf{x}) = 0$ on $\partial\Omega$. The aim is to introduce a partitioning of Ω domain into four nonoverlapping subdomains Ω^k ($1 \leq k \leq 4$) as suggested in Figure 4.6.6(b), and to apply a collocation approximation on each subdomains. The difficulty is to piece the subdomain approximations together.

There is some freedom in the choice of the GLJ grids at the subdomain level, but adjacent approximations must have the same polynomial degree along their common border. Here again, we assume that the polynomial degree is equal to N in both space directions and in all subdomains. Therefore, let $\mathcal{G}_{N,k}^{(\alpha,\beta)}$ denote the $(N + 1) \times (N + 1)$ GLJ collocation grid mapped from $\hat{\Omega}$ onto the subdomain Ω^k, and $\mathcal{G}_N^{(\alpha,\beta)} = \bigcup_k \mathcal{G}_{N,k}^{(\alpha,\beta)}$. With these conditions, an approximate solution $\hat{u}$ belongs to the polynomial space

$$\mathcal{P}_N(\Omega) := \{v \in C_0^0(\bar{\Omega}) | v_{|\Omega^k} \in \mathbb{P}_N(\Omega^k), k = 1\ldots, 4\}.$$

4.6. Collocation Discretizations

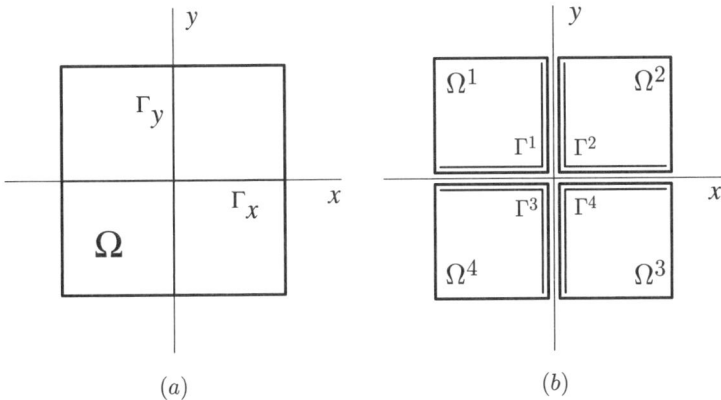

Figure 4.6.6. (*a*) The quadrangular integration domain Ω and (*b*) its partition into four nonoverlapping subdomains Ω^k.

and the collocation equations become

$$\mathcal{L}\hat{u}_k(\mathbf{x}) = f(\mathbf{x}) \qquad \forall \mathbf{x} \in \mathcal{G}_{N,k}^{(\alpha,\beta)} \cap \Omega^k,$$

$$\hat{u}_k(\mathbf{x}) = \hat{u}_l(\mathbf{x}) \qquad \forall \mathbf{x} \in \mathcal{G}_{N,k}^{(\alpha,\beta)} \cap \mathcal{G}_{N,l}^{(\alpha,\beta)}, \qquad (4.6.24)$$

$$p\frac{\partial \hat{u}_k}{\partial n^{(k)}}(\mathbf{x}) = -p\frac{\partial \hat{u}_l}{\partial n^{(l)}}(\mathbf{x}) \qquad \forall \mathbf{x} \in \mathcal{G}_{N,k}^{(\alpha,\beta)} \cap \mathcal{G}_{N,l}^{(\alpha,\beta)},$$

with $k, l = 1, \ldots, 4$. The outer normal derivative to the subdomain Ω^k at a boundary node $\mathbf{x}$ is denoted $n^{(k)}$. We face again the difficulty that two interface conditions apply at a common corner [see Figure 4.6.6(*a*)], which cannot be enforced at the same time. Just as with the Neumann boundary condition, this difficulty can be eased using the FE preconditioning algorithm.

We go back to the basic equation and remark that multiplying (4.6.1) by a test function $v(x) \in C_0^1(\Omega)$ and integrating over each subdomain gives

$$\int_{\Omega^k} (p\nabla u \cdot \nabla v + quv)\, d\mathbf{x} = \int_{\Omega^k} fv\, d\mathbf{x} + \int_{\Gamma^k} p\frac{\partial u}{\partial n} v\, dS, \qquad (4.6.25)$$

where Γ^k denotes the *internal* boundary of Ω^k shown in Figure 4.6.6(*b*). Adding all separate contributions, one gets

$$\int_{\Omega} (p\nabla u \cdot \nabla v + quv - fv)\, d\mathbf{x} = \int_{\Gamma_x} \left[p\frac{\partial u}{\partial y}\right] v\, dS + \int_{\Gamma_y} \left[p\frac{\partial u}{\partial x}\right] v\, dS. \qquad (4.6.26)$$

In this relationship, Γ_x and Γ_y are the subdomain internal boundaries as indicated on Figure 4.6.6(*a*), which is a simple rearrangement of the set $\{\Gamma^k\}$. The

214   4. Multidimensional Problems

quantities between brackets refer to the jumps in $p\frac{\partial u}{\partial y}$ and $p\frac{\partial u}{\partial x}$ across Γ_x and Γ_y, which should in fact vanish identically, as can be shown rigorously. The collocation equations (4.6.24) precisely deal with such a constraint.

The reasoning made about the Neumann boundary conditions may be repeated. A $\mathcal{Q}_1$ FE approximation $u_h \in H_0^1(\Omega)$ of the problem does *not* satisfy the variational interface condition

$$\int_{\Gamma_x} \left[p\frac{\partial u_h}{\partial y} \right] v \, dS + \int_{\Gamma_y} \left[p\frac{\partial u_h}{\partial x} \right] v \, dS = 0 \qquad \forall v \in H_0^1 \Omega. \qquad (4.6.27)$$

It would do so in the limit of a mesh size $h := (h_x, h_y)$ going to zero. With a nonzero mesh, interface residuals exist along Γ_x and Γ_y that must be handled by the collocation process.

The functional formulation of the FE preconditioned collocation scheme in the multidomain approach results from still another modification of the algorithm introduced in Section 2.7.3. Steps 3 to 5 of the algorithm become

- Step 3. Evaluate a first guess u_h^0 such that $\mathcal{A}_{\text{FE}} u_h^0 = f$, where the operator $\mathcal{A}_{\text{FE}}$ relates to the following problem: Find $u_h^0(\mathbf{x}) \in V_{N,0}$ such that, for all v in $V_{N,0} \subset H_0^1(\Omega)$,

$$\int_\Omega \left(p \nabla u_h^0 \cdot \nabla v + q u_h^0 v \right) d\mathbf{x} = \int_\Omega f v \, d\mathbf{x}.$$

- Step 4. Evaluate the spectral residuals r_N^0 on the interior nodes in the subdomains Ω^k, and the residuals $r_{N,y}^0, r_{N,x}^0$ associated with the interior boundary nodes along Γ_x and Γ_y, respectively:

$$r_N^0 := \mathcal{L} u_N^0 - f, \qquad r_{N,z}^0 := \left[p \frac{\partial u_N^0}{\partial z} \right] \qquad (z = x, y). \qquad (4.6.28)$$

The element $u_N^0 := I_N u^0$ is the spectral interpolation of the FE solution u_h^0 in $\mathcal{P}_N(\Omega)$.

- Step 5. Iterate till convergence:

$$u_N^{k+1} = u_N^k - \alpha I_N \mathcal{A}_{\text{FE}}^{-1} \left[I_h^0 r_N^k, J_x^0 r_{N,y}^k, J_y^0 r_{N,x}^k \right]. \qquad (4.6.29)$$

The interpolation operators J_x^0 and J_y^0 in (4.6.29) are quite similar to J_h^0 defined above. Here also it is possible to show that, asymptotically for k increasing to infinity, the partial differential equation is satisfied at all interior nodes, and the jumps in $p\frac{\partial u}{\partial n}$ across the interfaces are orthogonal to the trace of all $\mathcal{Q}_1$ finite elements along Γ_x and Γ_y. Canuto and Pietra discuss some other techniques for imposing the matching of the four spectral solutions at the interfaces.

4.6. Collocation Discretizations

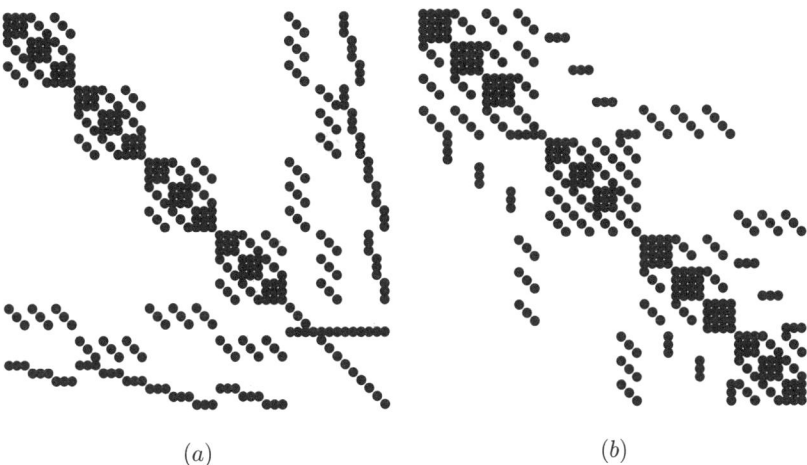

Figure 4.6.7. Two topological structures of L_C, the orthogonal collocation matrix on a 2×2 multidomain, depending on the ordering of the unknowns (see text). In this example, the polynomial expansion in each subdomain has $N = 5$.

Figure 4.6.7 displays two different topological structures of the collocation matrix L_C resulting from (4.6.24) on a 2×2 multidomain, depending on the ordering of the unknowns. The polynomial expansion in each subdomain has $N = 5$. In Fig. 4.6.7(a) the nodes are partitioned in two sets Ω and Γ, where Ω comprises the internal nodes in the subdomains Ω^k ($k = 1, \ldots, 4$), and Γ, the nodes belonging to the (inner) boundaries. Lexicographical order applies in each subdomain, and the subdomains are described from left to right and bottom to top. In Figure 4.6.7(b) lexicographical order applies again in each subdomain, this time including the nodes on the inner boundaries. The sweeping of the subdomains is the same. However, the reader should understand that the topological structure of L_C has no influence upon the inversion of the algebraic system, because this is done using FE–FD preconditioning. It is given here for the purpose of illustration.

Table 4.6.2, taken from [66], compares numerical results for different problems, obtained using Chebyshev collocation on monodomains and multidomains. The multidomain approach preserves the spectral accuracy. However, for a given number of degrees of freedom, the relative error in maximum norm is lower when collocation is used on a monodomain rather than on a multidomain. The multidomain collocation approach described in this section is particularly useful for complicated geometries that may be assembled from quadrangular domains. For a thorough discussion of the topic, the reader should consult [66].

Table 4.6.2. *Accuracy of the spectral-domain decomposition method: relative errors in the case of uniformly structured solutions of (4.6.1) (see [66]).*

	Relative error		
Solution	$u(x,y) = \sin 2\pi x \sin \pi y$	$u(x,y) = \cos 2\pi x \sin \pi y$	$u(x,y) = \cos 2\pi x \cos \pi y$
4 dom. (4 × 4)	0.14e + 00	0.54e + 00	0.62e + 00
1 dom. (8 × 8)	0.24e − 01	0.33e − 01	0.35e − 01
4 dom. (8 × 8)	0.13e − 03	0.11e − 02	0.12e − 02
1 dom. (16 × 16)	0.23e − 06	0.10e − 06	0.11e − 06
4 dom. (16 × 16)	0.27e − 11	0.43e − 10	0.49e − 10
1 dom. (32 × 32)	0.46e − 14	0.32e − 14	0.38e − 14

4.6.2 The Advection–Diffusion Case

To complete this section on collocation discretizations in multidimensional problems, we revisit the steady advection–diffusion problem already studied in Section 3.6, and we consider the BVP with homogeneous Dirichlet boundary conditions

$$\mathcal{L}_{\mathrm{AD}} u(\mathbf{x}) := -\nabla \cdot (\nu \nabla u) + \mathbf{c} \cdot \nabla u = f(\mathbf{x}), \quad \mathbf{x} \in \Omega, \\ u(\mathbf{x}) = 0, \quad \mathbf{x} \in \partial\Omega, \quad (4.6.30)$$

on the rectangular monodomain $\Omega := (0, L_1) \times (0, L_2) \subset \mathbb{R}^2$. The coefficients in the differential equations may be space-dependent. Physically, the vector $\mathbf{c} := (c_x, c_y)^T$ is the transport velocity of the quantity $u(\mathbf{x})$. Many important situations in pratice correspond to cases where diffusion is negligible compared to advection (i.e., $\nu \ll |\mathbf{c}|$). Here, we consider the constant-diffusion-coefficient case. We also assume that the domain Ω has been mapped onto the reference domain $\hat{\Omega}$, and we describe the algorithm on $\hat{\Omega}$.

Several attempts have been made to use the FE–FD preconditioning scheme, described above, for advection–diffusion problems [101, 143, 282]. As mentioned in Chapter 3, the most successful algorithm has been proposed by Funaro [143], who suggested collocating Equation (4.6.30) on a staggered grid rather than on the standard GLL grid. A similar approach was used a little later by Pinelli et al. for the Chebyshev collocation case [311].

The generation of the staggered grid in the one-dimensional case has been explained in Section 3.6.2. It is based on the property that, since in the pure diffusion case with operator $\mathcal{L}_D$ the GLL grid points are solutions to the polynomial equation $\mathcal{L}_D \chi_N^{(0,0)} = 0$, in the advection–diffusion case one might look

4.6. Collocation Discretizations

for an auxiliary grid, a polynomial solution to

$$\mathcal{L}_{AD}\chi_N^{(0,0)} := -\nu\frac{d^2\chi_N^{(0,0)}}{dx^2} + c\frac{d\chi_N^{(0,0)}}{dx} = 0. \tag{4.6.31}$$

Taking some basic properties of the Legendre polynomials into account, (4.6.31) reduces to

$$-\nu L'_N(x) + cL_N(x) = 0. \tag{4.6.32}$$

In two space dimensions, the generation of the staggered grid (also called the *upwind* grid) is based on the same principle. Here, the 2D equation $\mathcal{L}_{AD}\chi_N^{(0,0)}(x,y) = 0$, with

$$\chi_N^{(0,0)}(x,y) := (1-x^2)(1-y^2)L'_N(x)L'_N(y), \tag{4.6.33}$$

leads to two different relationships for the determination of the shifted node in the vicinity of $(\xi_i, \xi_j) \in \mathcal{G}_N^{(0,0)}$:

$$-\nu\left[2 - \xi_i^2 - \xi_j^2\right]L'_N(x) + c_x(\xi_i, \xi_j)(1 - \xi_j^2)L_N(x) = 0,$$
$$-\nu\left[2 - \xi_i^2 - \xi_j^2\right]L'_N(y) + c_y(\xi_i, \xi_j)(1 - \xi_i^2)L_N(y) = 0. \tag{4.6.34}$$

These equations are easily solved by using a few steps of a secant method, and the solution (θ_{ij}, τ_{ij}), with $1 \leq i, j \leq N-1$ satisfies the inequalities

$$\zeta_i < \theta_{ij} < \zeta_{i+1}, \qquad \zeta_j < \tau_{ij} < \zeta_{j+1},$$

where $\{\zeta\}$ designates the set of Gauss–Legendre (GL) quadrature nodes. Again, for a vanishing diffusion coefficient, the upwind grid gets close to the GL grid, whereas for a pure diffusion problem one recovers the classical GLL grid. The staggered grid $\mathcal{G}_{s,N}^{(0,0)}$ is usually rather unstructured, as shown on Figure 4.6.8 for a model problem. To simplify the notation we will drop the superscript on both grids, its absence implying the grid is based on Legendre polynomials.

Funaro's FD preconditioning scheme uses both the $\mathcal{G}_N$ and $\mathcal{G}_{s,N}$ grids. As previously, let L_C denote the collocation operator on $\mathcal{G}_N$. Although in the advection–diffusion case calculations are performed on the upwind grid $\mathcal{G}_{s,N}$, the solution approximation $u_N \in \mathbb{P}_N(\Omega)$ uses the values of the unknowns on the regular grid $\mathcal{G}_N$. An interpolation is then needed to switch from $\mathcal{G}_N$ to $\mathcal{G}_{s,N}$. We denote by S_N the associated matrix operator. As shown by Funaro, one reduces the computation cost entailed by the interpolation in 2D, with a simplified interpolation technique, to $O(N^3)$ instead of $O(N^4)$ for a regular interpolation scheme. For a polynomial function $r_N(x, y) \in \mathbb{P}_N(\Omega)$ defined on

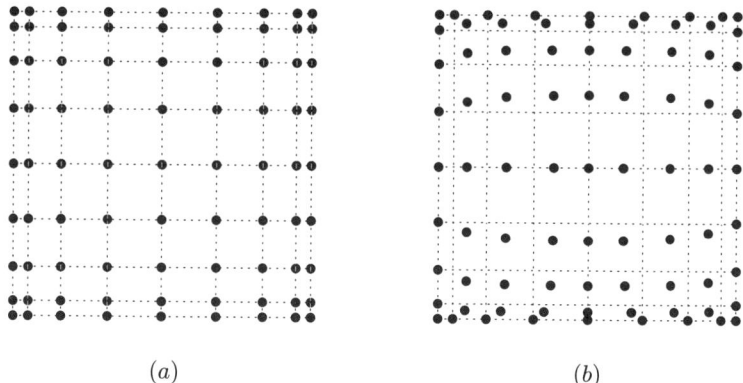

Figure 4.6.8. (a) The Gauss–Lobatto–Legendre grid with $N_x = N_y = 8$, and (b) the corresponding upwind grid for the problem $-\frac{1}{100}\Delta u + x\frac{\partial u}{\partial x} - y\frac{\partial u}{\partial y} = f$.

$\mathcal{G}_N$, the approximate value on the shifted node (θ_{ij}, τ_{ij}) is given by a Taylor series expansion limited to order 2:

$$r_N(\theta_{ij}, \tau_{ij}) \approx \sum_{k+\ell=0}^{2} \frac{(\theta_{ij} - \xi_i)^k}{k!} \frac{(\tau_{ij} - \xi_j)^\ell}{\ell!} (r_N)_{x,y}^{(k,\ell)}(\xi_i, \xi_j). \quad (4.6.35)$$

In practice however, the evaluation of $r_N(\theta_{ij}, \tau_{ij})$ uses a simplified Lagrange interpolation built on the coordinate lines crossing at (ξ_i, ξ_j). It is given by

$$r_N(\theta_{ij}, \tau_{ij}) = \sum_{\ell=0}^{N} r_N(\xi_\ell, \xi_j)\pi_\ell(\theta_{ij}) + \sum_{\ell=0}^{N} r_N(\xi_i, \xi_\ell)\pi_\ell(\tau_{ij}) - r_N(\xi_i, \xi_j),$$

$$(4.6.36)$$

where the functions π_ℓ are the basic Lagrange interpolation polynomials on the GLL grid. The multiplication cost of the operator S_N built on (4.6.36) by a vector is $O(N^3)$, and in the limit of pure diffusion problems, S_N tends to the identity operator.

A further important item to be dealt with is the preconditioning matrix L_P. Funaro's preconditioning scheme is such that in the limit of purely diffusive problems, the operator is built upon the classical five-point FD stencil on an irregular grid. The construction procedure for $L_P \equiv L_{\text{FD}}$ is based on the property that the truncation error associated with second-order finite differencing vanishes for a quadratic function. For instance, with the operator $\mathcal{L}_D$ given by (4.6.1), the biquadratic polynomial $\pi_2(x, y) \in \mathbb{P}_2(\Omega)$ uniquely defined by the set of values $\{u(\xi_{i+m}, \xi_{j+n})\}$ (with $-1 \leq m, n \leq 1$) at the nodes surrounding the internal node $(\xi_i, \xi_j) \in \mathcal{G}_N$ is such that the ℓ^*-component of the vector $L_{\text{FD}}\underline{u}$

4.6. Collocation Discretizations

is equal to

$$L_{FD}\underline{u}|_{\ell^*} = \mathcal{L}_D\pi_2(x, y)|_{(\xi_i, \xi_j)}, \quad 1 \le i, j \le N - 1. \quad (4.6.37)$$

One assumes lexicographic ordering of the coefficients in $\underline{u}$, and therefore the index $\ell^* = (j - 1)(N - 1) + i$, with $1 \le \ell^* \le (N - 1)^2$. Boundary conditions are enforced as usual in FD methods. By extension, and, taking into account previous remarks, the preconditioning operator in the advection–diffusion case, L_{FD}, is such that

$$L_{FD}\underline{u}|_{\ell^*} = \mathcal{L}_{AD}\pi_2(x, y)|_{(\theta_{ij}, \tau_{ij})}, \quad 1 \le i, j \le N - 1. \quad (4.6.38)$$

where $\mathcal{L}_{AD}$ is given in (4.6.30). The equation (4.6.38) leads to a nine-point stencil, the elements of which can be found in [143]. One verifies easily that for $|\mathbf{c}| = 0$, the five-point diffusion stencil is recovered.

Finally, the last item is the iterative scheme used to obtain the preconditioned collocation solution of the problem. Instead of the Richardson scheme (2.7.23), the preconditioning algorithm due to Funaro uses an iterative scheme derived from the DuFort–Frankel method described in Birkhoff and Lynch [44]:

$$\underline{u}^{(k+1)} = \frac{2\sigma_1}{1 + 2\sigma_1\sigma_2}\left[-L_{FD}^{-1}(L_C\underline{u}^{(k)} - \underline{f}) + 2\sigma_2\underline{u}^{(k)}\right] + \frac{1 - 2\sigma_1\sigma_2}{1 + 2\sigma_1\sigma_2}\underline{u}^{(k-1)}, \quad (4.6.39)$$

with $k = 1, 2, \ldots$. Every step of the iterative process involves three successive approximations, the algorithm being started with a guess $\underline{u}^{(0)} = \underline{u}^{(1)}$. The coefficients σ_1 and σ_2 depend on the eigenvalues of the iteration operator: $\sigma_1 = (\lambda_M\lambda_m)^{1/2}$ and $\sigma_2 = (\lambda_M + \lambda_m)/4$, λ_m and λ_M being respectively the minimum and maximum eigenvalues of $L_{FD}^{-1} \cdot L_C$. Strategies for the evaluation of these parameters can be found in the references.

The preconditioning-algorithm collocation solutions of advection–diffusion problems on a staggered grid may be summarized into a few steps.

- Step 1. Build the staggered collocation grid $\mathcal{G}_{s,N}$.
- Step 2. Assemble the collocation and the FD preconditioning matrices L_C and L_{FD}. The collocation matrix is built on $\mathcal{G}_N$. The FD matrix L_{FD} results from the enforcement of (4.6.38) at every internal node belonging to $\mathcal{G}_{s,N}$.
- Step 3. Assemble the shifting matrix S_N from $\mathcal{G}_N$ to $\mathcal{G}_{s,N}$.
- Step 4. Evaluate a first guess $\underline{u}^{(0)}$ such that $L_{FD}\underline{u}^{(0)} = S_N\underline{f}$, and set $\underline{u}^{(1)} = \underline{u}^{(0)}$.

Successively, for $k \ge 1$ till convergence:

- Step 5. Compute $L_C \underline{u}^{(k)}$, and evaluate the spectral residual on the staggered grid, $\underline{r}^{(k)} := S_N(L_C \underline{u}^{(k)} - \underline{f})$. If $\|\underline{r}^{(k)}\| \leq 10^{-n}$ (n given), then stop. If not, go to step 6.
- Step 6. Solve $L_{FD}\underline{v} = \underline{r}^{(k)}$, and update the solution $u_N(x, y)$ according to (4.6.39), that is,

$$\underline{u}^{(k+1)} = \frac{2\sigma_1}{1 + 2\sigma_1\sigma_2}\left(-\underline{v} + 2\sigma_2 \underline{u}^{(k)}\right) + \frac{1 - 2\sigma_1\sigma_2}{1 + 2\sigma_1\sigma_2}\underline{u}^{(k-1)},$$

and go to step 5;

In step 4, $\underline{f}$ contains the values of the r.h.s. of (4.6.30) at the nodes of the GLL grid in lexicographic ordering. We avoid detailing the selection of σ_1 and σ_2. Approximate evaluations of λ_m and λ_M (hence of σ_1 and σ_2) are indicated in [144].

Numerical results discussed in [143] show that the eigenspectrum of $L_{FD}^{-1} \cdot L_C$ is complex and located outside the unit circle. Spectral accuracy is reached in about 30 iterations, and, most important, evaluation of the residuals on the upwind grid eliminates the high-frequency oscillations polluting solutions obtained by collocation on the GLL grid. Pinelli et al. use almost the same algorithm (except for the Richardson scheme instead of the DuFort–Frankel) and get to the same conclusions [309, 310]. As a last comment we mention that the preconditioning algorithm discussed in [144] applies to a broader class of second-order BVPs than the problem (4.6.30) described here.

4.7 Parabolic Problems

We consider the solution of parabolic problems of the form

$$\frac{\partial u}{\partial t} = \nabla \cdot (p \nabla u) + f \quad \text{in } \Omega \times (0, T),$$
$$u(\mathbf{x}, t) = g(\mathbf{x}, t) \text{ on } \partial\Omega, \quad u(\mathbf{x}, 0) = u_0(\mathbf{x}), \quad (4.7.1)$$

which is a model for the unsteady heat equation with spatially varying conductivity, $p(\mathbf{x}) > 0$, and also constitutes one component of the momentum equation in viscous fluid flow. Initially, we consider $g \equiv 0$. Multiplying (4.7.1) by a test function v and integrating by parts, we arrive at the equivalent weak form: Find $u \in H_0^1$ such that

$$\frac{d}{dt}\int_\Omega v(\mathbf{x})u(\mathbf{x}, t)\, d\mathbf{x} = -\int_\Omega p \nabla v \cdot \nabla u\, d\mathbf{x} + \int_\Omega vf\, d\mathbf{x} \quad \forall v \in H_0^1. \quad (4.7.2)$$

4.7. Parabolic Problems

Here, the time derivative has been moved outside the first integral under the assumption that Ω is time-invariant. The moving-domain case is treated in Section 6.7.

Spatial discretization proceeds as in the elliptic case. We restrict the search for u to a finite-dimensional subspace $V_{N,0} \subset H_0^1$, representing u as

$$u(\mathbf{x}, t) = \sum_{j=1}^{\mathcal{N}} u_j(t)\psi_j(\mathbf{x}),$$

where $\{\psi_1, \ldots, \psi_\mathcal{N}\}$ is a basis for $V_{N,0}$, and $u_j(t)$ are unknown coefficients. We then insist that (4.7.2) hold for every $v = \psi_i, i = 1, \ldots, \mathcal{N}$, leading to the system of ODEs

$$\frac{d}{dt} M \underline{u} = -K \underline{u} + M \underline{f}, \qquad \underline{u}(0) = \underline{u}^0, \qquad (4.7.3)$$

where K and M are the stiffness and mass matrices introduced in Section 4.3, and $f(\mathbf{x})$ has been replaced by its interpolant, $\sum_j f_j \psi_j$.

As in the one-dimensional case considered in Section 3.4, several options are available for temporal discretization of (4.7.3), with an important distinction between the class of purely explicit methods, which have limited time-step sizes, and implicit methods, which require a system solve at each time step. It follows from the eigenvalue analyses (2.7.21)–(2.7.22) and (4.2.10) that, for tensor-product discretizations based on Nth-order collocation or Galerkin formulations, the spectral radius of $M^{-1}K$ scales as N^4, implying that the time step would need to scale as $\Delta t = O(N^{-4})$ to ensure a stable explicit formulation. Consequently, (4.7.3) is usually solved implicitly unless other terms in the governing problem mandate the use of a small time step.

Typical implicit formulations would be Crank–Nicolson or BDF2, with the latter given by

$$M \left(\frac{3\underline{u}^{n+1} - 4\underline{u}^n + \underline{u}^{n-1}}{2 \Delta t} \right) = -K \underline{u}^{n+1} + M \underline{f}^n.$$

This scheme is unconditionally stable and requires repeated solution of Helmholtz systems of the form

$$H \underline{u}^{n+1} = \underline{r}^{n+1}, \qquad H := \mathcal{M} Q^T \left(K_L + \frac{3}{2 \Delta t} M_L \right) Q \mathcal{M}. \qquad (4.7.4)$$

Here, $\underline{r}^{n+1}$ accounts for all the terms known prior to time t^{n+1}, $\mathcal{M}$ is the mask array enforcing homogeneous boundary conditions, Q is the connectivity matrix, and the subscript L denotes the unassembled stiffness and mass matrices.

In multiple space dimensions, it is generally too expensive in storage and operation count to solve (4.7.4) by direct factorization. Fortunately, the Helmholtz operator is typically diagonally dominant, and therefore Jacobi (diagonally) preconditioned Krylov methods converge quite rapidly. In particular, if a Galerkin scheme is used, then H is SPD and PCG is most effective. If (4.7.4) is recast into local form, the full system with inhomogeneous boundary conditions is

$$\mathcal{M}_L \Sigma' H_L \underline{u}_{0,L}^{n+1} = \underline{r}_L^{n+1}, \qquad (4.7.5)$$

$$\underline{r}_L^{n+1} := \mathcal{M}_L \Sigma' \left(\mathcal{M}_L \left[\frac{1}{2\Delta t} \mathcal{M}_L (4\underline{u}_L^n - \underline{u}_L^{n-1}) + \underline{f}_L^{n+1} \right] - H_L \underline{u}_{b,L}^{n+1} \right),$$

with the boundary-data splitting $\underline{u}_L^{n+1} := \underline{u}_{0,L}^{n+1} + \underline{u}_{b,L}^{n+1}$, where

$$\underline{u}_{b,L}^{n+1} := (I - \mathcal{M}_L)\underline{g}_L^{n+1} + \mathcal{M}_L \underline{u}_L^n. \qquad (4.7.6)$$

Equation (4.7.5) can be solved by using conjugate gradients (Algorithm 4.1). Note that the mask multiplying $4\underline{u}_L^n - \underline{u}_L^{n-1}$ is superfluous if $\mathcal{M}_L$ is diagonal, because $\mathcal{M}_L$ commutes with $\Sigma' \mathcal{M}_L$ and is idempotent. Also, the construction of $\underline{u}_{b,L}^{n+1}$ (4.7.6) implies that $\underline{u}_{0,L}^{n+1}$ satisfies homogeneous boundary conditions and, in the interior, accounts only for the *change* in the solution from time t^n to t^{n+1}. If the solution is smoothly varying in time, this can significantly reduce the work for an iterative solver, because the initial residual $\|\underline{r}_L^{n+1}\|$ may be quite small.

4.7.1 Time-Dependent Projection

Further improvements in the initial guess can be obtained by computing the *projection* of the solution onto the space of previous solutions, as suggested in [128]. Algorithm 4.2 (below) is a straightforward extension of Algorithm 2.1, with elements of the approximation space based on prior solutions, rather than elements of a Krylov subspace. It builds an H-orthonormal basis $\{\tilde{u}_1, \ldots, \tilde{u}_l\}$ satisfying $\tilde{u}_i^T H \tilde{u}_j = \delta_{ij}$, with $U := \text{span}\{\tilde{u}_1, \ldots, \tilde{u}_l\} = \text{span}\{\underline{u}_0^{n-l+1}, \ldots, \underline{u}_0^n\}$. Here $l \leq L$ is the number of prior solutions saved, and L is an upper bound determined from storage considerations. When this bound is exceeded, the approximation space is restarted with the most recent solution as the first entry. On the first pass ($l = 0$), there is no space to project onto, and the procedure simply generates a single element in U. Because $\tilde{\underline{u}}$ is H-orthogonal to U to within tolerance ϵ, the Gram–Schmidt procedure in step 5 constitutes a reorthogonalization of $\tilde{u}$,

4.7. Parabolic Problems

and stability is thus assured. The entire procedure requires only two additional matrix–vector products per time step. For ease of notation, we have reverted to the global vector representation and dropped the L subscripts. However, for local-storage implementations, all inner products must be effected with the multiplicity weighting, as in Algorithm 4.1.

Algorithm 4.2 (Time-dependent projection scheme) Given a r.h.s. $\underline{r}^{n+1}$, and initially $l = 0$, set

1. $\underline{\tilde{u}} = \sum_{i=1}^{l} \alpha_i \underline{\tilde{u}}_i$, $\alpha_i = \underline{\tilde{u}}_i^T \underline{r}^{n+1}$,
2. $\underline{\tilde{r}} = \underline{r}^{n+1} - H\underline{\tilde{u}}$,
3. solve $H\underline{\bar{u}} = \underline{\tilde{r}}$ to tolerance ϵ,
4. $\underline{u}_0^{n+1} = \underline{\bar{u}} + \underline{\tilde{u}}$,
5. If $l = L$ then
   $l = 1$
   $\underline{\tilde{u}}_l = \underline{u}_0^{n+1} / \|\underline{u}_0^{n+1}\|_H$,
   Else
   $l = l + 1$
   $\underline{\tilde{u}}_l = \underline{\bar{u}} - \sum_{i=1}^{l-1} \beta_i \underline{\tilde{u}}_i$, $\beta_i = \underline{\tilde{u}}_i^T H\underline{\bar{u}}$,
   $s = (\underline{\tilde{u}}_l^T H\underline{\tilde{u}}_l - \sum_{i=1}^{l-1} \beta_i^2)^{\frac{1}{2}}$,
   $\underline{\tilde{u}}_l = \underline{\tilde{u}}_l / s$,
   EndIf.

The table on the left in Figure 4.7.1 shows the results when Algorithm 4.2 is applied to (4.7.1) with $g = 0$ and f is a revolving Gaussian pulse:

$$f = e^{-2\tilde{r}^2}, \qquad \tilde{r}^2 := \frac{1}{100}[(x - \tilde{x})^2 + (y - \tilde{y})^2],$$

$$\tilde{x} := 0.5 + 0.25 \cos t,$$

$$\tilde{y} := 0.5 + 0.25 \sin t.$$

Three values of conductivity are considered, with the final case ($p = 10^3$) effectively corresponding to a Poisson equation to be solved at each step, similar to the pressure Poisson equation encountered in Chapter 6 [see (6.3.1) and (6.5.14)]. The table shows the average iteration counts over the interval $t \in [0, 2\pi]$ with (left) and without (right) projection. In these examples, $L = 20$. In each case, the projection approach provides a net reduction in iteration count, with more substantial reductions being obtained for smaller time steps. We note that for $p \leq 10^0$ and tol $\geq 10^{-6}$, each twofold reduction in Δt is

p	tol	$\Delta t = \pi/25$		$\pi/50$		$\pi/100$	
		W.	W.O.	W.	W.O.	W.	W.O.
10^{-3}	10^{-3}	4	4	2	3	1	3
	10^{-6}	9	10	5	8	2	6
	10^{-9}	14	15	9	12	4	9
10^{0}	10^{-3}	31	41	14	33	7	23
	10^{-6}	91	103	51	87	25	69
	10^{-9}	128	134	94	123	52	111
10^{3}	10^{-3}	33	45	17	38	8	30
	10^{-6}	101	113	66	107	36	102
	10^{-9}	136	142	111	138	78	135

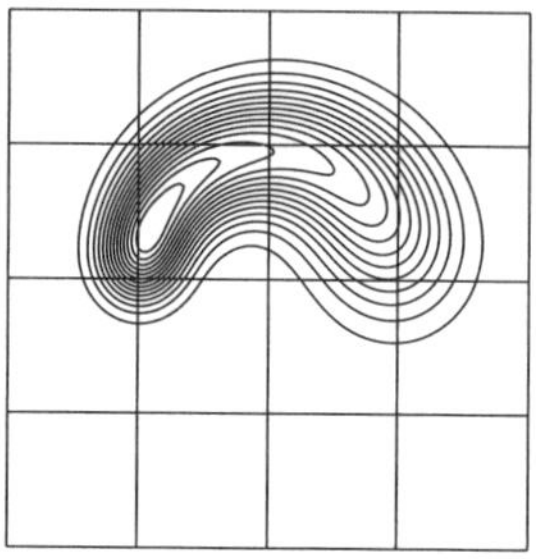

$u(\mathbf{x}, \pi), p = 10^{-3}$

Figure 4.7.1. Jacobi–PCG iteration counts with and without projection for an unsteady conduction problem in the unit square with $(E, N) = (16, 8)$.

accompanied by roughly a twofold reduction in iteration count, implying that the number of iterations required to integrate over the interval $t \in [0, 2\pi]$ is conserved by the projection method, independent of Δt. In these examples, the benefits are less significant when the iteration tolerance is tightened. This is in contrast to results given in [128]. There, and in [124], the projection method was seen to yield a two- to fourfold reduction in iteration count for the (consistent) pressure Poisson problem (6.3.1), preconditioned by domain decomposition methods.

4.7.2 Other Diffusion Systems

We consider the variational formulation of the parabolic problem

$$\frac{\partial u_i}{\partial t} = \frac{\partial}{\partial x_j} \mu \left(\frac{\partial u_i}{\partial x_j} + \frac{\partial u_j}{\partial x_i} \right), \qquad i = 1, \ldots, d, \qquad (4.7.7)$$

where, following standard indicial notation, summation over the repeated index j is implied. The system (4.7.7) is a model for the viscous contribution to the Navier–Stokes equations when the viscosity μ is spatially varying [see (1.6.2)]. In the constant-viscosity case, the second term on the right drops out as a result of the divergence-free constraint (1.2.10). Examples where variable viscosity is important are problems with large temperature gradients or in large-eddy simulations that use subgrid-scale models to account for local contributions to the Reynolds stress term (1.10.26).

4.7. Parabolic Problems

Following the standard weighted-residual approach, we multiply (4.7.7) by test functions v_i and integrate over Ω. After integration by parts we obtain

$$\frac{\partial}{\partial t}\int_\Omega v_i u_i\, d\mathbf{x} = -\int_\Omega \frac{\partial v_i}{\partial x_j}\mu\frac{\partial u_i}{\partial x_j}\,d\mathbf{x} - \int_\Omega \frac{\partial v_i}{\partial x_j}\mu\frac{\partial u_j}{\partial x_i}\,d\mathbf{x}. \quad (4.7.8)$$

The corresponding system matrices are generated after the usual choice of approximation spaces and bases, leading to the following system of ODEs for the velocity components (in $\mathbb{R}^2$):

$$\begin{bmatrix} M & 0 \\ 0 & M \end{bmatrix} \frac{d}{dt}\begin{pmatrix} \underline{u}_1 \\ \underline{u}_2 \end{pmatrix}$$
$$= -\left(\begin{bmatrix} K & 0 \\ 0 & K \end{bmatrix} + \begin{bmatrix} D_x^T \tilde{\mu} D_x & D_y^T \tilde{\mu} D_x \\ D_x^T \tilde{\mu} D_y & D_y^T \tilde{\mu} D_y \end{bmatrix}\right)\begin{pmatrix} \underline{u}_1 \\ \underline{u}_2 \end{pmatrix}, \quad (4.7.9)$$

where $K := D_x^T \tilde{\mu} D_x + D_y^T \tilde{\mu} D_y$ is the standard stiffness matrix, M is the mass matrix, $\tilde{\mu} := \mu M$ is the diagonally scaled mass matrix, and D_x and D_y are the standard derivative operators.

The difficulty introduced by variable viscosity is that $\underline{u}_1$ and $\underline{u}_2$ are now coupled by the full matrix on the r.h.s. of (4.7.9). A stable decoupling of this operator can be effected by moving diag$(2K)$ to the l.h.s. and using one of the semiimplicit splittings of Section 3.3, the simplest example being the Euler-implicit–Euler-explicit splitting

$$\begin{bmatrix} 2K + \frac{1}{\Delta t}M & 0 \\ 0 & 2K + \frac{1}{\Delta t}M \end{bmatrix}\begin{pmatrix} \underline{u}_1 \\ \underline{u}_2 \end{pmatrix}^{n+1}$$
$$= \frac{1}{\Delta t}\left(\begin{bmatrix} M & 0 \\ 0 & M \end{bmatrix} + \begin{bmatrix} D_y^T \tilde{\mu} D_y & -D_y^T \tilde{\mu} D_x \\ -D_x^T \tilde{\mu} D_y & D_x^T \tilde{\mu} D_x \end{bmatrix}\right)\begin{pmatrix} \underline{u}_1 \\ \underline{u}_2 \end{pmatrix}^n, \quad (4.7.10)$$

although higher-order splittings are also possible. This system requires only solution of the block-diagonal system on the left and can be shown to be unconditionally stable in $\mathbb{R}^2$ and $\mathbb{R}^3$ by a straightforward Rayleigh-quotient analysis. Note that if one moves only K to the left, rather than $2K$, stability is not obtained.

4.8 Hyperbolic Problems

We consider the multidimensional extension of the scalar transport problem (3.5.1),

$$\frac{\partial u}{\partial t} + \mathbf{c} \cdot \nabla u = f, \quad u = u_b \quad \text{on } \partial\Omega^{\text{in}}. \tag{4.8.1}$$

Here, $\mathbf{c}$ is a convecting field, and $\partial\Omega^{\text{in}} \subset \partial\Omega$ is the *inflow* boundary, where $\mathbf{c} \cdot \mathbf{n} < 0$, with $\mathbf{n}$ corresponding to the outward pointing normal on $\partial\Omega$. The weak formulation of this problem reads as follows: Find $u \in H_b^1$ such that

$$\int_\Omega v \left(\frac{\partial u}{\partial t} + \mathbf{c} \cdot \nabla u \right) dV = \int_\Omega v f \, dV \quad \forall v \in H_0^1, \tag{4.8.2}$$

where H_b^1 (H_0^1) is the space of all functions in H^1 satisfying the inflow (homogeneous) boundary conditions on $\partial\Omega^{\text{in}}$. As noted in Chapter 3, we do not strictly need both u and v in H^1. However, we assume that this holds, in anticipation of adding viscosity when we later consider the advection–diffusion and Navier–Stokes equations.

Discretization of (4.8.2) proceeds in the usual way by restricting the trial and test spaces to a finite-dimensional subspace V_N, representing u in terms of a suitable basis $\{\psi_j\}_1^N$ for V_N, and testing against every function in this basis, from which we arrive at the linear system

$$M \frac{d\underline{u}}{dt} = -C\underline{u} + M\underline{f}. \tag{4.8.3}$$

Here M is the usual mass matrix, and $\underline{u}$ the vector of unknown basis coefficients. The only new element in this equation is the convective operator $C_{ij} := \int_\Omega \psi_i \mathbf{c} \cdot \nabla \psi_j \, dV$.

Spectral-element construction of C proceeds by evaluating

$$\int_{\Omega^e} v \mathbf{c} \cdot \nabla u \, dV \tag{4.8.4}$$

element by element. We begin with the single-element case, dropping the superscript e for conciseness. Note that if $\mathbf{x}(\mathbf{r})$ is the mapping from $\hat{\Omega}$ to Ω, then

$$\mathbf{c} \cdot \nabla u := \sum_{i=1}^d c_i \frac{\partial u}{\partial x_i} = \sum_{i=1}^d c_i \sum_{j=1}^d \frac{\partial u}{\partial r_j} \frac{\partial r_j}{\partial x_i} = \sum_{j=1}^d \left(\sum_{i=1}^d c_i \frac{\partial r_j}{\partial x_i} \right) \frac{\partial u}{\partial r_j}.$$

4.8. Hyperbolic Problems

Defining $\tilde{c}_j := \sum_{i=1}^{d} c_i \, \partial r_j / \partial x_i$, $j = 1, \ldots, d$, we have

$$\mathbf{c} \cdot \nabla u := \sum_{j=1}^{d} \tilde{c}_j \frac{\partial u}{\partial r_j},$$

and the convective operator for a single element is thus

$$C = M \sum_{j=1}^{d} \tilde{c}_j D_j, \qquad (4.8.5)$$

where the mass matrix arises from the integration in (4.8.4). This is the multidimensional analogue of (3.5.5). When the form (4.8.5) is used, the computation of $C\underline{u}$ requires only d tensor-product evaluations $(D_j \underline{u})$ in the reference domain, rather than d^2 as would be required if each physical space component of ∇u were computed individually. For example, for a single element in $\mathbb{R}^2$, the product $\underline{w}^e = C^e \underline{u}^e$ is

$$w_{ij}^e = \rho_i \rho_j J_{ij}^e \left(\tilde{c}_{r,ij}^e u_{r,ij}^e + \tilde{c}_{s,ij}^e u_{s,ij}^e \right), \qquad (4.8.6)$$

where $\underline{u}_r^e := D_r \underline{u}^e = (I \otimes \hat{D}) \underline{u}^e$ and $\underline{u}_s^e := D_s \underline{u}^e = (\hat{D} \otimes I) \underline{u}^e$. The multi-domain extension of (4.8.6) is

$$\underline{w}_L = \Sigma' C_L \underline{u}_L,$$

assuming that the trial and test functions are continuous.

The form $\mathbf{c} \cdot \nabla u$ is the so-called convective form of the transport operator. Other forms, equivalent in the continuous case, include the conservation and skew-symmetric forms. From the identity

$$\nabla \cdot (\mathbf{c} u) = u \nabla \cdot \mathbf{c} + \mathbf{c} \cdot \nabla u \qquad (4.8.7)$$

we have, assuming $\nabla \cdot \mathbf{c} = 0$, the conservation form: $\nabla \cdot (\mathbf{c} u) = \mathbf{c} \cdot \nabla u$. Multiplying by a test function v and then integrating by parts yields

$$\int_\Omega v \nabla \cdot (\mathbf{c} u) \, dV = -\int_\Omega u \mathbf{c} \cdot \nabla v \, dV + \int_{\partial \Omega} v u \mathbf{c} \cdot \mathbf{n} \, dS. \qquad (4.8.8)$$

In situations where the surface integral on the right is vanishing, such as closed or periodic domains, we have

$$(v, \mathbf{c} \cdot \nabla u) = -(\mathbf{c} \cdot \nabla v, u), \qquad (4.8.9)$$

implying that the convection operator is skew-symmetric. The significance of this is that $\int u^2 \, dV$ is conserved (assuming $f \equiv 0$), since $(u, \mathbf{c} \cdot \nabla u) = 0$, as can be seen by setting $v = u$ in (4.8.9).

In the discrete case, it is possible to explicitly enforce skew-symmetry by defining

$$C_{\text{ss}} := \frac{1}{2} \sum_{i=1}^{d} \left(M c_i D_i - D_i^T c_i M \right).$$

The second term (the conservation form) is evaluated elementwise as

$$\begin{aligned}
\underline{w}^e &= C_c^e \underline{u}^e \\
&= D_1^T \left(\tilde{c}_1^e M^e \underline{u}^e \right) + D_2^T \left(\tilde{c}_2^e M^e \underline{u}^e \right) \\
&= \sum_{p=0}^{N} \hat{D}_{pi} \left(\tilde{c}_{1,pj}^e J_{pj}^e \rho_p \rho_j u_{pj}^e \right) + \sum_{q=0}^{N} \hat{D}_{qj} \left(\tilde{c}_{2,iq}^e J_{iq}^e \rho_i \rho_q u_{iq}^e \right).
\end{aligned}$$

The advantage of C_{ss} is that, since $C_{\text{ss}} = -C_{\text{ss}}^T$ by construction, its eigenvalues are purely imaginary, even if $\nabla \cdot \mathbf{c} \neq 0$, and no dissipation will result from the spatial discretization. Symmetric time-stepping schemes such as Crank–Nicolson will preserve this property, and any consistent time-stepping scheme will preserve it in the limit $\Delta t \longrightarrow 0$. Rønquist [331] demonstrated the potential advantages of the skew-symmetric form in Navier–Stokes simulations within closed domains. However, recent results by Wilhelm and Kleiser [412] have shown the potential for spurious instabilities in studies of Tollmien–Schlichting wave growth when using the skew-symmetric form in high-Reynolds-number applications. Moreover, the skew-symmetric form is more expensive and strictly applies only when the surface integral in (4.8.8) vanishes. If the surface integral in (4.8.8) is nonvanishing (e.g., due to outflow conditions), energy is not conserved in the continuous system, implying that there are eigenvalues with nontrivial real parts. One would consequently not expect the discrete operator to be skew-symmetric either.

To avoid implicit solution of the nonsymmetric system $(M + \Delta t \, C)$, one commonly advances (4.8.3) with explicit time-stepping schemes. AB3 and RK4 are appropriate choices, as their stability regions enclose a significant portion of the imaginary axis (see Figures 3.2.1 and 3.2.5). The CFL stability criterion can be determined by choosing Δt such that

$$\max_{\Omega} \Delta t \left(\frac{|\tilde{c}_1|}{\Delta r_1}, \ldots, \frac{|\tilde{c}_d|}{\Delta r_d} \right) \leq \frac{(\lambda h)^*}{S(N)},$$

where Δr_i and $\tilde{c}_i$ respectively refer to the local grid spacing and velocity in the ith direction in the reference domain $[-1, 1]^d$, $(\lambda h)^*$ is the maximum stable

4.8. Hyperbolic Problems

value in the λh plane (Figures 3.2.1 and 3.2.5), and $S(N)$ is the scaling factor given in Figure 3.5.2. Time advancement of (4.8.3) with, say, AB3 would be expressed in local form as

$$\underline{u}_L^{n+1} = \underline{u}_L^n - \mathcal{M}_L M_L^{-1} \Sigma' \left(\sum_{q=0}^{2} \frac{\gamma_{q+1}}{\underline{\gamma}} C_L^{n-q} \underline{u}_L^{n-q} \right). \quad (4.8.10)$$

Here, $\mathcal{M}_L$ is the local diagonal mask matrix, which is taken to be the identity everywhere except on $\partial\Omega^e \cap \partial\Omega^{in}$. The superscript on C_L^{n-q} refers to the time level in the case where **c** is a function of time. The AB3 coefficients $\underline{\gamma}$ and γ_q are given in Table 3.2.1.

A Two-Dimensional Example

We present a spectral-element example using the convective form coupled with AB3 time stepping (4.8.10). The problem is the convecting-cone problem first considered by Gottlieb and Orszag [163]. The convecting field is a plane rotation, $\mathbf{c} = (y - 0.5, 0.5 - x)$, on $\Omega = [0, 1]^2$. The initial condition is a cone of unit height with a base radius of 0.1 and center at $(0, .25)$. Periodic boundary conditions are applied in each direction. They are imposed by simply matching global node numbers on opposite sides of the domain; everything else carries through by the action of Σ' (see Figure 4.5.3).

Figure 4.8.1 shows the initial condition and the results for three different discretizations: $(E, N) = (16^2, 2)$, $(8^2, 4)$, and $(4^2, 8)$, all having essentially the

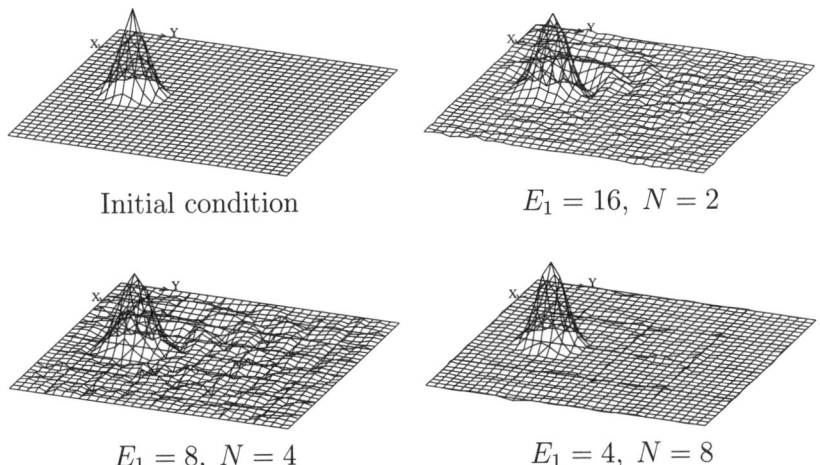

Figure 4.8.1. Convected-cone problem on a 32×32 grid comprising E_1^2 elements of order N.

same 32 × 32 resolution. The time step, $\Delta t = \pi/1000$, is the same in all cases and sufficiently small that spatial errors dominate. We observe in Figure 4.8.1 considerable dispersion for the lower-order cases after just a single revolution. The dispersion is significantly diminished for the moderately high-order case of $N = 8$, *despite the fact that the solution is not smooth*. The minima for the three respective cases are -0.1419, -0.1127, and -0.0371, and the maxima are 0.7693, 0.7413, and 0.8652. The results with RK4 are essentially identical.

4.9 Unsteady Advection–Diffusion Problems

This section introduces time-stepping schemes that are specific to mixed-physics problems in multiple space dimensions. In particular, we introduce the multidimensional extension of the operator-integration factor splitting (OIFS) scheme of Maday et al. [262]. This OIFS technique is also interpreted as a subcycling method where the advective part is integrated more often than the diffusive part because the time scale of advection is shorter and imposes more stringent stability conditions. We consider the multidimensional extension of the scalar transport problem (3.7.1),

$$\frac{\partial u}{\partial t} + \mathbf{c} \cdot \nabla u = \nu \Delta u, \qquad u = 0 \quad \text{on } \partial \Omega, \qquad (4.9.1)$$

where $\mathbf{c}$ is the given advection field. The weak formulation of this problem is as follows: Find $u \in H_0^1$ such that

$$\int_\Omega v \left(\frac{\partial u}{\partial t} + \mathbf{c} \cdot \nabla u \right) dV = -\nu \int_\Omega \nabla u \nabla v \, dV \qquad \forall v \in H_0^1. \qquad (4.9.2)$$

The discretization of Equation (4.9.2) produces the linear system for the basis coefficients:

$$M \frac{d\underline{u}}{dt} = -C \underline{u} - K \underline{u}. \qquad (4.9.3)$$

Recalling (3.9.1), one can write Equation (4.9.3) in terms of an integration factor $\mathcal{Q}_C^{t^*}(t)$:

$$\frac{d}{dt} \left[\mathcal{Q}_C^{t^*}(t) M \underline{u} \right] = \mathcal{Q}_C^{t^*}(t)(-K \underline{u}), \qquad (4.9.4)$$

which is defined here by

$$\frac{d}{dt} \mathcal{Q}_C^{t^*}(t) M = \mathcal{Q}_C^{t^*}(t) C, \qquad \mathcal{Q}_C^{t^*}(t^*) = I. \qquad (4.9.5)$$

4.9. Unsteady Advection–Diffusion Problems

Applying the BDFk scheme (3.2.29) to (4.9.4), one obtains

$$M\underline{u}^{n+1} - \sum_{j=1}^{k} a_j S_C\big(Q_C^{t^{n+1}}(t^{n+1-j})M\underline{u}^{n+1-j}\big) = -b_0 \Delta t\, K\underline{u}^{n+1} \quad (4.9.6)$$

The evaluation of the terms involving the operator integration factor, which is never constructed explicitly, is performed by using the relationship

$$S_C\big(Q_C^{t^{n+1}}(t^{n+1-j})M\underline{u}^{n+1-j}\big) = M\underline{\tilde{u}}_j^{n+1}, \quad j = 1, \dots, k, \quad (4.9.7)$$

where the unknowns $\underline{\tilde{u}}_j^{n+1}$ are computed by integration of the initial-value problem

$$M\frac{d}{ds}\underline{\tilde{u}}_j(s) = -C\underline{\tilde{u}}_j(s), \quad t^{n+1-j} \leq s \leq t^{n+1}, \quad (4.9.8)$$

with initial conditions $\underline{\tilde{u}}_j(t^{n+1-j}) = \underline{u}^{n+1-j}$. This last problem is solved with a time step $\Delta s = \Delta t/\eta$. The integer η represents the number of subcycles where explicit cheap evaluations of the advection term are computed.

The subcycling method is equivalent to the method of characteristics (cf. Quarteroni and Valli [319] and Ho et al. [206]), as the paper [262] on OIFS indicates. This can be shown by switching from Eulerian spatial description to Lagrangian material description. The equation (4.9.3) becomes

$$M\frac{d}{dt}\underline{u}(\mathbf{X}(\mathbf{x}, t), t) = -b_0 K\underline{u}(\mathbf{X}(\mathbf{x}, t), t), \quad (4.9.9)$$

where $\mathbf{X}(\mathbf{x}, t)$ represents the Lagrangian spatial coordinates. The l.h.s. of (4.9.9) corresponds now to the *particular derivative*, that is, the time derivative obtained by following a material point in the fluid. From continuum mechanics, we know that

$$M\frac{d}{dt}\underline{u}(\mathbf{X}(\mathbf{x}, t), t) = \left(M\frac{d}{dt} + C\right)\underline{u}(\mathbf{x}, t), \quad (4.9.10)$$

with $\underline{u}(\mathbf{x}, t)$, the scalar field in Eulerian coordinates. Applying BDF3 to Equation (4.9.9), we get

$$M\underline{u}(\mathbf{X}(\mathbf{x}, t^{n+1}), t^{n+1}) - \sum_{j=1}^{3} a_j M\underline{u}(\mathbf{X}(\mathbf{x}, t^{n+1-j}), t^{n+1-j})$$
$$= -\Delta t\, b_0 K\underline{u}(\mathbf{X}(\mathbf{x}, t^{n+1}), t^{n+1}). \quad (4.9.11)$$

In this relation, $\underline{u}(\mathbf{X}(\mathbf{x}, t^{n+1-j}), t^{n+1-j})$ is the value of $\underline{u}$ at time t^{n+1-j} at the foot of the characteristic originating from $\mathbf{x}$ at time t^{n+1}. By definition of the characteristics, $\mathbf{X}(\mathbf{x}, t^{n+1-j})$ is computed by the backward problem

$$\frac{\partial \mathbf{X}(\mathbf{x}, s)}{\partial s} = \underline{u}(\mathbf{X}(\mathbf{x}, s), s), \qquad s \in [t^{n+1-j}, t^{n+1}), \qquad (4.9.12)$$

$$\mathbf{X}(\mathbf{x}, t^{n+1}) = \mathbf{x}. \qquad (4.9.13)$$

When the condition (4.9.13) is taken into account, Equation (4.9.11) simplifies to

$$M\underline{u}^{n+1} - \sum_{j=1}^{3} a_j M\underline{u}(\mathbf{X}(\mathbf{x}, t^{n+1-j}), t^{n+1-j}) = -\Delta t\, b_0 K \underline{u}^{n+1}. \qquad (4.9.14)$$

The solution of the characteristics problem (4.9.12)–(4.9.13) by an appropriate time scheme like RK4 yields the values at the foot of the characteristics. The final step to prove the equivalence between subcycling and the method of characteristics is accomplished by the integration in time of Equation (4.9.10),

$$\int_{t^{n+1-j}}^{t^{n+1}} M \frac{d}{dt} \underline{u}(\mathbf{X}(\mathbf{x}, t), t)\, dt$$

$$= \int_{t^{n+1-j}}^{t^{n+1}} \left(M \frac{d}{dt} + C \right) \underline{u}(\mathbf{x}, t)\, dt, \qquad j = 1, \ldots, k,$$

which provides

$$M\underline{u}(\mathbf{X}(\mathbf{x}, t^{n+1-j}), t^{n+1-j}) = M\underline{u}^{n+1-j} - \int_{t^{n+1-j}}^{t^{n+1}} C\underline{u}(\mathbf{x}, t)\, dt, \qquad j = 1, \ldots, k. \qquad (4.9.15)$$

By use of the definition (4.9.7), we set

$$M\underline{u}(\mathbf{X}(\mathbf{x}, t^{n+1-j}), t^{n+1-j}) = Q_C^{(t^{n+1}, t^{n+1-j})} M\underline{u}(t^{n+1-j}), \qquad (4.9.16)$$

and this establishes that Equation (4.9.6) is equivalent to Equation (4.9.14).

4.10 Further Reading

Treatment of multidimensional elliptic, parabolic, and nonlinear hyperbolic equations by spectral methods is available in numerous texts. Theoretical treatments are presented in the books by Bernardi and Maday [39], Funaro [142], and Guo [177]. More applied approaches are presented in the books by Canuto

4.10. Further Reading

et al. [64], Funaro [144], and Trefethen [383]. Particular attention to spectral methods in axisymmetric geometries may be found in the texts by Boyd [48], Fornberg [137] and Bernardi et al. [34]. Recent advances in algorithms for hyperbolic problems, including weighted essentially nonoscillatory (WENO) and discontinuous Galerkin (DG) schemes, are given comprehensive treatment in Cockburn et al. [80].

5
Steady Stokes and Navier–Stokes Equations

In this chapter, we tackle the main subject of the monograph, namely, the numerical integration of the steady Navier–Stokes equations describing incompressible fluid flows subject to body forces $\mathbf{f}$ in the velocity–pressure formulation. We discard all description of high-order methods in the framework of the vorticity–stream-function formulation, because this approach lacks generality in terms of applications and ease of imposing various boundary conditions, especially when interfaces between two nonmixing fluids or free surfaces are present. We restrict ourselves here to the steady-state case, deferring the transient case to the next chapter.

5.1 Steady Velocity–Pressure Formulation

In dimensionless form, the steady-state Navier–Stokes equations are given by the following relationship [see Equations (1.7.1) and (1.2.5)]:

$$\mathbf{v} \cdot \nabla \mathbf{v} = -\nabla p + (Re)^{-1} \Delta \mathbf{v} + \mathbf{f} \quad \text{in } \Omega, \tag{5.1.1}$$

and the continuity equation

$$\text{div } \mathbf{v} = 0 \quad \text{in } \Omega, \tag{5.1.2}$$

where Ω denotes an open subset of $\mathbb{R}^d$, closed by the boundary $\partial \Omega$, which is Lipschitz-continuous, d being the spatial dimension. This set of equations is solved with appropriate boundary conditions. For the sake of simplicity, we will consider only homogeneous Dirichlet boundary conditions, unless otherwise explicitly stated. They are

$$\mathbf{v} = 0 \quad \text{on } \partial \Omega. \tag{5.1.3}$$

5.1. Steady Velocity–Pressure Formulation

Other possible conditions are expressed by imposing contact forces (1.3.2) on some part of the fluid boundary or its totality.

The steady Navier–Stokes equations are of advection–diffusion type, whereas the steady Stokes equations are linear and elliptic. In contrast with compressible fluid flows, where the pressure is given by the equation of state, incompressible fluids do not share this advantage. The pressure is the physical variable that responds to the continuity constraint (5.1.2), ensuring that the flow is incompressible everywhere in the computational domain. With the condition (5.1.3), the pressure in an incompressible fluid is known only up to a constant. Therefore, a reference pressure level has to be chosen. This is achieved by imposing a mean zero level over the domain:

$$\int_\Omega p\, dV = 0 \quad \text{on } \Omega. \tag{5.1.4}$$

In Section 5.2, the steady-state Stokes equations are analyzed. The weak formulation is first presented for the continuous case. The SEM is then described with the various possible discretizations leading to optimal versions free of spurious pressure modes. Legendre and Chebyshev collocation procedures are analyzed for single and staggered grids. Section 5.3 elaborates on the linear systems generated by the various discretization processes, and inspects their conditioning and the relevant algorithms to solve them efficiently with possible preconditioners. The Poisson pressure solver and the associated Green's-function technique are described in Section 5.4. Section 5.5 introduces divergence-free bases via the computation of a singular generalized Stokes eigenvalue problem. The stabilization of the $\mathbb{P}_N$–$\mathbb{P}_N$ approximation by the use of bubble functions is introduced in Section 5.6. These bubble functions are shown to bring to the discrete continuity equation an extra term that damps all oscillatory behavior in the pressure field. The discretization of the steady Stokes problem by hp methods is briefly discussed in Section 5.7. The steady Navier–Stokes equations are then presented in Section 5.8. The weak formulation for the continuous case is given. Some theoretical tools are introduced to allow further analysis of the collocation technique applied to the full Navier–Stokes equations. The question of convergence is, of course, of central importance. Iterative techniques are used to solve the problem, and the Newton method constitutes the natural algorithm to treat nonlinearities. In Section 5.9, a few relevant applications from both the numerical and the fluid-mechanics point of view are discussed. The chapter ends with Section 5.10, where further references to the literature are made, and some engineering considerations are raised in order to allow the reader to choose the best method for a specific application.

Readers interested in a detailed introduction to the numerical analysis of the steady Stokes problem are referred to Bernardi and Maday [38, 39].

5.2 Stokes Equations

In this section, we treat the dimensionless linear steady Stokes problem (1.7.6) and the associated divergence-free condition

$$\nabla p - \Delta \mathbf{v} = \mathbf{f} \quad \text{in } \Omega, \tag{5.2.1}$$

$$\text{div } \mathbf{v} = 0 \quad \text{in } \Omega, \tag{5.2.2}$$

with the homogeneous velocity boundary conditions (5.1.3).

A fundamental question in the development of the weak formulation of the Stokes problem is the appropriate choice of spaces for the velocity and pressure approximations.

5.2.1 The Weak Formulation

Equations (5.2.1) and (5.2.2) may be cast in the weak form. For this purpose, the functional spaces for the velocity and pressure are $H_0^1(\Omega)^d$ and $L_0^2(\Omega)$, respectively, with $H_0^1(\Omega)^d$ denoting the Sobolev space of vector functions with d components introduced in Section 4.3.1, satisfying homogeneous boundary conditions, and

$$L_0^2(\Omega) := \left\{ q \in L^2(\Omega) \,\Big|\, \int_\Omega q(\mathbf{x})\, d\mathbf{x} = 0 \right\}. \tag{5.2.3}$$

In this equation, the subscript indicates that the average pressure is set to zero. Then the weak form is as follows: Find $\mathbf{v}$ in $H_0^1(\Omega)^d$ and p in $L_0^2(\Omega)$ such that

$$\int_\Omega \nabla \mathbf{v} \cdot \nabla \mathbf{u}\, d\mathbf{x} - \int_\Omega p\, \text{div } \mathbf{u}\, d\mathbf{x} = \int_\Omega \mathbf{f} \cdot \mathbf{u}\, d\mathbf{x} \quad \forall \mathbf{u} \in H_0^1(\Omega)^d, \tag{5.2.4}$$

$$-\int_\Omega q\, \text{div } \mathbf{v}\, d\mathbf{x} = 0 \quad \forall q \in L_0^2(\Omega), \tag{5.2.5}$$

where $\mathbf{f}$ is given in $L^2(\Omega)^d$. In Equation (5.2.4), we recognize the terms corresponding to a stationary value of the energy functional for an elliptic problem plus the additional pressure term. In fact, the energy functional is modified in such a way that the pressure plays the role of a Lagrange multiplier for the continuity constraint [53, 71, 157, 176]. The total energy is given by the

5.2. Stokes Equations

functional

$$J(\mathbf{u}) = \frac{1}{2} \int_\Omega [\nabla \mathbf{u} \cdot \nabla \mathbf{u} - 2\mathbf{f} \cdot \mathbf{u}]\, dx. \quad (5.2.6)$$

To elaborate the Lagrange-multiplier formulation of the Stokes problem, one introduces the Lagrangian functional

$$L(\mathbf{u}, q) = J(\mathbf{u}) - \int_\Omega q\, \text{div}\, \mathbf{u}\, dx. \quad (5.2.7)$$

The classical minimization problem is now transformed into a saddle-point problem such that we look for a solution derived from the inf–sup problem

$$\inf_{\mathbf{u} \in H_0^1(\Omega)^d} \sup_{q \in L_0^2(\Omega)} \int_\Omega \left(\frac{1}{2} \nabla \mathbf{u} \cdot \nabla \mathbf{u} - \mathbf{f} \cdot \mathbf{u} - q\, \text{div}\, \mathbf{u}\right) dx. \quad (5.2.8)$$

Equivalently, the solution $(\mathbf{v}, p)$ of the previous saddle-point problem satisfies the inequalities

$$L(\mathbf{v}, q) \leq L(\mathbf{v}, p) \leq L(\mathbf{u}, p) \quad \forall \mathbf{u} \in H_0^1(\Omega)^d, \quad \forall q \in L_0^2(\Omega). \quad (5.2.9)$$

The term saddle point comes from the geometrical features of the Lagrangian functional, which is convex for the velocity minimization and concave for the pressure maximization, therefore showing a saddle form with double curvature.

The solution of the saddle-point problem possesses adequate stability and convergence properties if it complies with the well-known inf–sup condition, also called the Brezzi–Babuška condition [18, 51]. To this end, let us define the continuous bilinear forms $\mathcal{A}$ and $\mathcal{B}$:

$$\mathcal{A}: \quad H_0^1(\Omega)^d \times H_0^1(\Omega)^d \to \mathbb{R}, \quad \mathcal{A}(\mathbf{u}, \mathbf{v}) = (\nabla \mathbf{u}, \nabla \mathbf{v}), \quad (5.2.10)$$

$$\mathcal{B}: \quad H_0^1(\Omega)^d \times L_0^2(\Omega) \to \mathbb{R}, \quad \mathcal{B}(\mathbf{v}, q) = -(\text{div}\, \mathbf{v}, q), \quad (5.2.11)$$

where $(\cdot, \cdot)$ denotes the inner product in L^2 space. With these bilinear forms, the weak formulation (5.2.4)–(5.2.5) may be restated in the following abstract form: Find $(\mathbf{v}, p) \in H_0^1(\Omega)^d \times L_0^2(\Omega)$ such that

$$\mathcal{A}(\mathbf{u}, \mathbf{v}) + \mathcal{B}(\mathbf{u}, p) = \mathcal{F}(\mathbf{u}) \quad \forall \mathbf{u} \in H_0^1(\Omega)^d, \quad (5.2.12)$$

$$\mathcal{B}(\mathbf{v}, q) = 0 \quad \forall q \in L_0^2(\Omega). \quad (5.2.13)$$

With the norms defined in Section 4.3.1, we are now in a position to cite Brezzi's main result.

Theorem 5.1 *Under the assumptions*

1. *the continuous bilinear form $\mathcal{A}(\cdot, \cdot)$ is coercive on $H_0^1(\Omega)^d$ and satisfies an ellipticity condition for $\alpha > 0$, that is,*

$$\alpha \|\mathbf{u}\|_{H_0^1(\Omega)^d}^2 \leq \mathcal{A}(\mathbf{u}, \mathbf{u}) \qquad \forall \mathbf{u} \in H_0^1(\Omega)^d, \tag{5.2.14}$$

2. *the continuous bilinear form $\mathcal{B}(\cdot, \cdot)$ verifies an inf–sup condition for a constant $\beta > 0$, that is,*

$$\sup_{\mathbf{u} \in H_0^1(\Omega)^d} \frac{\mathcal{B}(\mathbf{u}, q)}{\|\mathbf{u}\|_{H_0^1(\Omega)^d}} \geq \beta \|q\|_{L_0^2(\Omega)} \qquad \forall q \in L_0^2(\Omega), \tag{5.2.15}$$

the problem (5.2.12)–(5.2.13) possesses a unique solution $(\mathbf{v}, p)$ with the following bounds:

$$\|\mathbf{v}\|_{H_0^1(\Omega)^d} \leq \frac{1}{\alpha} \|\mathbf{f}\|_{H^{-1}(\Omega)^d}, \tag{5.2.16}$$

$$\|p\|_{L_0^2(\Omega)} \leq \frac{1}{\beta}\left(1 + \frac{1}{\alpha}\right) \|\mathbf{f}\|_{H^{-1}(\Omega)^d}. \tag{5.2.17}$$

Related to the bilinear form (5.2.11), one introduces the operator B, corresponding to the continuity constraint:

$$B: \quad H_0^1(\Omega)^d \to \mathbb{R}, \quad B\mathbf{v} = -\mathrm{div}\,\mathbf{v}. \tag{5.2.18}$$

The image of B, denoted as Im B, coincides with $L_0^2(\Omega)$.

5.2.2 The Spectral-Element Method

As usual in the SEM, the first step consists in subdividing the domain $\bar{\Omega} = \Omega \cup \partial\Omega$ into E nonoverlapping rectilinear elements Ω^e, $e = 1, \ldots, E$, such that the intersection between two or more adjacent elements produces a face, an edge, or a vertex. Each element will involve a mesh constructed as a tensor product of one-dimensional grids. Although each space direction may be discretized independently of the others, without loss of generality we will consider only meshes obtained with the same number of nodes in each direction, denoted by $N + 1$, corresponding to the dimension of the space of Nth-order polynomials. To describe the discretization process accurately, we simplify the notations by defining

$$X := H_0^1(\Omega)^d, \quad Z := L_0^2(\Omega). \tag{5.2.19}$$

5.2. Stokes Equations

We define the approximate Stokes problem in its weak form: Find $(\mathbf{v}_h, p_h) \in (X_h, Z_h)$ such that

$$\mathcal{A}(\mathbf{u}_h, \mathbf{v}_h) + \mathcal{B}(\mathbf{u}_h, p_h) = \mathcal{F}_h(\mathbf{u}_h) \quad \forall \mathbf{u}_h \in X_h, \quad (5.2.20)$$

$$\mathcal{B}(\mathbf{v}_h, q_h) = 0 \quad \forall q_h \in Z_h. \quad (5.2.21)$$

where X_h and Z_h are finite-dimensional polynomial subspaces $X_h \subset X$ and $Z_h \subset Z$ in which the velocities and pressure will be approximated, respectively. Of course, in the sequel, the existence and uniqueness of the solution pair $(\mathbf{v}_h, p_h)$ are of primary concern. In finite elements, the subscript h is related to the typical mesh size, while in spectral elements, the parameter h refers to E, the number of elements and N, the polynomial degree.

The most obvious and simple choice for the velocity–pressure spectral-element discretization consists in resorting to the same polynomial space for both fields, namely, the so-called $\mathbb{P}_N$–$\mathbb{P}_N$ method where the discrete spaces, now denoted by X_N and Z_N, are defined as

$$X_N = X \cap \mathbb{P}_{N,E}^d, \quad (5.2.22)$$

$$Z_N = Z \cap \mathbb{P}_{N,E}, \quad (5.2.23)$$

with

$$\mathbb{P}_{N,E} = \{\phi | \phi \in L^2(\Omega); \phi|_{\Omega^e} \text{ is a polynomial of degree } \leq N\}, \quad (5.2.24)$$

where the superscript d in (5.2.22) reflects the fact that elements of X_N are d-dimensional. On the 3D parent element $\hat{\Omega} := [-1, 1]^3$, for example, the velocity field is approximated by

$$\mathbf{v}_N(\boldsymbol{\xi}) = \sum_{i,j,k=0}^{N} \mathbf{v}_{i,j,k} \, \pi_{i,j,k}^v \quad \forall \boldsymbol{\xi} \in \hat{\Omega} := [-1, 1]^3, \quad (5.2.25)$$

where the contracted sum extends over the three space dimensions in turn and refers to the indices i, j, k of the Lagrange Gauss–Lobatto–Legendre (GLL) grid nodes, and $\pi_{i,j,k}^v$ is the tensor product of elementary one-dimensional interpolant bases as introduced in (2.4.3). The GLL grid will be denoted by Ξ_N. Because the weak formulation is an integral form, quadrature rules are employed. The Legendre spectral elements take advantage of the special representation of the Lagrangian interpolants (5.2.25) and replace the continuous integrals by GLL rules made of tensor products of one-dimensional numerical quadratures. Consequently, the abstract problem (5.2.20)–(5.2.21) becomes:

Find $(\mathbf{v}_N, p_N) \in (X_N, Z_N)$ such that

$$\mathcal{A}_N(\mathbf{u}_N, \mathbf{v}_N) + \mathcal{B}_N(\mathbf{u}_N, p_N) = \mathcal{F}_N(\mathbf{u}_N) \quad \forall \mathbf{u}_N \in X_N, \quad (5.2.26)$$

$$\mathcal{B}_N(\mathbf{v}_N, q_N) = 0 \quad \forall q_N \in Z_N. \quad (5.2.27)$$

The bilinear forms appearing in (5.2.26)–(5.2.27) are given by the relationships

$$\mathcal{A}_N(\mathbf{u}_N, \mathbf{v}_N) := (\nabla \mathbf{u}_N, \nabla \mathbf{v}_N)_N, \quad (5.2.28)$$

$$\mathcal{B}_N(\mathbf{u}_N, q_N) := -(\text{div } \mathbf{u}_N, q_N)_N, \quad (5.2.29)$$

where the discrete scalar product $(\cdot, \cdot)_N$ is provided by the GLL quadrature rule. For the 3D case, one has on the reference cube

$$(f, g)_N := \sum_{i,j,k=0}^{N} f_{i,j,k} \, g_{i,j,k} \, \rho_{i,j,k}, \quad (5.2.30)$$

with $\rho_{i,j,k} := \rho_i \rho_j \rho_k$, denoting the product of three GLL weights (B.2.9).

From FE theory, it is well known that elements of the same polynomial degree for both primitive variables $\mathbf{v}$ and p suffer from the presence of spurious pressure modes. For example, the $\mathcal{Q}_1$–$\mathcal{Q}_1$ discretization on an equally spaced mesh triggers a pressure mode that, in two dimensions, is of the checkerboard type [342]. In fact, this element does not satisfy the inf–sup condition. For the $\mathbb{P}_N$–$\mathbb{P}_N$ method, the same conclusion holds (Bernardi and Maday [38]). This situation generates pressure solutions that satisfy the weak continuous relation

$$(\nabla p, \mathbf{u}) = 0 \quad \forall \mathbf{u} \in H_0^1(\Omega)^d. \quad (5.2.31)$$

From a practical point of view, the spurious pressure modes do not affect the velocity solution, although they do produce oscillatory pressure fields that should be filtered to recover meaningful physical solutions. We defer the analysis of these spurious pressure modes until Section 5.2.3, which is devoted to spectral collocation approximation.

Staggered Spectral Elements

A consistent and better choice for spectral-element discretizations that eliminates completely the spurious pressure modes relies on the idea of a staggered grid. It is well known in the FD or finite-volume context that this approach avoids the difficulties associated with collocated grids. By analogy with FE theory, where it is common to use pressure interpolants with polynomials of order one lower than for velocity, one might propose to work in $\mathbb{P}_N$–$\mathbb{P}_{N-1}$ spaces for the spectral-element discretization. Although this choice eliminates

5.2. Stokes Equations

the spurious modes in the case of a single spectral element, one can prove that it does not enforce the inf–sup condition for discretizations with several spectral elements. Therefore, Maday and Patera [261] developed an element based on $\mathbb{P}_N$–$\mathbb{P}_{N-2}$ spaces. The theory and its application to fluid flows were further reported by Maday et al. [260, 256] and Rønquist [332].

The functional spaces are defined as

$$X_N = X \cap \mathbb{P}^d_{N,E}, \tag{5.2.32}$$

$$Z_N = Z \cap \mathbb{P}_{N-2,E}. \tag{5.2.33}$$

The discrete problem (5.2.20)–(5.2.21) is still valid in this abstract setting. However, one must carefully consider the integration rules behind each term. As the velocities are in $\mathbb{P}_N$, the Lagrange interpolants are based on the GLL nodes, and the discrete scalar products in $\mathcal{A}_N(\cdot, \cdot)$ and $(\cdot, \cdot)_N$ are of the form (5.2.30). For the two remaining terms involving $\mathcal{B}_N(\cdot, \cdot)$, i.e. the pressure term and the divergence-free condition, a Gauss–Legendre (GL) integration rule is chosen (see Section B.2.2). On $\hat{\Omega}$, the 3D pressure Lagrange interpolant is written as

$$p_N(\zeta) = \sum_{i,j,k=1}^{N-1} p_{i,j,k} \pi^P_{i,j,k}(\zeta) \quad \forall \zeta \in \hat{\Omega}, \tag{5.2.34}$$

where the nodal values are attached to the GL grid denoted by $\mathcal{Z}_N$, and $\pi^P_{i,j,k}(\zeta) \in \mathbb{P}_{N-2}(\zeta)$ is the tensor product of the one-dimensional GL Lagrangian interpolants of degree $N-2$, that is,

$$\pi^P_l(\zeta) = \frac{L_{N-1}(\zeta)}{(\zeta - \zeta_l) L'_{N-1}(\zeta_l)}, \quad l \in \{i, j, k\} \quad \forall \zeta \in [-1, 1]. \tag{5.2.35}$$

Consequently, the discrete GL scalar product is

$$(f, g)_{N,\text{GL}} = \sum_{i,j,k=1}^{N-1} f_{i,j,k}\, g_{i,j,k}\, \omega_{i,j,k}, \tag{5.2.36}$$

where $\omega_{i,j,k} := \omega_i \omega_j \omega_k$ is the product of the GL weights (B.2.7). The reader should notice that we use a different notation for the weights from that used in the appendices. This is dictated by the presence of two quadrature rules in the weak formulation.

The $\mathbb{P}_N$–$\mathbb{P}_{N-2}$ Stokes problem becomes: Find $(\mathbf{v}_N, p_N) \in (X_N, Z_N)$ such that

$$\mathcal{A}_N(\mathbf{u}_N, \mathbf{v}_N) + \mathcal{B}_{N,\text{GL}}(\mathbf{u}_N, p_N) = \mathcal{F}_N(\mathbf{u}_N) \quad \forall \mathbf{u}_N \in X_N, \tag{5.2.37}$$

$$\mathcal{B}_{N,\text{GL}}(\mathbf{v}_N, q_N) = 0, \quad \forall q_N \in Z_N, \tag{5.2.38}$$

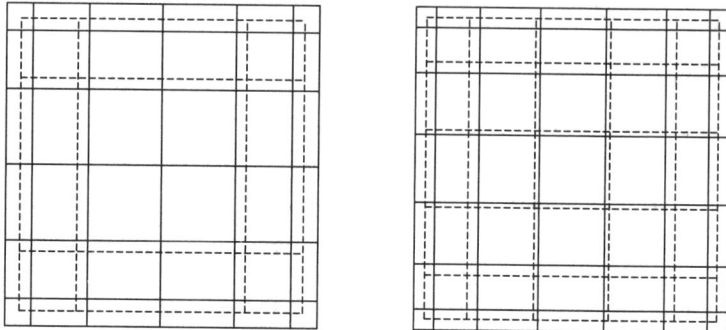

Figure 5.2.1. Staggered spectral element for $N = 6$ (left) and $N = 7$ (right).

with the bilinear forms in (5.2.37)–(5.2.38) defined by the relations

$$\mathcal{A}_N(\mathbf{u}_N, \mathbf{v}_N) := (\nabla \mathbf{u}_N, \nabla \mathbf{v}_N)_N, \tag{5.2.39}$$

$$\mathcal{B}_{N,GL}(\mathbf{v}_N, q_N) := -(\text{div } \mathbf{v}_N, q_N)_{N,GL}, \tag{5.2.40}$$

and the discrete scalar products given by (5.2.30) and (5.2.36), respectively. Note that the GL rule integrates exactly the pressure term and the continuity equation, as does the GLL rule for the other terms. Therefore, we can show easily that

$$\mathcal{B}_{N,GL}(\mathbf{u}_N, q_N) = \mathcal{B}(\mathbf{u}_N, q_N) \quad \forall \mathbf{u}_N \in X_N, \quad \forall q_N \in Z_N. \tag{5.2.41}$$

It is common to say the velocities in $\mathbb{P}_N$ are *conforming*, while the pressures are not. This means that the velocities are continuous along the element interfaces, while the pressure is *not* necessarily continuous. Figure 5.2.1 displays for one spectral element the staggered grid with an odd number (left) and an even number (right) of velocity nodes. The dashed lines represent the pressure grid, and the solid lines the velocity mesh. We notice that the two grids are intertwined and in the second example there is no pressure node in the central strips of the element.

Collocative Spectral Elements

An alternative to the staggered-grid approach of the $\mathbb{P}_N - \mathbb{P}_{N-2}$ formulation can also be derived. We will name this approach the collocative $\mathbb{P}_N - \mathbb{P}_{N-2}$ SEM. Taking into account the exactness of the GLL integration rule and the property (5.2.41), one deduces from Equation (5.2.37) the following collocation relation:

$$-\Delta \mathbf{v}_N + \nabla p_N = \mathbf{f}_N \quad \text{for} \quad \mathbf{x} \in \Xi_N. \tag{5.2.42}$$

5.2. Stokes Equations

Following the proposition of Azaïez and Coppoletta [17], we base the $\mathbb{P}_{N-2}$ discretization on the interior points of the GLL velocity mesh and thereby avoid interpolation between meshes, which is usually needed for the $\mathbb{P}_N$–$\mathbb{P}_{N-2}$ method. One can choose as 2D test functions

$$q_N(\zeta) := \hat{\pi}_{k,l}(\zeta) \in \mathbb{P}_{N-2}(\zeta), \qquad 1 \le k, l \le N-1, \qquad (5.2.43)$$

where $\hat{\pi}_{k,l}(\zeta)$ denotes the tensor product of one-dimensional Lagrangian interpolation polynomials associated with the zeros ξ_l ($l = 1, \ldots, N-1$) of L'_N:

$$\hat{\pi}_l(\zeta) = \frac{L'_N(\zeta)}{(\zeta - \xi_l) L''_N(\xi_l)}, \qquad 1 \le l \le N-1. \qquad (5.2.44)$$

Inserting the previous q_N in Equation (5.2.38) and fixing $\mathbf{x} = (\xi_i, \xi_j)$, one can show [17] that div $\mathbf{v}_N(\mathbf{x})$ is a linear combination of the values of div $\mathbf{v}_N$ evaluated at the four points $(-1, \xi_j)$, $(1, \xi_j)$, $(\xi_i, -1)$, $(\xi_i, 1)$. This is the reason that the $\mathbb{P}_N$–$\mathbb{P}_{N-2}$ method is not a full collocation method. Azaïez and Coppoletta call that method "quasi-collocation."

We note that when the pressure nodes coincide with the interior velocities, the abstract problem is again given by Equations (5.2.26)–(5.2.27) with GLL quadratures. The pressures are interpolated by the $\hat{\pi}$-polynomials. Numerical quadratures involving the pressure or the related test functions require extrapolation of these interpolants to the interface nodes (ξ_0, ξ_N).

It can be proven that the inf–sup condition yields

$$\sup_{\mathbf{u}_N \in X_N} \frac{\mathcal{B}_N(\mathbf{u}_N, q_N)}{\|\mathbf{u}_N\|_{H^1_0(\Omega)^d}} \ge C N^{(1-d)/2} \|q_N\|_{L^2_0(\Omega)} \qquad \forall q_N \in \mathbb{P}_{N-2} \cap L^2_0(\Omega). \qquad (5.2.45)$$

The expression in front of $\|q_N\|_{L^2_0(\Omega)}$ is called the inf–sup constant. It is denoted as β_N, and in this case we have $\beta_N := C N^{(1-d)/2}$. As a consequence of the inf–sup condition, one can prove the following theorem (see Maday et al. [263]).

Theorem 5.2 *Under the assumptions*

1. $N \ge 2$, with the pressure p_N in Z_N,
2. the solution $(\mathbf{v}, p)$ of the weak continuous problem (5.2.12)–(5.2.13) is in $H^s_0(\Omega)^d \times H^{s-1}(\Omega)$ for an integer $s \ge 1$ if $d = 2$, and $s \ge 3$ if $d = 3$; and the function $\mathbf{f}$ is in $H^t(\Omega)^d$ for an integer $t \ge d/2$,

the problem (5.2.37)–(5.2.38) possesses a unique solution $(\mathbf{v}_N, p_N)$ in $X_N \times Z_N$ with the following error bounds:

$$\|\mathbf{v} - \mathbf{v}_N\|_{H_0^1(\Omega)^d} \leq C\left[N^{1-s}\left(\|\mathbf{v}\|_{H_0^s(\Omega)^d} + \|p\|_{H^{s-1}(\Omega)}\right) + N^{-t}\|\mathbf{f}\|_{H^t(\Omega)^d}\right], \tag{5.2.46}$$

$$\|p - p_N\|_{L^2(\Omega)} \leq C_1\left[N^{(d+1)/2-s}\left(\|\mathbf{v}\|_{H_0^s(\Omega)^d} + \|p\|_{H^{s-1}(\Omega)}\right) + N^{\frac{d-1}{2}-t}\|\mathbf{f}\|_{H^t(\Omega)^d}\right], \tag{5.2.47}$$

with C and C_1 constants independent of N.

The last error estimate shows that the method is optimal for the velocity-field convergence, while the factor $N^{(d-1)/2}$ prevents optimality for the pressure. However, the overall convergence is nearly optimal, and the method is highly appealing because it constitutes a natural generalization of the finite-element technique. Studies of the convergence properties of the $\mathbb{P}_N$–$\mathbb{P}_{N-2}$ formulation in comparison with the $\mathbb{P}_N$–$\mathbb{P}_{N-1}$ formulation have been carried out by Shen [355].

Recently, Bernardi and Maday [40] have proposed a new pressure space to obtain an optimal inf–sup condition. The definition of Z_N is now for a single element $(E = 1)$

$$Z_N = Z \cap \mathbb{P}_{\lambda N}, \tag{5.2.48}$$

with $\lambda < 1$. The main result states that, for any real number λ, $0 < \lambda < 1$, there exists a positive constant β independent of N such that for any integer $N \geq 2/(1-\lambda)$ and all q_N in Z_N, the inf–sup condition is

$$\sup_{\mathbf{u}_N \in X_N} \frac{\mathcal{B}_N(\mathbf{u}_N, q_N)}{\|\mathbf{u}_N\|_{H_0^1(\Omega)^d}} \geq \beta \|q_N\|_{L_0^2(\Omega)}. \tag{5.2.49}$$

Furthermore, if Z_N is chosen as

$$Z_N = Z \cap \mathbb{P}_{\inf(\lambda N, N-2)}, \tag{5.2.50}$$

with inf the function giving the minimum of its two arguments, the constant β satisfies the inequality

$$\beta \geq C(1-\lambda)^{(d-1)/2}. \tag{5.2.51}$$

The pressure error is given by

$$\|p - p_N\|_{L^2(\Omega)} \leq C_1\left[N^{1-s}\left(\|\mathbf{v}\|_{H_0^s(\Omega)^d} + \lambda^{1-s}\|p\|_{H^{s-1}(\Omega)}\right) + N^{\frac{d-1}{2}-t}\|\mathbf{f}\|_{H^t(\Omega)^d}\right]. \tag{5.2.52}$$

5.2. Stokes Equations

The velocity error estimate in Bernardi and Maday [40] is optimal but it is measured in a divergence-free velocity space. As this is never achieved in a real computation, the convergence of the $\mathbb{P}_N$–$\mathbb{P}_{\lambda N}$ approximation for the velocity field is indeed affected by the choice of the λ-value.

In addition to the concept of spurious pressure modes, Vandeven [397] introduced the weakly spurious pressure (WSP) modes. These modes constitute a sequence such that

$$\lim_{N \to \infty} \beta_N \to 0. \qquad (5.2.53)$$

The existence of these WSP modes shows that the inf–sup condition, when it depends on the polynomial degree N, goes to zero when N goes to infinity. WSP modes will degrade the convergence rate of the iterative methods such as the CG algorithm.

5.2.3 Collocation Methods on Single and Staggered Grids

Collocation methods are inspired by the knowledge and experience acquired with the solution of elliptic problems (see Chapters 2 and 4). Specifically, these methods require that the residuals obtained by replacing the continuous variables with their numerical approximations vanish on a set of well-chosen points, the collocation nodes. Generally speaking, roots of any Jacobi polynomial are candidates for collocation. Nevertheless, numerical practice relies heavily on Chebyshev and Legendre-based collocation methods. We review them first in the context of single-grid collocation [36].

Legendre Single-Grid Collocation

If one chooses the GLL points as collocation nodes, the one-grid collocation technique can be viewed as the counterpart of the one-spectral-element approach ($E = 1$) in the framework of the $\mathbb{P}_N$–$\mathbb{P}_N$ method. It is based on velocity and pressure approximations of the form (5.2.25). The spaces are defined by the relationships

$$X_N = X \cap \mathbb{P}_N^d(\Omega), \qquad (5.2.54)$$

$$Z_N = Z \cap \mathbb{P}_N(\Omega), \qquad (5.2.55)$$

where $\mathbb{P}_N(\Omega)$ is the set of polynomials of degree $\leq N$ with respect to each of the independent variables. The abstract problem has the form (5.2.26)–(5.2.27), with the bilinear forms (5.2.28), (5.2.29) and the discrete scalar product (5.2.30).

It is possible to arrive at the collocation formulation of the abstract problem by using the exactness property of the GLL rule in conjunction with integration by parts. Omitting the details of such calculation, we restate the problem as follows: Find $\mathbf{v}_N \in \mathbb{P}_N^d(\Omega)$ and $p_N \in Z_N$ such that

$$-\Delta \mathbf{v}_N(\mathbf{x}) + \nabla p_N(\mathbf{x}) = \mathbf{f}_N(\mathbf{x}) \quad \text{for} \quad \mathbf{x} \in \Xi_N \cap \Omega, \quad (5.2.56)$$

$$\operatorname{div} \mathbf{v}_N(\mathbf{x}) = 0 \quad \text{for} \quad \mathbf{x} \in \Xi_N \cap \bar{Q}, \quad (5.2.57)$$

$$\mathbf{v}_N(\mathbf{x}) = 0 \quad \text{for} \quad \mathbf{x} \in \Xi_N \cap \partial\Omega. \quad (5.2.58)$$

As already mentioned, discretizations of velocity and pressure with interpolants of the same degree induce spurious pressure modes. These modes, say, $q_N \in \mathbb{P}_N$, constitute numerical artifacts and satisfy the weak formulation of Equation (5.2.31), that is,

$$\mathcal{B}_N(\mathbf{u}_N, q_N) = -(\operatorname{div} \mathbf{u}_N, q_N)_N = 0 \quad \forall \mathbf{u}_N \in X_N. \quad (5.2.59)$$

The vector subspace generated by these functions q_N will be denoted by Y_N; the corresponding discrete divergence operator is B_N, the discrete counterpart of B (5.2.18). The space $\operatorname{Im} X_N$ obtained through the divergence operator is the orthogonal subspace of Y_N in $\mathbb{P}_N$ with respect to the $L^2(\Omega)$ scalar product. Note also that unlike the continuous problem, $Y_N = \ker B_N^T$ is not reduced to constant functions (B^T is the gradient operator). Bernardi and Maday [38] show the space Y_N of spurious modes to be of dimension 8 for 2D problems. With the notation

$$L_{i,j} = L_i(x)L_j(y), \qquad L'_{i,j} = L'_i(x)L'_j(y), \quad (5.2.60)$$

Y_N can be identified as

$$Y_N = \operatorname{span}\{L_{0,0}, L_{0,N}, L_{N,0}, L_{N,N}, L'_{N,N}, xL'_{N,N}, yL'_{N,N}, xyL'_{N,N}\}.$$

One can also show that this pressure kernel comes from compatibility conditions between the discretized continuity constraint and the prescribed velocity boundary conditions (see Labrosse [234]). To get rid of these parasitic modes one eliminates, for example, the four corners of the computational box from the pressure grid. Furthermore, one eliminates four other points, arbitrarily chosen on $\partial\Omega$, that yield a nonvanishing determinant for the matrix formed by the values of the four polynomials

$$L_{0,0}, L_{0,N}, L_{N,0}, L_{N,N},$$

calculated at these particular nodes. Doing so enforces the inf–sup condition:

$$\sup_{\mathbf{u}_N \in X_N} \frac{\mathcal{B}_N(\mathbf{u}_N, q_N)}{\|\mathbf{u}_N\|_{H_0^1(\Omega)^d}} \geq CN^{-1} \|q_N\|_{L_0^2(\Omega)} \quad \forall q_N \in Z_N, \quad (5.2.61)$$

5.2. Stokes Equations

where the space Z_N is complementary to Y_N and is chosen to be such that

$$\mathbb{P}_N(\Omega) = Z_N \oplus Y_N. \tag{5.2.62}$$

In Equation (5.2.62), the symbol $\oplus$ denotes the *direct sum* of the two spaces (Section A.4.5). This decomposition is unique, and the space Z_N, called the *complement* of Y_N, is constructed in such a way that

$$\dim Y_N + \dim Z_N = \dim \mathbb{P}_N(\Omega), \qquad Y_N \cap Z_N = \{0\}. \tag{5.2.63}$$

For 3D applications, the space Y_N contains $12N + 4$ spurious pressure modes (see Bernardi et al. [41]), among which $12N - 4$ are generated at the nodes lying on the edges of Ω and the remaining ones are in the space

$$\text{span}\{L_{0,0,0}, L_{N,0,0}, L_{0,N,0}, L_{0,0,N}, L_{N,N,0}, L_{N,0,N}, L_{0,N,N}, L_{N,N,N}\}. \tag{5.2.64}$$

With the pressure space (5.2.62), one can prove the following convergence results (see [41]).

Theorem 5.3 *Under the assumptions*

1. $N \geq 2$, with the pressure p_N in Z_N,
2. *the solution* $(\mathbf{v}, p)$ *of the weak continuous problem (5.2.12)–(5.2.13) is in* $H_0^s(\Omega)^d \times H^{s-1}(\Omega)$ *for an integer* $s \geq 1$, *and the function* $\mathbf{f}$ *is in* $H^t(\Omega)^d$ *for an integer* $t \geq 2$,

the problem (5.2.26)–(5.2.27) possesses a unique solution $(\mathbf{v}_N, p_N)$ *in* $X_N \times Z_N$ *with the following bounds for the errors:*

$$\|\mathbf{v} - \mathbf{v}_N\|_{H_0^1(\Omega)^d} \leq C\big(N^{1-s}\|\mathbf{v}\|_{H_0^s(\Omega)^d} + N^{-t}\|\mathbf{f}\|_{H^t(\Omega)^d}\big), \tag{5.2.65}$$

$$\|\mathbf{v} - \mathbf{v}_N\|_{L^2(\Omega)^d} \leq C_1\big(N^{-s}\|\mathbf{v}\|_{H_0^s(\Omega)^d} + N^{-t}\|\mathbf{f}\|_{H^t(\Omega)^d}\big), \tag{5.2.66}$$

$$\|p - p_N\|_{L_0^2(\Omega)} \leq C_2\big[N^{2-s}\big(\|\mathbf{v}\|_{H_0^s(\Omega)^d} + \|p\|_{H^{s-1}(\Omega)}\big) + N^{1-t}\|\mathbf{f}\|_{H^t(\Omega)^d}\big], \tag{5.2.67}$$

with C, C_1, *and* C_2 *constants independent of* N.

Chebyshev Single-Grid Collocation

The Chebyshev collocation grid is obtained from the Gauss–Lobatto–Chebyshev (GLC) nodes ζ_i^c, $i = 0, \ldots, N$, which are the roots of the equation

$$(1 - \zeta^2) T_N' = 0, \qquad \zeta \in [-1, 1], \tag{5.2.68}$$

where T_N' is the derivative of the Chebyshev polynomial of first kind and of degree N. Those collocation points (see Figure 2.7.1) form the grid $\mathbf{Z}_N$ and are written in closed form as

$$\zeta_i^c = \cos\frac{\pi i}{N}, \qquad 0 \le i \le N. \tag{5.2.69}$$

Each Chebyshev grid $\mathbf{Z}_N$ in a d-dimensional problem is built up by a tensor product of one-dimensional GLC grids:

$$\mathbf{Z}_N = \otimes\, \zeta_{N_{x_i}}^c, \qquad i = 1, \ldots, d. \tag{5.2.70}$$

With the same choice for X_N and Z_N as in Equations (5.2.54)–(5.2.55), the discrete problem may be presented as follows: Find $\mathbf{v}_N \in \mathbb{P}_N^d(\Omega)$ and $p_N \in Z_N$ such that

$$-\Delta \mathbf{v}_N(\mathbf{x}) + \nabla p_N(\mathbf{x}) = \mathbf{f}_N(\mathbf{x}) \qquad \forall \mathbf{x} \in \mathbf{Z}_N \cap \Omega, \tag{5.2.71}$$

$$\operatorname{div} \mathbf{v}_N(\mathbf{x}) = 0 \qquad \forall \mathbf{x} \in \mathbf{Z}_N \cap \bar{\Omega}, \tag{5.2.72}$$

$$\mathbf{v}_N(\mathbf{x}) = 0 \qquad \forall \mathbf{x} \in \mathbf{Z}_N \cap \partial\Omega. \tag{5.2.73}$$

A variational statement of the Chebyshev collocation process may be obtained if we generalize the saddle-point formulation according to Bernardi et al. [35]. This is necessary to describe the Stokes problem within the abstract setting of weighted Sobolev spaces employing the Chebyshev weight function w_d defined for a d-dimensional problem as the product of one-dimensional Chebyshev weights in each (continuous) variable ζ_i:

$$w_d(\mathbf{Z}) = \prod_{i=1}^{d} w_i(\zeta), \qquad \mathbf{Z} \in \hat{\Omega}, \tag{5.2.74}$$

$$w_i(\zeta) = (1 - \zeta^2)^{-\frac{1}{2}}, \qquad \zeta \in [-1, 1]. \tag{5.2.75}$$

Let us introduce the inner product based on the GLC integration rule for the 3D case:

$$(f, g)_{N,C} = \sum_{i,j,k=0}^{N} f_{i,j,k}\, g_{i,j,k}\, \omega_{i,j,k}^c, \tag{5.2.76}$$

with $\omega_{i,j,k}^c := \omega_i^c \omega_j^c \omega_k^c$ denoting the product of the corresponding GLC weights (B.2.13). The discrete abstract problem becomes: Find $(\mathbf{v}_N, p_N) \in (X_N, Z_N)$ such that

$$\mathcal{A}_{N,C}(\mathbf{v}_N, \mathbf{u}_N) + \mathcal{B}_{1,N,C}(\mathbf{u}_N, p_N) = \mathcal{F}_N(\mathbf{u}_N)_{N,C} \qquad \forall \mathbf{u}_N \in X_N, \tag{5.2.77}$$

$$\mathcal{B}_{2,N,C}(\mathbf{v}_N, q_N) = 0 \qquad \forall q_N \in Z_N. \tag{5.2.78}$$

5.2. Stokes Equations

The bilinear forms appearing in (5.2.77)–(5.2.78) are given by the relations

$$\mathcal{A}_{N,C}(\mathbf{v}_N, \mathbf{u}_N) = \left(\nabla \mathbf{v}_N, \nabla(\mathbf{u}_N w_d) w_d^{-1}\right)_{N,C}, \quad (5.2.79)$$

$$\mathcal{B}_{1,N,C}(\mathbf{u}_N, q_N) = -\left(\text{div}(\mathbf{u}_N w) w_d^{-1}, q_N\right)_{N,C}, \quad (5.2.80)$$

$$\mathcal{B}_{2,N,C}(\mathbf{u}_N, q_N) = -(\text{div } \mathbf{u}_N, q_N)_{N,C}. \quad (5.2.81)$$

The spaces $Y_{i,N,C}$, $i = 1, 2$, of spurious modes are, as in the Legendre case, of dimension 8 for 2D problems. They correspond to the definitions

$$Y_{i,N,C} = \{q_N \in \mathbb{P}_N; \ \forall \mathbf{u}_N \in X_N, \ \mathcal{B}_{i,N,C}(\mathbf{u}_N, q_N) = 0\}. \quad (5.2.82)$$

With the notation

$$T_{i,j} = T_i(x) T_j(y), \ T'_{i,j} = T'_i(x) T'_j(y), \quad (5.2.83)$$

the space $Y_{1,C,N}$ is generated by

$$Y_{1,N,C} = \text{span}\{T_{0,0}, T_{0,N}, T_{N,0}, T_{N,N}, T'_{N,N}, xT'_{N,N}, yT'_{N,N}, xyT'_{N,N}\}, \quad (5.2.84)$$

and the space $Y_{2,N,C}$ by

$$Y_{2,N,C} = \text{span}\{q_N^*(x) q_N^*(y), q_N^*(x) T_N(y), T_N(x) q_N^*(y), T_{N,N},$$
$$T'_{N,N}, xT'_{N,N}, yT'_{N,N}, xyT'_{N,N}\}. \quad (5.2.85)$$

In the previous relation, the polynomial q_N^* is defined implicitly by the equation

$$\forall \varphi \in \mathbb{P}_N(-1, 1), \quad (q_N^*, \varphi)_{N,C} = \int_{-1}^{+1} \varphi(\zeta) \, d\zeta \quad (5.2.86)$$

and is the discrete orthogonal projection of $w^{-1}(\zeta)$ onto $\mathbb{P}_N(-1, 1)$ for the scalar product (5.2.76).

If one chooses the test functions for (5.2.81) in the supplementary space of $Y_{2,N,C}$, an inf–sup condition results with a constant that is dependent on N but that does not show up in the error bounds. For the bilinear form $\mathcal{B}_{1,N,C}$, one can prove that for every $q_N \in \mathbb{P}_N(\Omega)$ orthogonal to $Y_{1,C,N}$, there exists a constant C independent of N such that

$$\sup_{\mathbf{u}_N \in X_N} \frac{\mathcal{B}_{1,N,C}(\mathbf{u}_N, q_N)}{\|\mathbf{u}_N\|_{H_C^1(\Omega)^2}} \geq CN^{-2} \|q_N\|_{L_C^2(\Omega)}. \quad (5.2.87)$$

Consequently, the next theorem provides the error estimates.

Theorem 5.4 *Under the assumptions*

1. *$N \geq 2$, with the pressure p_N in Z_N the supplementary space of $Y_{1,N,C} \in \mathbb{P}_N(\Omega)$,*
2. *the solution $(\mathbf{v}, p)$ of the weak continuous problem (5.2.12)–(5.2.13) is in $H_0^s(\Omega)^d \times H^{s-1}(\Omega)$ for an integer $s \geq 1$, and the function $\mathbf{f}$ is in $H^t(\Omega)^d$ for an integer $t \geq 2$,*

the problem (5.2.77)–(5.2.78) possesses a unique solution $(\mathbf{v}_N, p_N)$ in $X_N \times Z_N$ with the following bounds for the errors:

$$\|\mathbf{v} - \mathbf{v}_N\|_{H_C^1(\Omega)^d} \leq C\left(N^{1-s}\|\mathbf{v}\|_{H_C^s(\Omega)^d} + N^{-t}\|\mathbf{f}\|_{H^t(\Omega)^d}\right), \tag{5.2.88}$$

$$\|p - p_N\|_{L_C^2(\Omega)} \leq C_1\left[N^{3-s}\left(\|\mathbf{v}\|_{H_C^s(\Omega)^d} + \|p\|_{H^{s-1}(\Omega)}\right) + N^{2-t}\|\mathbf{f}\|_{H^t(\Omega)^d}\right], \tag{5.2.89}$$

with C and C_1 constants independent of N.

Legendre Staggered-Grid Collocation

In the same spirit as the FD staggered-mesh (MAC) method [185] and with the goal of getting rid of the spurious pressure modes arising in the single-grid collocation approach, Bernardi and Maday [37] proposed a staggered Legendre scheme. This procedure may also be used with a Chebyshev discretization. As usual in a staggered grid, the horizontal velocity components are attached to the mid-nodes of the left and right faces of the computational cell, and the vertical velocities are defined at the mid-nodes of the bottom and top faces. The pressures are cell-centered. Therefore, the 2D grid on the reference square is built up as follows (where, for the Legendre case, Ξ and ζ are the respective GLL and GL grids):

$$G_u = \Xi_{N_x} \otimes \zeta_{N_y}, \tag{5.2.90}$$

$$G_v = \zeta_{N_x} \otimes \Xi_{N_y}, \tag{5.2.91}$$

$$G_p = \zeta_{N_x} \otimes \zeta_{N_y}. \tag{5.2.92}$$

Figure 5.2.2 shows this grid for the case $N_x = N_y = 7$. The dots, the triangles, and the stars are the respective nodes for the pressure and for the x- and y-velocity components. Because of the staggered velocity nodes, the approximation spaces are defined by

$$X_{N,u} = X \cap \{\mathbb{P}_N(x) \otimes \mathbb{P}_{N+1}(y)\}, \tag{5.2.93}$$

$$X_{N,v} = X \cap \{\mathbb{P}_{N+1}(x) \otimes \mathbb{P}_N(y)\}, \tag{5.2.94}$$

5.2. Stokes Equations

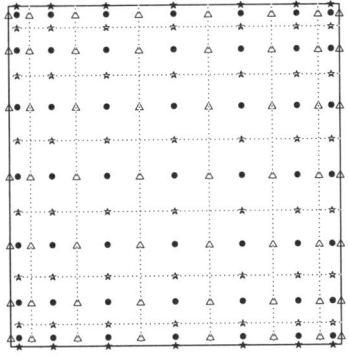

Figure 5.2.2. Legendre staggered collocation grid for $N_x = N_y = 7$.

and the pressures are in

$$Z_N = Z \cap \mathbb{P}_{N-1}(\Omega). \tag{5.2.95}$$

We note that the spaces $\mathbb{P}_{N+1}$ involved in the definitions (5.2.93)–(5.2.94) are needed to impose the homogeneous boundary conditions on the velocity field, although the corresponding grids contain only $N-1$ points.

The collocation process yields the following problem: Find $\mathbf{v}_N \in X_N = (X_{N,u} \times X_{N,v})$ and $p_N \in Z_N$ such that

$$\left(-\Delta u_N + \frac{\partial p_N}{\partial x}\right)(\mathbf{x}) = f_x(\mathbf{x}) \qquad \forall \mathbf{x} \in G_u \cap \Omega, \tag{5.2.96}$$

$$\left(-\Delta v_N + \frac{\partial p_N}{\partial y}\right)(\mathbf{x}) = f_y(\mathbf{x}) \qquad \forall \mathbf{x} \in G_v \cap \Omega, \tag{5.2.97}$$

$$\operatorname{div} \mathbf{v}_N(\mathbf{x}) = 0 \qquad \forall \mathbf{x} \in G_p \cap \Omega. \tag{5.2.98}$$

The 3D generalization of this collocation method is straightforward, with obvious extensions such as

$$G_u = \Xi_{N,x} \otimes \zeta_{N,y} \otimes \zeta_{N,z} \tag{5.2.99}$$

and

$$X_{N,u} = X \cap \{\mathbb{P}_N(x) \otimes \mathbb{P}_{N+1}(y) \otimes \mathbb{P}_{N+1}(z)\}. \tag{5.2.100}$$

To recast the above collocation problem into a variational formulation, one uses the polynomial bases defined as follows:

$$g^u_{i,j} = \pi_i(x)(1-y^2)\tilde{\pi}_j(y), \qquad 1 \le i \le N-1, \quad 1 \le j \le N, \tag{5.2.101}$$

$$g^v_{i,j} = (1-x^2)\tilde{\pi}_i(x)\pi_j(y), \qquad 1 \le i \le N, \quad 1 \le j \le N-1, \tag{5.2.102}$$

formed by the GLL Lagrangian interpolants $\pi_i \in \mathbb{P}_N(\hat{\Omega})$ given by (2.4.3) and the GL Lagrangian interpolants $\tilde{\pi}_i \in \mathbb{P}_{N-1}(\hat{\Omega})$ given by

$$\tilde{\pi}_i(\zeta) = \frac{L_N(\zeta)}{(\zeta - \zeta_i)L'_N(\zeta_i)}, \qquad 1 \leq i \leq N-1. \qquad (5.2.103)$$

In addition, an appropriate factor ensures the satisfaction of the boundary conditions.

Let us define the bilinear form for the 2D functions **f** and **g**:

$$(\mathbf{f}, \mathbf{g})_N = \sum_{i=0}^{N} \sum_{j=1}^{N} f_x(\xi_i, \zeta_j) g_x(\xi_i, \zeta_j) \rho_i \omega_j + \sum_{i=1}^{N} \sum_{j=0}^{N} f_y(\zeta_i, \xi_j) g_y(\zeta_i, \xi_j) \omega_i \rho_j. \qquad (5.2.104)$$

Multiplying through the momentum equations (5.2.96)–(5.2.97) by $g^u_{i,j} \rho_i \omega_j$ and $g^v_{i,j} \omega_i \rho_j$, respectively, summing these relations, and taking the exactness of the GLL and GL integration formulae into account to perform integration by parts, we arrive at the following abstract problem: Find $\mathbf{v}_N \in X_N$ and $p_N \in Z_N$ such that

$$\mathcal{A}_N(\mathbf{v}_N, \mathbf{u}_N) + \mathcal{B}_N(\mathbf{u}_N, p_n) = \mathcal{F}(\mathbf{g})_N \qquad \forall \mathbf{u}_N \in X_N, \quad (5.2.105)$$

$$\mathcal{B}_N(\mathbf{v}_N, q_N) = 0 \qquad \forall q_N \in Z_N, \quad (5.2.106)$$

where the forms are

$$\mathcal{A}_N(\mathbf{v}_N, \mathbf{u}_N) = \left(\frac{\partial \mathbf{v}_N}{\partial x}, \frac{\partial \mathbf{u}_N}{\partial x}\right)_N + \left(\frac{\partial \mathbf{v}_N}{\partial y}, \frac{\partial \mathbf{u}_N}{\partial y}\right)_N, \qquad (5.2.107)$$

$$\mathcal{B}_N(\mathbf{u}_N, q_N) = -\sum_{i=1}^{N} \sum_{j=1}^{N} \operatorname{div} \mathbf{u}_N(\zeta_i, \zeta_j) q_N(\zeta_i, \zeta_j) \omega_i \omega_j. \qquad (5.2.108)$$

One can prove that the discrete problem has no spurious pressure mode. Furthermore, after lengthy developments [37], one can prove that the form $\mathcal{A}_N$ satisfies continuity and ellipticity properties. However, the ellipticity constant is not independent of N and varies like CN^{-1}. Likewise, the form $\mathcal{B}_N$ is also continuous and is characterized by the inf–sup condition

$$\sup_{\mathbf{u}_N \in X_N} \frac{\mathcal{B}_N(\mathbf{u}_N, q_N)}{\|\mathbf{u}_N\|_{H^1(\Omega)^2}} \geq CN^{-1/2} \|q_N\|_{L^2(\Omega)} \qquad \forall q_N \in Z_N. \quad (5.2.109)$$

Because the ellipticity constant for $\mathcal{A}_N$ is not bounded independently of N and although the inf–sup constant, $\beta_N := CN^{-1/2}$, is better in this case than in single-grid collocation, one obtains the following error estimates, which are not optimal for either the velocities or the pressures.

5.3. Linear Systems, Algorithms, and Preconditioners

Theorem 5.5 *Under the assumptions*

1. $N \geq 2$, with the pressure p_N in Z_N given by (5.2.95),
2. the solution $(\mathbf{v}, p)$ of the weak continuous problem with $d = 2$ (5.2.12)–(5.2.13) is in $H_0^s(\Omega)^d \times H^{s-1}(\Omega)$ for an integer $s \geq 1$, and the function $\mathbf{f}$ is in $H^t(\Omega)^d$ for an integer $t \geq 2$,

the problem (5.2.105)–(5.2.106) possesses a unique solution $(\mathbf{v}_N, p_N)$ *in* $X_N \times Z_N$ *with the following error bounds:*

$$\|\mathbf{v} - \mathbf{v}_N\|_{H^1(\Omega)^d} + N^{-1/2}\|p - p_N\|_{L^2(\Omega)}$$
$$\leq C \big(N^{2-s} \big(\|\mathbf{v}\|_{H^s(\Omega)^d} + \|p\|_{H^{s-1}(\Omega)} \big) + N^{2-t} \|\mathbf{f}\|_{H^t(\Omega)^d} \big),$$

with C a constant independent of N.

We note that in practical implementations, especially with collocation calculations, the pressure is not in $L_0^2(\Omega)$. Instead, a reference pressure is fixed by pinning a reference level at a well-chosen point. With iterative methods, one can enforce $p \in L_0^2(\Omega)$ through projection [see (5.1.4) and (2.7.40)].

5.3 Linear Systems, Algorithms, and Preconditioners

Having gathered the results available from numerical analysis, we are now ready to consider the linear systems obtained from the various techniques, the available solvers, the conditioning of the matrices, and the design of efficient preconditioners.

5.3.1 Spectral-Element Methods and Uzawa Algorithm

In the classical staggered $\mathbb{P}_N$–$\mathbb{P}_{N-2}$ method [261, 332] the pressure gradients when evaluated are interpolated on the velocity grid Ξ_N. In the collocative formulation, where the internal GLL points constitute the pressure nodes, this interpolation process is not needed, thereby reducing the operation count of the method. The inf–sup constant β_N, which is related to the pressure discretization, is important because it has to be positive in order to obtain a generalized inverse of the discrete velocity–pressure system. The pressure approximation is optimal when β_N is independent of N. If this is not the case, the pressure incurs a loss of accuracy of order β_N^{-1}. In addition, the number of iterations in the Uzawa algorithm [9] will be proportional to β_N^{-1}.

Let us examine the matrix involved in the discrete form of the incompressibility condition (5.2.38). For a single element in $\mathbb{R}^2$ we find

$$(q_N, \operatorname{div} \mathbf{v}_N)_{N,\mathrm{GL}} = \sum_{l=1}^{2} \sum_{i,j=1}^{N-1} q_N(\zeta_i, \zeta_j) \frac{\partial v_{l,N}}{\partial x_l}(\zeta_i, \zeta_j) \omega_i \omega_j. \quad (5.3.1)$$

The contribution from q_N presents no difficulty, because q_N is represented by the Lagrangian interpolants π_i^p (5.2.35). However, the velocity derivatives must be interpolated from GLL to GL grids, giving rise to the following matrix form:

$$(q_N, \operatorname{div} \mathbf{v}_N)_{N,\mathrm{GL}} = \underline{q}^T (D_1 \underline{v}_1 + D_2 \underline{v}_2), \quad (5.3.2)$$

where $\underline{v}_i$, $i = 1, \ldots, d$, constitute the discrete set of velocity components. When dealing with rectilinear affine mappings, one defines the local derivative matrices as

$$D_1 = \frac{L_2}{2} \tilde{I} \otimes \tilde{D}, \qquad D_2 = \frac{L_1}{2} \tilde{D} \otimes \tilde{I}, \quad (5.3.3)$$

where the matrix $\tilde{I}$ with the components

$$\tilde{I}_{i,j} = \omega_i \pi_j^v(\zeta_i) \quad (5.3.4)$$

takes care of the weighted interpolation between the two grids, and

$$\tilde{D}_{i,j} = \omega_i \left. \frac{d\pi_j^v}{dr} \right|_{\zeta_i} \quad (5.3.5)$$

is the weighted one-dimensional derivative matrix, interpolated on the GL grid. It is left as an exercise for the reader to show that the pressure gradient matrix coming from the discretization of the term $\mathcal{B}_{N,\mathrm{GL}}(\mathbf{u}_N, p_N)$ in (5.2.37) yields matrices that are the transpose of the matrices involved in the divergence operator.

Using test functions based on the Lagrangian polynomials of the velocity approximation, the projection procedure of Equations (5.2.37)–(5.2.38) leads to a symmetric, positive semidefinite system. Using $\underline{p}$ to denote the set of discrete unknowns of the pressure, and K, D^T, D, to denote the discrete Laplace, gradient, and divergence operators, respectively, one finds

$$K \underline{v}_i - D_i^T \underline{p} = M \underline{f}_i, \qquad i = 1, \ldots, d, \quad (5.3.6)$$

$$-\sum_{i=1}^{d} D_i \underline{v}_i = 0. \quad (5.3.7)$$

5.3. Linear Systems, Algorithms, and Preconditioners

The r.h.s. $\underline{f}_i$ corresponds to the discretization of the data **f**, and M is the diagonal mass matrix. Because the underlying bilinear form $\mathcal{A}_N$ is symmetric, K is also symmetric. Furthermore, D_i and D_i^T are transposed matrices, as they are adjoint operators in the continuous formulation, a property that is conserved at the discrete level. The ellipticity property of $\mathcal{A}_N$ implies that K is invertible. We can therefore proceed with a block Gaussian elimination. From (5.3.6), the velocity components are

$$\underline{v}_i = K^{-1} D_i^T \underline{p} + K^{-1} M \underline{f}_i, \qquad i = 1, \ldots, d. \tag{5.3.8}$$

Inserting this last expression in the continuity condition (5.3.7), one gets

$$\sum_{i=1}^{d} D_i K^{-1} D_i^T \underline{p} = \sum_{i=1}^{d} -D_i K^{-1} M \underline{f}_i. \tag{5.3.9}$$

Consequently, solving this last equation for the pressure, one computes the velocity components from (5.3.6).

The key question now is how to solve (5.3.9). Let us denote the discrete pressure operator, called the *Uzawa operator*, as

$$S = \sum_{i=1}^{d} D_i K^{-1} D_i^T. \tag{5.3.10}$$

This matrix is symmetric positive semidefinite. Formally, the continuous counterpart of the pressure operator is close to identity. Indeed, it corresponds to the chain of differentiation (grad), two integrations (the inverse Laplacian), and differentiation again (div). Hence, we expect that the condition number of S will be close to unity.

This condition number may be analyzed by solving the generalized eigenvalue problem

$$S\underline{\omega}^k = \lambda_k \tilde{M} \underline{\omega}^k, \qquad k = 1, \ldots, r, \tag{5.3.11}$$

where $\tilde{M}$ is the diagonal pressure mass matrix, $\underline{\omega}^k$ is the set of generalized pressure eigenvectors, and r is the rank of $\tilde{M}$. Fortin [138] proved that the inf–sup constant β appearing in (5.2.15) is related to the former problem by the result

$$\beta = (\lambda_{\min})^{1/2}, \tag{5.3.12}$$

where $\lambda_{\min}$ denotes the minimum (nonvanishing) eigenvalue of the spectrum $\{\lambda_k\}$, and the square root of the maximum eigenvalue yields the norm of the bilinear form $\mathcal{B}(\cdot,\cdot)$. We note that the spurious pressure modes are the eigenvectors of the Uzawa matrix corresponding to the vanishing eigenvalues of this operator.

In the discrete situation, however, S is no longer close to the identity matrix, but to $\tilde{M}$, the variational equivalent of the identity operator. This suggests that $\tilde{M}^{-1}$ is an efficient preconditioner of the CG technique. The elliptic character of the bilinear forms involved in the problem motivates application of the preconditioned conjugate gradient (PCG) method to the solution of the pressure equation (5.3.9). The algorithm proceeds as in Algorithm 2.2. Here, the unknown $\underline{x}$ is the pressure $\underline{p}$. With $\underline{p}_0$ given, the initial residual $\underline{r}_0$ is computed by the relation $\underline{r}_0 = D_i K^{-1} M \underline{f}_i + S \underline{p}_0$. Note that the matrix A of the PCG is now S. As can be seen, the inner iteration implies the evaluation of $S\underline{w}$. This calculation can be carried out with the next stages,

$$\underline{y}_i = D_i^T \underline{w}, \quad i = 1, \ldots, d, \tag{5.3.13}$$

$$K \underline{z}_i = \underline{y}_i, \quad i = 1, \ldots, d, \tag{5.3.14}$$

$$S \underline{w} = \sum_{i=1}^{d} D_i \underline{z}_i. \tag{5.3.15}$$

The intermediate step involves the solution of d elliptic equations, which can be treated again by the PCG method or by direct solvers such as the fast diagonalization method [86]. We note that the residual $\underline{r}$ in the outer CG iteration is the discrete divergence $-D_i \underline{v}_i$. Its norm can serve as a stopping criterion in the iterations. Algorithm 2.2 in Section 2.7 and Equations (5.3.13)–(5.3.15) together form the Uzawa algorithm [9].

The number of iterations required for PCG is proportional to the square root of the condition number of the system matrix. Bernardi and Maday [38] and Maday et al. [256] prove that the maximum eigenvalue $\lambda_{\max}$ of $\tilde{M}^{-1} S$ is of the order of unity, while the minimum eigenvalue $\lambda_{\min}$ scales like β_N^2. The β_N expression is the inf–sup constant given in (5.2.45), namely, $N^{(1-d)/2}$. As a consequence, the condition number of the pressure matrix is such that

$$\kappa(\tilde{M}^{-1} S) \sim \beta_N^{-2} = N^{d-1}. \tag{5.3.16}$$

This theoretical estimate is numerically confirmed (see, for example, [355]). The PCG method will converge in $O(N^{\frac{1}{2}})$ iterations for 2D problems and in $O(N)$ iterations for 3D ones. However, this asymptotic behavior does not match what is observed when N is very low, a situation close to classical FE discretizations. In this case, the inf–sup condition is independent of the characteristic spatial resolution, $h \sim N^{-1}$, as long as the discrete spaces are compatible. Practical

5.3. Linear Systems, Algorithms, and Preconditioners

computations also indicate that the condition number (5.3.16) is too pessimistic and that the Uzawa algorithm converges more rapidly.

5.3.2 Collocation Methods

Legendre single-grid collocation is easily implemented if the $\mathbb{P}_N$–$\mathbb{P}_N$ method is adopted. The number of velocity and pressure nodes is the same, a fact that eases the coding. The spurious pressure modes are eliminated, for example, in a postprocessing step, where the pressures are obtained on the right subspace by an orthogonal projection (see Métivet [275]). The linear system associated with this approach has basically the same form as the one coming from spectral-element discretization. In the framework of Chebyshev collocation, Phillips and Roberts [307] remove the spurious pressure modes by applying a singular value decomposition (SVD) [367, 384] to the linear-algebraic pressure system.

For staggered mesh collocation, Le Quéré [237] discusses details of the implementation for Chebyshev collocation. The tensor-product property of the polynomial bases is thoroughly exploited to compute gradients or divergence through one-dimensional derivatives in the required space directions.

We will concentrate on Chebyshev single-grid collocation because it has been heavily practiced over the past decade. Orszag [288] and Morchoisne [280] early recognized that the conditioning of Chebyshev collocation would impede any direct solution of the system matrix generated by this method. Therefore, a preconditioning technique is indispensable. Among various possibilities, the FEM has provided a powerful and efficient tool. Demaret and Deville [94] analyze different 2D elements, namely, the nine-node Lagrangian Q_2–Q_1 element, the Q_1–Q_0 element, and a stabilized Q_1–Q_1 element due to Brezzi and Pitkäranta [54]. As usual, the vertices of the FE grid coincide with the GLC nodes.

The Chebyshev collocation matrix system is

$$K^c \underline{v}_i - G_i^c \underline{p} = \underline{f}_i, \quad i = 1, \ldots, d, \quad (5.3.17)$$

$$-\sum_{i=1}^{d} D_i^c \underline{v}_i = 0, \quad (5.3.18)$$

where K^c, G_i^c, and D_i^c are the discrete Chebyshev Laplace, gradient, and divergence operators, respectively. From the variational equivalent of the collocation problem, we know that K^c is asymmetric; furthermore, the discrete gradient and divergence operators are no longer transposes of each other because of the Chebyshev weight influence in the integration by parts. We denote by L_S^c the full Chebyshev Stokes matrix corresponding to Equations (5.3.17)–(5.3.18). The FE preconditioner produces the matrix L_S in the l.h.s. for the Stokes operator, and M_S in the r.h.s.

If a Richardson iteration method is set up, the essential question is the convergence of the method, which is governed by the spectral radius of the Richardson iteration matrix R^S defined by the equation

$$R^S = I - \alpha P^{-1} L_S^c. \qquad (5.3.19)$$

The spectral radius has to be less than one to ensure convergence. In this equation, α is a relaxation factor and P the preconditioning matrix defined as

$$P = (M_S)^{-1}(L_S). \qquad (5.3.20)$$

Even though $P^{-1}L_S^c$ is not a symmetric matrix, one customarily considers $\kappa(P^{-1}L_S^c) = \lambda_{max}/\lambda_{min}$ as an indicator of its condition number, λ_{max} and λ_{min} being the maximum and minimum eigenvalues of $P^{-1}L_S^c$, respectively. Demaret and Deville showed that $\kappa(P^{-1}L_S^c)$ is close to 2 when the Q_2–Q_1 element is used with Dirichlet boundary conditions and approaches $\pi^2/4$ with natural boundary conditions. Consequently, the optimum value of α [see (2.7.28)] is 2/3. Using this preconditioner, Demaret and Deville [94] noticed the systematic absence of spurious pressure modes in their numerous calculations. It seems that the FE, which satisfies the Brezzi–Babuška condition, has no spurious pressure mode and plays the role of an adequate filter for the collocation approximation.

For the Q_1–Q_0 preconditioner, the optimum value of α degrades to a value close to 0.4. This behavior is due to the slower spatial convergence of the Q_1–Q_0 FE pair, which possesses a pressure checkerboard mode to be filtered in the iteration procedure. These results on the preconditioned pressure matrix have been confirmed by the theoretical Fourier eigenspectrum analysis carried out by Deville and Mund [102] applied to a semiperiodic Stokes problem with FE preconditioning.

The stabilized Q_1–Q_1 element produces dismal performance, because incorporating the Laplace operator in the weak continuity equation with a multiplicative factor proportional to the square of the local dimension of the element induces λ_{min} to go to zero for increasing values of the discretization parameter N. Therefore, this element is not recommended.

Besides quadrilateral elements, CFD practitioners use triangles because they provide the user with more flexibility to handle complicated geometries. Deville et al. [103] use the triangular counterpart of the Q_2–Q_1 element, namely, the P_2–P_1 element. This last notation is used by Ciarlet [79] for triangles. In order to reach a prescribed level of accuracy on a Stokes analytical problem, the triangles need almost four times more iterations than do the quadrilateral elements. However, the extension to tetrahedra for 3D applications, replacing the expensive triquadratic–trilinear parallelepipeds, might be more rewarding in computational efficiency. This is still to be demonstrated.

5.4 Poisson Pressure Solver and Green's-Function Technique

In compressible fluid flow problems, a state equation – that is, a prognostic relation – is available to handle the pressure. For incompressible fluid flows, a diagnostic equation has to be generated instead.

5.4.1 General Considerations

In the continuum, the diagnostic relationship is obtained by applying the divergence operator to Equation (5.1.1) and taking the continuity requirement (5.1.2) into account. This produces a Poisson equation for the pressure:

$$\Delta p = \text{div}(-\mathbf{v} \cdot \nabla \mathbf{v} + \mathbf{f}) \quad \text{in } \Omega. \tag{5.4.1}$$

This second-order elliptic equation can be solved if we provide boundary conditions. For that purpose, the most natural choice, guided by the inviscid Euler case, consists of a Neumann condition that is itself derived from the Navier–Stokes equations by use of the normal pressure gradient. If we denote by $\mathbf{n}$ the unit vector normal to the boundary $\partial \Omega$, the condition reads

$$\mathbf{n} \cdot \nabla p = \mathbf{n} \cdot [-\mathbf{v} \cdot \nabla \mathbf{v} + (Re)^{-1} \Delta \mathbf{v} + \mathbf{f}] \quad \text{on } \partial \Omega. \tag{5.4.2}$$

If the Reynolds number goes to infinity and if the homogeneous Dirichlet velocity boundary condition (5.1.3) is imposed, the Neumann condition (5.4.2) reduces to a homogeneous one (neglecting the body-force term),

$$\mathbf{n} \cdot \nabla p = 0 \quad \text{on } \partial \Omega, \tag{5.4.3}$$

which is consistent with the integration of the Euler equations.

Another possible choice of boundary condition for the pressure Poisson equation (5.4.1) resorts to the tangential projection of the pressure gradient. With τ the unit vector tangent to $\partial \Omega$, the pressure condition is

$$\tau \cdot \nabla p = \tau \cdot [-\mathbf{v} \cdot \nabla \mathbf{v} + (Re)^{-1} \Delta \mathbf{v} + \mathbf{f}] \quad \text{on } \partial \Omega. \tag{5.4.4}$$

This condition is in fact a Dirichlet condition because the pressure can be integrated along $\partial \Omega$. It is discussed by Gresho [167, 166]. This boundary condition is no longer employed, however, because one can prove that Equations (5.4.1) and (5.4.2) imply Equation (5.4.4).

5.4.2 The Green's-Function Method

For the steady Stokes problem, Equations (5.2.1)–(5.2.2) may be combined into

$$\Delta p = \text{div}\, \mathbf{f} \quad \text{in } \Omega, \tag{5.4.5}$$

with a nonhomogeneous Neumann boundary condition

$$\mathbf{n} \cdot \nabla p = \mathbf{n} \cdot (\Delta \mathbf{v} + \mathbf{f}) \quad \text{on } \partial \Omega. \tag{5.4.6}$$

We observe that because of this last boundary condition the velocity and pressure are still coupled, leading to a full solve. In 3D problems this situation is unacceptable because the (discrete) matrix problem may not fit into main memory.

One way around this difficulty is to decouple the pressure and velocity calculations by the Green's-function method, also called the influence-matrix technique, where the pressure is imposed as a Dirichlet condition in such a way that the incompressibility constraint is satisfied everywhere, boundaries included.

One knows from continuum considerations that the solution $(\mathbf{v}, p)$ of Equation (5.2.1) satisfies at the same time, on the boundary $\partial \Omega$, the condition (5.4.6) and the incompressibility condition (5.2.2). In order to avoid the use of the Neumann condition for the pressure problem, it is tempting to impose a Dirichlet pressure condition expressing that the incompressibility constraint holds at the boundary. Indeed, by solving Equation (5.2.1) with homogeneous velocity boundary conditions and taking (5.4.5) into account with (5.2.2) on $\partial \Omega$, it is simple to conclude that

$$-\Delta \,\text{div}\, \mathbf{v} = 0 \quad \text{in } \Omega. \tag{5.4.7}$$

Therefore, the solution is such that $\text{div}\, \mathbf{v} = 0$ everywhere in Ω and the pair $(\mathbf{v}, p)$ is a solution of the Stokes problem (5.2.1). Unfortunately, the incompressibility condition at the boundary is not directly applicable to the Poisson pressure equation. The Green's-function procedure helps to circumvent that difficulty.

Let us first solve the pressure Poisson equation (5.4.5) with the boundary condition

$$p = p_b \quad \text{on } \partial \Omega, \tag{5.4.8}$$

with p_b a function to be specified later. The associated velocity field resulting from the solution of Equation (5.2.1) with homogeneous velocity boundary conditions and the previous incorrect pressure boundary field is not

5.4. Poisson Pressure Solver and Green's-Function Technique

divergence-free. To achieve that goal, we use the Helmholtz–Weyl decomposition [145]: any velocity field can be decomposed into a solenoidal part and an irrotational part:

$$\mathbf{v} = \mathbf{v}_S + \mathbf{v}_I, \tag{5.4.9}$$

such that

$$\text{div } \mathbf{v}_S = 0, \quad \mathbf{curl } \mathbf{v}_I = 0. \tag{5.4.10}$$

Because of these conditions, we may rewrite the decomposition, taking advantage of vector identities, as

$$\mathbf{v} = \mathbf{curl } \psi + \nabla \phi, \tag{5.4.11}$$

where ψ and ϕ are the vector and scalar potentials, respectively. To fix the reference level of this scalar potential, we assume that ϕ is vanishing on the boundary. Applying the divergence operator to (5.4.11), one obtains the Poisson problem

$$\Delta \phi = \text{div } \mathbf{v} \quad \text{in } \Omega \tag{5.4.12}$$

$$\phi = 0 \quad \text{on } \partial \Omega. \tag{5.4.13}$$

Enforcing Equation (5.4.7) on the potential ϕ leads to the biharmonic relation

$$\Delta^2 \phi = 0, \tag{5.4.14}$$

where $\Delta^2 = \Delta \Delta$. This equation implies through (5.4.12), (5.2.1), and the boundary condition (5.4.8) that

$$\Delta^2 \phi = \text{div}(\nabla p - \mathbf{f}) = 0. \tag{5.4.15}$$

Since $\text{div}(\nabla p - \mathbf{f})$ depends on p_b, we may infer that if we construct p_b in such a way that the homogeneous Neumann condition

$$\mathbf{n} \cdot \nabla \phi = 0 \quad \text{on } \partial \Omega \tag{5.4.16}$$

is satisfied, the biharmonic equation with both homogeneous Dirichlet and Neumann conditions has a trivial solution, implying that $\text{div } \mathbf{v} = 0$ over all the domain.

Using the linear superposition principle, one can split the variables as follows:

$$p = \hat{p} + p_0, \quad \mathbf{v} = \hat{\mathbf{v}} + \mathbf{v}_0. \tag{5.4.17}$$

Each new variable will be the solution of a Poisson problem. First, one computes the solution (v_0, p_0) of a problem with homogeneous pressure conditions

$$\Delta p_0 = \operatorname{div} \mathbf{f} \quad \text{in } \Omega, \tag{5.4.18}$$

$$p_0 = 0 \quad \text{on } \partial\Omega, \tag{5.4.19}$$

and

$$\nabla p_0 - \Delta \mathbf{v}_0 = \mathbf{f} \quad \text{in } \Omega, \tag{5.4.20}$$

$$\mathbf{v}_0 = 0 \quad \text{on } \partial\Omega. \tag{5.4.21}$$

Then, the other parts $(\hat{\mathbf{v}}, \hat{p})$ are obtained with inhomogeneous pressure conditions through

$$\Delta \hat{p}(p_b) = 0 \quad \text{in } \Omega, \tag{5.4.22}$$

$$\hat{p} = p_b \quad \text{on } \partial\Omega, \tag{5.4.23}$$

and

$$\nabla \hat{p}(p_b) - \Delta \hat{\mathbf{v}}(p_b) = 0 \quad \text{in } \Omega, \tag{5.4.24}$$

$$\hat{\mathbf{v}}(p_b) = 0 \quad \text{on } \partial\Omega. \tag{5.4.25}$$

Let us recall Green's formula: $\forall v \in H^1(\Omega)$ and $\mathbf{q} \in H^1(\Omega)^d$,

$$\int_\Omega \operatorname{div} \mathbf{q}\, v\, dV + \int_\Omega \mathbf{q} \cdot \nabla v\, dV = \int_{\partial\Omega} \mathbf{q} \cdot \mathbf{n}\, v\, dS, \tag{5.4.26}$$

where v in the last integral denotes the trace of v on the boundary $\partial\Omega$.

To impose the boundary condition (5.4.16) on ϕ, we use Green's formula (5.4.26), setting $\mathbf{q} = \nabla\phi$. With Equations (5.4.22)–(5.4.23) and (5.4.12)–(5.4.13), one obtains, for each boundary pressure function p_b that is a "trial" function belonging to the class of functions p_b,

$$\int_{\partial\Omega} \mathbf{n} \cdot \nabla\phi\, p_b\, dS = \int_\Omega \Delta\phi\, \hat{p}\, dV + \int_\Omega \nabla\phi \cdot \nabla\hat{p}\, dV$$

$$= \int_\Omega \operatorname{div} \mathbf{v}\, \hat{p}\, dV - \int_\Omega \phi \Delta\hat{p}\, dV$$

$$= \int_\Omega \operatorname{div} \mathbf{v}\, \hat{p}(p_b)\, dV. \tag{5.4.27}$$

As the potential ϕ satisfies a homogeneous Neumann condition (5.4.16), the pressure boundary function p_b is the solution of the problem

$$\int_\Omega \operatorname{div} \mathbf{v}(p_b)\, \hat{p}(p_b)\, dV = 0. \tag{5.4.28}$$

5.4. Poisson Pressure Solver and Green's-Function Technique

Applying Green's formula to Equation (5.4.28) and using the relationships (5.4.24) and (5.4.17), one easily gets the weak formulation of the Green's-function technique: Find p_b such that

$$\int_\Omega \nabla \hat{v}(p_b) \nabla \hat{v}(\mathsf{p}_b) \, dV = -\int_\Omega \nabla v_0 \nabla \hat{v}(\mathsf{p}_b) \, dV. \qquad (5.4.29)$$

This equation is represented by a positive definite, symmetric matrix, a candidate for a direct Cholesky solver or the CG method. However, Equation (5.4.29) is not yet in an efficient computational form.

5.4.3 Implementation

We will explain the algorithmic implementation for a 2D problem. For that purpose, let C_i ($i = 1, \ldots, \mathcal{N}_{bc}$) denote the number of boundary nodes resulting from the discretization and lying on $\partial\Omega$. The first step constructs the influence matrix, denoted by Q, which converts the pressure information on $\partial\Omega$ in the evaluation of div **v** at the same boundary points. If the problem is solved with N and M spatial points in the x and y directions respectively, the matrix Q has dimension $\mathcal{N}_{bc} = 2(N + M)$. If one discards the corners of the computational box because div **v** $= 0$ is achieved there identically, then $\mathcal{N}_{bc} = 2(N + M - 2)$. The $\mathcal{N}_{bc}$ pressure distributions used to set up the matrix Q are obtained by solving $\mathcal{N}_{bc}$ problems (5.4.22) and (5.4.23) with unit pressure loads such that

$$\hat{p}_b(C_i) = \delta(\mathbf{x}(C_i)), \qquad i = 1, \ldots, \mathcal{N}_{bc}. \qquad (5.4.30)$$

The resulting velocity field $\hat{\mathbf{v}}$ is computed by Equations (5.4.24)–(5.4.25). The influence matrix is built up by storing columnwise the values of $[\text{div } \hat{\mathbf{v}}(C_i)]_k$.

Once the influence matrix has been evaluated, one solves Equations (5.4.18)–(5.4.21). This computation allows one to obtain div $\mathbf{v}_0 = \bar{d}$ at the boundary points C_i. Indeed, we want div **v** $= 0$ at each boundary point C_i. Taking (5.4.17) into account, we obtain

$$\text{div } \hat{\mathbf{v}} = -\text{div } \mathbf{v}_0. \qquad (5.4.31)$$

The solution of the linear system

$$p_b = -Q^{-1}\bar{d} \qquad (5.4.32)$$

yields the pressure that must be enforced on the boundary to compute the correct pressure field by Equations (5.4.5) and (5.4.8). The velocity is obtained by solving

$$\nabla p - \Delta \mathbf{v} = \mathbf{f} \quad \text{in } \Omega, \tag{5.4.33}$$

$$\mathbf{v} = 0 \quad \text{on } \partial\Omega. \tag{5.4.34}$$

The extension of this algorithm to 3D applications is conceptually simple. However, its implementation on a computer is memory-intensive. Indeed, if N, M, and L are the numbers of points in the three space directions, one has to solve $O(2(NM + NL + ML))$ Stokes problems, and the matrix Q has dimension $2(NM + NL + ML)$.

For a FE presentation, we refer the reader to the paper by Glowinski and Pironneau [158] or the book by Girault and Raviart [157]. For spectral methods, an algorithm has been proposed by Kleiser and Schumann [226] with a Chebyshev approximation in one direction and Fourier series in the others; and a generalization for 2D problems was presented by Le Quéré and Alziary de Roquefort [238]. The use of this method for spectral-element calculations does not present any difficulty, as it is a mere extension of the FE approach.

We comment here about the so-called Poisson equation obtained from (5.3.10), where the stiffness matrix K is replaced by the diagonal mass matrix M. The relationship reads

$$A = D_i M^{-1} D_i^T. \tag{5.4.35}$$

We note that no pressure conditions are necessary, since the mass matrix imposes the velocity boundary conditions. With respect to the Uzawa pressure operator, the pseudo-Poisson approach avoids the solution of d Helmholtz equations and can be considered as a preconditioning technique of the full Uzawa method. We will return to this point in the next chapter when we analyze the pressure computation in unsteady problems.

5.5 Divergence-Free Bases

Key to successfully integrating the Stokes equations is the modification of the weak formulation through the use of a constrained Sobolev space, which we denote H_{div} and define by the relation

$$H_{\text{div}}(\Omega) = \left\{ \mathbf{u} \in H_0^1(\Omega)^d \,\middle|\, \text{div}\,\mathbf{u} = 0 \right\}. \tag{5.5.1}$$

This new space is clearly a subspace of the usual Sobolev space. Choosing the test functions in H_{div}, we may restate the weak formulation as follows: Find

5.5. Divergence-Free Bases

$\mathbf{v} \in H_{\text{div}}(\Omega)$ such that

$$\mathcal{A}(\mathbf{u}, \mathbf{v}) = \mathcal{F}(\mathbf{u}) \qquad \forall \mathbf{u} \in H_{\text{div}}(\Omega), \tag{5.5.2}$$

with the bilinear form $\mathcal{A}$ defined by the relation

$$\mathcal{A}: H_{\text{div}}(\Omega) \times H_{\text{div}}(\Omega) \to \mathbb{R}, \qquad \mathcal{A}(\mathbf{u}, \mathbf{v}) = (\nabla \mathbf{u}, \nabla \mathbf{v}). \tag{5.5.3}$$

One can easily show that the abstract form (5.5.2) is equivalent to the one given in Equations (5.2.12)–(5.2.13). See, for example, Girault and Raviart [157].

From this abstract form, it is simple to obtain a Galerkin approximation of the following problem: Find $\mathbf{v}_h \in H_{\text{div},h}$ such that

$$\mathcal{A}(\mathbf{u}_h, \mathbf{v}_h) = \mathcal{F}_h(\mathbf{u}_h) \qquad \forall \mathbf{u}_h \in H_{\text{div},h}, \tag{5.5.4}$$

where $H_{\text{div},h}$ is a finite-dimensional subspace of H_{div}. Here again the notation h may correspond to a finite or a spectral-element discretization. It is fairly easy to prove that

$$\|\mathbf{v} - \mathbf{v}_h\| \leq C \inf_{\mathbf{u}_h \in H_{\text{div},h}} \|\mathbf{v} - \mathbf{u}_h\|. \tag{5.5.5}$$

Although the proposed method is simple to write down in its abstract setting, a computer implementation is far from trivial. The essential task is the elaboration of a basis function that is inherently divergence-free. In the spectral frame, the first attempt was designed by Moser et al. [281] and further analyzed by Pasquarelli et al. [296] and Pasquarelli [295]. However, this new basis was proposed with trigonometric polynomials in two periodic directions and Chebyshev polynomials in the remaining direction. The generalization to the full nonhomogeneous 3D case is not simple. For this reason, Batcho and Karniadakis [24] introduce a generalized Stokes operator and compute the set of Stokes eigenfunctions in order to solve the Navier–Stokes equations.

The continuous Stokes problem (5.2.1) is generalized and reformulated as the following self-adjoint eigenproblem:

$$\nabla p - \text{div}[\rho(\mathbf{x})\nabla \mathbf{v}] + q(\mathbf{x})\mathbf{v} = \lambda \mathbf{v} \qquad \text{in } \Omega, \tag{5.5.6}$$

$$\text{div } \mathbf{v} = 0 \qquad \text{in } \Omega, \tag{5.5.7}$$

with the homogeneous boundary condition (5.1.3). The forcing term $\mathbf{f}$ has been replaced by $\lambda \mathbf{v}$, and the viscous term contains a nonconstant coefficient. The velocity field is sought in H_{div}. We note that the generalized Stokes problem reduces to the standard Stokes eigensystem if $\rho(\mathbf{x}) = 1$ and $q(\mathbf{x}) = 0$.

If one concentrates on the standard Stokes eigenproblem, the velocity field may be approximated by a series expansion of Stokes eigenfunctions such that

$$\mathbf{v} = \sum_{i=0}^{N} a_i \mathbf{w}_i, \qquad (5.5.8)$$

each eigenfunction being the solution of the eigensystem

$$\nabla \Pi - \Delta \mathbf{w}_i = \lambda_i \mathbf{w}_i \quad \text{in } \Omega, \qquad (5.5.9)$$

$$\text{div } \mathbf{w}_i = 0 \quad \text{in } \Omega, \qquad (5.5.10)$$

$$\mathbf{w}_i = 0 \quad \text{on } \partial\Omega. \qquad (5.5.11)$$

The eigenspectrum is built up in such a way that λ_0 designates the smallest eigenvalue. The scalar quantity Π in Equation (5.5.9) enforces the incompressibility constraint on each eigenfunction. The coefficients in the series (5.5.8) are obtained as the scalar products

$$a_i = (\mathbf{v}, \mathbf{w}_i) = \int_\Omega \mathbf{v} \mathbf{w}_i \, d\mathbf{x}. \qquad (5.5.12)$$

Inserting Equation (5.5.9) in (5.5.12) and using the Green's identity twice for the integration by parts [the scalar $\Pi(\mathbf{x})$ is eliminated in this procedure] with Equation (5.5.11) taken into account, one obtains

$$a_i = -\frac{1}{\lambda_i} \int_\Omega \Delta \mathbf{v} \mathbf{w}_i \, d\mathbf{x}. \qquad (5.5.13)$$

Using (5.5.9) again and integrating by parts twice, one writes the final result:

$$a_i = \frac{1}{\lambda_i^2} \left((\mathbf{w}_i, \Delta^2 \mathbf{v}) - \int_{\partial\Omega} \Delta \mathbf{v} \cdot (\mathbf{n} \cdot \nabla \mathbf{w}_i) \, dS \right). \qquad (5.5.14)$$

Inspection of this last equation shows that one can reach convergence rates higher than $O(1/\lambda_i^2)$ only if the boundary term vanishes, indicating that the Laplacian of the velocity field must go to zero when one approaches the boundary $\partial\Omega$. This is the case only for very peculiar situations, for example, in the presence of periodic boundary conditions. Following Batcho and Karniadakis, the rate of convergence is therefore $O(1/\lambda_i^2)$. Since the spectrum scales as $\lambda_n = O(n^{2/d})$ in $\mathbb{R}^d$ [82], the convergence rate for the Stokes eigenfunction series is $O(n^{-4/d})$, showing quadratic convergence for 2D problems ($d = 2$).

To improve the convergence rate, Batcho and Karniadakis, following the theory of Sturm–Liouville operators, propose to deal with a singular Stokes

5.5. Divergence-Free Bases

system. The convergence properties are enhanced through the use of a Laplacian operator with variable coefficient vanishing at the boundaries, that is, if one has

$$\int_{\partial\Omega} \rho(\mathbf{x})[\nabla^p \mathbf{v} \cdot (\mathbf{n} \cdot \nabla \mathbf{w}_i)] \, dS = 0. \tag{5.5.15}$$

For regular functions such that $\mathbf{v} \in C^\infty$, the eigenfunctions converge exponentially fast, whereas for nonsmooth functions the convergence rate is governed by the highest derivative order p for which the condition (5.5.15) holds. The eigenspectrum scales as $a_n = O(\lambda_n^{-p})$.

Singular systems with homogeneous Dirichlet eigenfunctions are designed by adjusting the diagonal coefficient $q(\mathbf{x})$ in (5.5.6) to alter the behavior of the eigenfunctions when one approaches the boundaries. A possible choice consists in selecting $q(\mathbf{x}) \sim \hat{\rho}^{-1}(\mathbf{x})$ near $\partial\Omega$ so that the eigenfunctions are forced to go to zero when $\hat{\rho}(\mathbf{x})$ vanishes. The scalar function $\hat{\rho}(\mathbf{x})$ is chosen to be positive definite in the interior and subjected to homogeneous conditions. The singular Stokes system then is written as follows:

$$\nabla p - \mathrm{div}[\rho(\mathbf{x})\nabla\mathbf{v}] + \frac{\mathbf{v}}{\hat{\rho}(\mathbf{x})} = \lambda \mathbf{v} \quad \text{in } \Omega, \tag{5.5.16}$$

$$\mathrm{div}\,\mathbf{v} = 0 \quad \text{in } \Omega, \tag{5.5.17}$$

$$\mathbf{v} = 0 \quad \text{on } \partial\Omega. \tag{5.5.18}$$

Batcho and Karniadakis report that the choice of $q(\mathbf{x}) = -4/\rho(\mathbf{x})$ yields excellent results if the scalar $\rho(\mathbf{x})$ is computed as the solution of a Poisson problem with positive forcing $C > 0$, that is,

$$-\Delta\rho(\mathbf{x}) = C \quad \text{in } \Omega, \tag{5.5.19}$$

$$\rho(\mathbf{x}) = 0 \quad \text{on } \partial\Omega. \tag{5.5.20}$$

For simple geometries such as the plane channel or the eddy-promoter domain, $\rho(\mathbf{x})$ is C^∞, but for a backward-facing step or a grooved channel, $\rho(\mathbf{x})$ is only C^1 because of the presence of corner singularities in the geometry (see Figure 5.5.1).

The algebraic problem associated with Equations (5.5.16)–(5.5.17) may be cast in the following form:

$$H\underline{v}_i - D_i^T p = M\lambda\underline{v}_i, \quad i = 1,\ldots,d, \quad \text{in } \Omega \tag{5.5.21}$$

$$-D_i\underline{v}_i = 0 \quad \text{in } \Omega, \tag{5.5.22}$$

where H is the discrete counterpart of the operator $\mathrm{div}[\rho(\mathbf{x})\nabla] + q(\mathbf{x})$, D_i^T the discrete gradient, M the mass matrix, and D_i the discrete divergence. All

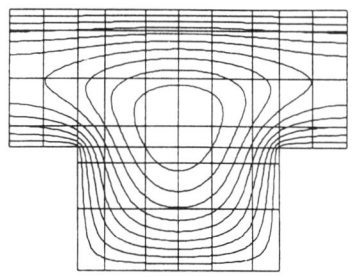

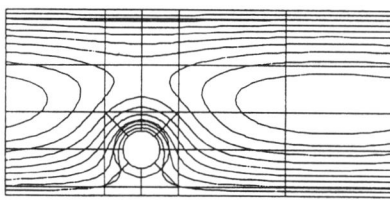

Figure 5.5.1. Countours of computed scalar function $\rho(\mathbf{x})$ for the grooved channel and eddy-promoter geometries. The spectral-element discretization is based on $N_x = N_y = 13$.

discrete terms may be obtained through the $\mathbb{P}_N$–$\mathbb{P}_{N-2}$ method. The Uzawa algorithm leads to the set of equations

$$Sp = D_i H^{-1} M \lambda \underline{v}_i, \qquad (5.5.23)$$

$$H\underline{v}_i = D_i^T p + M\lambda \underline{v}_i, \qquad i = 1, \ldots, d, \qquad (5.5.24)$$

where S is defined by the relation

$$S = D_i H^{-1} D_i^T. \qquad (5.5.25)$$

The iterative solution of Equations (5.5.23)–(5.5.24) for $d = 2$ is best computed if this system is rewritten in the following way:

$$\begin{pmatrix} H & 0 \\ 0 & H \end{pmatrix} \begin{pmatrix} \underline{v}_1 \\ \underline{v}_2 \end{pmatrix} = \begin{pmatrix} S_{11} + B & S_{12} \\ S_{21} & S_{22} + B \end{pmatrix} \begin{pmatrix} \lambda \underline{v}_1 \\ \lambda \underline{v}_2 \end{pmatrix}, \qquad (5.5.26)$$

where the block matrices are given by

$$S_{ij} = D_i^T S^{-1} D_j H^{-1} M, \qquad i, j = 1, 2. \qquad (5.5.27)$$

The matrix system is solved by an iterative procedure: either the inverse orthogonal method or the Lanczos–Uzawa algorithm. Full details are given in Batcho and Karniadakis [24].

5.6 Stabilization of the $\mathbb{P}_N$–$\mathbb{P}_N$ Approximation by Bubble Functions

Spurious pressure modes can be prevented by using bubble functions. The idea was introduced first in the FE context by Crouzeix and Raviart [88] and Arnold et al. [8]. The trick consists in adding a supplementary space to the standard FE space where the discretized velocities are sought. The extra bases and test functions are spanned by local functions having their support in one element, the so-called bubble functions. This technique was extended to spectral discretizations by Canuto and Van Kemenade [65].

For the sake of simplicity, we present the method for one spectral element covering the reference domain $\Omega = (-1, 1)^d$. The space of discrete velocities is

$$X_N = H_0^1(\Omega)^d \cap \mathbb{P}_N^d, \qquad (5.6.1)$$

while the pressure space is

$$Z_N = L_0^2(\Omega) \cap \mathbb{P}_N. \qquad (5.6.2)$$

Covering $\hat{\Omega} := [-1, 1]^d$ with a GLL grid, one obtains a decomposition into C cells or FEs, the vertices of which are neighbors, so as to build up quadrilateral cells in two dimensions and bricks in three dimensions. Typically, the cell $C_{i,j}$ in a 2D problem is built up on the four grid points $\xi_{i,j}$, $\xi_{i+1,j}$, $\xi_{i,j+1}$, and $\xi_{i+1,j+1}$. This grid will be denoted by G_h. On each of the C elements, one defines a bubble function b_C vanishing on ∂C and outside C. Such a function may result, for example, as the tensor product of d parabolas, vanishing at the end points of the parent element and taking the value unity at the origin. The space of bubble functions is denoted by B_N^d. The augmented space for discrete velocities is defined by the relation

$$X_N^b = X_N \oplus B_N^d. \qquad (5.6.3)$$

The approximate problem is now the following: Find $(\mathbf{w}_N, p_N) \in (X_N^b, Z_N)$ such that

$$\mathcal{A}(\mathbf{u}_N^b, \mathbf{w}_N) + \mathcal{B}(\mathbf{u}_N^b, p_N) = \mathcal{F}_N(\mathbf{u}_N^b) \quad \forall \mathbf{u}_N^b \in X_N^b, \qquad (5.6.4)$$

$$\mathcal{B}(\mathbf{w}_N, q_N) = 0 \quad \forall q_N \in Z_N, \qquad (5.6.5)$$

with the definitions

$$\mathcal{A}(\mathbf{u}_N, \mathbf{v}_N) = (\nabla \mathbf{u}_N, \nabla \mathbf{v}_N), \qquad (5.6.6)$$

$$\mathcal{B}(\mathbf{u}_N, q_N) = -(\operatorname{div} \mathbf{u}_N, q_N). \qquad (5.6.7)$$

Decomposing $\mathbf{w}_N$ as $\mathbf{w}_N = \mathbf{v}_N + \mathbf{v}_b \in X_N^b$, one can write the previous problem in a split form: Find $\mathbf{v}_N \in X_N$, $\mathbf{v}_b \in B_N^d$, $p_N \in Z_N$ such that

$$\mathcal{A}(\mathbf{u}_N, \mathbf{v}_N) + \mathcal{A}(\mathbf{u}_N, \mathbf{v}_b) + \mathcal{B}(\mathbf{u}_N, p_N) = \mathcal{F}_N(\mathbf{u}_N) \quad \forall \mathbf{u}_N \in X_N, \quad (5.6.8)$$

$$\mathcal{A}(\mathbf{u}_b, \mathbf{v}_N) + \mathcal{A}(\mathbf{u}_b, \mathbf{v}_b) + \mathcal{B}(\mathbf{u}_b, p_N) = \mathcal{F}_N(\mathbf{u}_b) \quad \forall \mathbf{u}_b \in B_N^d, \quad (5.6.9)$$

$$\mathcal{B}(\mathbf{v}_N, q_N) + \mathcal{B}(\mathbf{v}_b, q_N) = 0 \quad \forall q_N \in Z_N. \quad (5.6.10)$$

Let us now look at the interaction of the bubbles with the global polynomial functions. To this end, as a first step, we choose $\mathbf{u}_N$ as $\mathbf{u}_h$, the FE linear interpolation of the high-order polynomial at the nodes of G_h. This is applied to the second term on the l.h.s. of (5.6.8); similarly, q_N is replaced by q_h in the second term of (5.6.10). Through this localization procedure, we obtain

$$(\nabla \mathbf{v}_b, \nabla \mathbf{u}_h) = \sum_{C \in G_h} (\nabla \mathbf{v}_b, \nabla \mathbf{u}_h)_C = - \sum_{C \in G_h} (\mathbf{v}_b, \Delta \mathbf{u}_h)_C = 0, \quad (5.6.11)$$

since $\mathbf{v}_b = 0$ on ∂C and $\Delta \mathbf{u}_h$ is identically zero on C. The notation $(\cdot, \cdot)_C$ indicates the usual L^2 inner product on the element C.

In the second step, we introduce the projection operator J_h from $C^0(\bar{\Omega})$ functions into the space of piecewise constant function on the cells of G_h. The easiest way to implement this is to approximate a function f in the cell C by its constant value $f(C)$ evaluated at the cell barycenter. Applying this operator J_h to every term in Equations (5.6.8)–(5.6.10), where bubbles interact with the global polynomial or with its piecewise linear interpolant, leads to the modified Galerkin method: Find $\mathbf{v}_N \in X_N$, $\mathbf{v}_b \in B_N^d$ and $p_N \in Z_N$ such that

$$\mathcal{A}(\mathbf{u}_N, \mathbf{v}_N) + \mathcal{B}(\mathbf{u}_N, p_N) = \mathcal{F}_N(\mathbf{u}_N) \quad \forall \mathbf{u}_N \in X_N, \quad (5.6.12)$$

$$-(J_h(\Delta \mathbf{v}_N), \mathbf{u}_b) + \mathcal{A}(\mathbf{u}_b, \mathbf{v}_b) + (J_h(\nabla p_N), \mathbf{u}_b) = (J_h \mathbf{f}_N, \mathbf{u}_b) \quad \forall \mathbf{u}_b \in B_N^d, \quad (5.6.13)$$

$$\mathcal{B}(\mathbf{v}_N, q_N) - (\mathbf{v}_b, J_h(\nabla q_h)) = 0 \quad \forall q_N \in Z_N. \quad (5.6.14)$$

From Eq. (5.6.13), we extract the bubble component $\mathbf{v}_b$ as follows. Let

$$\mathbf{v}_b = \sum_{C \in G_h} \mathbf{v}_C b_C, \quad (5.6.15)$$

where $\mathbf{v}_C$ is the (unknown) vector weighting factor in the element C. Introducing the residual

$$\mathbf{r}_N := -\Delta \mathbf{v}_N + \nabla p_N - \mathbf{f}_N \quad (5.6.16)$$

5.6. Stabilization of the $\mathbb{P}_N$–$\mathbb{P}_N$ Approximation by Bubble Functions

and the test function $\mathbf{u}_b = b_C$ successively in all cells, we readily obtain

$$\mathbf{v}_C = -\frac{(b_C, 1)_C}{(\nabla b_C, \nabla b_C)_C}(J_h \mathbf{r}_N)|_C \qquad \forall C \in G_h, \tag{5.6.17}$$

where, as mentioned above, $(J_h \mathbf{r}_N)|_C$ denotes the value of the projection of the residual at the barycenter of cell C. Inserting (5.6.15) in the l.h.s. of (5.6.14) and taking (5.6.17) into account, we get

$$-(\mathbf{v}_b, J_h(\nabla q_h)) = \sum_{C \in G_h} \frac{(b_C, 1)_C^2}{(\nabla b_C, \nabla b_C)_C} \frac{1}{|C|}(J_h \mathbf{r}_N, J_h(\nabla q_h))_C. \tag{5.6.18}$$

In this relation, $|C|$ represents a measure of the element C, namely, its area. Therefore, the problem (5.6.12)–(5.6.14) reduces to the following: Find $\mathbf{v}_N \in X_N$ and $p_N \in Z_N$ such that

$$\mathcal{A}(\mathbf{u}_N, \mathbf{v}_N) + \mathcal{B}(\mathbf{u}_N, p_N) = \mathcal{F}_N(\mathbf{u}_N) \qquad \forall \mathbf{u}_N \in X_N, \tag{5.6.19}$$

$$\mathcal{B}(\mathbf{v}_N, q_N) + \sum_{C \in G_h} \tau_C (J_h \mathbf{r}_N, J_h(\nabla(q_h)))_C = 0 \qquad \forall q_N \in Z_N. \tag{5.6.20}$$

The modified Galerkin method introduces into the continuity equation a stabilizing term to damp the oscillatory pressure modes. The choice of τ_C is crucial to render the method efficient. This stabilization parameter, defined by the relation

$$\tau_C = \frac{(b_C, 1)_C^2}{(\nabla b_C, \nabla b_C)_C} \frac{1}{|C|}, \tag{5.6.21}$$

depends on the cell size and the chosen bubble. Referring to Canuto and Van Kemenade [65] for details, we estimate τ_C by the formula

$$\tau_C = \frac{h_C}{2|\mathbf{v}|_C} f(Re_C) = \begin{cases} \frac{C' h_C^2}{8\nu}, & 0 \leq Re_C \leq 1, \\ \frac{h_C}{2|\mathbf{v}|_C}, & Re_C > 1, \end{cases} \tag{5.6.22}$$

where h_C is a typical mesh size, $|\mathbf{v}|_C$ the magnitude of the velocity field (like the Euclidian norm at the cell center), ν the kinematic viscosity, C' a constant (typically 1/3), and Re_C the cell Reynolds number.

In [70], Canuto et al. generalize the previous method to gain flexibility in the computation of the stabilization parameter. The bubble is still defined as in (5.6.15). Choosing again $\mathbf{u}_b = b_C$ and inserting this expression and (5.6.15) into Equation (5.6.13), one obtains with the residual (5.6.16) the relation

$$\mathcal{A}(b_C, \mathbf{v}_C) + (\mathbf{r}_C, b_C) = 0, \tag{5.6.23}$$

where $\mathbf{r}_C = J_h \mathbf{r}_N|_C$. The variational statement (5.6.23) corresponds to the strong problem

$$-\Delta \mathbf{v}_C = \mathbf{r}_C, \qquad (5.6.24)$$

to be solved in each C. As the residual $\mathbf{r}_C$ is constant over each C, we may set

$$\mathbf{v}_C = \mathbf{r}_C \boldsymbol{\varphi}_C. \qquad (5.6.25)$$

Using this relation as the basis for $\mathbf{v}_C$ and diagonalizing the 2×2 matrix corresponding to $\boldsymbol{\varphi}_C$ as $\boldsymbol{\varphi}_C = \text{diag}(\varphi_C, \varphi_C)$, we can solve the following problem for φ_C:

$$-\Delta \varphi_C = 1, \qquad \varphi_C \in H_0^1(C), \qquad (5.6.26)$$

for every C. By the same steps as before, we evaluate the expression analogous to Equation (5.6.18). The variational problem is still of the form (5.6.19)–(5.6.20), but the cell stabilization parameter is now

$$\tau_C = \frac{1}{|C|} \int_C \varphi_C \, dV. \qquad (5.6.27)$$

In [70], all the details are provided to solve Equation (5.6.26) on the parent element and to compute the parameter τ_C.

5.7 *hp*-Methods for Stokes Problems

For completeness, we introduce in this section the *hp*-method, which has been in use since the early eighties. The d-dimensional bases for the approximated fields are obtained as tensor products of one-dimensional polynomials of degree p as in Equation (2.3.20). This strategy is used for meshes composed of quadrilaterals and hexahedra. The book by Szabó and Babuška [370] lists all possible combinations and presents them in the form of hierarchical bases, by associating the shape functions with the topological objects of the FE, namely, the vertex modes, the edge modes, and the interior nodes (the bubble functions). We observe that the *p*-methodology is of spectral type, the approximation space being $\mathbb{P}_N$ ($p = N$). It is common practice for classical fluid-mechanics problems to resort to a $\mathbb{P}_N$–$\mathbb{P}_{N-2}$ or p–($p-2$) discretization for velocity and pressure, respectively (see, for example, Oden et al. [287] and Warichet and Legat [407]).

In the frame of a mixed FE approach, Jensen and Zhang [212] show that the error on $\mathbf{v}_N$ is still of spectral convergence. We have the following estimate.

Theorem 5.6 *Under the assumptions*

1. $N \geq 2$,
2. *the solution* $(\mathbf{v}, p)$ *of the weak continuous problem (5.2.12)–(5.2.13) is in* $H_0^s(\Omega)^d \times H^{s-1}(\Omega)$ *with s the smoothness of the exact solution,*

the problem (5.2.37)–(5.2.38) possesses a unique solution $(\mathbf{v}_N, p_N)$ *in* $X_N \times Z_N$ *with the following error bound:*

$$\|\mathbf{v} - \mathbf{v}_N\|_{H_0^1(\Omega)^d} \leq Ch^{\min(N,s)} N^{-s} \|\mathbf{v}\|_{H_0^{s+1}(\Omega)^d}, \qquad (5.7.1)$$

with the constant C independent of N.

High-order interpolation bases were developed for triangles and tetrahedra in [370]. Recently, however, Sherwin and Karniadakis [358, 359, 221] used modal expansions with these elements (see Chapter 7) in a splitting formulation of the unsteady Navier–Stokes equations. This numerical approximation brings more flexibility in coping with difficult geometrical problems. Specifically, the elements allow local h-refinements where FEs are subdivided into smaller pieces and local p-enrichments through the use of bubble functions. Schwab [348, 349] has covered many theoretical issues related to the use of p and hp methods. In [349], the Stokes problem is revisited and the results are compatible with the previous theorem. Recently, Schwab and Suri [350] analyzed thoroughly mixed hp-methods for parallelograms and triangles. They show that exponential convergence for both the velocity and the pressure can be achieved for geometric meshes that combine quadrilaterals and triangles and use graded refinement near singularities.

5.8 Steady Navier–Stokes Equations

We now consider the steady dimensionless Navier–Stokes equations in a form that reduces to the Stokes equations when the Reynolds number vanishes (1.7.5):

$$Re\,(\mathbf{v} \cdot \nabla \mathbf{v}) = -\nabla p + \Delta \mathbf{v} + \mathbf{f} \qquad \text{in } \Omega, \qquad (5.8.1)$$

$$\text{div } \mathbf{v} = 0 \qquad \text{in } \Omega, \qquad (5.8.2)$$

with the homogeneous velocity boundary conditions (5.1.3). One notices that the Navier–Stokes problem becomes stiffer when Re increases. Some algorithms exploit the fact that for low or moderate Reynolds number, the l.h.s. of (5.8.1) is a perturbation of the former Stokes problem. At higher Re-values, however, another approach must be undertaken. Indeed, when Re reaches higher values, the physics may change drastically. Experiments show that one may switch from steady-state solutions to time-dependent ones. In this case, the dynamics, and therefore the temporal behavior of the Navier–Stokes equations, becomes essential.

To be able to treat nonlinear operators like the Navier–Stokes equations from the abstract point of view, we introduce the Fréchet derivative (see Appendix A for the definition), denoted by $D_{NS}(\mathbf{v}, p)$ to indicate that it operates with respect to velocity and pressure. One has

$$D_{NS} = Re\,[(\cdot) \cdot \nabla \mathbf{v} + \mathbf{v} \cdot \nabla(\cdot)] + \nabla(\cdot) - \Delta(\cdot). \qquad (5.8.3)$$

The symbol $(\cdot)$ represents a dummy variable, either the velocity or the pressure, made clear by context.

5.8.1 Weak Formulation

With the same Hilbert spaces introduced in Section 5.2, the previous problem may be restated in its weak form: Find $\mathbf{v}$ in $H_0^1(\Omega)^d$ and p in $L_0^2(\Omega)$ such that

$$\mathcal{A}(\mathbf{u}, \mathbf{v}) + Re\,\mathcal{C}(\mathbf{v}; \mathbf{v}, \mathbf{u}) + \mathcal{B}(\mathbf{u}, p) = \mathcal{F}(\mathbf{u}) \qquad \forall \mathbf{u} \in H_0^1(\Omega)^d, \qquad (5.8.4)$$

$$\mathcal{B}(\mathbf{v}, q) = 0 \qquad \forall q \in L_0^2(\Omega), \qquad (5.8.5)$$

where the continuous forms $\mathcal{A}$ and $\mathcal{B}$ have been defined by (5.2.10) and (5.2.11). The new term in (5.8.4) corresponds to the trilinear form associated with the nonlinear term. Its definition is given by the relation

$$\mathcal{C} : H_0^1(\Omega)^d \times H_0^1(\Omega)^d \times H_0^1(\Omega)^d \to \mathbb{R},$$

$$\mathcal{C}(\mathbf{v}; \mathbf{w}, \mathbf{u}) = \int_\Omega [(\mathbf{v} \cdot \nabla)\mathbf{w}]\,\mathbf{u}\,dV = \left(v_j \frac{\partial w_i}{\partial x_j}, u_i\right). \qquad (5.8.6)$$

We note that no integration by parts is performed on the nonlinear term. This approach contrasts with the practice in compressible flows, where the integration by parts provides momentum fluxes, which are more easily treated in the presence of shocks in the flow.

The abstract problem (5.8.4)–(5.8.5) is an instantiation of a more general problem: Find $\mathbf{x} = (\mathbf{v}, p)$ in $X = H_0^1(\Omega)^d \times L_0^2(\Omega)$ such that

$$F(\mathbf{x}, Re) := \mathbf{x} + TG(\mathbf{x}, Re) = 0, \qquad (5.8.7)$$

where T is a linear operator and G is a C^∞ mapping. The linear operator acts on the body force $\mathbf{f}$ and produces the solution

$$T\mathbf{f} = \mathbf{x} \qquad (5.8.8)$$

5.8. Steady Navier–Stokes Equations

of a Stokes problem with homogeneous Dirichlet velocity conditions

$$\mathcal{A}(\mathbf{v}, \mathbf{u}) + \mathcal{B}(\mathbf{u}, p) = \mathcal{F}(\mathbf{u}) \quad \forall \mathbf{u} \in H_0^1(\Omega)^d, \quad (5.8.9)$$

$$\mathcal{B}(\mathbf{v}, q) = 0 \quad \forall q \in L_0^2(\Omega). \quad (5.8.10)$$

The C^∞ mapping is defined by the relation

$$G(\mathbf{x}, Re) := Re\,(\mathbf{v} \cdot \nabla \mathbf{v} - Re^{-1}\mathbf{f}). \quad (5.8.11)$$

Notice that the mapping G does not depend on the pressure p. With the relations (5.8.8), (5.8.11), the Navier–Stokes problem (5.8.7) has the pair $\mathbf{x}$ as a solution. Inspection of the expression (5.8.3) for D_{NS} shows that through (5.8.7), it yields the relation

$$D_{NS} = D_{\mathbf{x}}F = 1 + T(D_{\mathbf{x}}G). \quad (5.8.12)$$

The numerical solution of the Navier–Stokes equations requires a finite-dimensional approach. We will present the collocation approximation, which was thoroughly analyzed by Bernardi et al. [36].

5.8.2 Collocation Approximation of the Navier–Stokes Equations

Nonsingular solutions of the Navier–Stokes equations can be obtained only if the Fréchet derivative D_{NS} exists or, to be more precise, if the Fréchet derivative of the map F with respect to $\mathbf{x}$ is an isomorphism of X (i.e., a bijective mapping that preserves distances; see Appendix A). This is achieved if the Fréchet derivative is bounded from below. Therefore, there exists a constant $C > 0$ such that $\forall \mathbf{y} \in X$

$$\|(D_{\mathbf{x}}F)\mathbf{y}\|_X = \|\mathbf{y} + T(D_{\mathbf{x}}G)\mathbf{y}\|_X \geq C\|\mathbf{y}\|_X. \quad (5.8.13)$$

In practical terms, satisfaction of this last inequality requires working at small or moderate Reynolds numbers. However, our numerical computations have shown that collocation or even spectral-element methods are robust in this respect, that is, yield converged solutions for Reynolds numbers where other standard numerical techniques such as finite elements or finite volumes fail completely. One can consequently obtain numerical results that are nonsingular solutions of the (discrete) Navier–Stokes equations but that are meaningless from the physics standpoint, since the steady-state assumption is physically implausible.

Suppose that G is a C^2 map from $X \times \mathbb{R}_+$ into the dual space of $H_0^1(\Omega)^d$, with a bounded second-order derivative D^2G. Assume furthermore that T is a compact operator. One can prove that the discretization error is bounded as follows:

$$\|x(Re) - x_N(Re)\|_X \leq C(\|x(Re) - \Pi_N x(Re)\|_X + \|(T - T_N)G(\mathbf{x}, Re)\|_X),$$
(5.8.14)

where $C > 0$ is independent of both Re and N. Equation (5.8.14) shows that the spatial discretization error is bounded by the interpolation error and by the quality of the approximation T_N of the linear operator T. This last consideration is linked to what has been analyzed under the heading of Stokes equations. Nonetheless, when dealing with the discrete Navier–Stokes equations within a Legendre or Chebyshev collocation procedure on a single grid, the abstract problem becomes: Find $\mathbf{x}_N = (\mathbf{v}_N, p_N)$ in $X_N = [H_0^1(\Omega)^d \cap \mathbb{P}_N^d(\Omega)] \times [L_0^2(\Omega) \cap \mathbb{P}_N(\Omega)]$ such that

$$F_N(\mathbf{x}_N, Re) := \mathbf{x}_N + T_N G_N(\mathbf{x}_N, Re) = 0,$$
(5.8.15)

where T_N will obviously result from the Stokes problem approximation and G_N is the discretization of Equation (5.8.11). This term does not raise any special hardship because we are in the collocation framework, especially if the nonlinear term has been written under conservative form.

The theorems proved by Bernardi et al. [36] generalize the previous Theorems 5.3 and 5.4 for Legendre and Chebyshev collocation, respectively, of the Stokes problem provided that the discrete Fréchet derivative $D_{\mathbf{x}_N} F_N = 1 + T_N(DG_N)$ is an isomorphism of X_N. In the corresponding appropriate norms, one obtains the following theorem, with the convention $\alpha = 0$ for the Legendre case and $\alpha = -1/2$ for the Chebyshev case:

Theorem 5.7 *Under the assumptions*

1. *N is large enough,*
2. *the solution $(\mathbf{v}, p)$ of the weak continuous problem (5.8.4)–(5.8.5) is in $H_0^s(\Omega)^d \times H^{s-1}(\Omega)$ for an integer $s \geq 1$, and the function $\mathbf{f}$ is in $H^t(\Omega)^d$ for an integer $t \geq 2$,*

the problem (5.8.15) possesses a solution $(\mathbf{v}_N, p_N)$ in X_N with the following error bounds:

$$\|\mathbf{v} - \mathbf{v}_N\|_{H^1(\Omega)^d} \leq c_1(\mathbf{v}) N^{1-s} + c_2(\mathbf{f}) N^{1+2\alpha-t},$$
(5.8.16)

$$\|p - p_N\|_{L^2(\Omega)} \leq c_3(\mathbf{v}, p) N^{3-s} + c_2(\mathbf{f}) N^{3+2\alpha-t}$$
(5.8.17)

with $c_1(\mathbf{v})$, $c_3(\mathbf{v}, p)$, $c_2(\mathbf{f})$ constants independent of N.

5.8. Steady Navier–Stokes Equations

The constants c_i, $i = 1, 3$, involved in this theorem depend upon the velocity and pressure fields. It is indeed difficult to get better estimates for the nonlinear problem, because the advection effects play the role of a source term. This point of view will directly affect the solution algorithms.

5.8.3 Solution Algorithms: Iterative and Newton Methods

A nonlinear operator is best treated by iterative techniques, with the question of convergence being one of the key issues. The simplest algorithm consists of the so-called incremental loading based on the structure of Equation (5.8.1). Using the upper index k to denote the iteration counter and dropping the lower index related to the discretization for the sake of simplicity, we can state the iterative method as follows: Find $\mathbf{x}^{k+1} := (\mathbf{v}^{k+1}, p^{k+1}) \in X$ such that $\forall (\mathbf{u}, q) \in X$

$$\mathcal{A}(\mathbf{u}, \mathbf{v}^{k+1}) + \mathcal{B}(\mathbf{u}, p^{k+1}) = -Re\,\mathcal{C}(\mathbf{v}^k; \mathbf{v}^k, \mathbf{u}) + (\mathbf{f}, \mathbf{u}), \qquad (5.8.18)$$

$$\mathcal{B}(\mathbf{v}^{k+1}, q) = 0. \qquad (5.8.19)$$

First, one solves the Stokes problem obtained by setting $Re = 0$ in Equation (5.8.18); then, the Navier–Stokes problem is handled by increasing Re incrementally, using the last solution computed at the previous Re-value as the initial guess of the new iteration. As mentioned previously, this iterative process is robust. The source of stiffness is the nonlinearity in the r.h.s., making it diverge at moderate Reynolds numbers. This stiffness depends not only on Re itself, but also on the geometrical complexity of the application at hand. The convergence of this method is evidently related to the quality of the Stokes solver and its preconditioner, the increment value, and to the presence of bifurcation points in the solution family of the abstract problem (5.8.15). Because it is an explicit method in which the nonlinearity is generated at the old iterate level, it is less stable than other techniques, implying some implicitness of the operator. Therefore, if we want to cope with high Reynolds numbers and still use the plain Navier–Stokes equations with no turbulence modeling, it is better to resort to more sophisticated algorithms.

The Newton method applied to the discrete collocation Navier–Stokes problem (5.8.15) requires solving the relation

$$D_{\text{NS}}(\mathbf{x}^k, Re)(\mathbf{x}^{k+1} - \mathbf{x}^k) = -F(\mathbf{x}^k, Re), \qquad (5.8.20)$$

where we recognize a generalization of the classical Newton algorithm. The Newton method is quadratically convergent, meaning that the error, defined as the difference between $\mathbf{x}^k$ and the exact solution $\mathbf{x}$, at iteration k is bounded from above by the square of the error at the preceding iteration. This quadratic convergence is spectacular: it doubles the number of significant figures at each iteration, once the iterative process has hooked a solution branch.

In some numerical procedures, one does not want to reconstruct (at non-negligible expense) the operator $D_{NS}(\mathbf{x}^k, Re)$ at each iteration. Therefore, one freezes it as $D_{NS}(\mathbf{x}^0, Re)$, obtaining the numerical algorithm known as the Picard iterative method, which converges only linearly. Let us apply the Newton method (5.8.20) to the weak formulation (5.8.4)–(5.8.5). One writes the following: Find $\mathbf{x}^{k+1} \in X$ such that $\forall (\mathbf{u}, q) \in X$

$$\mathcal{A}(\mathbf{u}, \mathbf{v}^{k+1}) + Re\,[\mathcal{C}(\mathbf{v}^{k+1}; \mathbf{v}^k, \mathbf{u}) + \mathcal{C}(\mathbf{v}^k; \mathbf{v}^{k+1}, \mathbf{u})] + \mathcal{B}(\mathbf{u}, p^{k+1})$$
$$= -Re\,\mathcal{C}(\mathbf{v}^k; \mathbf{v}^k, \mathbf{u}) + \mathcal{F}(\mathbf{u}), \quad (5.8.21)$$
$$\mathcal{B}(\mathbf{v}^{k+1}, q) = 0. \quad (5.8.22)$$

From the assumptions on the map G, one can show that the Fréchet derivative is Lipschitz continuous, thereby inducing the quadratic convergence of the Newton method. In order to converge, the Newton method needs an initial guess $\mathbf{x}^0$ close to the solution. Using Picard iterations at the beginning of the algorithm may help to bring the sequence of iterates closer to the convergence domain of the Newton method.

5.9 Applications

In this section, we illustrate our subject by describing a few relevant problems. Some of them will be more related to numerical considerations and algorithms; the others will exemplify fluid-mechanics applications.

5.9.1 Stokes Problems

Square-Cavity Problem

The problem of choosing the proper spectral solution method for the Stokes equations was thoroughly examined by Schumack et al. [347]. They considered the use of staggered versus nonstaggered grid, weak formulation versus collocation, and velocity–pressure discretizations of the same order or with pressure approximations of lower order than that for the velocity bases. Ten different formulations were set up and compared on three test problems. Among these problems was the well-known driven-cavity problem.

This problem is solved on the reference square $[-1, 1]^2$; the velocity boundary conditions are homogeneous on the three fixed lower walls, while the top lid

5.9. Applications

is sliding at a unit horizontal velocity component. The problem is ill posed because the boundary conditions at the upper corners drive the velocity field to be singular and make the divergence of the velocity field blow up to infinite values. As a consequence, the pressure is singular and also goes to infinite values. This effect is clearly demonstrated when one refines meshes where the computed solutions are affected more and more by the presence of these singularities.

There are several ways to cope with singular problems. Schumack et al. subtracted the singularity using a singular solution provided in Batchelor [23]. From their numerical results, we can draw the following conclusions:

1. Collocation is, practically speaking, as good as the SEM as far as spatial accuracy for the velocity field is concerned. Collocation is easier to program but has less flexibility and less generality for dealing with complicated geometries and 3D problems.
2. The use of a pressure approximation of order two less than that for the velocity helps to avoid the spurious pressure modes. The pressure nodes defined on the interior velocity nodes constitute the best choice. This procedure needs no interpolation between grids and provides both efficiency and accurate quadratures.
3. The removal of singularities improves the convergence by two or three orders of magnitude (see Figure 5.9.1). The special treatment of corner singularities

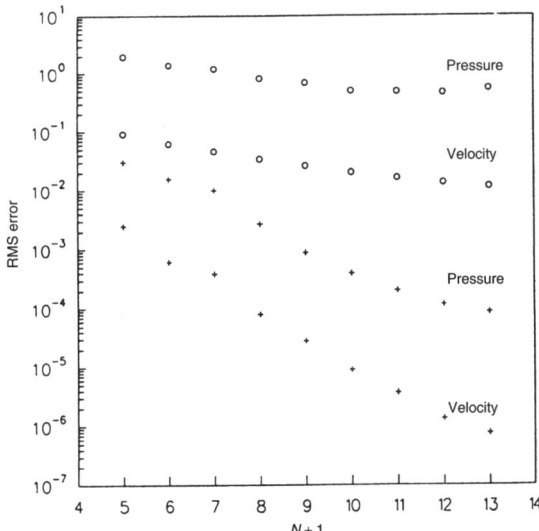

Figure 5.9.1. Square-cavity problem: velocity and pressure error with and without special treatment for top-corner singularities. Crosses indicate special treatment; circles indicate no special treatment.

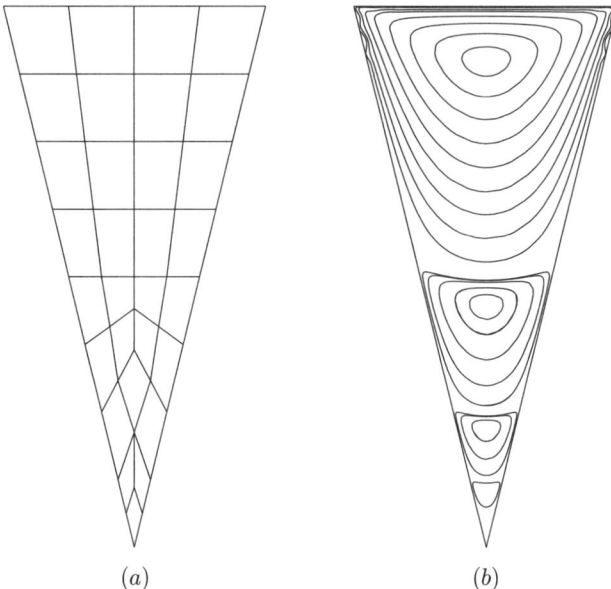

(a)   (b)

Figure 5.9.2. Creeping flow in a wedge [331]. Velocity boundary conditions are no slip on the sides and unit horizontal velocity on the top. The $E = 30$, $N = 8$ spectral-element discretization is shown in (a), and the sequence of decaying eddies, in the form of streamlines, in (b). The vortex strength decays by a factor of roughly 400 from one eddy to the next.

is, moreover, preferable to a domain decomposition that attempts to resolve them by using smaller domains near the corners.

Flow in a Wedge

Let us now turn to the Stokes flow in a wedge. This problem was solved by Rønquist [332] using the SEM. As the top lid moves at unit velocity (Figure 5.9.2), a series of Moffatt eddies [277] is generated. These eddies are stacked from top to bottom in an infinite cascade to the tip. The wedge shown has an aperture angle of $28.5°$. The asymptotic ratio of successive eddy intensities is 405. With the discretization shown in the figure, one obtains four eddies. The ratio of the strength of two successive eddies is from top to bottom 386, 406, and 411.

Grooved Channel

Batcho and Karniadakis [24] have shown that the standard Stokes eigenfunctions, when used to compute high-Reynolds-number flows, fail to produce good accuracy. We will concentrate here on the singular Stokes eigenproblem. Figure 5.9.3 displays for the grooved-channel problem of Figure 5.5.1 the

5.9. Applications

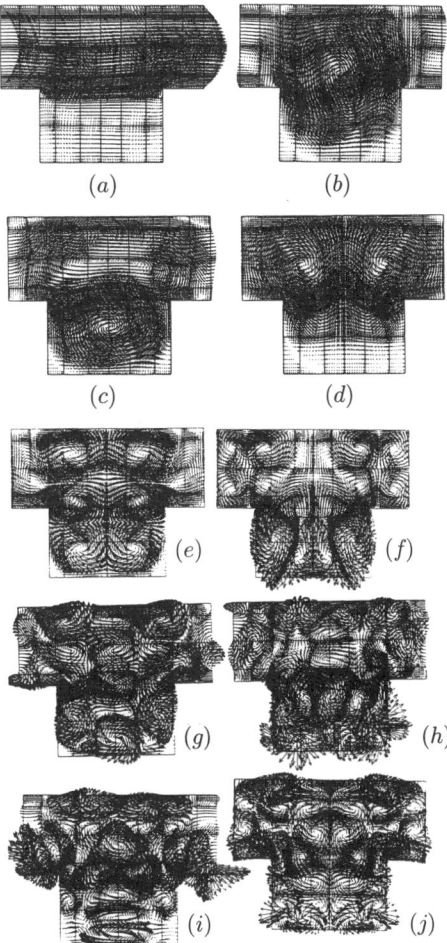

Figure 5.9.3. Grooved-channel flow: velocity-vector plots of the first four singular Stokes eigenfunctions. In each spectral element, $N_x = N_y = 13$. The first, second, third, and fourth eigenfunctions are shown in (a)–(d), respectively. In (e)–(j) are given the velocity-vector plots of eigenfunctions 10, 20, 30, 40, 50, and 60, respectively.

velocity-vector plots of the first four leading eigenfunctions and selected higher eigenmodes. The discretization is based on $N = 13$ and $\|\rho(\mathbf{x})\|_\infty = 1.5$. The level of the discrete divergence for these eigenmodes is $\sim 10^{-2}$ to $\sim 10^{-3}$. If one compares the solution obtained by the divergence-free basis with 100 eigenfunctions to the solution obtained by the standard SEM, the Stokes eigenproblem yields an L^2 error on the velocity of order 10^{-3}, whereas the singular Stokes problem loses one order of magnitude at a level of order 10^{-2}.

282   5. Steady Stokes and Navier–Stokes Equations

Stabilization by Bubble Functions

The stabilization of the $\mathbb{P}_N-\mathbb{P}_N$ method by bubble functions is very efficient. Canuto and Van Kemenade [65] compute the Stokes flow in the regularized driven-cavity problem ($0 \leq x, y \leq 1$), where the prescribed velocity profile on the top lid is given by $u(x, 1) = 16x^2(1-x)^2$. Even with a coarse discretization like $N_x = N_y = 8$, the bubble-stabilized spectral method is able to reproduce two digits of the "exact" solution obtained with $N_x = N_y = 28$. Furthermore, the computed pressure field is very smooth and does not show any oscillations due to lack of resolution.

Wannier–Stokes Flow

The Wannier–Stokes flow, a 2D Stokes flow in a semiinfinite plane, is set into motion by a horizontal wall, coinciding with the horizontal axis, moving with constant velocity U in its own plane. A fixed cylinder of radius R is located with its center at a distance d above the wall. Figure 5.9.4 gives a sketch of the geometry of the problem. Wannier [404] derived an analytical solution in

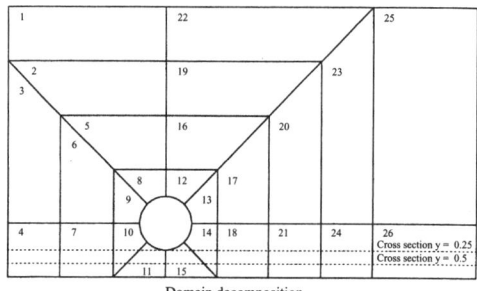

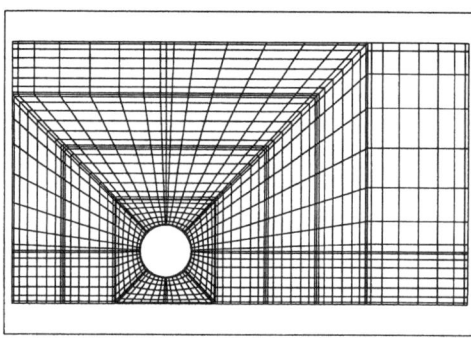

Figure 5.9.4. Wannier–Stokes flow: domain decomposition (top) and corresponding mesh with $N = 9$ in each subdomain (bottom).

the framework of lubrication theory. The Wannier–Stokes solution allows for a reliable study of the errors induced by nonrectangular geometries. The exact solution for the velocity components is

$$u = U - \frac{2(A+Fy)}{K_1}\left((s+y) + \frac{K_1}{K_2}(s-y)\right) - \frac{B}{K_1}\left((s+2y) - \frac{2y(s+y)^2}{K_2}\right)$$
$$- \frac{C}{K_2}\left((s-2y) - \frac{2y(s-y)^2}{K_2}\right) - F\ln\left(\frac{K_1}{K_2}\right), \quad (5.9.1)$$

$$v = -\frac{2x(A+Fy)}{K_1 K_2}(K_2 - K_1) - \frac{2Bxy(s+y)}{K_1^2} - \frac{2Cxy(s-y)}{K_2^2}, \quad (5.9.2)$$

where

$$A = -\frac{Ud}{\ln(\Gamma)}, \quad B = \frac{2U(d+s)}{\ln(\Gamma)}, \quad C = \frac{2U(d-s)}{\ln(\Gamma)}, \quad F = \frac{U}{\ln(\Gamma)},$$

$$\Gamma = \frac{d+s}{d-s}, \quad K_1 = x^2 + (s+y)^2, \quad K_2 = x^2 + (s-y)^2, \quad s^2 = d^2 - R^2.$$

We choose $d = 2R, U = 1$. As the wall movement induces the fluid to be pushed through the gap between the wall and the cylinder, Schneidesch and Deville [345] decompose the geometry into 26 subdomains, with some refinement near the cylinder. A typical mesh is shown in Figure 5.9.4 for a Chebyshev collocation with $N_x = N_y = 9$. Essential boundary conditions are imposed at the boundaries of the computational domain from the exact solution (5.9.1)–(5.9.2). The velocity components, pressure, and vorticity fields are shown from top to bottom in Figure 5.9.5. We observe the smoothness of the various quantities and, especially, that of the pressure. Even though C^1 continuity is not enforced by the method, inspection of this figure shows that the collocation method produces smooth results if a certain degree of regularity exists in the approximated solution. Figure 5.9.6 provides the pressure distribution computed at different horizontal cross sections in the geometry (see Figure 5.9.4 for the definition of these cross sections). Even with a coarse mesh, acceptable results are obtained. These results demonstrate the powerful capabilities of the spectral discretization.

5.9.2 Navier–Stokes Problems

Kovasznay Flow

Kovasznay [227] gives an analytical steady-state solution to the Navier–Stokes equations that is similar to the 2D flow field behind a periodic array of

284   5. Steady Stokes and Navier–Stokes Equations

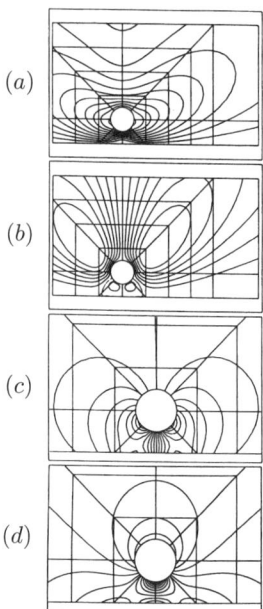

Figure 5.9.5. Isolines for the Wannier–Stokes flow of the velocity components (a, b), pressure (c), and vorticity (d).

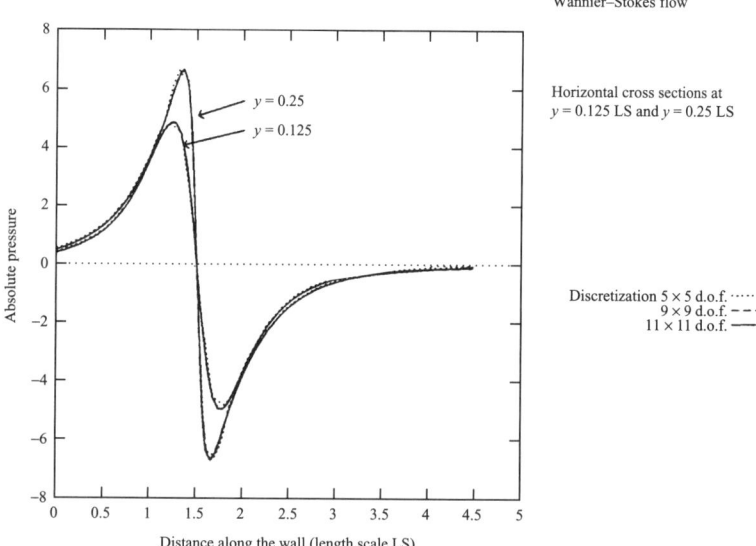

Figure 5.9.6. Evolution of the pressure along horizontal lines located under the cylinder.

5.9. Applications

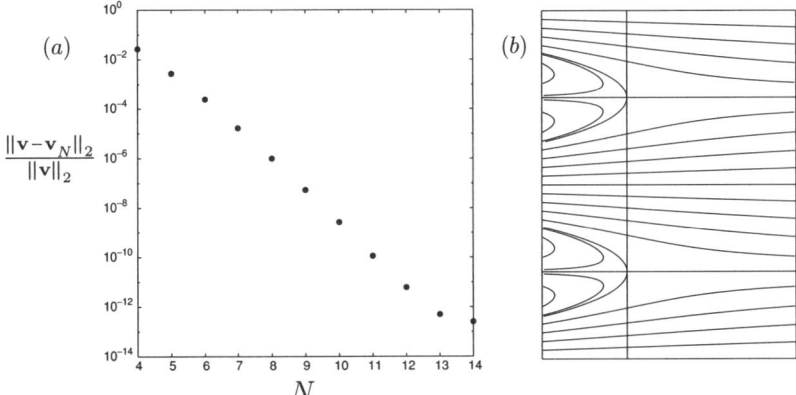

Figure 5.9.7. Kovasznay flow: spectral-element convergence (*a*) and streamlines (*b*) for $Re = 40$.

cylinders,

$$u_x = 1 - e^{\lambda x} \cos 2\pi y,$$

$$v = \frac{\lambda}{2\pi} e^{\lambda x} \sin 2\pi y,$$

$$\lambda := \frac{Re}{2} - \sqrt{\frac{Re^2}{4} + 4\pi^2},$$

where Re is the Reynolds number based on mean flow velocity and separation between vortices. Because this solution incorporates nonlinear effects (unlike Poiseuille flow), it is a good test for full Navier–Stokes solution algorithms. The results here are computed by marching to a steady state, but could equally well be computed using a steady-state solver. Figure 5.9.7(*a*) shows exponential convergence for $Re = 40$ using the $\mathbb{P}_N - \mathbb{P}_{N-2}$ SEM with $E = 8$ elements and N varying from 4 to 15. Periodic boundary conditions are applied in the *y*-direction, and the exact solution is prescribed on the left and right boundaries. Streamlines for this case are shown in Figure 5.9.7(*b*). The domain is $\Omega = [-0.5, 1.0] \times [-0.5, 1.5]$.

Grooved Channel

We compare the divergence-free basis method [24], applied to the grooved-channel flow at $Re = 100$, with the SEM. In this case, the singular Stokes eigenfunctions yield a fairly consistent decay of the L^2 error norm with respect to the Stokes eigenfunctions, which give more erratic behavior when the dimension

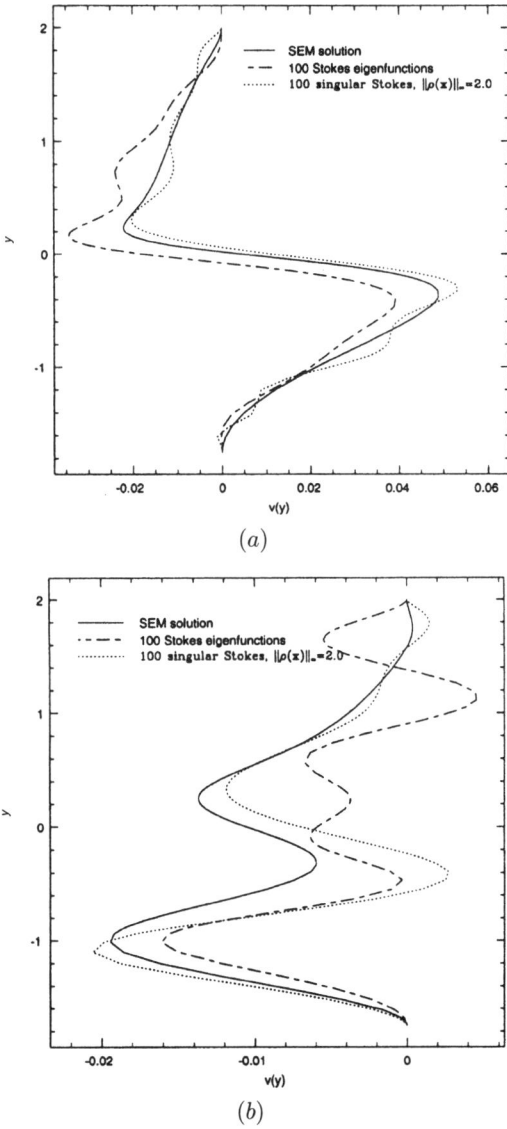

Figure 5.9.8. Vertical velocity profile for the grooved channel at $Re = 100$: (a) $x = 1.3$; (b) $x = 3.1$.

of the system increases. Figure 5.9.8 compares the three possible approaches: SEM, Stokes, and singular Stokes Galerkin projections. The locations $x = 1.3$ and 3.1 are at the core of the recirculation zone in the groove region. If we take the spectral-element computation as reference, the singular Stokes eigensystem is able to resolve the SEM velocity solution better than the Stokes system. This

5.9. Applications

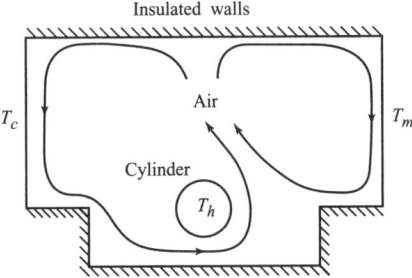

Figure 5.9.9. The cooled-hot-cylinder problem: geometry.

shows that the method is robust and capable of reproducing complicated profiles with low-dimensional expansions.

Cooled Hot Cylinder

We consider the natural convection of air in a complicated cavity depicted in Figure 5.9.9. The center of the hot cylinder of radius R is placed on the vertical axis of symmetry of the cavity at a distance from the bottom wall equal to the diameter. The overall dimensions of the cavity are $12R \times 8R$. The cylinder is hot at temperature T_h, while the vertical walls are set at two different temperatures T_c and T_m such that $T_c < T_m < T_h$. All other boundaries are adiabatic. The fluid is set into motion by heat exchange from the hot cylinder in the middle of the cavity. Convective cells appear on each side of the cylinder as the fluid is cooled by the vertical walls. The problem is not symmetric with respect to the vertical axis of symmetry, because of the temperature difference between the two vertical walls. The Navier–Stokes equations and the temperature equation are coupled via the Boussinesq approximation (see Chapter 1). The computed temperature field is obtained for a Rayleigh number of 500,000. Eighteen subdomains with an 11×11 Chebyshev discretization are used. The isotherms in Figure 5.9.10 reveal a strong gradient on the southwest side of the cylinder, indicating the

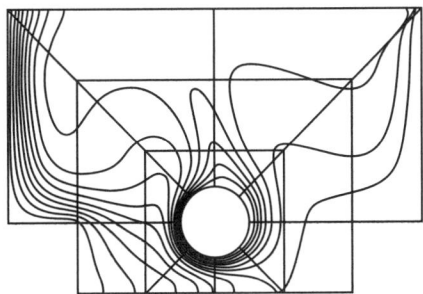

Figure 5.9.10. The cooled-hot-cylinder problem: isothermal lines (see [346]).

sharp temperature variation in the bottom left corner of the cavity. We observe also a hot plume rising from the cylinder to about two-thirds of the domain height.

5.10 Complements and Engineering Considerations

A Chebyshev collocation method was designed by Ku et al. [231] with the aim of enforcing the continuity condition everywhere in the domain, on the boundaries as well as in the interior. The method generates a Poisson pressure equation that satisfies the velocity boundary conditions and the incompressibility constraint on the boundaries.

Because we resolved to focus on primitive variables, we have not considered in this chapter the vorticity–stream-function approach. However, numerous studies have been devoted to that topic and to the biharmonic stream-function problem where vorticity as a variable is eliminated. We refer to Chapter 5 of the book by Bernardi and Maday [38] for fourth-order problems.

Heinrichs [189] discusses the splitting procedure for the biharmonic problem in a system of two equations involving the Laplacian. The Chebyshev collocation method is used. The vorticity is expressed in the polynomial space $\mathbb{P}_N$, while the stream function is discretized in $\mathbb{P}_{N+2}$ in order to automatically satisfy the Dirichlet boundary conditions. Finite differences are applied to precondition the linear systems, which are solved with a multigrid method. Excellent results are produced on analytical solutions.

Owens and Phillips [291] reformulate the Stokes equations in terms of stream function and Airy stress function. This formulation is ideal for mass and momentum conservation. With these new variables, one has to solve biharmonic equations. The boundary conditions are generated by means of a least-squares formulation. A domain decomposition based on Chebyshev collocation is worked out and applied to the planar stick–slip problem. Good results are obtained.

Instead of the classical velocity–pressure formulation, another possible choice is based on the three-field approach, namely the velocity–pressure–stress formulation. This is used often in rheology, where constitutive relationships are dealing with viscoelastic fluids. The constitutive relation for an incompressible fluid may be written as

$$\sigma = -p\mathbf{I} + \tau, \qquad (5.10.1)$$

where τ is called the extra-stress tensor by rheologists and the shear stress tensor by fluid mechanicists. The Navier–Stokes equations become

$$\rho \frac{D\mathbf{v}}{Dt} = -\nabla p + \operatorname{div} \tau + \rho \mathbf{f}. \qquad (5.10.2)$$

5.10. Complements and Engineering Considerations

Reducing (5.10.2) to the steady Stokes problem of a viscous, Newtonian, incompressible fluid, we obtain the velocity–pressure–stress formulation

$$\nabla p - \operatorname{div} \boldsymbol{\tau} = \rho \mathbf{f}, \tag{5.10.3}$$

$$\operatorname{div} \mathbf{v} = 0, \tag{5.10.4}$$

$$\boldsymbol{\tau} = 2\mu \mathbf{d}, \tag{5.10.5}$$

where the rate of deformation tensor **d** is defined by Equation (1.5.4). Gerritsma and Phillips [155] show that the spectral element approximation of the three-field formulation ($\mathbf{v}, p, \boldsymbol{\tau}$) is compatible if the functional spaces for one element are chosen as $\mathbb{P}_N, \mathbb{P}_{N-2}, \mathbb{P}_N$. This means that the velocity–pressure fields satisfy the same compatibility condition as before, while the stress components are in the same space as the velocity. However, when the discussion of the discretization is extended to the case of several spectral elements, one has to recognize that the stress components are in $L^2(\Omega)$. Consequently, we require that the discrete space for $\boldsymbol{\tau}_N$ contain the discrete space for the velocity gradient. Gerritsma and Phillips [156] propose that this may be achieved by a discontinuous approximation of $\boldsymbol{\tau}$ between the elements.

The three-field approach is very appealing as a generic methodology to treat all cases of viscous incompressible fluids. Indeed, the tensor $\boldsymbol{\tau}$ may represent the Newtonian constitutive equation or any viscoelastic fluid. Furthermore, this tensor may take the turbulence phenomena into account by incorporating the Reynolds stress tensor (1.10.9) or the LES subgrid scale tensor (1.10.25). More generally, the velocity–pressure–stress formulation is ideal for properly treating the interaction between the flow of a viscous incompressible fluid and a solid wall. The unknowns expressed by the stress components are the appropriate ones for the interface conditions.

From the implementation point of view, all methods are not equivalent. Collocation preconditioned by FEs is probably a good choice when a FE code is available. Our experience shows that symbolic manipulation packages are also a useful route to quick development of a computing tool. However, extending collocation to domain decomposition requires the treatment of patching conditions, as will be explained in Chapter 7, and this procedure reduces the robustness of the algorithms. The SEM does not suffer this lack of robustness. Since it is a generalization of the FEM, it can cope with complicated geometries, nonlinearities, and nonconstant coefficients. It is probably the best choice from a numerical and fluid-mechanics point of view.

A final comment is due with respect to the accuracy attained with spectral methods. A reference benchmark solution was produced in the eighties by de Vahl Davis [93] for a thermal convection problem in a square cavity filled with air ($Pr = 0.71$). The Navier–Stokes and energy equations (1.6.6)–(1.6.8) are integrated by using the Boussinesq approximation. No-slip boundary conditions are applied to all boundaries. The cavity horizontal walls are insulated while given differential temperatures are prescribed on the vertical walls. The benchmark solution is obtained by extrapolating to the limit second-order FD results obtained on successively refined grids from 21×21 to 81×81 points. Excellent agreement, to three decimal places, is obtained by Chebyshev collocation [96] with two to three times fewer points in each space direction. In particular, for a nonlinear problem in a simple geometry, high-order methods, compared with FD or FE schemes, save at least an order of magnitude in the number of degrees of freedom, while maintaining the same accuracy.

6
Unsteady Stokes and Navier–Stokes Equations

The physics involved in fluid flow phenomena is three-dimensional in essence and, for most cases in nature or in technology transient with a complex temporal behavior. Consequently, stable and accurate (in space and time) algorithms are required in order to investigate and analyze these complicated situations.

6.1 Unsteady Velocity–Pressure Formulation

Let us consider the numerical integration of the Navier–Stokes equations describing transient incompressible fluid flows in primitive variables subject to body forces $\mathbf{f}$. In dimensionless form, where the characteristic time is the inertial flow time L/U and the reference pressure the dynamic pressure, the Navier–Stokes equations are given by the following relationship [see Equation (1.7.1)]:

$$\frac{D\mathbf{v}}{Dt} = -\nabla p + (Re)^{-1}\Delta\mathbf{v} + \mathbf{f} \quad \text{in } \Omega \quad \text{for } t \geq 0, \qquad (6.1.1)$$

and the continuity equation

$$\text{div } \mathbf{v} = 0 \quad \text{in } \Omega \quad \text{for } t \geq 0. \qquad (6.1.2)$$

This set of equations is solved with appropriate initial and boundary conditions. As in the preceding chapter, we will consider only homogeneous Dirichlet boundary conditions, unless otherwise explicitly stated. They are

$$\mathbf{v} = 0 \quad \text{on } \partial\Omega \quad \text{for } t \geq 0. \qquad (6.1.3)$$

Initial conditions will provide an incompressible velocity field at the beginning of the time integration. With the superscript 0 for this initial field, we have according to Equations (1.6.18)–(1.6.19)

$$\mathbf{v}(\mathbf{x}, t = 0) = \mathbf{v}^0(\mathbf{x}) \quad \text{in } \Omega \tag{6.1.4}$$

$$\text{div } \mathbf{v}^0 = 0 \quad \text{in } \Omega. \tag{6.1.5}$$

We will assume that the time integration is carried out over a finite time interval $(0, T)$.

In the transient Navier–Stokes equations, the nonlinear advective terms account for some of the difficulties in the mathematical analysis of their solutions and in the numerical integration process, because they generate bifurcations (qualitative change of the solutions in the phase space), transition, and, eventually, turbulence. The existence and uniqueness proofs for the solutions of the Navier–Stokes equations have been investigated by physicists, fluid dynamicists, and mathematicians since the early nineteenth century, when those equations were derived. The books by Galdi [145, 146, 147, 148] and Témam [375, 376] summarize the state of the art in this domain and also introduce the reader to challenging open questions. Indeed, although some results are well established, the advent of the theory of dynamical systems and chaos (notions of strange attractors and extreme sensitivity to the initial conditions) at the end of the twentieth century has changed considerably the issues raised by such analysis (see Bergé et al. [32]).

The problem defined by (6.1.1)–(6.1.5) is well posed. However, integrating transient 3D problems in the velocity–pressure formulation constitutes a formidable computational task. Therefore, a pressure Poisson equation is produced by applying the divergence operator to Equation (6.1.1) and taking the continuity requirement (6.1.2) into account. The pressure Poisson equation is given by

$$\Delta p = \text{div}\left(-\frac{D\mathbf{v}}{Dt} + \mathbf{f}\right) \quad \text{in } \Omega \quad \text{for } t \geq 0, \tag{6.1.6}$$

with the boundary condition

$$\mathbf{n} \cdot \nabla p = \mathbf{n} \cdot \left(-\frac{D\mathbf{v}}{Dt} + (Re)^{-1} \Delta \mathbf{v} + \mathbf{f}\right) \quad \text{on } \partial\Omega. \tag{6.1.7}$$

The reference pressure level is imposed by the relation

$$\int_\Omega p \, dV = 0 \quad \text{on } \Omega. \tag{6.1.8}$$

For $Re = \infty$ and for the homogeneous Dirichlet velocity boundary condition (6.1.3), the Neumann condition (6.1.7) reduces to a homogeneous one (neglecting the body-force term):

$$\mathbf{n} \cdot \nabla p = 0 \quad \text{on } \partial\Omega. \tag{6.1.9}$$

This is the pressure boundary condition for an inviscid fluid flow.

In Section 6.2, the unsteady Stokes equations are analyzed. The weak formulation is first presented for the continuous case. We apply the SEM for space discretization, and we propose various possible time schemes. The key issue of splitting errors induced by the velocity–pressure-decoupled computation is addressed in Section 6.3. Section 6.4 presents the weak continuous formulation of the Navier–Stokes equations. Then, the nonlinear advective term is treated, and various time discretization methods are given, including extrapolation and operator-integration-factor splitting techniques (OIFS). Section 6.5 analyzes another possible way to implement the velocity–pressure decoupling through the projection method applied to the continuous equations. The stabilization of unsteady flows by a filtering technique is described in Section 6.6. The direct simulation of free-surface flows is a difficult nonlinear problem. It is presented in Section 6.7 with the method of arbitrary Lagrangian–Eulerian formulation, which enables the treatment of moving domains. We close in Section 6.8 with some unsteady applications. Suggestions for further reading and engineering considerations are given in Section 6.9.

6.2 Unsteady Stokes Equations

In this section, we treat the dimensionless linear unsteady Stokes problem (1.7.6) and the associated divergence-free condition

$$\frac{\partial \mathbf{v}}{\partial t} + \nabla p - \Delta \mathbf{v} = \mathbf{f} \quad \text{in } \Omega \quad \text{for } t \geq 0, \tag{6.2.1}$$

$$\text{div } \mathbf{v} = 0 \quad \text{in } \Omega, \tag{6.2.2}$$

with the homogeneous velocity boundary conditions (6.1.3) and an initial divergence-free velocity field $\mathbf{v}^0$.

We will deal with the weak formulation of the unsteady Stokes continuous problem, because this provides a variational approach to be discretized in space and time.

6.2.1 The Weak Formulation

Equations (6.2.1) and (6.2.2) are cast in the weak form. For this purpose, the functional spaces for the velocity and pressure are $H_0^1(\Omega)^d$ and $L_0^2(\Omega)$, as they were defined in Section 5.2.1. They will be designated here by X and Z, respectively. Based on the bilinear forms $\mathcal{A}$ and $\mathcal{B}$ given by (5.2.10) and (5.2.11), respectively, the weak form is as follows: Find $(\mathbf{v}(t), p(t)) \in X \times Z$ such that for almost every $t \in (0, T)$

$$\frac{d}{dt}(\mathbf{v}(t), \mathbf{u}) + \mathcal{A}(\mathbf{v}(t), \mathbf{u}) + \mathcal{B}(\mathbf{u}, p(t)) = \mathcal{F}(t, \mathbf{u}) \qquad \forall \mathbf{u} \in X, \quad (6.2.3)$$

$$\mathcal{B}(\mathbf{v}(t), q) = 0 \qquad \forall q \in Z, \quad (6.2.4)$$

$$\mathbf{v}(0) = \mathbf{v}^0. \quad (6.2.5)$$

The expression "almost every t" rules out the occurrence of wild temporal behavior, for example, a step function in the velocity field, which is always detrimental in that it produces a numerical transient state with no physical meaning.

Before time-discretizing the former abstract problem, we can proceed with the space discretization investigated in Chapter 5. For ease of development in the sequel, we will use the $\mathbb{P}_N$–$\mathbb{P}_{N-2}$ SEM, although collocation could also be used. With the definitions of the discrete spaces given by Equations (5.2.32)–(5.2.33), the approximate abstract problem becomes the following: Find $(\mathbf{v}_N(t), p_N(t)) \in X_N \times Z_N$ such that for almost every $t \in (0, T)$

$$\frac{d}{dt}(\mathbf{v}_N(t), \mathbf{u}_N) + \mathcal{A}(\mathbf{v}_N(t), \mathbf{u}_N) + \mathcal{B}(\mathbf{u}_N, p_N(t)) = \mathcal{F}_N(t, \mathbf{u}_N) \qquad \forall \mathbf{u}_N \in X_N,$$
$$(6.2.6)$$

$$\mathcal{B}(\mathbf{v}_N(t), q) = 0 \qquad \forall q_N \in Z_N, \quad (6.2.7)$$

$$\mathbf{v}_N(0) = \mathbf{v}_N^0. \quad (6.2.8)$$

With the same matrix notation as in Sections 4.7 and 5.3, and the Einstein summation convention over repeated indices, the semidiscrete problem is

$$M\frac{d\underline{v}_i}{dt} + K\underline{v}_i - D_i^T \underline{p} = M\underline{f}_i, \qquad i = 1, d, \quad (6.2.9)$$

$$-D_i \underline{v}_i = 0, \quad (6.2.10)$$

$$\underline{v}_i(0) = \underline{v}_i^0. \quad (6.2.11)$$

It produces a set of ordinary differential equations.

For the time discretization, we may resort to any of the time-marching schemes presented in Chapter 3. From this plentiful choice, we have selected a few algorithms with the intent of achieving high-order (greater than two) time accuracy. Indeed, as was noted in the introductory chapter, more and

6.2. Unsteady Stokes Equations

more industrial applications consider 3D transient calculations by DNS or LES methodologies. In order to derive and extract meaningful statistics from these large databases, accurate time integration is clearly needed.

Explicit time-marching schemes applied to the viscous term are subjected to stability restrictions on the time step Δt, imposing $\Delta t \leq (Re\, C)/N^4$ [see Equation (3.4.12)]. For moderate values of the Reynolds number, this condition may be stringent, and it is customary to rely on implicit backward differentiation formulas of order k (BDFk). When (3.2.29) is applied to (6.2.9), one gets the following, with the superscript indicating the time level:

$$M\left(\beta_k \underline{v}_i^{n+1} + \beta_{k-1} \underline{v}_i^n + \cdots + \beta_0 \underline{v}_i^{n+1-k}\right)$$
$$= \Delta t\left(-K\underline{v}_i^{n+1} + D_i^T \underline{p}^{n+1} + M\underline{f}_i^{n+1}\right), \quad i = 1, \ldots, d, \quad (6.2.12)$$

where the coefficients β_k are obtained by the relationships

$$\beta_k = \frac{1}{b_0}, \quad \beta_{k-j} = \frac{-a_j}{b_0}, \quad j = 1, \ldots, k. \quad (6.2.13)$$

The coefficients b_0 and a_j are those of the BDFk formula given in Equation (3.2.29) with the corresponding numerical values given in Table 3.2.3. We note that the number of implicit relations to be solved remains constant on increasing order of the BDF. Hence, the computational cost increases only very moderately with the order.

6.2.2 Uzawa Algorithm

We begin by redefining the source term $M\mathbf{f}^{n+1}$ to include all explicit contributions coming from the l.h.s. of the BDFk (6.2.12) for $\underline{v}_i^{n+1}$. The discrete problem thus is as follows: Solve for $\underline{v}_i^{n+1}$ and $\underline{p}^{n+1}$

$$H\underline{v}_i^{n+1} - D_i^T \underline{p}^{n+1} = M\underline{f}_i^{n+1}, \quad i = 1, \ldots, d, \quad (6.2.14)$$

$$-D_i \underline{v}_i^{n+1} = 0, \quad (6.2.15)$$

$$\underline{v}_i(0) = \underline{v}_i^0. \quad (6.2.16)$$

In the relation (6.2.14), the matrix H is obtained as

$$H = \frac{\beta_k}{\Delta t} M + K. \quad (6.2.17)$$

The Uzawa algorithm consists in carrying through the block Gaussian elimination on the system (6.2.14)–(6.2.15) [256]. From (6.2.14), the velocity components are

$$\underline{v}_i^{n+1} = H^{-1} D_i^T \underline{p}^{n+1} + H^{-1} M \underline{f}_i^{n+1}, \qquad i = 1, \ldots, d. \quad (6.2.18)$$

Applying the continuity constraint (6.2.15) to the previous relationship, one obtains the pressure equation

$$S_{\Delta t} \underline{p}^{n+1} = -D_i H^{-1} M \underline{f}_i^{n+1}, \qquad (6.2.19)$$

with

$$S_{\Delta t} = D_i H^{-1} D_i^T. \qquad (6.2.20)$$

The discrete pressure operator $S_{\Delta t}$ is time-dependent (hence the subscript Δt), as opposed to the steady Stokes operator (5.3.10). Once the pressure is known, one can compute the velocity field by solving Helmholtz problems:

$$H \underline{v}_i^{n+1} = D_i^T \underline{p}^{n+1} + M \underline{f}_i^{n+1}, \qquad i = 1, \ldots, d. \quad (6.2.21)$$

6.2.3 Splitting and Decoupling Algorithms

For the unsteady Stokes operator, the Uzawa algorithm is known to converge slowly at large computational expense. Therefore, much effort has been devoted to decoupling the pressure from the velocity-field calculation. This induces a time error called the *splitting error*, which should be of the same order as the global time scheme and also of the order of each element of the split marching procedure. The simplest, most popular decoupling method is the fractional-step method due to Chorin [78] and Témam [374]. It has a first-order time error, which can be improved to second order by the pressure correction algorithm (see Van Kan [396]). Initially, most of these projection methods were applied to the continuous equations and analyzed only with respect to the time discretization technique, thereby ignoring the influence of space discretization. The analyses require an additional boundary condition for the pressure computation, and this topic was the subject of research and intense discussion during the eighties. The Neumann pressure boundary condition has to be time-approximated at the same order as the decoupling error (see Orszag et al. [289], Karniadakis et al. [219]). This is no trivial task and calls for careful analysis. These splitting methods exhibit error layers near the walls of the computational domain, where the accuracy is reduced. The layers have a thickness that is $O(\sqrt{\nu \Delta t})$. Inside a layer, the pressure error may be of the order of unity, and the divergence of the velocity field may also be of the order of unity or even worse.

6.2. Unsteady Stokes Equations

In this section, we examine the decoupling error of various methods resulting from the space and time discretization of the Stokes equations. Since the velocity boundary conditions are incorporated in the discrete operators, no additional pressure boundary condition is needed. The analysis reported here has been inspired by the illuminating papers of Perot [305, 306] and the thesis of Couzy [83]. We use the shorthand notation $\mathbf{H}$ for the block-diagonal matrix containing d blocks, H, on the diagonal, and $\mathbf{D}$ for the matrix composed of the relevant derivative matrices given by Equation (5.3.3) to build up the discrete divergence operator. In matrix form, the equations (6.2.14) and (6.2.15) yield

$$\begin{pmatrix} \mathbf{H} & -\mathbf{D}^T \\ -\mathbf{D} & 0 \end{pmatrix} \begin{pmatrix} \underline{v}^{n+1} \\ \underline{p}^{n+1} \end{pmatrix} = \begin{pmatrix} \mathbf{M}\underline{f}^{n+1} \\ 0 \end{pmatrix}. \tag{6.2.22}$$

We will use block LU factorization to carry out the splitting-error study. We generalize the previous matrix formulation by introducing an arbitrary matrix $\mathbf{Q}$. We obtain

$$\begin{pmatrix} \mathbf{H} & -\mathbf{H}\mathbf{Q}\mathbf{D}^T \\ -\mathbf{D} & 0 \end{pmatrix} \begin{pmatrix} \underline{v}^{n+1} \\ \underline{p}^{n+1} \end{pmatrix} = \begin{pmatrix} \mathbf{M}\underline{f}^{n+1} \\ 0 \end{pmatrix} + \begin{pmatrix} \underline{r}^{n+1} \\ 0 \end{pmatrix}. \tag{6.2.23}$$

The term $\underline{r}^{n+1}$ is a residual accounting for the difference between (6.2.22) and (6.2.23); it will be discarded in the LU decomposition. We will take full advantage, however, of the presence of the matrix $\mathbf{Q}$, which will give us flexibility in choosing the projection method (defined later). The solution of the block LU decomposition of (6.2.23) is a two-step procedure:

$$\begin{pmatrix} \mathbf{H} & 0 \\ -\mathbf{D} & -\mathbf{D}\mathbf{Q}\mathbf{D}^T \end{pmatrix} \begin{pmatrix} \underline{v}^* \\ \underline{p}^{n+1} \end{pmatrix} = \begin{pmatrix} \mathbf{M}\underline{f}^{n+1} \\ 0 \end{pmatrix} \tag{6.2.24}$$

and

$$\begin{pmatrix} \mathbf{I} & -\mathbf{Q}\mathbf{D}^T \\ 0 & \mathbf{I} \end{pmatrix} \begin{pmatrix} \underline{v}^{n+1} \\ \underline{p}^{n+1} \end{pmatrix} = \begin{pmatrix} \underline{v}^* \\ \underline{p}^{n+1} \end{pmatrix}. \tag{6.2.25}$$

The intermediate velocity $\underline{v}^*$ is not divergence-free. The second step (6.2.25) is the projection of the non-divergence-free $\underline{v}^*$ onto the divergence-free velocity $\underline{v}^{n+1}$. This is the origin of the so-called *projection method*. The system (6.2.24)–(6.2.25) covers a large class of projection techniques depending on the choice of $\mathbf{Q}$.

If we set

$$\mathbf{Q} = \mathbf{H}^{-1}, \tag{6.2.26}$$

we get the preceding Uzawa algorithm, and no decoupling error is introduced, because the two-step projection is formally equivalent to the original system (6.2.22). If we choose

$$\mathbf{Q} = \frac{\Delta t}{\beta_k} \mathbf{M}^{-1}, \qquad (6.2.27)$$

we obtain the fractional-step method, although this name is also given to the space-continuous splitting (Section 6.5). With the latter choice, the residual is

$$\underline{r}^{n+1} := (\mathbf{HQ} - \mathbf{I})\mathbf{D}^T \underline{p}^{n+1} = \frac{\Delta t}{\beta_k} \mathbf{KM}^{-1}\mathbf{D}^T \underline{p}^{n+1} = O(\Delta t). \qquad (6.2.28)$$

From the definition (6.2.17) we can see that the velocity in Equation (6.2.14) is multiplied by $(\Delta t)^{-1}$, and the local time error of the fractional-step method is therefore of second order, thereby producing a globally first-order time-accurate scheme. Furthermore, the decoupling error does not vanish when the problem reaches a steady-state solution. As we mentioned earlier, the numerical methods in fluid mechanics aim at 3D transient computations by DNS or LES to obtain the correct temporal dynamics. A first-order time-accurate scheme is insufficient in that respect; second-order accuracy is the minimum required; and in the spectral computations, higher-order accuracy in time should ideally be achieved to avoid spoiling the spatial accuracy by a poor time-marching scheme.

A better, second-order approximation of $\mathbf{H}^{-1}$ is obtained by incorporating the next term in the Taylor series of $\mathbf{H}^{-1}$:

$$\mathbf{Q} = \frac{\Delta t}{\beta_k} \mathbf{M}^{-1} - \left(\frac{\Delta t}{\beta_k}\right)^2 \mathbf{M}^{-1} \mathbf{K} \mathbf{M}^{-1}. \qquad (6.2.29)$$

The second-order decoupling error is evaluated from the relation

$$\underline{r}^{n+1} = \left(\frac{\Delta t}{\beta_k}\right)^2 (\mathbf{KM}^{-1})^2 \mathbf{D}^T \underline{p}^{n+1} = O(\Delta t^2). \qquad (6.2.30)$$

As noted by Couzy [83], this method does not work efficiently, because (6.2.29) is positive definite only for small values of Δt.

Increasing the accuracy to third order, we get

$$\mathbf{Q} = \frac{\Delta t}{\beta_k} \mathbf{M}^{-1} - \left(\frac{\Delta t}{\beta_k}\right)^2 \mathbf{M}^{-1}\mathbf{K}\mathbf{M}^{-1} + \left(\frac{\Delta t}{\beta_k}\right)^3 (\mathbf{KM}^{-1})^2 \mathbf{M}^{-1}. \qquad (6.2.31)$$

This matrix is always positive definite and produces the decoupling error

$$\underline{r}^{n+1} = \left(\frac{\Delta t}{\beta_k}\right)^3 (\mathbf{KM}^{-1})^3 \mathbf{D}^T \underline{p}^{n+1} = O(\Delta t^3). \qquad (6.2.32)$$

This method will be called P3.

6.2. Unsteady Stokes Equations

Up to now, all splitting schemes have nonvanishing errors for steady-state solutions. These may be detrimental, because converged solutions might depend on the time step used and change with it. To escape from this dismal performance and remove the decoupling error for steady problems, we can modify the fractional-step method as follows:

$$\begin{pmatrix} \mathbf{H} & -\mathbf{HQD}^T \\ -\mathbf{D} & 0 \end{pmatrix} \begin{pmatrix} \underline{v}^{n+1} \\ \underline{p}^{n+1} - \underline{p}^n \end{pmatrix} = \begin{pmatrix} \mathbf{M}\underline{f}^{n+1} + \mathbf{D}^T \underline{p}^n \\ 0 \end{pmatrix} + \begin{pmatrix} \underline{r}^{n+1} \\ 0 \end{pmatrix}. \tag{6.2.33}$$

The block LU factorization of (6.2.33) with the matrix $\mathbf{Q}$ given by (6.2.27) leads to the pressure correction (PC) method, which is a second-order scheme, as proven by Van Kan [396]. The decoupling error is obtained by the relationship

$$\begin{aligned} \underline{r}^{n+1} &:= (\mathbf{HQ} - \mathbf{I})\mathbf{D}^T(\underline{p}^{n+1} - \underline{p}^n) \\ &= \frac{\Delta t}{\beta_k}\mathbf{KM}^{-1}\mathbf{D}^T(\underline{p}^{n+1} - \underline{p}^n) = O(\Delta t^2). \end{aligned} \tag{6.2.34}$$

We notice that this decoupling error vanishes for a steady solution because, in this case, $\underline{p}^{n+1} - \underline{p}^n = 0$. The same trick based on adding $\mathbf{D}^T \underline{p}^n$ to the r.h.s. of Equation (6.2.22) and computing a correction term for the pressure can be applied to the P3 scheme. One obtains the P3 PC scheme with a fourth-order decoupling error, namely,

$$\left(\frac{\Delta t}{\beta_k}\right)^3 (\mathbf{KM}^{-1})^3 \mathbf{D}^T(\underline{p}^{n+1} - \underline{p}^n) = O(\Delta t^4), \tag{6.2.35}$$

which will also vanish for steady solutions.

Following the same strategy, we could derive high-order (fifth, sixth) projection methods. Nonetheless, the practical interest is limited. Indeed, the global time accuracy of the split scheme will be $\min(k, l)$, where k is the order of the BDF algorithm and l the error order of the decoupling. In increasing the accuracy, the BDF scheme will bring much of the improvement. Finally, we note that the projection scheme will provide a divergence-free velocity at each time step, implying that even if the initial condition is not divergence-free, the solution will be after the first step.

The LU factorization recommended by Perot [305] can be approximated by the use of inexact factorization. Typically, the matrix $\mathbf{H}^{-1}$ is approximated by replacing $\mathbf{Q}$ by $\mathbf{Q}_1$ in the L block and by $\mathbf{Q}_2$ in the U block. New factorization schemes are produced by the various available choices for $\mathbf{Q}_1$ and $\mathbf{Q}_2$. The Yosida method proposed by Quarteroni et al. [318] sets $\mathbf{Q}_1 = (\Delta t/\beta_k)\mathbf{M}^{-1}$

and $\mathbf{Q}_2 = \mathbf{H}^{-1}$. The two-step process becomes

$$\begin{pmatrix} \mathbf{H} & 0 \\ -\mathbf{D} & -\mathbf{DQ}_1\mathbf{D}^T \end{pmatrix} \begin{pmatrix} \underline{v}^* \\ \underline{p}^{n+1} \end{pmatrix} = \begin{pmatrix} \mathbf{M}\underline{f}^{n+1} \\ 0 \end{pmatrix} \quad (6.2.36)$$

and

$$\begin{pmatrix} \mathbf{I} & -\mathbf{Q}_2\mathbf{D}^T \\ 0 & \mathbf{I} \end{pmatrix} \begin{pmatrix} \underline{v}^{n+1} \\ \underline{p}^{n+1} \end{pmatrix} = \begin{pmatrix} \underline{v}^* \\ \underline{p}^{n+1} \end{pmatrix}. \quad (6.2.37)$$

The full matrix is now

$$\begin{pmatrix} \mathbf{H} & -\mathbf{D}^T \\ -\mathbf{D} & \mathbf{D}(\mathbf{Q}_2 - \mathbf{Q}_1)\mathbf{D}^T \end{pmatrix}. \quad (6.2.38)$$

We notice that the inexact factorization produces a perturbation term in the incompressibility constraint, which may be viewed as a quasicompressible regularization. In (6.2.36), the matrix $\mathbf{DQ}_1\mathbf{D}^T = (\Delta t/\beta_k)\mathbf{DM}^{-1}\mathbf{D}^T$ computing the pressure is the pressure pseudo-Laplacian matrix, which will be used in the next section for preconditioning purposes. The perturbation term in (6.2.38) yields

$$-\mathbf{D}(\mathbf{Q}_1 - \mathbf{Q}_2)\mathbf{D}^T = -\mathbf{D}\left(\frac{\Delta t}{\beta_k}\mathbf{M}^{-1}\right)\left[\mathbf{I} - \left(\mathbf{I} + \frac{\Delta t}{\beta_k}\mathbf{M}^{-1}\mathbf{K}\right)^{-1}\right]\mathbf{D}^T$$

$$= -\left(\frac{\Delta t}{\beta_k}\mathbf{M}^{-1}\right)^2 \mathbf{DYD}^T, \quad (6.2.39)$$

where the matrix $\mathbf{Y}$ is given by

$$\mathbf{Y} = \left(\frac{\beta_k}{\Delta t}\mathbf{M}\right)\left[\mathbf{I} - \left(\mathbf{I} + \frac{\Delta t}{\beta_k}\mathbf{M}^{-1}\mathbf{K}\right)^{-1}\right]. \quad (6.2.40)$$

The Yosida matrix corresponds to the Yosida regularization of the Laplace operator $\mathbf{K}$. We emphasize that this method has been tested successfully with FEs, but not in the spectral-element context.

6.3 Pressure Preconditioning

The matrix $S_{\Delta t}$ in Equation (6.2.20) is full, and Equation (6.2.19) requires an iterative approach as in the steady Stokes case. However, whereas the steady pressure matrix S was preconditioned by the inverse of the pressure mass matrix $\tilde{M}$ (5.3.11), this is no longer true for $S_{\Delta t}$. For large time steps, H tends to the

6.3. Pressure Preconditioning

limit K, and $\tilde{M}^{-1}S_{\Delta t}$ is well conditioned. For small time steps with $\Delta t \to 0$, the pressure matrix $S_{\Delta t}$ goes to the pseudo-Laplacian matrix A [see (5.4.35) in the steady case], defined by

$$A = D_i M^{-1} D_i^T. \tag{6.3.1}$$

We note that the matrix A arises in the splitting equation (6.2.24) when Q is given by Equation (6.2.27). Here, M^{-1} is defined on the GLL grid and enforces the velocity boundary conditions. The matrix A corresponds to the discrete consistent Poisson pressure solver resulting from the spectral-element discretization of the unsteady Stokes problem (6.2.1)–(6.2.2) treated by an explicit algorithm. The conditioning of the system is as bad as the one for a Laplace equation. A preconditioner for $S_{\Delta t}$ that works for small and large time steps is given by

$$P^{-1} = \tilde{M}^{-1} + \frac{\beta_k}{\Delta t} A^{-1}, \tag{6.3.2}$$

where $\tilde{M}$ is the diagonal pressure mass matrix. For both cases, the product $P^{-1} S_{\Delta t}$ is close to identity, as shown by Cahouet and Chabard [60]. The difficulty with this preconditioner is that because A itself is ill conditioned, it too needs to be preconditioned to obtain the fast convergence of the iterative method used to solve for A.

Rønquist [333] points out that the pressure Poisson operator (6.3.1) is more difficult to solve than the classical Laplace equation because the first one is in L^2 and the second one in H^1. For this reason, Rønquist proposes to solve for the pressure in two different subspaces. The first set of equations computes the constant level of the pressure in each element by building up a global coarse approximation (subscript c). The second subspace represents the local pressure variation with respect to the previous piecewise constant frame (subscript f for fine). Using the notation of Chapter 5, we decompose the pressure space into two disjoint parts Z_c and $Z_{N,E}$ defined as

$$Z_c = L_0^2(\Omega) \cap \mathbb{P}_{0,E}(\Omega), \tag{6.3.3}$$

$$Z_{N,E} = L_0^2(\Omega) \cap \mathbb{P}_{N-2,E}(\Omega), \tag{6.3.4}$$

such that the full pressure results from

$$\underline{p} = J\underline{p}_c + \underline{p}_f \in Z_N = Z_c \oplus Z_{N,E}. \tag{6.3.5}$$

Here, $\underline{p}_c$ accounts for the coarse pressure and $\underline{p}_f$ for the local variation. The operator J maps (by interpolation) the local pressure level onto the space Z_N. Combining Equation (6.2.19) with (6.3.5), we write

$$S_{\Delta t}(J\underline{p}_c + \underline{p}_f) = -D_i H^{-1} M \underline{f}_i^{n+1}. \tag{6.3.6}$$

Multiplying (6.3.6) by the transpose J^T, we obtain for $\underline{p}_c$

$$\underline{p}_c = (J^T S_{\Delta t} J)^{-1} J^T \left(-D_i H^{-1} M \underline{f}_i^{n+1} - S_{\Delta t} \underline{p}_f \right). \tag{6.3.7}$$

The definition $S_c = J^T S_{\Delta t} J$ yields the coarse-grid operator, and $J\underline{p}_c$ is the projection (in the S-norm) of $\underline{p}$ onto $\mathcal{R}(J)$. Substituting (6.3.7) in (6.3.6), we arrive at the following expression for $\underline{p}_f$:

$$\left(I - S_{\Delta t} J S_c^{-1} J^T\right) S_{\Delta t} \underline{p}_f = -\left(I - S_{\Delta t} J S_c^{-1} J^T\right) D_i H^{-1} M \underline{f}_i^{n+1}. \tag{6.3.8}$$

The fine-grid operator S_f given by

$$S_f = \left(I - S_{\Delta t} J S_c^{-1} J^T\right) S_{\Delta t} \tag{6.3.9}$$

allows us to rewrite Equations (6.3.7) and (6.3.8) as

$$S_c \underline{p}_c = J^T \left(-D_i H^{-1} M \underline{f}_i^{n+1} - S_{\Delta t} \underline{p}_f\right), \tag{6.3.10}$$

$$S_f \underline{p}_f = -\left(I - S_{\Delta t} J S_c^{-1} J^T\right) D_i H^{-1} M \underline{f}_i^{n+1}. \tag{6.3.11}$$

The former scheme is referred to as *deflation*, by analogy with the standard deflation technique used in root finding [284]. The coarse-grid equation (6.3.10) involves E unknowns. The moderate size of this system leads to the choice of a direct solver. On the contrary, an iterative method such as preconditioned conjugate gradients is preferable for tackling the fine-grid problem (6.3.11).

An efficient preconditioner corresponds to the elemental local matrix $S_{\Delta t}$, where homogeneous Dirichlet boundary conditions are imposed through H^{-1} on all boundaries and interfaces. A major advantage of this procedure comes from the fact that the preconditioner is such that

$$\ker(\text{block}(S_{\Delta t})) = \ker(S_f) \tag{6.3.12}$$

and is completely decoupled per element. This is a characteristic feature amenable to parallel processing. In practical computations, however, it is difficult to build $\text{block}(S_{\Delta t})$. To overcome this drawback, one uses instead $\tilde{M}^{-1} + (\beta_k/\Delta t)\,\text{block}(A)^{-1}$, which is spectrally close according to Equation (6.3.2).

Fischer and Rønquist [132] use a FE preconditioner to solve for the matrix block(A). Couzy and Deville [85] suggest using the fast diagonalization method to compute block(A)$^{-1}$. They also introduce a two-stage preconditioning technique in which the first stage preconditions the operator $S_{\Delta t}$ by (6.3.2) and the second stage applies the same technique to A as has been applied to $S_{\Delta t}$. In other words, the evaluation of Equation (6.3.2) will require the solution of systems like $A\underline{q} = $ r.h.s. This equation is written as $A(J\underline{q}_c + \underline{q}_f) = $ r.h.s. Coarse-grid (A_c) and fine-grid (A_f) operators are introduced with definitions similar to S_c and S_f, leading to two algebraic systems. Again, the matrix A_f will be preconditioned by block(A).

In Couzy [83], the homogeneous Dirichlet boundary conditions of the local preconditioner are replaced by Neumann velocity conditions at the element interfaces. These help to partially circumvent the L^2-specific feature of the pressure operator. In the same spirit, Fischer [127] uses the additive Schwarz method, which will be described in full detail in the next chapter. There, the interelement coupling is improved by using overlapping subdomains in conjunction with a continuous piecewise-linear approximation for the coarse-grid problem.

6.4 Unsteady Navier–Stokes Equations

If we refer to the diffusive time L^2/ν and normalize the pressure with respect to the viscous stress $\mu U/L$, the dimensionless Navier–Stokes equations take the form

$$\frac{\partial \mathbf{v}}{\partial t} + Re\,(\mathbf{v} \cdot \nabla \mathbf{v}) = -\nabla p + \Delta \mathbf{v} + \mathbf{f} \quad \text{in } \Omega \quad \text{for } t \geq 0, \quad (6.4.1)$$

together with the mass conservation equation (6.1.2), the homogeneous boundary conditions, and an initial incompressible velocity field. With respect to the Stokes equations (6.2.1), the nonlinear term multiplied by the Reynolds number brings all the stiffness to the problem. This advective nonlinearity not only will complicate the physics of the fluid flow but also will add to the difficulties of the numerical simulation.

6.4.1 Weak Formulation

With the same spaces as those defined earlier in Sections 5.2 and 6.2, the weak form of the Navier–Stokes equations is: Find $(\mathbf{v}(t), p(t)) \in X \times Z$ such that for almost every $t \in (0, T)$

$$\frac{d}{dt}(\mathbf{v}(t), \mathbf{u}) + \mathcal{A}(\mathbf{v}(t), \mathbf{u}) + \mathcal{B}(\mathbf{u}, p(t))$$
$$= -\text{Re}\,\mathcal{C}(\mathbf{v}(t); \mathbf{v}(t), \mathbf{u}) + \mathcal{F}(t, \mathbf{u}) \quad \forall \mathbf{u} \in X, \quad (6.4.2)$$
$$\mathcal{B}(\mathbf{v}(t), q) = 0 \quad \forall q \in Z, \quad (6.4.3)$$
$$\mathbf{v}(0) = \mathbf{v}^0, \quad (6.4.4)$$

where $\mathcal{C}$ has been defined in Chapter 5 [see Equation (5.8.6)].

The system resulting from the discretization of the problem (6.4.2) is nonlinear and nonsymmetric. A common practice consists in using an explicit treatment of the advection term to avoid the use of sophisticated linearization procedures like the Newton method combined with nonsymmetric iterative solvers like GMRES [340]. The explicit time advancement for the nonlinear term avoids the linearization and reduces the computational burden to the solution of a Stokes problem with a given source term. As a penalty for that practice, however, a stability condition on the maximum allowed time step emerges.

6.4.2 Advection Treatment

Different choices are possible for the advection operator. Three forms can be written:

$$\mathbf{v} \cdot \nabla \mathbf{v} \quad \text{(convective form)}, \quad (6.4.5)$$
$$\nabla \cdot \mathbf{v}\mathbf{v} \quad \text{(conservative form)}, \quad (6.4.6)$$
$$\frac{1}{2}(\mathbf{v} \cdot \nabla \mathbf{v} + \nabla \cdot \mathbf{v}\mathbf{v}) \quad \text{(skew-symmetric form)}. \quad (6.4.7)$$

These three forms are equivalent for the continuous equations (with homogeneous Dirichlet boundary conditions only) because of the incompressibility condition. However, this is no longer true for the discrete operators. Couzy [83] showed numerically the clear superiority of the skew-symmetric form with respect to the conservative and convective forms, and this observation was confirmed by Rønquist [334]. The reason for this optimal performance is that the eigenvalues of the discrete skew-symmetric operator lie on the imaginary axis of the complex plane. Both other forms have real parts. Recently, however, Wilhelm and Kleiser [412] noticed an unstable behavior in Navier–Stokes simulations, depending on the formulation of the advection operator. The numerical

6.4. Unsteady Navier–Stokes Equations

scheme is stable for the convective form and unstable for both other forms. They demonstrated that this instability comes from the staggered-grid discretization. Although the continuity constraint is enforced at the GL pressure points, the divergence errors may still grow exponentially elsewhere in the computational domain, especially at the GLL velocity points.

The Galerkin method applied to the convective form chooses the test functions in the same polynomial bases as the velocity approximation. According to the SEM, the discretization process for the reference element $\hat{\Omega}$ with a superscript e gives

$$C^e \underline{v}^e = \rho_l \rho_m \rho_n \sum_{q=0}^{N} \left(v^e_{1,lmn} D^e_{lq} v^e_{qmn} + v^e_{2,lmn} D^e_{mq} v^e_{lqn} \right.$$
$$\left. + v^e_{3,lmn} D^e_{nq} v^e_{lmq} \right), \quad l, m, n \in [0, N]. \quad (6.4.8)$$

Let $V^e_1 = \text{diag}(v^e_{1,\hat{i}})$ be the diagonal matrix having entries $v^e_{1,\hat{i}} := v^e_{1,lmn}$ such that $\hat{i} = 1 + l + (N+1)m + (N+1)^2 n$ is the index map that defines a natural ordering for the tensor product of the GLL quadrature points. Analogous definitions hold for V^e_2 and V^e_3. We note from (6.4.8) that the evaluation of the nonlinear term involves only pointwise multiplication (or collocation) of the nodal values of the velocity components with the spatial derivatives of $\mathbf{v}^e$. If we define for an $L^e_x \times L^e_y \times L^e_z$ element Ω^e

$$D^e_1 := (I \otimes I \otimes \hat{D}) \frac{2}{L^e_x}, \quad D^e_2 := (I \otimes \hat{D} \otimes I) \frac{2}{L^e_y}, \quad D^e_3 := (\hat{D} \otimes I \otimes I) \frac{2}{L^e_z},$$

for the derivative operators and

$$M^e := \frac{L^e_x L^e_y L^e_z}{8} \hat{M} \otimes \hat{M} \otimes \hat{M} \quad (6.4.9)$$

for the mass matrix, the nonlinear term evaluation may be cast in the tensor-product form

$$C^e \underline{v}^e = M^e \left(V^e_1 D^e_1 + V^e_2 D^e_2 + V^e_3 D^e_3 \right) \underline{v}^e. \quad (6.4.10)$$

The full product $C\underline{v}$ is obtained by the direct stiffness summation (4.5.8) of all elemental contributions:

$$C\underline{v} = \Sigma' C^e \underline{v}^e. \quad (6.4.11)$$

The semidiscrete Navier–Stokes problem becomes

$$M\frac{d\underline{v}_i}{dt} + Re\, C\underline{v}_i + K\underline{v}_i - D_i^T \underline{p} = M\underline{f}_i, \quad i = 1,\ldots,d, \quad (6.4.12)$$

$$-D_i\underline{v}_i = 0. \quad (6.4.13)$$

$$\underline{v}_i(0) = \underline{v}_i^0. \quad (6.4.14)$$

In order to avoid the iterative solution of a nonsymmetric, nonlinear system that occurs when the advection terms are treated implicitly, the only remaining choice is to treat them explicitly. The family of Adams–Bashforth (AB) schemes uses a multistep procedure. The k-step method is of global order k. The advection operator being nondissipative, its stability region should embrace the largest possible portion of the imaginary axis. As was shown in Figure 3.2.1, the AB method of order two (AB2) has no intersection of its stability region with the imaginary axis. This is not the case for the AB3 method, however. We select this scheme to compute $C\underline{v}$. The splitting method with an implicit treatment of the viscous term (as before) and an explicit scheme for advection leads, for example, to the choice BDF1/AB3 for (6.4.12)–(6.4.13). Assuming for the sequel that the index i will always be in $[1, d]$, we obtain

$$\left(\frac{M}{\Delta t} + K\right)\underline{v}_i^{n+1} - D_i^T \underline{p}^{n+1}$$

$$= \frac{M}{\Delta t}\underline{v}_i^n + M\underline{f}_i^{n+1} - Re\left(\tfrac{23}{12}C\underline{v}_i^n - \tfrac{4}{3}C\underline{v}_i^{n-1} + \tfrac{5}{12}C\underline{v}_i^{n-2}\right), \quad (6.4.15)$$

$$-D_i\underline{v}_i^{n+1} = 0. \quad (6.4.16)$$

The local error analysis shows that this scheme is $O(\Delta t^2)$, providing a global error of order one. To achieve higher-order accuracy in time is no trivial task. The combination BDF2/AB3 may be written as

$$\left(\frac{3M}{2\Delta t} + K\right)\underline{v}_i^{n+1} - D_i^T \underline{p}^{n+1}$$

$$= \frac{M}{\Delta t}\left(2\underline{v}_i^n - \tfrac{1}{2}\underline{v}_i^{n-1}\right) + M\underline{f}_i^{n+1} - 2\,Re\left(\tfrac{23}{12}C\underline{v}_i^n - \tfrac{4}{3}C\underline{v}_i^{n-1} + \tfrac{5}{12}C\underline{v}_i^{n-2}\right),$$

$$(6.4.17)$$

$$-D_i\underline{v}_i^{n+1} = 0. \quad (6.4.18)$$

The global error is still of order Δt [83] and is due to time splitting between explicit advective and implicit diffusive parts of the Navier–Stokes equations.

Another simple method, used by Karniadakis et al. [219], consists in extrapolating the nonlinear terms. More precisely, a BDFk (here BDF3) is applied to

6.4. Unsteady Navier–Stokes Equations

both linear and nonlinear terms:

$$\left(\frac{11}{6\Delta t}M + K\right)\underline{v}_i^{n+1} - D_i^T \underline{p}^{n+1} = \frac{M}{\Delta t}\left(3\underline{v}_i^n - \tfrac{3}{2}\underline{v}_i^{n-1} + \tfrac{1}{3}\underline{v}_i^{n-2}\right)$$
$$+ M\underline{f}_i^{n+1} - Re\, C\underline{v}_i^{n+1}, \quad (6.4.19)$$

$$-D_i \underline{v}_i^{n+1} = 0. \quad (6.4.20)$$

To cope with the nonlinear term at the new time level and avoid solving an implicit relation, one uses an extrapolation scheme (EX3) to determine $C\,\underline{v}_i^{n+1}$:

$$C\,\underline{v}_i^{n+1} = 3C\underline{v}_i^n - 3C\underline{v}_i^{n-1} + C\underline{v}_i^{n-2} + O(\Delta t^3). \quad (6.4.21)$$

Combining (6.4.19) with (6.4.21) gives

$$\left(\frac{11}{6\Delta t}M + K\right)\underline{v}_i^{n+1} - D_i^T \underline{p}^{n+1}$$
$$= \frac{M}{\Delta t}\left(3\underline{v}_i^n - \tfrac{3}{2}\underline{v}_i^{n-1} + \tfrac{1}{3}\underline{v}_i^{n-2}\right) + M\underline{f}_i^{n+1} - Re\left(3C\underline{v}_i^n - 3C\underline{v}_i^{n-1} + C\underline{v}_i^{n-2}\right).$$
$$(6.4.22)$$

For this scheme, there is no time-splitting error. Furthermore, it is easy to show that the BDF3/EX3 scheme provides global third-order accuracy.

As indicated in Chapter 3, operator-integration-factor splitting (OIFS) techniques help to produce efficient and more stable numerical schemes. The basic idea is to separate the linear viscous term from the nonlinear advection contributions to the flow. The latter constitute the essential part of the dynamics for increasing Reynolds numbers. The Reynolds number is interpreted here as the ratio of the viscous diffusive time to the inertial time of the flow. OIFS techniques are tailored to quickly advance the time integration of the advection term using an explicit scheme, while solving less often for the implicit Stokes operator, which imposes incompressibility. The method was applied to the Navier–Stokes equations by Maday et al. [262], Couzy [84], and Timmermans [379]. With the spectral-element discretization, Equation (6.4.12) is written in terms of an integration factor $\mathcal{Q}_C^{t^*}(t)$:

$$\frac{d}{dt}[\mathcal{Q}_C^{t^*}(t)\mathbf{M}\underline{v}] = \mathcal{Q}_C^{t^*}(t)(-\mathbf{K}\underline{v} + \mathbf{D}^T\underline{p} + \mathbf{M}\underline{f}), \quad (6.4.23)$$

which is defined here by

$$\frac{d}{dt}[\mathcal{Q}_C^{t^*}(t)\mathbf{M}] = Re\,\mathcal{Q}_C^{t^*}(t)C(\underline{v}), \quad \mathcal{Q}_C^{t^*}(t^*) = I. \quad (6.4.24)$$

Applying BDFk to Equation (6.4.23), one obtains [see Equation (3.3.15)]

$$\mathbf{M}\beta_k \underline{v}^{n+1} + \sum_{j=1}^{k} \beta_{k-j} S_C\left(\mathcal{Q}_C^{t^{n+1}}(t^{n+1-j})\mathbf{M}\underline{v}^{n+1-j}\right)$$
$$= \Delta t(-\mathbf{K}\underline{v}^{n+1} + \mathbf{D}^T \underline{p}^{n+1} + \mathbf{M}\underline{f}^{n+1}). \quad (6.4.25)$$

The terms involving the operator integration factor, which is never constructed explicitly, are evaluated using the relationship

$$S_C\left(\mathcal{Q}_C^{t^{n+1}}(t^{n+1-j})\mathbf{M}\underline{v}^{n+1-j}\right) = \mathbf{M}\underline{\tilde{v}}_j^{n+1}, \quad j=1,\ldots,k, \quad (6.4.26)$$

where the unknowns $\underline{\tilde{v}}_j^{n+1}$ are computed by integration of the initial-value problem

$$\mathbf{M}\frac{d}{ds}\underline{\tilde{v}}_j(s) = -\operatorname{Re} C(\underline{\tilde{v}}_j(s))\underline{\tilde{v}}_j(s), \quad t^{n+1-j} \leq s \leq t^{n+1}, \quad (6.4.27)$$

with initial conditions $\underline{\tilde{v}}_j(t^{n+1-j}) = \underline{v}^{n+1-j}$. This last problem is solved with a time step $\Delta s = \Delta t/\gamma$. The integer γ represents the number of subcycles where explicit cheap evaluations of the nonlinear terms are computed.

Because stability is at stake, we choose the fourth-order RK scheme, since it embraces a large portion of the imaginary axis of the complex plane (as shown in Figure 3.2.5) and does not require initial data as a multistep method as AB3 would. For the Stokes operator, we use a BDFk scheme. Therefore, an OIFS BDF3/RK4 scheme (see Section 3.3.2) for the Navier–Stokes equations is

$$\left(\frac{11}{6\Delta t}\mathbf{M} + \mathbf{K}\right)\underline{v}^{n+1} - \mathbf{D}^T \underline{p}^{n+1} = \frac{\mathbf{M}}{\Delta t}\left(3\underline{\tilde{v}}_1^{n+1} - \tfrac{3}{2}\underline{\tilde{v}}_2^{n+1} + \tfrac{1}{3}\underline{\tilde{v}}_3^{n+1}\right) + \mathbf{M}\underline{f}^{n+1},$$
$$(6.4.28)$$

$$-\mathbf{D}\underline{v}^{n+1} = 0. \quad (6.4.29)$$

The RK4 scheme is applied to (6.4.27), which is solved in turn

- on the interval $[t^n, t^{n+1}]$, $\underline{\tilde{v}}_1^n = \underline{v}^n$, to compute $\underline{\tilde{v}}_1^{n+1}$,
- on the interval $[t^{n-1}, t^{n+1}]$, $\underline{\tilde{v}}_2^{n-1} = \underline{v}^{n-1}$, to compute $\underline{\tilde{v}}_2^{n+1}$, and
- on the interval $[t^{n-2}, t^{n+1}]$, $\underline{\tilde{v}}_3^{n-2} = \underline{v}^{n-2}$, to compute $\underline{\tilde{v}}_3^{n+1}$.

For $\Delta t = \gamma \Delta s$, we require 24γ evaluations of the nonlinear term. Typically, one chooses Δs such that the CFL condition based on this time step is less than one. This is more expensive than the extrapolation method. However, the stability of the algorithm is better, and the final decision in choosing between extrapolation and OIFS should be based on accuracy and stability considerations. We point out here that OIFS induces a nonvanishing splitting error for steady-state problems.

6.5 Projection Methods

The linear algebra problem we face if we discretize the full 3D, transient Navier–Stokes equations in space and time is so immense that, inevitably, we come to the conclusion that some velocity–pressure decoupling procedure has to be designed. This may be carried out first in the continuum framework, thereby avoiding all subsequent considerations of compatibility between discrete spaces like the inf–sup condition. However, the pressure boundary conditions have to be handled in the right fashion in order to keep the time discretization error consistent with each piece of the split scheme. The methods presented in this section have been analyzed and thoroughly discussed by Orszag et al. [289], Tomboulides et al. [382], and Karniadakis et al. [219]. Projection methods aim at the decoupled computation of the velocity and pressure fields. Most of the algorithms compute first a tentative or intermediate velocity that is not divergence-free. This field is then projected on the subspace of solenoidal vector functions.

Let us write the Navier–Stokes equations (6.4.1), (6.1.2) as

$$\frac{\partial \mathbf{v}}{\partial t} = -\nabla p + L(\mathbf{v}) + N(\mathbf{v}) + \mathbf{f}, \tag{6.5.1}$$

$$\mathcal{D} := \operatorname{div} \mathbf{v} = 0, \tag{6.5.2}$$

where the linear term $L(\mathbf{v})$ is given by

$$L(\mathbf{v}) = \Delta \mathbf{v} = \nabla \operatorname{div} \mathbf{v} - \nabla \times (\nabla \times \mathbf{v}) \tag{6.5.3}$$

and the advection term is given by

$$N(\mathbf{v}) = -(Re)\mathbf{v} \cdot \nabla \mathbf{v} \tag{6.5.4}$$

Here, the nonlinear term is expressed by its convective form, but it might also be chosen among the various other forms (6.4.6)–(6.4.7). Time integration of Equation (6.5.1) between t^n and t^{n+1} gives

$$\mathbf{v}^{n+1} - \mathbf{v}^n = -\int_{t^n}^{t^{n+1}} \nabla p \, dt + \int_{t^n}^{t^{n+1}} L(\mathbf{v}) \, dt + \int_{t^n}^{t^{n+1}} N(\mathbf{v}) \, dt + \int_{t^n}^{t^{n+1}} \mathbf{f} \, dt. \tag{6.5.5}$$

Commuting the space and time operators acting on the pressure, we write

$$\int_{t^n}^{t^{n+1}} \nabla p \, dt = \nabla \int_{t^n}^{t^{n+1}} p \, dt = \Delta t \, \nabla p^{n+1}, \tag{6.5.6}$$

so that the pressure p^{n+1} ensures the incompressibility of the velocity field at the new time level. The new-time-level pressure indicates that this value may be obtained, for example, over the time step Δt by a quadrature method such as the (right-hand) rectangular rule. The remaining integrals in (6.5.5) are dealt with by an ABk scheme for the advective term and a BDFk scheme for the linear term. Following this approach for the Navier–Stokes equations, we get

$$\beta_k \mathbf{v}^{n+1} + \sum_{j=1}^{k} \beta_{k-j} \mathbf{v}^{n+1-j}$$
$$= \Delta t \left(-\nabla \bar{p}^{n+1} + \sum_{j=1}^{k} b_j N(\mathbf{v}^{n+1-j}) + L(\mathbf{v}^{n+1}) + \mathbf{f}^{n+1} \right), \quad (6.5.7)$$

where the coefficients β_j are given by Equation (6.2.13) and the b_j are the standard coefficients of the AB scheme, Equation (3.2.22).

6.5.1 Fractional-Step Method

The fractional-step method is a splitting technique (as introduced in Chapter 3) that decouples the nonlinear, divergence-free, and viscous terms in the Navier–Stokes equations into separate subproblems. It requires solution of the pressure Poisson equation (6.1.6) with the Neumann boundary condition (6.1.7). Formally, since the pressure field has to enforce continuity at the new time level, the former boundary condition must be expressed at t^{n+1}. This couples velocity and pressure. To avoid this link, which would need some (costly) iterations, we proceed in three steps to split the computation.

The first step evaluates an intermediate field by the ABk scheme

$$\hat{\mathbf{v}} = -\sum_{j=1}^{k} \beta_{k-j} \mathbf{v}^{n+1-j} + \Delta t \left[\sum_{j=1}^{k} b_j N(\mathbf{v}^{n+1-j}) + \mathbf{f}^{n+1} \right] \quad (6.5.8)$$

The second step renders the velocity field incompressible by adding the pressure gradient

$$\hat{\hat{\mathbf{v}}} = \hat{\mathbf{v}} - \Delta t \, \nabla p^{n+1}. \quad (6.5.9)$$

The third and final step is the viscous step:

$$\beta_k \mathbf{v}^{n+1} = \hat{\hat{\mathbf{v}}} + \Delta t L(\mathbf{v}^{n+1}). \quad (6.5.10)$$

For more generality, we will assume that the flow satisfies inhomogeneous boundary conditions

$$\mathbf{v}^{n+1} = \mathbf{g} \quad \text{on } \partial\Omega. \quad (6.5.11)$$

6.5. Projection Methods

The splitting scheme is built on two further hypotheses: First, the velocity field $\hat{\mathbf{v}}$ satisfies the incompressibility condition, and therefore

$$\operatorname{div} \hat{\mathbf{v}} = 0. \tag{6.5.12}$$

Second, we assume that the same field $\hat{\mathbf{v}}$ satisfies the inhomogeneous Dirichlet condition in the normal direction $\mathbf{n}$ to the boundary:

$$\hat{\mathbf{v}} \cdot \mathbf{n} = \mathbf{g} \cdot \mathbf{n}. \tag{6.5.13}$$

The resulting Poisson equation becomes

$$\Delta p = \operatorname{div}\left(\frac{\hat{\mathbf{v}}}{\Delta t}\right), \tag{6.5.14}$$

with the boundary condition obtained from (6.5.9),

$$\Delta t \, \nabla p^{n+1} \cdot \mathbf{n} = -(\mathbf{g} - \hat{\mathbf{v}}) \cdot \mathbf{n} \quad \text{on } \partial\Omega. \tag{6.5.15}$$

This scheme is prone to time-splitting errors, because the Poisson pressure equation is solved with the incorrect boundary condition (6.5.15). If we assume furthermore that the wall is fixed and impermeable, the boundary condition reduces to $\Delta t \, \nabla p^{n+1} \cdot \mathbf{n} = 0$, which is appropriate for an inviscid fluid but far away from physical reality for a viscous flow. This way of solving the pressure may lead to errors of $O(1)$. The splitting error induces a boundary layer for the numerical divergence.

Let us estimate the boundary-layer thickness. Applying the divergence to Equation (6.5.7) and recalling the notation introduced in Equation (6.5.2) for the divergence of the velocity field, we obtain

$$\beta_k \mathcal{D}^{n+1} + \sum_{j=1}^{k} \beta_{k-j} \mathcal{D}^{n+1-j}$$
$$= \Delta t \left[-\Delta p^{n+1} + \operatorname{div}\left(\sum_{j=1}^{k} b_j N(\mathbf{v}^{n+1-j}) + L(\mathbf{v}^{n+1}) + \mathbf{f}^{n+1} \right) \right]. \tag{6.5.16}$$

Combination of the three steps (6.5.8)–(6.5.10) shows that the pressure equation takes all explicit terms into account, while at the new time level the divergence equation is

$$\beta_k \mathcal{D}^{n+1} - \Delta t \, \Delta \mathcal{D}^{n+1} = 0. \tag{6.5.17}$$

We may conclude that there exists a numerical boundary layer of thickness $l = O(\sqrt{\Delta t})$ or, in dimensional form, $l = O(\sqrt{\nu \Delta t})$. One can show that the velocity error at the wall is proportional to l^2, that is, to Δt. If the pressure computation is failing by an error $O(1)$ at the wall, the velocity is still of first order in time. If now the pressure computation is of first order in time, the velocity will be of second order.

If we again assume that the boundary is fixed and impermeable, the boundary condition (6.1.7) reduces to

$$\mathbf{n} \cdot \nabla p = \mathbf{n} \cdot (\Delta \mathbf{v} + \mathbf{f}) \qquad \text{on } \partial\Omega. \tag{6.5.18}$$

Decomposing as in (5.4.9) the velocity field into an irrotational part $\mathbf{v}_I$ and a solenoidal part $\mathbf{v}_S$ such that

$$\nabla \times \mathbf{v}_I = 0, \tag{6.5.19}$$

$$\text{div } \mathbf{v}_S = 0, \tag{6.5.20}$$

we can rewrite the pressure boundary condition as

$$\mathbf{n} \cdot \nabla p = \mathbf{n} \cdot [\nabla \text{ div } \mathbf{v}_I - \nabla \times (\nabla \times \mathbf{v}_S) + \mathbf{f}] \qquad \text{on } \partial\Omega. \tag{6.5.21}$$

If we impose the condition that $\mathbf{v}_I$ vanishes in an incompressible flow, the pressure boundary condition becomes

$$\mathbf{n} \cdot \nabla p = \mathbf{n} \cdot [-\nabla \times (\nabla \times \mathbf{v}) + \mathbf{f}] \qquad \text{on } \partial\Omega. \tag{6.5.22}$$

Following Karniadakis et al. [219], we combine Equation (6.5.15) with the previous condition (6.5.22) and approximate the r.h.s. by an ABk scheme to obtain

$$\mathbf{n} \cdot \nabla p^{n+1} = -\mathbf{n} \cdot \left(\frac{1}{\Delta t}(\mathbf{g} - \hat{\mathbf{v}}) + \sum_{j=1}^{k} b_j \nabla \times (\nabla \times \mathbf{v}^{n+1-j}) - \mathbf{f}^{n+1} \right) \qquad \text{on } \partial\Omega. \tag{6.5.23}$$

The usual practice is to use a $\mathbb{P}_N$–$\mathbb{P}_N$ approximation for velocity and pressure, thereby specifying the same discretization for every step of the splitting scheme with the same solver and preconditioner. This approach eases the programming effort dramatically. We note that each time step requires only four elliptic solves.

In summary, the algorithm proceeds as follows:

- Compute the intermediate velocity $\hat{\mathbf{v}}$ by (6.5.8).
- Solve for the pressure (6.5.14) with boundary condition (6.5.23), and add the pressure gradient to the velocity $\hat{\mathbf{v}}$, obtaining (6.5.9).
- Solve the Helmholtz equations (6.5.10) to produce the final velocity at time t^{n+1}.

A deep and careful analysis of the Stokes solver associated with the previous algorithm was carried out by Leriche and Labrosse [242] for Chebyshev collocation. They showed by computing the eigenvalue spectra that the boundary condition (6.5.22) keeping the rotational part of the velocity field does preserve the elliptic character of the Stokes problem. The explicit time treatment of the rotational boundary term by first- and second-order time schemes leads to stable schemes with the viscous term being implicitly integrated. For higher- (third- and fourth-) order schemes, we recover a time-step restriction of $O(N^{-4})$.

6.5.2 Pressure Correction Method

The pressure correction method is basically a predictor–corrector procedure between the velocity and the pressure fields. The idea resides in the incorporation of the old-time-level pressure gradient to integrate the momentum equations. The resulting velocity field is obviously not divergence-free. In a correction step, a pressure correction that is the difference between the new- and the old-time-level pressures is computed and added to the preceding velocity to enforce the continuity constraint. The pressure correction was first introduced in the FD framework by Goda [159]. Unlike the splitting scheme, this pressure correction scheme yields a divergence-free field at the end of the time step.

The pressure correction scheme advances in two steps. First, we compute an intermediate velocity field $\hat{\mathbf{v}}$ by

$$\beta_k \hat{\mathbf{v}} + \sum_{j=1}^{k} \beta_{k-j} \mathbf{v}^{n+1-j} = \Delta t \left(-\nabla p^n + \sum_{j=1}^{k} b_j N(\mathbf{v}^{n+1-j}) + L(\hat{\mathbf{v}}) + \mathbf{f}^{n+1} \right), \quad (6.5.24)$$

with the old time-level pressure p^n. Then, defining the pressure correction $\delta p = p^{n+1} - p^n$ and subtracting (6.5.24) from (6.5.7), we obtain

$$\beta_k (\mathbf{v}^{n+1} - \hat{\mathbf{v}}) = \Delta t \{-\nabla \delta p + [L(\mathbf{v}^{n+1}) - L(\hat{\mathbf{v}})]\}. \quad (6.5.25)$$

Applying the divergence to this relation and assuming that the linear operators commute (this might be not the case for the discrete operators), we obtain a Poisson equation for the pressure correction:

$$\Delta \delta p = \text{div} \left(\frac{\beta_k \hat{\mathbf{v}}}{\Delta t} - L(\hat{\mathbf{v}}) \right). \quad (6.5.26)$$

The final velocity is given by

$$\mathbf{v}^{n+1} = \hat{\mathbf{v}} - \frac{\Delta t}{\beta_k} \nabla \delta p. \quad (6.5.27)$$

The integration of the momentum equation (6.5.24) is performed with the boundary condition $\hat{\mathbf{v}} = \mathbf{g}$. If we take this velocity condition into account, the pressure correction boundary condition comes from (6.5.27) as $\mathbf{n} \cdot \nabla \delta p = 0$. This condition does not allow much change in the boundary layer and therefore does not ensure the correct dynamics in that location.

Timmermans et al. [379, 381] overcame the disadvantages of the previous method by the following modifications. The first step still integrates (6.5.24) with the boundary condition $\hat{\mathbf{v}} = \mathbf{g} = \mathbf{v}^{n+1}$. The second step computes a scalar correction field q (replacing δp; note that q is not the pressure). The relationship for q is generated by the application of the divergence to

$$\mathbf{v}^{n+1} = \hat{\mathbf{v}} - \frac{\Delta t}{\beta_k} \nabla q. \tag{6.5.28}$$

Imposing the incompressibility constraint at time t^{n+1}, we obtain

$$\Delta q = \mathrm{div}\left(\frac{\beta_k \hat{\mathbf{v}}}{\Delta t}\right). \tag{6.5.29}$$

The gradient of the correction is added through Equation (6.5.28). If we compare Equation (6.5.28) with Equation (5.4.9), we see that the gradient of the correction corresponds to the irrotational part of the velocity. Consequently, Equation (6.5.25), which is still valid, can be written with the help of (6.5.28) and (5.4.9) as

$$\nabla q = \nabla(p^{n+1} - p^n) - \nabla \mathrm{div}\, \mathbf{v}^{n+1} + \nabla \times \nabla \mathbf{v}^{n+1} + \nabla \mathrm{div}\, \hat{\mathbf{v}} - \nabla \times \nabla \hat{\mathbf{v}}. \tag{6.5.30}$$

From this expression, the final pressure field is calculated as

$$p^{n+1} = p^n + q - \mathrm{div}\, \hat{\mathbf{v}}. \tag{6.5.31}$$

The last term of the r.h.s. of Equation (6.5.31) contributes to the pressure adaptation in the boundary layer.

The OIFS method is applied to the Navier–Stokes equations in order to decouple the advection problem from the viscous part, which will include the pressure gradient at the previous time level. So far, only the BDF2 scheme has been applied to the implicit Stokes operator.

The algorithm evolves by successive steps:

- Solve the initial-value problem (6.4.27) by a three-step explicit Taylor–Galerkin method, which is a generalization of the two-step method introduced in Section 3.10. This scheme does not suffer from the weak instability suffered by the two-step scheme.

6.6. Stabilizing Unsteady Flows

- Solve the OIFS version of Equation (6.5.24) via a BDF2 method, and obtain $\hat{\mathbf{v}}$.
- Solve the Poisson Equation (6.5.26) for the pressure correction.
- Use Equation (6.5.27) to compute the final incompressible velocity field.
- Update the pressure by Equation (6.5.31).

Timmermans et al. [381] showed that the correction method is second-order accurate in time for the velocity computation. The performance of the algorithm for the pressure is not as good, however. The pressure time accuracy is between first and second order, depending on the geometrical complexity of the problem.

Heinrichs [192, 194] proposed a third-order pressure correction scheme for the Stokes problem that can be extended to the Navier–Stokes equations. Equation (6.5.24) is computed with a BDF3/AB3 scheme by setting $k = 3$, and the pressure gradient ∇p^n is replaced by $\nabla \bar{p}^{n+1}$, where the pressure $\bar{p}^{n+1}$ is obtained by second-order time extrapolation:

$$\bar{p}^{n+1} = 2p^n - p^{n-1}. \qquad (6.5.32)$$

The pressure correction is now defined by $\delta p = p^{n+1} - \bar{p}^{n+1}$ and results from the Poisson equation (6.5.26). The new velocity is given by (6.5.27).

6.6 Stabilizing Unsteady Flows

In Chapters 3, 4, and 5 we have discussed the use of stabilization techniques based on bubble functions and shifted grids to control spurious oscillations in convection-dominated problems and in the Stokes problem. Another approach to stabilization is based on the idea of filtering. This technique has been widely used in high-order applications to hyperbolic problems [106, 254] and, more recently, in spectral simulations of incompressible flows [47, 130]. The basic idea is to suppress at the end of each time step any spurious modes that may pollute an otherwise accurate solution.

In [130], a stabilizing filter was developed for unsteady incompressible simulations based on the $\mathbb{P}_N$–$\mathbb{P}_{N-2}$ SEM (5.2.32)–(5.2.33). Using any of the techniques developed earlier in the chapter [e.g., (6.2.33)], one begins by computing an intermediate solution, $(\tilde{\underline{\mathbf{v}}}, \underline{p}^{n+1})$, to the unsteady Stokes problem

$$\mathbf{H}\tilde{\underline{\mathbf{v}}} - \mathbf{D}^T \underline{p}^{n+1} = \mathbf{M}\underline{\mathbf{f}}^n, \qquad \mathbf{D}\tilde{\underline{\mathbf{v}}} = 0, \qquad (6.6.1)$$

where $\underline{\mathbf{f}}^n$ represents the contribution of the nonlinear terms. The resultant velocity field is then filtered to generate the solution at time t^{n+1},

$$\underline{v}^{n+1} = F_\alpha \underline{\tilde{v}}. \tag{6.6.2}$$

The advantage of filter-based stabilization is that one can use all of the technology developed for the symmetric unsteady Stokes problem (6.6.1) without change to either discretization or solver.

For reasons of efficiency, it is desirable that the action of the filter be local, yet preserve interelement continuity. In [130], these conditions are met by the following construction. For any integer N let $\pi_{N,j}$ represent the set of Lagrangian interpolation polynomials associated to the GLL quadrature rule with $N+1$ nodes $\{\xi_{N,k}\}$ given by (B.2.9). Here, we slightly modify the notation to indicate the polynomial degree (N), because we are going to use GLL schemes with different orders. Let I_N^M denote the $(M+1) \times (N+1)$ matrix operator with components

$$\left(I_N^M\right)_{ij} = \pi_{N,j}(\xi_{M,i}), \qquad 0 \leq i \leq M, \quad 0 \leq j \leq N. \tag{6.6.3}$$

Consider, in the same spirit, the $(N+1) \times (N+1)$ matrix operator

$$\Pi_{N-1} = I_{N-1}^N I_N^{N-1}, \tag{6.6.4}$$

where the elements of the matrix product are defined by (6.6.3).

Any polynomial $u_N \in \mathbb{P}_N(\hat{\Omega})$ may be uniquely decomposed into

$$u_N(x) = \sum_{k=0}^N u_k \, \pi_{N,k}(x), \qquad u_k = u_N(\xi_{N,k}).$$

Therefore, the vector $\Pi_{N-1}\underline{u}$ with $\underline{u} := (u_0, \ldots, u_N)^T$ gives the nodal values on the $\{\xi_{N,k}\}$ grid of the polynomial $I_{N-1}u_N$, where I_{N-1} denotes the Lagrangian interpolation operator of degree $N-1$ on $\{\xi_{N-1,k}\}$. This interpolation automatically eliminates some high-frequency contribution as discussed below; hence the filtering process. One may view Π_{N-1} as a pseudoprojector, in that in any case $\Pi_{N-1}^2 \neq \Pi_{N-1}$. The one-dimensional filter is a combination of the pseudoprojector (6.6.4) with the identity matrix, that is,

$$\hat{F}_\alpha := \alpha \Pi_{N-1} + (1-\alpha) I_N^N. \tag{6.6.5}$$

In higher space dimensions, one simply uses the tensor-product form within each element, $F_\alpha^e := \hat{F}_\alpha \otimes \cdots \otimes \hat{F}_\alpha$. The free parameter α determines the strength of the filter, with $\alpha = 1$ corresponding to a full pseudoprojection

6.6. Stabilizing Unsteady Flows

onto $\mathbb{P}_{N-1}$. The choice $0.05 < \alpha < 0.30$ typically yields a smoother result [see Figure 6.8.2(c, d)]. The interpolation-based procedure ensures that interelement continuity is preserved, and because the interpolation error $\|\tilde{v} - \Pi_{N-1}\tilde{v}\|$ tends to zero exponentially fast as $N \to \infty$, spectral convergence is not compromised.

Heuristically, because the points $\{\xi_{N,i}\}$ interlace $\{\xi_{N-1,i}\}$, the pseudoprojector tends to suppress high-frequency oscillations. To quantify this, consider $v(\xi) := u(\xi) - I_{N-1}u(\xi)$ for $u(\xi) \in \mathbb{P}_N(\hat{\Omega})$. Because $v(\xi) \in \mathbb{P}_N$ and $v(\xi_{N-1,i}) = 0$ $(i = 0, \ldots, N-1)$, it follows that

$$v(\xi) = \tilde{u}_N (1 - \xi^2) L'_{N-1}(\xi) = \hat{u}_N [L_N(\xi) - L_{N-2}(\xi)] =: \hat{u}_N \phi_N(\xi),$$

where $\tilde{u}_N$ and $\hat{u}_N$ are constants. Part of this relationship may be explained taking into account the properties (B.1.11) and (B.1.20) of Legendre polynomials. Here, ϕ_N is an element in the basis

$$\phi_0 = \frac{1-\xi}{2}, \qquad \phi_1 = \frac{1+\xi}{2}, \qquad \phi_k = L_k - L_{k-2}, \quad k \geq 2, \quad (6.6.6)$$

which has been used by several authors (e.g., [355, 47]), and which has been presented in Section 2.3.2 devoted to the p-version of finite elements; $\hat{u}_N$ is the associated basis coefficient. One can change from (6.6.6) to the usual Lagrangian basis by recognizing that, for any $u \in \mathbb{P}_N$,

$$u_i := u(\xi_{N,i}) = \sum_{k=0}^{N} \hat{u}_k \phi_k(\xi_{N,i}), \qquad i = 1, \ldots, N. \quad (6.6.7)$$

Hence, setting $(Z)_{ik} := \phi_k(\xi_{N,i})$, one has $\underline{u} = Z\underline{\hat{u}}$ as the one-dimensional transformation from wave (or modal) space to physical (or nodal) space.

As suggested by Boyd [47], the basis (6.6.6) can be used to construct an arbitrary filter that scales each $\hat{u}_k$, $k = 2, \ldots, N$. With $\Sigma = \text{diag}(\sigma_k)$, with $\sigma_0 = \sigma_1 = 1$, and with σ_k as the desired reduction at each wavenumber $k \geq 2$, the general filter is given by $\hat{F}_\Sigma := Z \Sigma Z^{-1}$. With the linear combination (6.6.5), $\hat{F}_\alpha$ acts to scale $\hat{u}_N$ by $1 - \alpha$, leaving the other coefficients unchanged.

The stabilizing role of the filter can be understood by considering its application to the one-dimensional unsteady advection–diffusion problem (1.9.6) with $c = 1$ and the addition of a forcing term f,

$$u_t + u_x = \nu u_{xx} + f, \qquad u(0) = u(1) = 0. \quad (6.6.8)$$

If we assume that f is time-independent, u eventually evolves to the solution of the steady-state problem (3.6.1), which has well-known stability problems as

$\nu \to 0$ (e.g., see [62]). Discretizing (6.6.8) by the SEM in space and AB3CN in time, and then coupling with the filter, we have

$$H\underline{\tilde{u}} = H_R \underline{u}^n - C\left(\tfrac{23}{12}\underline{u}^n - \tfrac{16}{12}\underline{u}^{n-1} + \tfrac{5}{12}\underline{u}^{n-2}\right) + M\underline{f}, \qquad \underline{u}^{n+1} = F_\alpha \underline{\tilde{u}}, \tag{6.6.9}$$

where $H = (\nu/2)K + (1/\Delta t)M$ and $H_R = -(\nu/2)K + (1/\Delta t)M$ are discrete Helmholtz operators and C is the convection operator.

Equation (6.6.9) can be viewed as a fixed-point iteration with a solution satisfying

$$\left[\nu K + C + H\left(F_\alpha^{-1} - I\right)\right]\underline{u}^\infty = M\underline{f}. \tag{6.6.10}$$

For any Galerkin formulation, C is skew-symmetric and therefore singular if the number of variables is odd (the spurious mode in the single spectral element case being $L_N - L_0$). On the other hand, the eigenvalues of $F_\alpha^{-1} - I$ are $\{0, 0, \ldots, 0, \tfrac{\alpha}{1-\alpha}\}$ [the nonzero eigenmode being $\phi_N(x) := L_N(x) - L_{N-2}(x)$]. The stabilizing term $H(F_\alpha^{-1} - I)$ controls the growth of the component involving L_N and prevents (6.6.10) from blowing up as $\nu \to 0$.

As noted above, one can easily suppress more elements in this basis in order to construct smoother filters, as suggested, for example, in [46, 399]. However, experiences in [130] and asymptotic analysis [$\nu \to 0$ in (6.6.10)] indicate that a slight suppression of just the Nth mode is sufficient to stabilize the $\mathbb{P}_N$–$\mathbb{P}_{N-2}$ method at moderate to high Reynolds numbers.

6.7 Arbitrary Lagrangian–Eulerian Formulation and Free-Surface Flows

Most of the time, fluid mechanics problems are expressed by using the Eulerian formulation, whereas solid mechanics problems rely on the Lagrangian formulation. However, in some cases, such as free-surface flows or fluid–structure interaction, neither of these approaches is completely adequate because of their intrinsic limitations.

In the Eulerian (or spatial) formulation, the reference system is fixed in space. The numerical methods use elements also fixed in space. If the boundary of the fluid domain deforms (as often occurs in materials-processing systems) and does not coincide with an element side or face, the application of the boundary conditions at the right location becomes difficult and needs some local interpolation on the geometry and on the conditions themselves.

6.7. Arbitrary Lagrangian–Eulerian Formulation

In the Lagrangian (or material) formulation, the reference system is attached to material points of the fluid (or, more generally, of the continuous medium). Within the discretization method, the elements deform with the material and can be excessively distorted. If the aspect ratio of the elements reaches values preventing good spatial convergence, some reconstruction technique is called for to regenerate a suitable mesh.

6.7.1 ALE *Formulation*

The arbitrary Lagrangian–Eulerian (ALE) formulation attempts to combine the advantages of both techniques and to avoid the drawbacks of each of them. The reference system is not fixed a priori in space or attached to the material. It may be considered as a computational reference system, which can be chosen quite arbitrarily. The method was first proposed in the FD context by the Los Alamos group (Hirt et al. [205], Pracht [315]). In the eighties, the method was applied within the FE community (Hughes et al. [211], Donea et al. [108], Ramaswamy and Kawahara [322]). More recently, Ho and Patera [207, 208] brought the ALE technique into the spectral-element frame. Before we discuss the discretization, we will first extend the concept of material time derivative in an arbitrary frame of reference and then present the variational formulation of the free-surface problem.

The material derivative describes the time evolution of variables attached to material points followed in their motion. As was already mentioned, it is interesting to track the evolution of variables or quantities at points characterized by a proper velocity field $\mathbf{w}(\mathbf{x}, t)$ different from the velocity $\mathbf{v}(\mathbf{x}, t)$ of the material particles. Typically, in a practical computation, $\mathbf{w}$ will be the mesh velocity. Let us denote by $\delta/\delta t$ the time derivative when a fictitious particle moving with the velocity $\mathbf{w}(\mathbf{x}, t)$ is followed; this derivative will be called the ALE derivative. It suffices to replace D/Dt and $\mathbf{v}$ by $\delta/\delta t$ and $\mathbf{w}$, respectively, in the definition of the material derivative (1.2.5) to obtain for the function f

$$\frac{\delta f}{\delta t} = \frac{\partial f}{\partial t} + \mathbf{w} \cdot \nabla f. \tag{6.7.1}$$

If $\mathbf{w} = 0$, the mesh is fixed, and we recover the Eulerian description where $\delta/\delta t$ is the classical partial derivative with respect to time. If $\mathbf{w} = \mathbf{v}$, we obtain the Lagrangian description, and $\delta/\delta t$ is the particle or substantial derivative. Finally, if $\mathbf{w} \neq \mathbf{v} \neq 0$, the mesh moves with its own velocity, and the time derivative takes that information into account. Following Truesdell and Toupin [387], the associated Reynolds transport theorem for the time change of a volume integral is

$$\frac{\delta}{\delta t} \int_{\Omega(t)} f \, dV = \int_{\Omega(t)} \left(\frac{\delta f}{\delta t} + f \operatorname{div} \mathbf{w} \right) dV = \int_{\Omega(t)} \left(\frac{\partial f}{\partial t} + \operatorname{div}(f\mathbf{w}) \right) dV. \quad (6.7.2)$$

The interpretation of the last equality needs some development and comment. Indeed, according to Truesdell and Toupin, one can use the divergence theorem to write

$$\frac{\delta}{\delta t} \int_{\Omega(t)} f \, dV = \int_{\Omega} \frac{\partial f}{\partial t} dV + \int_{\partial \Omega} f(\mathbf{w} \cdot \mathbf{n}) \, dS. \quad (6.7.3)$$

Here, the first term of the right-hand side acts on the spatial volume Ω, which at time t corresponds to the configuration of the material volume $\Omega(t)$. The notation $\delta/\delta t$ reminds us that the volume of integration is material with respect to the velocity $\mathbf{w}$.

At this stage, let us introduce the relative velocity $\mathbf{c}$ of the continuous media with respect to the motion of the reference frame defined by $\mathbf{w}$ such that

$$\mathbf{v} = \mathbf{c} + \mathbf{w}. \quad (6.7.4)$$

We are able to determine the relation between the two time derivatives D/Dt and $\delta/\delta t$. We obtain

$$\frac{Df}{Dt} = \frac{\delta f}{\delta t} + \mathbf{c} \cdot \nabla f. \quad (6.7.5)$$

The kinematics of volume integrals defined on the same material volume but with respect to different reference frames produce the relation

$$\frac{d}{dt} \int_{\Omega(t)} f \, dV = \frac{\delta}{\delta t} \int_{\Omega(t)} f \, dV + \int_{\partial \Omega(t)} f (\mathbf{c} \cdot \mathbf{n}) \, dS. \quad (6.7.6)$$

6.7.2 Free-Surface Conditions

A free surface [see Equation (1.6.11)] is a special case of an interface. When two immiscible fluids are in contact, an interface is created that, for 3D problems, is a general surface whose geometry may be quite complicated. Since the interface is in mechanical equilibrium, the interface condition imposes that the vector sum of the contact forces of the two fluids with respective subscripts I and II vanishes:

$$\mathbf{t}_I + \mathbf{t}_{II} = 0. \quad (6.7.7)$$

With the definition (1.3.2) and because $\mathbf{n}_I = -\mathbf{n}_{II}$, we have

$$\sigma_I \mathbf{n}_I = \sigma_{II} \mathbf{n}_I. \quad (6.7.8)$$

6.7. Arbitrary Lagrangian–Eulerian Formulation

In this relation the stress tensor σ defines the rheological behavior of the fluids. At an interface, both of them are viscous Newtonian.

When a free surface is present, one of the fluids (II) is assumed to be air, considered as an inviscid fluid. Recall the constitutive relation for a viscous Newtonian incompressible fluid (1.6.1) for I, at a free surface with the normal unit vector **n** pointing out of the viscous fluid. The interface condition is

$$(-p_\mathrm{I}\mathbf{I} + 2\mu\mathbf{d}_\mathrm{I})\mathbf{n} = -p_\mathrm{II}\mathbf{n}. \tag{6.7.9}$$

The free-surface conditions for a 2D problem are obtained by projection of the previous relation on the normal **n** and on the tangent vector τ to the surface:

$$-p_\mathrm{I} + 2\mu(\mathbf{d}_\mathrm{I}\mathbf{n}) \cdot \mathbf{n} = -p_\mathrm{II}, \tag{6.7.10}$$

$$2\mu(\mathbf{d}_\mathrm{I}\mathbf{n}) \cdot \tau = 0. \tag{6.7.11}$$

If, furthermore, surface tension effects play an important role, as they do in coating flows for example, a generalization of the Laplace relation gives

$$-p_\mathrm{I} + 2\mu(\mathbf{d}_\mathrm{I}\mathbf{n}) \cdot \mathbf{n} = -p_\mathrm{II} - \gamma\left(R_1^{-1} + R_2^{-1}\right), \tag{6.7.12}$$

where in this section, γ denotes the surface tension coefficient, and R_1 and R_2 are the principal radii of curvature of the surface. The quantity between parentheses multiplying the surface tension coefficient is equal to $2\,\Xi$, with Ξ denoting the mean curvature of the surface, in this section. The sign of Ξ depends on the concavity ($-$) or the convexity ($+$) of the surface. More elaborate formulae are given by Ericksen in an appendix to Truesdell and Toupin's presentation.

The relevant dimensionless number when surface tension effects are important is the Weber number

$$We := \frac{\gamma}{\rho L U^2}, \tag{6.7.13}$$

which compares the surface tension force with inertia.

In the past decade, the surface tension effects on fluid flow have been incorporated in the continuum model itself. The developments are due to Brackbill et al. [49] and Lafaurie et al. [235]. The state of the art of the DNS of free-surface flows is reviewed by Scardovelli and Zaleski [343]. The capillary force acting on a free surface is modeled by an additional term in the Navier–Stokes equations (1.6.3), expressed by $2\,\Xi\gamma\delta_S\mathbf{n}$, where δ_S is a Dirac delta function applied only on the surface S. The mean curvature is obtained as the divergence of the normal,

$$2\Xi = -\nabla_S \cdot \mathbf{n}, \tag{6.7.14}$$

with ∇_S being the gradient operator restricted to the surface. This approach leads to the definition of the capillary tensor

$$\boldsymbol{\Gamma} = -\gamma(\mathbf{I} - \mathbf{n} \otimes \mathbf{n})\delta_S \qquad (6.7.15)$$

such that

$$2\,\Xi\gamma\delta_S\mathbf{n} = -\operatorname{div}\boldsymbol{\Gamma}. \qquad (6.7.16)$$

If a conservative form of the momentum equations is needed, then this last expression of the surface tension effects may be easily incorporated. This way of dealing with the free surface opens the door to the level-set approach, which is recognized as a powerful alternative to existing methods (see the book by Sethian [353]).

We note that if the 2D free surface may be explicitly represented by the relation

$$F(x, y, t) = y - f(x, t) = 0, \qquad (6.7.17)$$

then the components of $\mathbf{n}$ may be written as

$$n_x = \frac{1}{(1 + f'^2)^{1/2}} \quad \text{and} \quad n_y = \frac{-f'}{(1 + f'^2)^{1/2}}, \qquad (6.7.18)$$

with f' denoting the partial derivative of f with respect to x. The use of (6.7.17) requires that the free surface be represented by a uniform function. For example, this excludes the case of a breaking wave on a slanted beach, where the free-surface is obviously multiform. For transient free-surface flows, some algorithms take advantage of the kinematical constraint expressing that the free surface remains a material surface containing the same fluid particles as time elapses. Using (6.7.17), we get

$$\frac{dF}{dt} = 0, \quad \text{i.e.} \quad \frac{\partial f}{\partial t} + u\frac{\partial f}{\partial x} = v. \qquad (6.7.19)$$

6.7.3 Variational Formulation of Free-Surface Flows

The variational formulation is built up by the use of the linear momentum conservation law, Equation (1.3.1), recast in the ALE frame. Indeed, if we proceed as usual, then (as in Section 6.4) we will run into trouble because the appropriate functional spaces will be defined on $\Omega(t)$, a time-dependent domain, and the time derivative will involve the spatial frame of reference. The technical difficulty comes from the fact that the test functions cannot be chosen invariable with respect to time, because they should vanish on the moving part of the

6.7. Arbitrary Lagrangian–Eulerian Formulation

boundary, where essential conditions are applied. Therefore, we define them also in the ALE frame for full consistency of the method. The test functions **u** for the velocity field are independent of time in the reference system, and we can infer that in the current configuration they satisfy the relation

$$\frac{\delta \mathbf{u}}{\delta t} = 0. \tag{6.7.20}$$

Let us denote by u a scalar test function obeying the condition (6.7.20). The Reynolds transport theorem (6.7.2) allows us to write successively

$$\int_{\Omega(t)} u \left[\frac{\delta f}{\delta t} + f \operatorname{div} \mathbf{w} \right] dV$$

$$= \int_{\Omega(t)} \left(\frac{\delta(uf)}{\delta t} + uf \operatorname{div} \mathbf{w} \right) dV = \frac{\delta}{\delta t} \int_{\Omega(t)} uf \, dV. \tag{6.7.21}$$

We may generalize the previous relation by replacing the function f by vector fields like the velocity **v** and the test function u by its vector counterpart **u**.

The weak ALE formulation of the transient Navier–Stokes equations for free-surface flows is obtained through the following steps. The integral form of the momentum equation is given by the use of Equations (6.7.6) and (1.3.2) and of the divergence theorem as

$$\frac{d}{dt} \int_{\Omega(t)} \rho \mathbf{v} \, dV = \frac{\delta}{\delta t} \int_{\Omega(t)} \rho \mathbf{v} \, dV + \int_{\Omega(t)} \rho \operatorname{div}(\mathbf{vc}) \, dV$$

$$= \int_{\Omega(t)} (\operatorname{div} \boldsymbol{\sigma} + \rho \mathbf{f}) \, dV. \tag{6.7.22}$$

The functional space for the velocity is a subset of $H_0^1(\Omega(t))^d$, where the test functions satisfy (6.7.20). For the pressure, the test function q will be in $L^2(\Omega(t))$, since the reference pressure level will be imposed by the normal free-surface condition. Multiplying under each integral by the test functions and taking the Reynolds theorem (6.7.21) in its vector form into account and (6.7.4), we have for the weak form: Find $(\mathbf{v}(t), p(t)) \in H_0^1(\Omega(t))^d \times L^2(\Omega(t))$ such that for almost every $t \in (0, T)$

$$\frac{\delta}{\delta t} \int_{\Omega(t)} \mathbf{u} \cdot \mathbf{v} \, dV + \int_{\Omega(t)} \mathbf{u} \cdot \operatorname{div}(\mathbf{vv} - \mathbf{vw})$$

$$= \int_{\Omega(t)} \left(\mathbf{f} \cdot \mathbf{u} + \frac{p}{\rho} \operatorname{div} \mathbf{u} - 2\nu \mathbf{d} \cdot \nabla \mathbf{u} \right) dV + \int_{\partial \Omega_s(t)} \frac{\gamma}{\rho} 2 \Xi \mathbf{n} \cdot \mathbf{u} \, dS$$

$$\forall \mathbf{u} \in H_0^1(\Omega(t))^d, \tag{6.7.23}$$

$$- \int_{\Omega(t)} q \operatorname{div} \mathbf{v} \, dV = 0 \quad \forall q \in L^2(\Omega(t)). \tag{6.7.24}$$

In the above relations, $\partial\Omega_S(t)$ denotes the free surface. Upon integration of this term by parts, one obtains an expression involving the contact angles of the free surface with solid walls. We could also deal with gradients of the surface tension coefficient as they appear in the Marangoni effect.

The free surface is obviously a material surface. We require there that the normal mesh velocity be that of the interface:

$$\mathbf{w} \cdot \mathbf{n} = \mathbf{v} \cdot \mathbf{n}. \tag{6.7.25}$$

That kinematical condition does not require anything of the tangential mesh velocity. However, imposition of the condition

$$\mathbf{w} \cdot \boldsymbol{\tau} = 0 \tag{6.7.26}$$

turns out to be computationally interesting because it minimizes the mesh deformation. On solid fixed walls, the mesh velocity is set to

$$\mathbf{w} = 0. \tag{6.7.27}$$

Several approaches are possible for computing the mesh velocity inside the domain. Ho and Patera [207, 208] proposed to solve an elliptic problem of the form

$$\mathcal{E}\mathbf{w} = 0, \tag{6.7.28}$$

where the operator $\mathcal{E}$ comes from an elasticity or a Stokes problem. More specifically, the steady equilibrium equations for linear infinitesimal elasticity are given by

$$\operatorname{div} \sigma_E + \rho \mathbf{f} = 0. \tag{6.7.29}$$

The constitutive relationship for the stress tensor implies the deformation tensor ϵ, which is expressed in terms of the displacement gradient tensor

$$\sigma_E = \lambda \operatorname{tr} \epsilon \, \mathbf{I} + 2\mu \, \epsilon, \tag{6.7.30}$$

$$\epsilon = \frac{1}{2}[\nabla \mathbf{u} + (\nabla \mathbf{u})^T], \tag{6.7.31}$$

where the vector $\mathbf{u}$ represents the displacement, the material parameters λ and μ being the Lamé coefficients. Setting $\mathbf{f} = 0$ and replacing the displacement $\mathbf{u}$ by the velocity $\mathbf{w}$ in the elasticity problem, we are led to the solution of (6.7.28). We note that the elasticity solution in fact depends on only *one* material parameter,

6.7. Arbitrary Lagrangian–Eulerian Formulation

since the Lamé coefficients can be rewritten by using Young's modulus E and the Poisson coefficient ν:

$$\mu = \frac{E}{2(1+\nu)}, \quad \lambda = \frac{\nu E}{(1+\nu)(1-2\nu)}, \tag{6.7.32}$$

where $0 \leq \nu < 1/2$. For the incompressible elastic case ($\nu = 0.5$), the coefficient λ goes to infinity. Nevertheless, we can introduce the deviatoric part of the infinitesimal strain tensor

$$\epsilon^D = \epsilon - \frac{1}{3}\operatorname{tr}\epsilon. \tag{6.7.33}$$

The stress tensor becomes

$$\sigma_E = \sigma^D + p\,\mathbf{I}, \tag{6.7.34}$$

with the corresponding definitions

$$p = (3\lambda + 2\mu)\operatorname{div}\mathbf{u}, \tag{6.7.35}$$

$$\sigma^D = 2\mu\epsilon. \tag{6.7.36}$$

The scalar field is the pressure that corresponds to the incompressible character of the material. We observe that this incompressible elastic problem is related to the Stokes incompressible flow. For that reason it is better to use steady Stokes solvers, since, in their deformation, spectral elements have to conserve their volume.

When the mesh velocity has been obtained, we can next evaluate the mesh movement by integrating the equation

$$\frac{d\mathbf{x}}{dt} = \mathbf{w}, \tag{6.7.37}$$

where $\mathbf{x}$ is the position of the mesh points. Other strategies are also possible. For example, Formaggia and Nobile [136] worked directly on $\mathbf{x}$, solving a diffusion equation or a harmonic extension problem.

6.7.4 Space and Time Discretization

The spatial discretization of the free-surface problem is based on the $\mathbb{P}_N$–$\mathbb{P}_{N-2}$ SEM. We will capitalize on former developments and skip all unnecessary details. The semidiscrete equations derived from Equations (6.7.23)–(6.7.24) are

$$\frac{d}{dt}(\mathbf{M}\underline{\mathbf{v}}) = \mathbf{C}\underline{\mathbf{v}} - \mathbf{K}\underline{\mathbf{v}} + \mathbf{D}^T\underline{p} + \underline{\mathbf{F}}, \tag{6.7.38}$$

$$-\mathbf{D}\underline{\mathbf{v}} = 0. \tag{6.7.39}$$

Note that we have come back to the classical notation for the time derivative.

The convective term $\mathbf{C}\underline{\mathbf{v}}$ corresponds to the nonlinear term incorporating the mesh velocity. The source term $\underline{\mathbf{F}}$ is composed of the body force and the surface-tension contribution. All other quantities are standard. The nodal mesh coordinates are updated by using the mesh velocity

$$\frac{d}{dt}\underline{\mathbf{x}} = \underline{\mathbf{w}}. \tag{6.7.40}$$

The viscous part and pressure are computed implicitly; all other variables are evaluated explicitly. For example, if we resort to a BDF2/AB3 scheme as in (6.4.17), Equation (6.7.38) becomes

$$\beta_2 \mathbf{M}^{n+1}\underline{\mathbf{v}}^{n+1} + \beta_1 \mathbf{M}^n\underline{\mathbf{v}}^n + \beta_0 \mathbf{M}^{n-1}\underline{\mathbf{v}}^{n-1}$$
$$= \Delta t \left(\sum_{j=1}^{3} b_j (\mathbf{C}\mathbf{v})^{n+1-j} - \mathbf{K}^{n+1}\underline{\mathbf{v}}^{n+1} + \mathbf{D}^{T,n+1}\underline{p}^{n+1} + \underline{\mathbf{F}}^{n+1} \right). \tag{6.7.41}$$

All quantities at the new time level $n+1$ are computed in the updated geometry (hence the time superscript for the matrices); the other variables are evaluated in the previous geometrical configurations. The time integration of Equation (6.7.40) may be performed by any stable scheme proposed in Chapter 3 (e.g., the AB2 scheme).

6.8 Unsteady Applications

In this section, we consider successively three applications that demonstrate the spectral-element capabilities. The first application is a free-surface problem related to materials processing, namely, the extrusion of a 3D die. Although the solution is steady-state, the free-surface methodology relies on transient calculations. We start from a given geometrically simple initial condition, and, through a time marching procedure, we obtain the final configuration enforcing the free-surface boundary conditions.

The second and third problems are transient cases. The vortex-sheet roll-up offers a rich and complex physics; it is a perfect example of the help brought by the filtering process in achieving higher-Reynolds-number flows. The last application computes unsteady blood flow through an arteriovenous graft.

6.8.1 Extrusion from a Die

Rønquist [335] considered the time evolution to steady-state conditions for a material being extruded from a square die. Applying symmetry conditions, one needs to examine only a quarter of the die. The problem is characterized by the Reynolds and Weber numbers $Re = 20$ and $We = 2$. The reference

6.8. Unsteady Applications

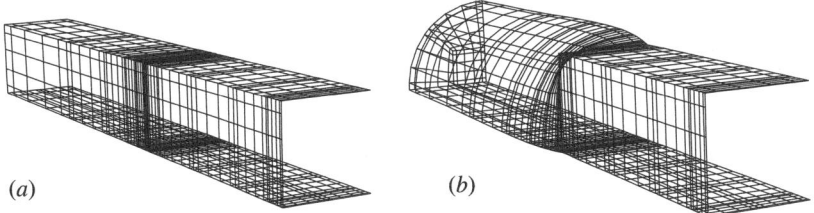

Figure 6.8.1. Free-surface die problem (due to Rønquist [335]).

length L is the die cross-section height, and the velocity U is based on the inlet profile. The boundary conditions specify the inlet profile, no-slip wall conditions inside the extruder, free-surface conditions along the product sides outside the extruder, and outflow (stress-free) conditions at the exit. Symmetry conditions are imposed on symmetry planes.

The computational domain comprises $E = 33$ spectral elements with $N = 5$ for each of them. The initial mesh is displayed in Figure 6.8.1(a), where the right half of the mesh corresponds to the extruder interior, and the left half to the product. The steady-state configuration is displayed in (b), which shows the final free surface of the product when the stresses have relaxed. As is well known from low-Reynolds-number flows with surface tension, the product swells away from the corners of the die exit and contracts near the corner. The product also changes its shape from a square cross section at the die exit to a circular one at the outlet section of the computational domain.

6.8.2 Vortex-Sheet Roll-Up

The vortex-sheet roll-up problem has been studied by a number of authors as a means of testing the stability of numerical methods on underresolved grids (e.g., [25, 56]). Doubly periodic boundary conditions are applied on $\Omega := [0, 1]^2$, with initial conditions

$$u = \begin{cases} \tanh[\rho(y - 0.25)] & \text{for} \quad y \leq 0.5, \\ \tanh[\rho(0.75 - y)] & \text{for} \quad y > 0.5, \end{cases} \quad v = 0.05 \sin(2\pi x).$$

The perturbed vortex sheets of initial thickness ρ are continuously stretched until the thickness is well below the grid resolution. Ideally, the sheets should break up and dissipate, allowing the calculation to continue the advection and diffusion of the remaining vortex structures.

Figure 6.8.2 shows several spectral-element results for the shear-layer roll-up and illustrates the effectiveness of the stabilizing filter described in Section 6.6. Each case consists of a 16×16 array of elements, save for (e), which is 32×32. The time step size is $\Delta t = 0.002$ in all cases, corresponding to CFL numbers in the range of 1 to 5. Without filtering, the SEM fails to simulate this difficult problem at any reasonable resolution.

Figure 6.8.2(a) shows the results just prior to blowup for the unfiltered case with $N = 16$, which corresponds to an $\mathcal{N} \times \mathcal{N}$ grid with $\mathcal{N} = 256$. Resolutions

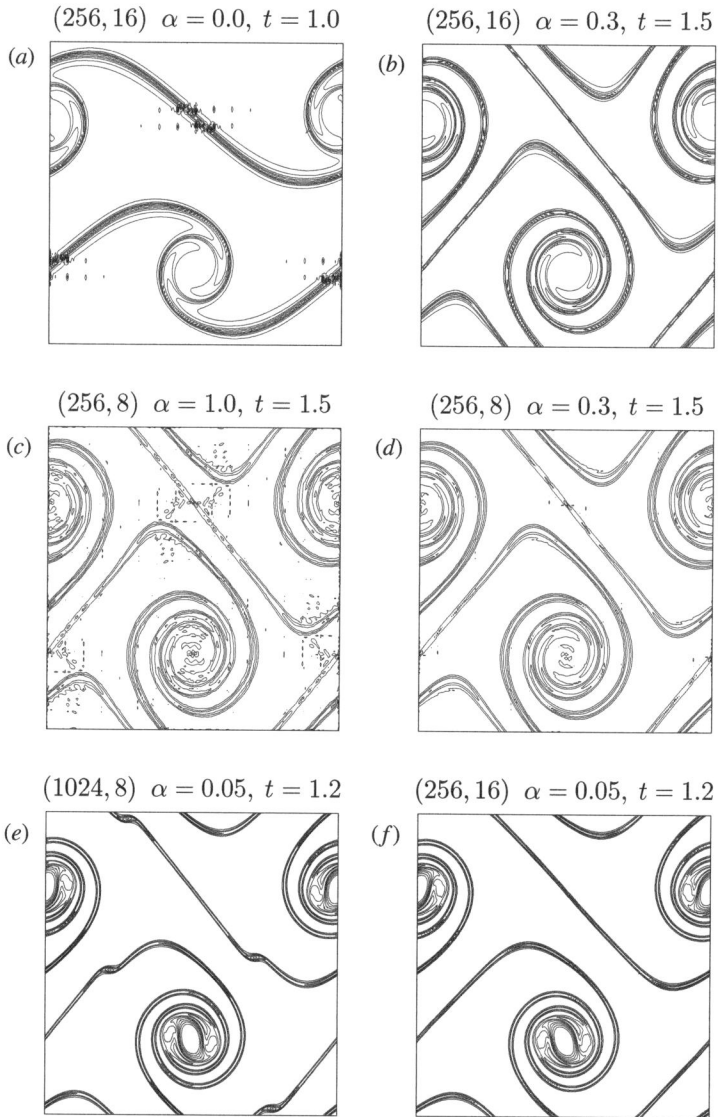

Figure 6.8.2. Vorticity contours for different (E, N) pairings: $(a$–$d)$ thick shear layer, $\rho = 30$, $Re = 10^5$, contours from -70 to 70 by $140/15$; $(e$–$f)$ thin shear layer, $\rho = 100$, $Re = 40{,}000$, contours from -36 to 36 by $72/13$ (cf. Figure 3c in [56]).

6.9. Further Reading and Engineering Considerations

of $N=8$ ($\mathcal{N}=128$) and $N=32$ ($\mathcal{N}=512$) yield similar results. Filtering with $\alpha = 0.3$ yields a dramatic improvement for $\mathcal{N}=256$ (b) and $\mathcal{N}=128$ (d). Although full projection ($\alpha=1$) is also stable, it is clear by comparing (c) and (d) that partial filtering ($\alpha < 1$) is preferable. Finally, (e) and (f) correspond to the difficult thin-shear-layer case of $\rho = 100$, $Re = 40{,}000$ [56]. The spurious vortices observed in (e) are eliminated in (f) by increasing the order to $N = 16$ and reducing the element array to 8×8. Since the resolution is effectively the same in each case ($\mathcal{N}=256$), there is a significant advantage to using higher order for this particular application.

6.8.3 Unsteady Flow in Arteriovenous Grafts

Figure 6.8.3 shows spectral-element simulations of flow in an arteriovenous graft model, as studied in [247]. The simulations permit the study of stress distributions in the vein wall downstream of the graft, where disease formation occurs. The geometry in (a) consists of a graft of inlet diameter D connected to a vein of diameter $0.625D$. In the experimental–numerical comparisons shown, steady Poiseuille profiles are specified at the graft and vein inlets (left), with respective volume flow rates Q and $Q/9$. The Reynolds number is $Re := 4Q/\nu\pi D$. Steady-state centerplane velocity distributions at $Re = 1060$ are shown in (b) for $(E, N) = (3840, 5)$, along with experimentally measured profiles from [10]. At higher Reynolds numbers, the flow transitions to an unsteady state, as illustrated in (c) by a snapshot of the vorticity contours at $Re = 1820$. The experimental and numerical mean velocity and u_{rms} profiles are shown for $Re = 1820$ in (d) and (e), respectively. The spectral-element simulations at $Re = 1820$ were computed with $N = 9$, and all the calculations employed the filter of Section 6.6 with $\alpha = 0.05$.

6.9 Further Reading and Engineering Considerations

Free-surface flows for inviscid, incompressible fluid were investigated by Robertson and Sherwin [329] using hp-spectral elements with the ALE formulation. The influence of the mesh construction (symmetric or asymmetric) was analyzed. Robertson and Sherwin showed that the asymmetric mesh induces numerical instabilities by generating spurious eigenmodes. Some remedies were proposed. One of them consists in adding a diffusive term with a viscosity coefficient that is compatible with the hp-accuracy of the spectral-element discretization.

A $\mathbb{P}_N$–$\mathbb{P}_{N-2}$ Chebyshev collocation based on GLC nodes for the velocity and on the interior velocity nodes for the pressure was proposed by Heinrichs [191] to integrate the unsteady Stokes problem. The splitting technique uses

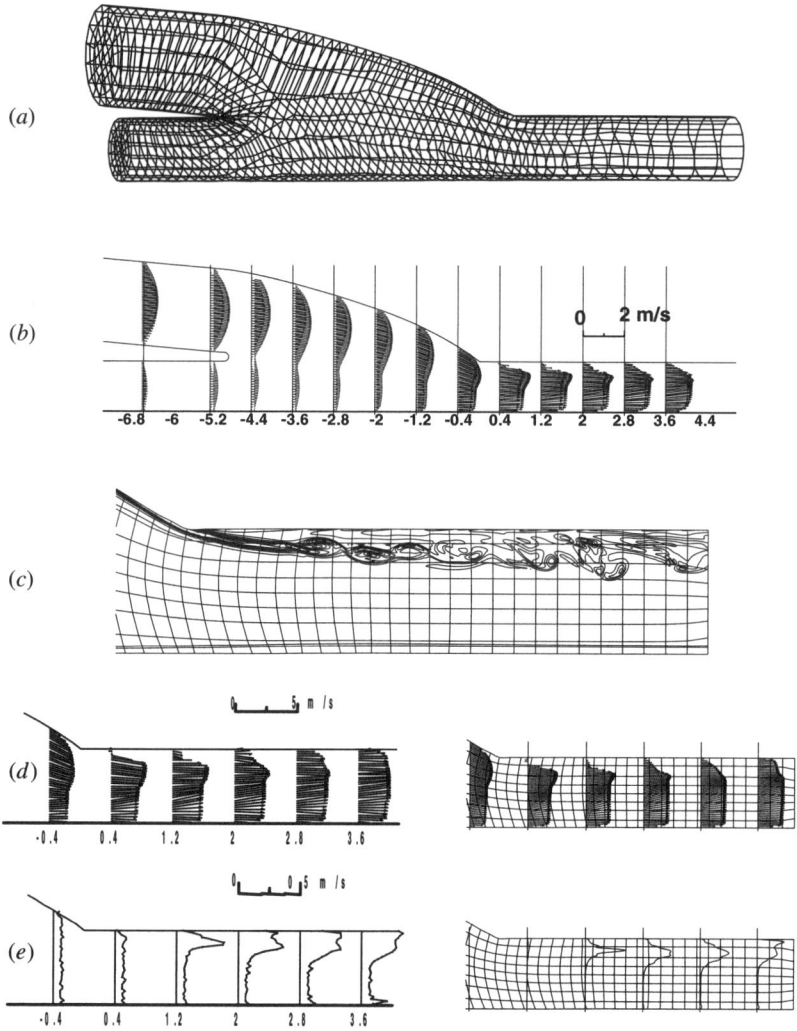

Figure 6.8.3. Flow in an arteriovenous graft model.

the pressure correction method. The solution is computed by spectral multigrid techniques preconditioned by finite differences.

The divergence-free basis of eigenfunctions computed from the singular Stokes operator by Batcho and Karniadakis [24] was used in the bifurcation study for the eddy-promoter geometry of Figure 5.5.1 and was fully reported in the paper by Bangia et al. [20]. The Galerkin approximation generates a nonlinear system of ordinary differential equations (ODEs) by the use of solenoidal

6.9. Further Reading and Engineering Considerations

test functions. Obviously, the pressure term is removed. Available packages for ODEs are applied for the time evolution. The dynamics of the flow shows a Hopf bifurcation from steady to oscillatory flow at a Reynolds number of 350. A second frequency appears at $Re \approx 890$. The multiple frequency dynamics complicates the flow behavior by bringing a substantial increase in the spatial and temporal scales. It turns out that the divergence-free basis compares favorably with the proper orthogonal decomposition (POD) of Lumley [249, 33].

The hp-SEM has also been used in biomedical flow. Sherwin et al. [360] analyzed the blood flow in a fully 45° distal end-to-side anastomosis. The bypass vessel is out of the plane of symmetry of the host vessel. This situation creates a nonplanar 3D flow. The computed results were compared with magnetic resonance imaging data.

A moving SEM was designed to compute the dynamics inside a journal bearing [180, 246]. The staggered SEM is used with a preconditioned Uzawa method. The time integration is made by a first-order backward Euler scheme. The bearing movement is integrated by the equation of motion, which couples the acceleration of the position vector of the center of the journal to the applied load and the force exerted by the viscous fluid on the journal.

The projection–diffusion (PRDI) method is a fractional-step method involving two stages. Let us introduce in the frame of the Stokes problem the acceleration vector

$$\mathbf{a} = \frac{\partial \mathbf{v}}{\partial t} - \Delta \mathbf{v}. \tag{6.9.1}$$

The first step solves the problem

$$\mathbf{a} + \nabla p = 0 \quad \text{in } \Omega, \tag{6.9.2}$$

$$\operatorname{div} \mathbf{a} = 0 \quad \text{in } \Omega, \tag{6.9.3}$$

$$\mathbf{n} \cdot \mathbf{a} = \mathbf{n} \cdot \left(\frac{\partial \mathbf{v}}{\partial t} - \Delta \mathbf{v} \right) \quad \text{on } \partial\Omega. \tag{6.9.4}$$

The second is a diffusive viscous step:

$$\frac{\partial \mathbf{v}}{\partial t} - \Delta \mathbf{v} = \mathbf{a} \quad \text{in } \Omega, \tag{6.9.5}$$

$$\mathbf{v} = \mathbf{g} \quad \text{on } \Omega, \tag{6.9.6}$$

where $\mathbf{g}$ is the prescribed velocity boundary condition.

The first step is a Darcy-type problem, which was analyzed by Azaïez et al. [16, 15] in the framework of variational spectral discretizations. The Darcy problem involves conditions on normal fluxes through some parts of the boundary.

One of the interesting characteristics of the method is that the decoupling is independent of the time scheme [see Equation (6.9.2)]. Leriche and Labrosse [242] carried out the full study of the temporal behavior of the PRDI decoupling, which gave the expected time accuracy for the analyzed time schemes. However, the PRDI method is close to the usual fractional-step method, which should be chosen for the more general boundary conditions that are tractable. The time-dependent calculations performed by Azaïez et al. [15] confirm the analysis of Leriche and Labrosse.

With the view of integrating the LES equations (1.10.23)–(1.10.26), Karamanos and Sherwin [217] generalize the splitting scheme of Karniadakis et al. [219] by decomposing the variable viscosity into two pieces. The first one is independent of time and treated implicitly; the second one is instantaneous, temporally and spatially varying, and treated explicitly. The stability analysis is performed on a one-dimensional diffusion equation and leads to a mild stability condition.

Finally, we mention the Chebyshev collocation element method based on a fractional-step method and the enforcement of the incompressibility constraint everywhere, up to the walls. This method was performed by Ku, Hirsh, Taylor and colleagues and reported in a series of papers that appeared over a decade [231, 232, 233, 230]. The Poisson solver incorporates the boundary continuity equation, and the matrix solver enforces C^1 continuity across interfaces.

Many CFD practitioners compare the computational efficiency of their algorithms and Navier–Stokes solvers using the amount of CPU time spent per degree of freedom (the number of degrees of freedom is the number of grid nodes multiplied by the numbers of variables) and per time step. This exercise is always difficult, because it implies rescaling those values to get a figure for the computer that is considered as the standard tool at the time of comparison. Furthermore, this kind of procedure will conclude to the detriment of spectral methods. Since they are global methods, each computational node is coupled to more neighbors than in standard finite-difference, finite-volume, or finite-element methods and therefore imply more computational work per grid point. A sensible and fair rule consists in fixing a priori the level of accuracy (5%, say, for engineering purposes) required to achieve a long-term and meaningful integration. We will have to specify the spatial and time accuracy of the numerical method in order to meet such a requirement. In real computations, the user should give the numerical error bar of the simulation in order to promote better and long-lasting results (see [218]). We consider that this will bring an enormous improvement in assessing fluid-flow computational results.

7
Domain Decomposition

Domain decomposition for numerical solution of PDEs has been an active area of research, with a well-organized international conference series held annually since 1987. The area encompasses preconditioning of linear systems, discretizations, and solution of hybrid systems (e.g., coupled Navier–Stokes and Euler problems). We consider the first two of these in the context of high-order methods.

7.1 Introduction

Domain decomposition (DD) alleviates the solution complexity associated with the full problem in a complicated geometry. Broadly, the aim of DD consists in formulating independent problems in separate subdomains whose union constitutes the whole. The decomposition of the domain may be motivated by differing physics within different subdomains, by the availability of fast solvers for each subdomain, by the desire to partition the computational effort across separate processors, or by the inherent heterogeneity of the discretization. We discuss in Section 7.2 preconditioning methods such as substructuring, Schwarz overlapping, and multigrid techniques. Section 7.3 describes the mortar element method, which encompasses both functional and geometrical nonconforming discretizations; suggestions for implementation are given. Section 7.4 reviews the coupling between finite and spectral elements, provides some theoretical considerations about adaptivity near geometrical singularities, and considers for the 2D case hp-spectral triangular elements, which open the way for the coupling between quadrilaterals and triangles.

7.2 Preconditioning Methods

Domain-decomposition preconditioning has gained much attention over the past decade, both in theory and in practice. The underlying strategy is to precondition the linear system arising from the discretization of a PDE in Ω by solving smaller local problems in subdomains Ω^s of Ω with $\bar{\Omega} = \bigcup \bar{\Omega}^s$. Variations arise in how the local solutions are recombined, whether one uses exact or inexact solves for the local problems, and whether one includes a global coarse-grid problem or not. Fortunately, the strategy is robust enough to tolerate many shortcuts to facilitate a given implementation. While much of the renewed interest in domain decomposition resulted from the desire to increase parallelism, the method has also proven to be an effective serial preconditioning strategy. General references on this topic include the books by Smith et al. [363] and Quarteroni and Valli [320].

Section 7.2.1 presents a technique known as substructuring, in which subdomain interior degrees of freedom are eliminated by direct means, and the remaining interface variables are computed either directly or iteratively. Section 7.2.2 describes overlapping methods, and Section 7.2.3 reviews multigrid methods for spectral elements.

7.2.1 Substructuring and the Steklov–Poincaré Operator

In this subsection, we treat a very simple elliptic problem in order to introduce a few concepts based on the continuous representation of the substructuring method and the associated Steklov–Poincaré operator. We then show how these concepts translate in a discrete spectral-element representation.

At this stage it is important to specify the notation used with regard to topology. As was the case in previous chapters, Ω is a bounded open domain in $\mathbb{R}^d$. All superscripts refer to a subdomain, a finite or spectral element, depending on the context. Closed domains (or subdomains) are indicated by a bar (e.g., $\bar{\Omega}$). In Section 7.3 on the mortar element method there is a need to associate two superscripts to the subdomain boundaries $\partial \Omega^{e,s}$: the first one (e) refers to an element, while, conventionally, the second one (s) refers to a particular edge ($d = 2$) or face ($d = 3$) inside the element. Finally, there is also a need to specifically denote interior boundaries or interfaces between subdomains. These are designated by Γ.

The Continuous Presentation

Suppose we want to solve the Poisson equation

$$-\Delta u = f \quad \text{in } \Omega, \quad u = 0 \quad \text{on } \partial \Omega. \tag{7.2.1}$$

7.2. Preconditioning Methods

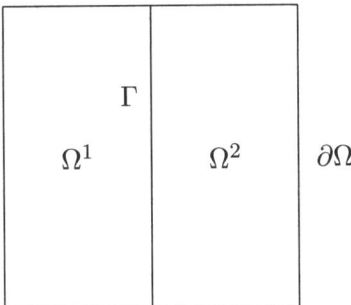

Figure 7.2.1. Subdomain decomposition of a bounded domain $\Omega \subset \mathbb{R}^2$.

The DD method begins with a subdivision of Ω into S subdomains Ω^s, $s = 1, \ldots, S$. For the sake of simplicity, we initially will restrict ourselves to the case of two nonoverlapping adjacent subdomains Ω^s, $s = 1, 2$ (see Figure 7.2.1).

At the interface Γ two unit normal vectors $\mathbf{n}_s$ on $\partial\Omega^s$ are pointing outward. However, as $\mathbf{n}_1 = -\mathbf{n}_2$, we will refer to $\mathbf{n}_1$ as n. The one-domain problem (7.2.1) is reformulated as the multidomain form ($s = 1, 2$) with u_s the solution in domain s. Assuming $f \in L^2$, we have

$$-\Delta u_s = f_s \quad \text{in } \Omega^s, \tag{7.2.2}$$

$$u_s = 0 \quad \text{on } \partial\Omega^s \cap \partial\Omega, \tag{7.2.3}$$

$$u_1 = u_2 \quad \text{on } \Gamma, \tag{7.2.4}$$

$$\frac{\partial u_1}{\partial n} = \frac{\partial u_2}{\partial n} \quad \text{on } \Gamma, \tag{7.2.5}$$

where $\partial/\partial n$ is the normal derivative. We remark that if f were only in H^{-1}, there could be a nontrivial jump in the normal derivative on Γ, and (7.2.5) would need to be modified accordingly. The conditions (7.2.4)–(7.2.5) express the C^1 continuity matching conditions. If u were the temperature, these conditions would impose continuity across the interface of the temperature field and of the heat flux. In fact, the DD method engenders local independent subproblems, which can be solved with a direct or an iterative procedure. The global solution is obtained by combining the local solutions at the interface through an iterative process. This method will depend strongly on the interface equation, which originates from the interface problem defined by the so-called Steklov–Poincaré operator (see Quarteroni and Valli for full details [320]). Let us call v the unknown value of u on Γ. We solve two Dirichlet problems with $s = 1, 2$:

$$-\Delta w_s = f_s \quad \text{in } \Omega^s, \tag{7.2.6}$$

$$w_s = 0 \quad \text{on } \partial\Omega^s \cap \partial\Omega, \tag{7.2.7}$$

$$w_s = v \quad \text{on } \Gamma. \tag{7.2.8}$$

Because these problems are linear, we can obtain the full solution

$$w_s = w_{L,s} + w_{P,s}, \tag{7.2.9}$$

from the following problems:

$$-\Delta w_{L,s} = 0 \quad \text{in } \Omega^s, \tag{7.2.10}$$

$$w_{L,s} = 0 \quad \text{on } \partial\Omega^s \cap \partial\Omega, \tag{7.2.11}$$

$$w_{L,s} = v \quad \text{on } \Gamma, \tag{7.2.12}$$

and

$$-\Delta w_{P,s} = f_s \quad \text{in } \Omega^s, \tag{7.2.13}$$

$$w_{P,s} = 0 \quad \text{on } \partial\Omega^s \cap \partial\Omega, \tag{7.2.14}$$

$$w_{P,s} = 0 \quad \text{on } \Gamma. \tag{7.2.15}$$

The solutions $w_{L,s}$ result from a Laplace operator and are called the harmonic extension of v in Ω^s, while the solutions $w_{P,s}$ are obtained from a Poisson equation satisfying homogeneous boundary conditions. The solution w_s given by the problem (7.2.6)–(7.2.8) will be equal to u_s of the problem (7.2.2)–(7.2.5) iff

$$\frac{\partial w_1}{\partial n} = \frac{\partial w_2}{\partial n} \quad \text{on } \Gamma. \tag{7.2.16}$$

We can now write the equation for the unknown value v on Γ:

$$\mathcal{S}v = \theta \quad \text{on } \Gamma, \tag{7.2.17}$$

with the definitions

$$\theta = \frac{\partial w_{P,2}}{\partial n} - \frac{\partial w_{P,1}}{\partial n} \tag{7.2.18}$$

and the Steklov–Poincaré operator

$$\mathcal{S}v := \frac{\partial w_{L,1}}{\partial n} - \frac{\partial w_{L,2}}{\partial n}. \tag{7.2.19}$$

7.2. Preconditioning Methods

We note that the Steklov–Poincaré operator is built on the jump of the normal derivatives at the interface. Of course, this equation will be handled by a weak formulation, as are other elliptic equations. As soon as we know the solution v, the problems (7.2.10)–(7.2.12) can be solved independently and in parallel, as the interface condition is given.

The Discrete Presentation

Let us decompose the full domain of Figure 7.2.1 in two spectral elements coinciding with the two subdomains. The unknowns generated by the weak form of the problem (7.2.1) are collected by gathering first the interior variables of the two elements and then the interface variables. In matrix form, we obtain

$$\begin{pmatrix} K_{11} & 0 & K_{1\Gamma} \\ 0 & K_{22} & K_{2\Gamma} \\ K_{\Gamma 1} & K_{\Gamma 2} & K_{\Gamma\Gamma} \end{pmatrix} \begin{pmatrix} \underline{u}_1 \\ \underline{u}_2 \\ \underline{u}_\Gamma \end{pmatrix} = \begin{pmatrix} \underline{f}_1 \\ \underline{f}_2 \\ \underline{f}_\Gamma \end{pmatrix}. \qquad (7.2.20)$$

The notation is defined as follows. The matrix K_{ss} is the stiffness matrix of subdomain s and concerns the internal nodes only. The matrix $K_{s\Gamma}$ represents the coupling between the unknowns of subdomain s and the interface, whereas $K_{\Gamma\Gamma}$ gives the coupling of the interface variables with themselves. The block diagonal matrices are SPD; one can show that $K_{s\Gamma} = K_{\Gamma s}^T$. If we eliminate the unknowns related to the interior nodes, we obtain the system

$$\mathcal{S}_N \underline{u}_\Gamma = \underline{\theta}_\Gamma, \qquad (7.2.21)$$

with

$$\underline{\theta}_\Gamma := \underline{f}_\Gamma - K_{\Gamma 1} K_{11}^{-1} \underline{f}_1 - K_{\Gamma 2} K_{22}^{-1} \underline{f}_2 \qquad (7.2.22)$$

and

$$\mathcal{S}_N := K_{\Gamma\Gamma} - K_{\Gamma 1} K_{11}^{-1} K_{1\Gamma} - K_{\Gamma 2} K_{22}^{-1} K_{2\Gamma}. \qquad (7.2.23)$$

The matrix $\mathcal{S}_N$ is the Schur complement system. It is the spectral-element realization of the Steklov–Poincaré operator $\mathcal{S}$ introduced in (7.2.19). Once (7.2.23) is solved, the interior variables may be computed from

$$K_{ss}\underline{u}_s = -K_{s\Gamma}\underline{u}_\Gamma + \underline{f}_s. \qquad (7.2.24)$$

Suppose now we have four subdomains Ω^s, $s = 1, \ldots, 4$, lined up like a sandwich, with three interfaces Γ^{12}, Γ^{23}, and Γ^{34}, where Γ^{ij} is the interface between the subdomains Ω^i and Ω^j. It is left as an exercise for the reader to show that the Schur complement problem becomes

$$\begin{pmatrix} S_{11} & S_{12} & \\ S_{21} & S_{22} & S_{23} \\ & S_{32} & S_{33} \end{pmatrix} \begin{pmatrix} \underline{u}_{\Gamma^{12}} \\ \underline{u}_{\Gamma^{23}} \\ \underline{u}_{\Gamma^{34}} \end{pmatrix} = \begin{pmatrix} \underline{f}_{\Gamma^{12}} \\ \underline{f}_{\Gamma^{23}} \\ \underline{f}_{\Gamma^{34}} \end{pmatrix} \qquad (7.2.25)$$

with obvious definitions of S_{ij} resulting from the generalization of Equation (7.2.23). The block Schur complement (7.2.25) has a sparse structure much like a typical FD or FE matrix and can be solved directly or iteratively. The off-diagonal blocks represent the interface–interface interactions that arise from the elimination of the interior degrees of freedom.

When the Schur matrices are explicitly computed, the present method, where all interior variables are eliminated and their contribution reported on the interfaces, is referred to as *substructuring*, or static condensation. It has been employed in engineering applications for some time, especially in solid mechanics problems, and was first used in the spectral-element context by Patera [298].

Couzy and Deville [86] study the preconditioning of the Schur complement matrix (7.2.25). Two preconditioners are examined. The first one is the inverse of the diagonal of (7.2.25), and the second is the block-diagonal matrix that has as entries the inverses of the diagonal blocks of (7.2.25). This latter (*block-Jacobi*) method yields a condition number of the preconditioned system practically independent of N and gives good performance on Stokes test cases. It is possible to extend the block-Jacobi preconditioner by applying the overlapping Schwarz techniques introduced in the next section to the interface variables. This approach has been studied by a number of researchers, including Rønquist [334, 335], Pavarino and Widlund [301], Pavarino and Warburton [300], and Casarin [73, 72, 74]. A related approach that also addresses the Stokes Schur-complement system is the balancing Neumann–Neumann method developed by Pavarino and Widlund [302].

7.2.2 Overlapping Schwarz Procedures

The overlapping Schwarz approach to DD differs from the substructuring method in that the domain-to-domain information exchange is effected through regions of overlap, rather than through transmission conditions at the interface. We will take as our example the Poisson equation (7.2.1) in $\mathbb{R}^2$. However, the method readily extends to more general equations and higher space dimensions. See [363, 320] for more details.

7.2. Preconditioning Methods

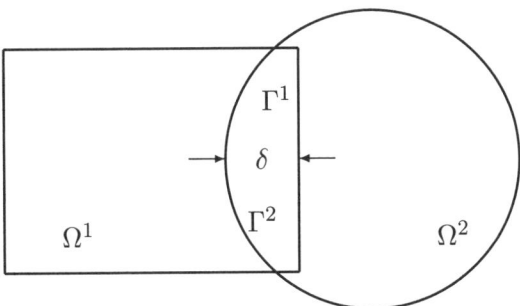

Figure 7.2.2. Domain Ω decomposed into overlapping domains Ω^1 and Ω^2 having boundaries $\partial\Omega^1$ and $\partial\Omega^2$.

To motivate the development, we begin with a brief discussion of the classical Schwarz alternating method. Assume that Ω is partitioned into the regions Ω^1 and Ω^2, with respective interior boundaries $\Gamma^s := \partial\Omega^s \setminus (\partial\Omega \cap \partial\Omega^s)$ and characteristic overlap width δ, as depicted in Figure 7.2.2. The Schwarz iteration alternately solves the Poisson equation in each subdomain, using the most recent solution in one domain to provide boundary conditions for the other. That is, setting $u^0 = 0$, we find for $k > 0$

$$-\Delta u^{k+\frac{1}{2}} = f \quad \text{in } \Omega^1, \qquad u^{k+\frac{1}{2}} = u^k \quad \text{on } \bar{\Omega}\setminus\Omega^1,$$
$$-\Delta u^{k+1} = f \quad \text{in } \Omega^2, \qquad u^{k+1} = u^{k+\frac{1}{2}} \quad \text{on } \bar{\Omega}\setminus\Omega^2. \tag{7.2.26}$$

Note that the imposition of the solution on $\bar{\Omega}\setminus\Omega^s$ provides both the boundary condition on $\partial\Omega^s$ and the extension of the solution from Ω^s to $\bar{\Omega}$. Subtracting (7.2.1) from (7.2.26) yields the equation for the error $e^k := u^k - u$:

$$-\Delta e^{k+\frac{1}{2}} = 0 \quad \text{in } \Omega^1, \qquad e^{k+\frac{1}{2}} = e^k \quad \text{on } \bar{\Omega}\setminus\Omega^1,$$
$$-\Delta e^{k+1} = 0 \quad \text{in } \Omega^2, \qquad e^{k+1} = e^{k+\frac{1}{2}} \quad \text{on } \bar{\Omega}\setminus\Omega^2. \tag{7.2.27}$$

Convergence follows from the maximum principle. For each iteration, $|e^{k+\frac{1}{2}}|$ has its maximum on Γ^1 and decays to zero on $\partial\Omega$. Increased separation δ between Γ^2 and Γ^1 leads to a decrease in $\max_{\Gamma^2} |e^{k+\frac{1}{2}}|$, which in turn bounds $|e^{k+1}|$, and therefore to an increase in the rate of convergence. This is illustrated explicitly by the one-dimensional example of Figure 7.2.3. Starting with a uniform error of $\|e^0\|_\infty$, the error at the end of each substep is [from (7.2.27)] given by the straight line segments shown. A simple geometric argument reveals

$$\|e^k\|_\infty = \left(\frac{1-\delta/2}{1+\delta/2}\right)^{2k} \|e^0\|_\infty, \tag{7.2.28}$$

from which the advantage of increased overlap is evident.

340   7. Domain Decomposition

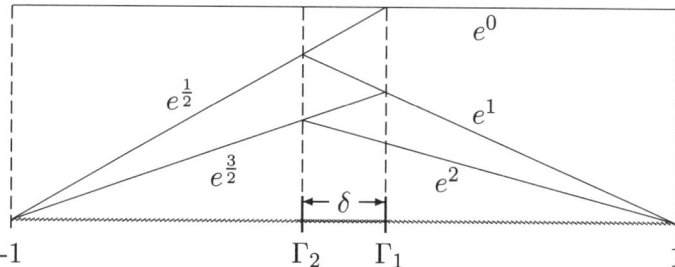

Figure 7.2.3. Error behavior for the one-dimensional alternating Schwarz method on $\Omega = (-1, 1)$ with overlap $\delta = 0.25$ with respect to $x = 0$.

Notice that the convergence result (7.2.28) depends only on the Green's functions (7.2.27) and the geometry of the overlap. This is typical of Schwarz methods and has the desirable consequence that the convergence rate does not depend on mesh discretization parameters such as h or N.

Discrete Formulation

We derive the algebraic formulation of (7.2.26) by discretizing (7.2.1) with a Galerkin formulation. Let V_0 be the Q_1 Lagrangian FE functions spanned by $\underline{\psi}(\mathbf{x}) := (\psi_1(\mathbf{x}), \ldots, \psi_\mathcal{N}(\mathbf{x}))^T$, built upon a given triangulation of Ω and vanishing on $\partial\Omega$. Let $V_0^s \subset V_0$, $s = 1, 2$, comprise the subset of these functions that vanish on $\bar{\Omega} \setminus \Omega^s$. Denote by $\mathcal{I}_s$ the index set associated with each subdomain such that $j \in \mathcal{I}_s \iff \psi_j \in V_0^s$, and assume for convenience that the ψ_j's are ordered with elements in $\mathcal{I}_1 \setminus \mathcal{I}_2$ first, $\mathcal{I}_1 \cap \mathcal{I}_2$ second, and $\mathcal{I}_2 \setminus \mathcal{I}_1$ last. In other words, one considers those elements that do not vanish on Ω^1 first, then those that do not vanish on $\Omega^1 \cap \Omega^2$, and finally those that do not vanish on Ω^2.

The weak formulation of (7.2.1) is: Find $u \in V_0$ such that

$$\int_\Omega \nabla v \cdot \nabla u \, dV = \int_\Omega v f \, dV \quad \forall v \in V_0. \tag{7.2.29}$$

Inserting the basis for V_0 gives rise to the $\mathcal{N} \times \mathcal{N}$ system

$$K\underline{u} = M\underline{f}, \quad K_{ij} := \int_\Omega \nabla \psi_i \cdot \nabla \psi_j \, dV, \quad M_{ij} := \int_\Omega \psi_i \psi_j \, dV,$$

where $\underline{u} \in \mathbb{R}^\mathcal{N}$ is the vector of unknown basis coefficients and f has been replaced by its Lagrangian interpolant.

7.2. Preconditioning Methods

Now consider the following subproblems for $s = 1, 2$:

$$u_s := \delta u_s + u_s^b, \tag{7.2.30}$$

$$\delta u_s \in V_0^s : \quad \int_{\Omega^s} \nabla v_s \cdot \nabla \delta u_s \, dV = \int_{\Omega^s} v_s f \, dV - \int_{\Omega^s} \nabla v_s \cdot \nabla u_s^b \, dV$$

$$\forall v_s \in V_0^s.$$

Here, u_s has been split into δu_s, satisfying homogeneous Dirichlet boundary conditions on $\partial \Omega^s$, and $u_s^b \in V_0$, satisfying $u_s^b = u_s$ on Γ^s. Setting ($u_1^b := u^k$, $u_2^b := u^{k+\frac{1}{2}}$) and ($u^{k+\frac{1}{2}} := u_1$, $u^{k+1} := u_2$), we recover (7.2.26).

To cast the alternating Schwarz method into matrix form, we observe that, since $v_s \in V_0^s$ vanishes outside Ω^s, the domain of integration in (7.2.30) can be changed to Ω. Moreover, for any $\underline{v}_s \in \mathbb{R}^{\mathcal{N}_s}$, we can define $v_s := \underline{\psi}^T R_s^T \underline{v}_s \in V_0^s$, where R_s^T is the *prolongation* matrix given in the current example by

$$R_1^T := (\underline{\hat{e}}_1, \ldots, \underline{\hat{e}}_{\mathcal{N}_1}), \quad R_2^T := (\underline{\hat{e}}_{\mathcal{N}-\mathcal{N}_2+1}, \ldots, \underline{\hat{e}}_{\mathcal{N}}).$$

Here, $\underline{\hat{e}}_j$ denotes the jth column of the $\mathcal{N} \times \mathcal{N}$ identity matrix, and $\mathcal{N}_s$ denotes the cardinality of $\mathcal{I}_s$. R_s is referred to as the *restriction* matrix.

Substituting $\underline{v}_s^T R_s \underline{\psi}$ and $\underline{\psi}^T R_s^T \delta \underline{u}_s$ into (7.2.30) for v_s and δu_s, respectively, we have the equations for the basis coefficients:

$$\underline{u}_s = \underline{u}_s^b + R_s^T \delta \underline{u}_s,$$

$$R_s K R_s^T \delta \underline{u}_s = R_s (M \underline{f} - K \underline{u}_s^b).$$

Setting ($\underline{u}_1^b := \underline{u}^k$, $\underline{u}_2^b := \underline{u}^{k+\frac{1}{2}}$), ($\underline{u}^{k+\frac{1}{2}} := \underline{u}_1$, $\underline{u}^{k+1} := \underline{u}_2$), and defining $K_s := R_s K R_s^T$, we arrive at the matrix form of (7.2.26):

$$\underline{u}^{k+\frac{1}{2}} = \underline{u}^k + R_1^T K_1^{-1} R_1 (M \underline{f} - K \underline{u}^k), \tag{7.2.31}$$

$$\underline{u}^{k+1} = \underline{u}^{k+\frac{1}{2}} + R_2^T K_2^{-1} R_2 (M \underline{f} - K \underline{u}^{k+\frac{1}{2}}). \tag{7.2.32}$$

As illustrated in Figure 7.2.4, K_1 corresponds to a leading principal submatrix of K, while K_2 corresponds to a trailing principal submatrix (because of the particular vertex ordering chosen for this example).

Combining (7.2.31)–(7.2.32) with $M \underline{f} = K \underline{u}$ yields the fixed-point iteration for the multiplicative Schwarz procedure,

$$\underline{u}^{k+1} = \underline{u}^k + P_1(\underline{u} - \underline{u}^k) + P_2(\underline{u} - \underline{u}^k) - P_2 P_1(\underline{u} - \underline{u}^k)$$
$$= \underline{u} + (I - P_2)(I - P_1)(\underline{u}^k - \underline{u}), \tag{7.2.33}$$

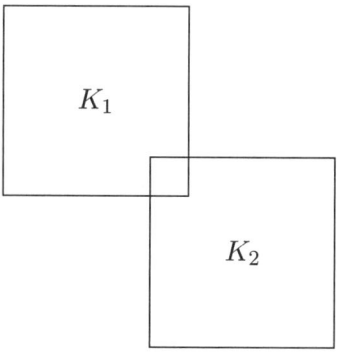

Figure 7.2.4. Overlapping subdomain operators $K_s = R_s^T K R_s$ corresponding to principal submatrices of K.

where $P_s := R_s^T K_s^{-1} R_s K$. Recall from (2.7.36) that P_s is a projector in the K-norm onto $\mathcal{R}(R_s^T)$. Each iterate is thus equal to the solution plus successive projections of the current error onto $\mathcal{R}^\perp(R_s^T)$, the orthogonal complement of $\mathcal{R}(R_s^T)$.

Schwarz Preconditioners

It is clear that the fixed-point iteration (7.2.33) corresponds to the Richardson scheme (2.7.23) applied to the preconditioned operator

$$P^{-1}K := P_2 + P_1 - P_2 P_1.$$

This algorithm is significantly improved by being embedded in a Krylov-based projection procedure such as GMRES. Recall that the preconditioning step in each iteration of Algorithm 2.3 calls for the solution of a system of the form $P\underline{z} = \underline{r}$, where $P \approx K$. Schwarz preconditioning simply involves applying one round of (7.2.31)–(7.2.32) to the problem $K\underline{z} = \underline{r}$, with initial guess $\underline{z} = 0$. Generalized to S subdomains, this gives the following:

$$\begin{aligned}
&\underline{z}_0 := 0 \\
&\text{do } s = 1, S \\
&\quad \underline{z}_s := \underline{z}_{s-1} + R_s^T K_s^{-1} R_s (\underline{r} - K\underline{z}_{s-1}) \\
&\text{enddo}
\end{aligned} \qquad (7.2.34)$$

In practice, the intermediate vectors $\underline{z}_s$ are replaced by $\underline{z}$, which is simply overwritten. Also, restriction and prolongation are effected through the use of the index sets $\mathcal{I}_s$, rather than by explicitly forming R_s and R_s^T. In contrast to the Schur-complement (substructuring) approach, where the local solves must be exact, one can approximate the operators K_s^{-1} and K in (7.2.34), because here DD plays only the role of a preconditioner, rather than a reformulation of the problem. One can, for instance, solve the subproblems on grids different from the originating grid, in which case R_s^T becomes an interpolation operator. It is nonetheless necessary that P^{-1} have full rank, implying, for example, that the total number of degrees of freedom in the subdomain problems must at least equal the number of degrees of freedom in the original problem.

The scheme (7.2.34) is a polynomial in the projectors P_s, with an error propagator of the form

$$(I - P_S) \cdots (I - P_2)(I - P_1).$$

This algorithm is consequently referred to as *multiplicative* Schwarz preconditioning. As it stands, it lacks parallelism because the intermediate residual must be recomputed for each subdomain. However, because $P_s P_{s'} = 0$ whenever $\Omega^s \cap \Omega^{s'} = \emptyset$, parallelism can be recovered through the use of a multicolor ordering of the subdomains, as noted by Dryja and Widlund [112].

To further increase parallelism, Dryja and Widlund [111] developed the *additive* Schwarz preconditioner, which is based upon the simplest possible polynomial in the projectors, namely,

$$P_{AS}^{-1} K := P_1 + P_2 + \cdots + P_S. \tag{7.2.35}$$

The additive Schwarz preconditioning step takes the simple form

$$\underline{z} := P_{AS}^{-1} \underline{r} = \sum_{s=1}^{S} R_s^T K_s^{-1} R_s \underline{r}.$$

Since P_{AS} is symmetric positive definite, PCG is the iterative method of choice. [The multiplicative algorithm can also be symmetrized by repeating (7.2.34) in reverse order.] Note that additive Schwarz is a generalization of block-Jacobi preconditioning (with blocks as illustrated in Figure 7.2.4), while multiplicative Schwarz is a generalization of block Gauss–Seidel. Although multiplicative Schwarz generally converges in fewer iterations (see, e.g., [363]), additive Schwarz is quite popular because it is fully parallel and because it does not require reevaluation of any part of the residual. This latter point is of particular importance for high-order methods where formation of the stiffness matrix is impractical.

A Brief Analysis

Since the number of iterations for PCG scales as $\sqrt{\kappa(P^{-1}K)}$, it is important that the condition number κ be bounded independent of problem size. For the additive Schwarz method, the upper bound is determined by using the projection properties of P_s. From a basic property of the spectral radius (see Section 2.7), (7.2.35), and the triangle inequality, one has

$$\lambda_{\max} := \rho \leq \|P_{\mathrm{AS}}^{-1} K\|_K \leq \|P_1\|_K + \|P_2\|_K + \cdots + \|P_S\|_K = S.$$

The final equality follows from the identity $\|P_s\|_K = 1$ for any projector P_s. Recognizing that $P_s + P_{s'}$ is also a projector whenever $\Omega^s \cap \Omega^{s'} = \emptyset$, we can improve this estimate to $\lambda_{\max} \leq \mathcal{C}$, where $\mathcal{C}$ is the minimum number of colors required to color each subdomain so that no two domains of the same color overlap. For regular partitions, $\mathcal{C}$ is bounded independent of S.

The lower bound is more difficult to determine. A key point is that for any $u \in V_0$, it should be possible to bound the minimum energy of the decomposition of u by the energy of u. Let C_0^2 be the smallest constant such that, $\forall u \in V_0$,

$$\mu(u) := \min_{\substack{u_s \in V_0^s \\ u = \sum_s u_s}} \sum_{s=1}^S \int_{\Omega^s} \nabla u_s \cdot \nabla u_s \, dV \leq C_0^2 \int_\Omega \nabla u \cdot \nabla u \, dV = C_0^2 (\underline{u}, \underline{u})_K.$$

(7.2.36)

It is a straightforward algebraic argument to show that $(\underline{u}, \underline{u})_P = \mu(u)$ (see [363, 339]), from which it follows immediately that

$$\frac{1}{C_0^2} = \lambda_{\min} := \min_{\underline{u} \in \mathbb{R}^N} \frac{(\underline{u}, \underline{u})_K}{(\underline{u}, \underline{u})_P}.$$

The constant C_0^2 is estimated from FE theory by using the inequality (7.2.36). The key to reducing C_0^2 is to increase the domain overlap, which allows elements of the decomposition $u = \sum_s u_s$ to decay smoothly to zero in the overlap region, thus lowering $\mu(u)$.

Two-Level Preconditioners

If there are many subdomains, it is generally necessary to augment DD-based preconditioners with a coarse-grid space to efficiently eliminate low-wavenumber components in the residual. Widlund [410] argues that, without such a space, information propagation is limited by subdomain-to-subdomain interactions, and the condition number of the Schwarz algorithms must grow at least as fast as $(L/H)^2$, where H is the characteristic diameter of the subdomains and L the characteristic diameter of Ω, resulting in an iteration count that is dependent on problem size.

Addition of a coarse-grid operator amounts to constructing a new approximation to (7.2.1) using a space $V_0^c = \mathrm{span}\{\psi_1^c(\mathbf{x}), \ldots, \psi_{n_c}^c(\mathbf{x})\}$, where the basis

7.2. Preconditioning Methods

functions are based on a relatively coarse triangulation of Ω. One then forms the system

$$K_0 \underline{u}_0 = M_0 \underline{f}_0, \quad (K_0)_{ij} := \int_\Omega \nabla \psi_i^c \cdot \nabla \psi_j^c \, dV, \quad (M_0)_{ij} := \int_\Omega \psi_i^c \psi_j^c \, dV.$$

In addition, a prolongation operator, R_0^T, is needed to interpolate the coarse-grid solution back to the original fine mesh. The preconditioner then is

$$P_{AS}^{-1} := \sum_{s=0}^{S} R_s^T K_s^{-1} R_s,$$

with a similar change for the multiplicative case.

There is considerable flexibility in building the coarse-grid operator. For instance, the interpolation and solution bases need not be the same. The key consideration is that the coarse mesh spacing should be proportional to (or finer than) the diameters of the subdomains.

With the addition of a coarse-grid approximation, the condition number for the additive Schwarz algorithm satisfies

$$\kappa\left(P_{AS}^{-1} K\right) \leq C \left(1 + \max_i \frac{H_i}{\delta_i}\right), \tag{7.2.37}$$

where H_i is the local coarse-mesh size, and δ_i is the subdomain overlap in the vicinity of the associated coarse-mesh element. The estimate (7.2.37) is independent of the number of subdomains and holds in either two or three dimensions (see [363]).

7.2.3 Schwarz Preconditioners for High-Order Methods

To develop the overlapping Schwarz method for SEMs, we start with the elements Ω^e and extend these by an amount δ in each direction to create overlapping subdomains $\tilde{\Omega}^e$. This approach provides a well-defined coarse-grid space in the form of the underlying spectral-element mesh and allows the local subdomain solves to be element-oriented. To precondition the spectral-element Laplacian K, Pahl [292] suggests using local operators $\tilde{K}_e$ built on linear FEs triangulated at the GLL points, as this allows for $\delta < H$ while ensuring that the basis functions $\psi \in V_0^e$ vanish outside $\tilde{\Omega}^e$, a key assumption in the Schwarz theory. Using the spectral equivalence between the FEM and SEM bases (see Section 2.7), Casarin [72] proves that this approach does achieve the bound (7.2.37), provided that the preconditioner is coupled with a coarse-grid space and that the spectral elements are shape-regular. He also proves a similar bound if one uses

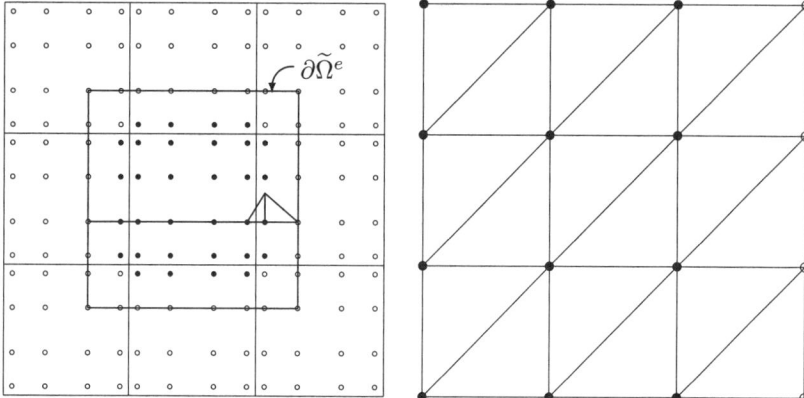

Figure 7.2.5. Overlapping Schwarz degrees of freedom (closed circles) for fine and coarse subproblems with $E = 9$ subdomains. The local system (left) is built on the Gauss points, using a tensor product of linear FE functions. The global coarse-grid problem (right) is based on linear triangles, with homogenous Dirichlet boundary conditions along the outflow boundary on the left.

the spectral-element basis with degrees of freedom restricted to nodes interior to $\tilde{\Omega}^e$.

A similar approach was employed by Fischer and coworkers [127, 129] to precondition the consistent (L^2) Poisson operator, $A := \mathbf{D}\mathbf{M}^{-1}\mathbf{D}^T$, which governs the pressure in the $\mathbb{P}_N$–$\mathbb{P}_{N-2}$ spectral-element formulation when the pressure-correction scheme (6.2.33) is used to advance the incompressible Navier–Stokes equations. The local and coarse-grid operators are based on FE approximations to the standard (H^1) Laplacian K. Numerical evidence in [127] indicates that K is spectrally equivalent to A.

Fischer et al. [129] construct the local operators $\tilde{K}_e$ from a tensor product of one-dimensional FEs based on the pressure (Gauss) nodes, as illustrated in Figure 7.2.5. This construction allows the use of the fast diagonalization method (FDM) (4.2.15) to solve the local problems, which is particularly important in the 3D case. For deformed elements, the tensor-product Laplacian is approximated by using one-dimensional operators based on the *average* separation of the sides in each direction, as suggested in [83]. Additional geometric flexibility is obtained by excluding, during restriction and prolongation, nodes in any element not sharing a full face with Ω^e, such as those denoted by the open circles in Figure 7.2.5 (left). This strategy allows an arbitrary number of elements to be connected at the spectral-element vertices without affecting the local tensor-product topology, which is critical to the FDM. We remark that subdomain data exchanges can be effected by using the routines that implement Σ' (4.5.8) (see Section 8.5.2).

7.2. Preconditioning Methods

Table 7.2.1. *Performance of the additive Schwarz algorithm.*

	FDM		$N_o = 0$		$N_o = 1$		$N_o = 3$		$K_0 = 0$		Deflation	
E	Iter.	CPU	Iter.	CPU	Iter.	CPU	Iter.	CPU	Iter.	CPU	Iter.	CPU
93	67	4.4	121	10	64	5.9	49	5.6	169	19	126	17
372	114	37	203	74	106	43	73	39	364	193	216	125
1488	166	225	303	470	158	274	107	242	802	1798	327	845

The coarse-grid operator, K_0, is based on linear elements with Dirichlet boundary conditions applied at the outflow boundary, as depicted in Figure 7.2.5. The coarse-grid prolongation, R_0^T, employs bilinear tensor-product interpolation to map the coarse-grid solution to the Gauss points, which is both memory-efficient and fast. Ensuring a correct implementation of the restriction operator R_0 requires some care, however. Since R_0^T *copies* vertex values to each element, R_0 must *sum* contributions from neighboring elements onto the coarse-grid vertex values. This issue is particularly subtle in the H^1 case if the residual is stored in its local form. In this case, $\underline{r}_L$ contains redundantly stored interface values that must be premultiplied by the inverse multiplicity $\mathcal{W}$ (4.5.17) to allow for the multiple copies generated by R_0^T.

Table 7.2.1 contrasts the performance of the additive Schwarz preconditioner with the deflation-based scheme (6.3.10)–(6.3.11) when PCG is used to compute the pressure on the first step of impulsively started flow past a cylinder at $Re_D = 5000$. Three spectral-element discretizations are considered, with $E = 93, 372$, and 1488 (see Figure 7.2.6), and $N = 7$ in all cases. The residual reduction is 10^{-5}, and CPU times (in seconds) are for an SGI Onyx 2 workstation.

Two local discretizations are considered for the additive Schwarz case. The first, denoted FDM, employs the tensor-product discretization of Figure 7.2.5.

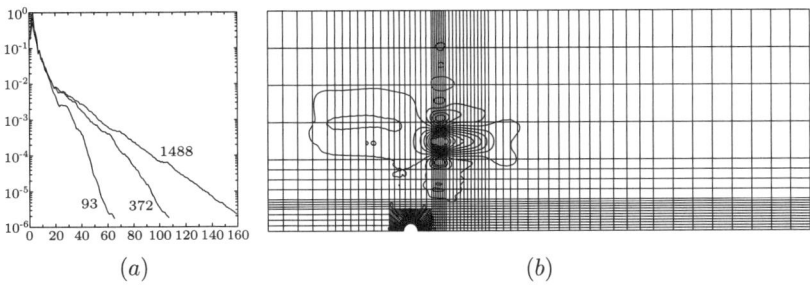

Figure 7.2.6. *(a)* Convergence histories for $N_0 = 1$ case of Table 7.1 with $E = 93, 372$, and 1488. *(b)* Contours of $\underline{p}^{\text{final}} - \underline{p}^{25}$ for $E = 1488$.

The second, denoted N_o, employs an unstructured triangulation of the Gauss points, as described in [127]. N_o indicates the number of rows of FEs attached to Ω^e to form the overlap region $\tilde{\Omega}^e$, with $N_o = 0$ corresponding to block-Jacobi augmented with a coarse-grid solve. As expected, increasing the overlap from 0 to 3 reduces the iteration count. Nevertheless, the FDM scheme with minimal overlap (corresponding to $N_o = 1$) is faster because of the efficiency of the tensor-product solves. (This effect is even more dramatic in $\mathbb{R}^3$, where other approaches are not competitive.) The deflation scheme requires two products in A per iteration, which increases its solution time. The importance of the coarse grid is illustrated by the $K_0 = 0$ column, which shows a five- to eightfold increase in iteration count for the $N_o = 3$ case when the coarse-grid solver is excluded.

Unfortunately, the iteration counts in Table 7.2.1 are not bounded with respect to the mesh refinement E. While the initial convergence rate is identical for all resolutions, the performance begins to degrade after about 20 iterations for the higher-resolution cases, as seen in Figure 7.2.6(a). In [127], the cause of this is traced to the high-aspect-ratio elements in the domain, which violate the shape-regularity assumptions in the Schwarz theory. Of the 158 iterations required to compute the solution in the $N_o = 1, E = 1488$ case, the last 133 are used to eliminate the mode shown in Figure 7.2.6(b). It is clear that this mode is a by-product of the high-aspect-ratio elements in the far field and has nothing to do with the physics of the problem. Possible remedies are discussed in [127] and include either increasing the overlap of the high-aspect-ratio elements or eliminating them completely by using nonconforming elements, as described in Section 7.3. In addition, one can also consider using a finer coarse-grid space.

7.2.4 Spectral-Element Multigrid

Suppose that K is an $\mathcal{N} \times \mathcal{N}$ SPD system arising from discretization of an elliptic problem such as (7.2.1), and consider Richardson iteration for solving $K\underline{u} = \underline{f}$, with an SPD preconditioner B and initial guess $\underline{\hat{u}}^i$:

$$\underline{u}^0 = \underline{\hat{u}}^i, \qquad (7.2.38)$$

$$\underline{u}^l = \underline{u}^{l-1} + B^{-1}(\underline{f} - K\underline{u}^{l-1}), \quad l = 1, \ldots, m. \qquad (7.2.39)$$

The equation for the corresponding error, $\underline{e}^m := \underline{u} - \underline{u}^m$, is

$$\underline{e}^m = \underline{e}^{m-1} - B^{-1}K\underline{e}^{m-1} = (I - B^{-1}K)\underline{e}^{m-1} = (I - B^{-1}K)^m \underline{e}^0.$$

Let $\{\underline{z}_k\}$ be the set of B-orthonormal eigenvectors satisfying $K\underline{z}_k = \lambda_k B \underline{z}_k$, with B scaled so that $0 < \lambda_1 \leq \lambda_2 \leq \cdots \leq \lambda_\mathcal{N} := 1$. If $\underline{e}^m = \sum_k \hat{e}_k^m \underline{z}_k$, it follows

7.2. Preconditioning Methods

that

$$\hat{e}_k^m = (1 - \lambda_k)^m \hat{e}_k^0, \quad 1 \le k \le \mathcal{N}. \tag{7.2.40}$$

Because of the scaling of B, the high-wavenumber ($\lambda_k \to 1$) error components are damped quickly, while the low-wavenumber components are reduced less. The idea behind multigrid is to use the *smoothing* step (7.2.38)–(7.2.39) to reduce the error in the upper part of the spectrum, followed by a coarse-grid correction to reduce the error in the lower part.

Now, consider a coarse grid with p degrees of freedom. If the $p \times \mathcal{N}$ matrix R is a restriction operator mapping the fine to the coarse grid, then the coarse-grid correction for a two-level scheme is computed as

$$\begin{aligned}\delta \underline{u} &= R^T K_c^{-1} R(\underline{f} - K \underline{u}^m) \\ \tilde{\underline{u}} &= \underline{u}^m + \delta \underline{u}.\end{aligned} \tag{7.2.41}$$

As in the Schwarz case, R^T is the prolongation operator, which is usually taken to be the interpolant from the coarse to the fine mesh. If $K_c := RKR^T$, the corresponding error,

$$\tilde{\underline{e}} = \left(I - R^T K_c^{-1} R K\right) \underline{e}^m,$$

is the K-orthogonal projection of $\underline{e}^m$ onto $\mathcal{R}^\perp(R)$ [see (2.7.36)]. If the range of R is the span of the p first eigenvectors of $B^{-1}K$, $\mathcal{R}(R) = \text{span}\{\underline{z}_1, \ldots, \underline{z}_p\}$, then the error spectrum after (7.2.41) is

$$\hat{\tilde{e}}_k = \begin{cases} 0, & 1 \le k \le p, \\ (1 - \lambda_k)^m \hat{e}_k^0, & p < k \le n, \end{cases} \tag{7.2.42}$$

and, because of the orthogonality of the $\underline{z}_k$'s, $\|\tilde{\underline{e}}\|_K \le (1 - \lambda_{p+1})^m \|\underline{e}^0\|_K$.

Usually, the coarse-grid space is not perfectly coincident with the low-wavenumber eigenfunctions, and the errors introduced by the coarse-grid correction must be corrected by further fine-grid smoothing, that is, repeating (7.2.38)–(7.2.39), setting $\underline{u}^0 := \tilde{\underline{u}}$ on each successive round. In addition, it is possible to implement the coarse grid correction (7.2.41) as a smooth–correct–smooth sequence on another grid pair, which can be recursively nested, giving rise to the classic multigrid V- or W-cycles.

Figure 7.2.7 illustrates the V-cycle. At the coarsest grid level the linear system is usually solved by a direct method, because the number of unknowns is small. Assuming (7.2.42) does hold, the nominal error reduction per V-cycle would be $\rho^{2m} := (1 - \lambda_{p+1})^{2m}$. The quantity ρ is referred to as the multigrid convergence factor.

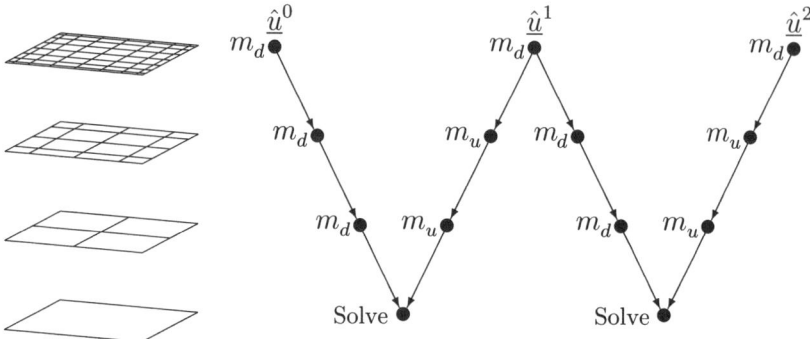

Figure 7.2.7. Multigrid V-cycle with m_d and m_u smoothings on the downward and upward sweeps, respectively.

In FD applications, (7.2.39) is usually based upon Gauss–Seidel relaxation. However, in spectral methods, the matrix elements are not explicitly available, and it is necessary to consider alternatives. In [337, 331], Rønquist and Patera implement multigrid for the SEM using Jacobi smoothing. They choose

$$B := \sigma \operatorname{diag}(K_{ii}), \qquad \sigma := \max_{\underline{x}\in\mathbb{R}^N,\ \underline{x}\neq 0} \frac{\underline{x}^T K \underline{x}}{\underline{x}^T \operatorname{diag}(K_{ii})\underline{x}},$$

with K_{ii} given by (4.4.13)–(4.4.14) and σ computed in a preprocessing step using a simple power iteration. The coarse-to-fine mapping is effected by the interpolation operator

$$(R^T)_{ij} = \pi_j^{\tilde{N}}(\xi_i^N), \qquad (7.2.43)$$

where $\tilde{N} < N$ (corresponding to $p < \mathcal{N}$) is the order of the coarse approximation in one space dimension. In the multidimensional multielement case, (7.2.43) is applied to the residual in a tensor-product form on an element-by-element basis. In addition, the coarse-grid stiffness matrix K_c is based upon a discretization of (7.2.1) with a polynomial of degree $\tilde{N}$, rather than through the explicit construction RKR^T. In one dimension the two forms are identical. In higher dimensions, this identity ceases to hold because of the approximate quadrature (4.3.18) used for $\hat{M}$ in, for example, the form $K^e = \hat{M} \otimes \hat{K} + \hat{K} \otimes \hat{M}$.

Maday and Muñoz [257, 258, 283] analyze this algorithm for the one- and two-dimensional cases with $\tilde{N} = N/2$. In one dimension with a single element, $E = 1$, they establish that (7.2.42) holds with $\lambda_{p+1} \approx 1/4$ and $p + 1 = \tilde{N}$. For $E > 1$, (7.2.42) ceases to hold. Following the theoretical framework of Bank and Douglas [21], they show however that the convergence factor $\rho \sim 0.75$ still

7.2. Preconditioning Methods

holds, independent of E and N, provided $m \leq 3$. These results are confirmed by the numerical experiments of Rønquist and Patera [337, 331, 259].

In two dimensions, the results are no longer independent of N. Instead, one has $\rho \sim 1 - C/N$, where C is a constant. To improve upon this, Rønquist and Patera suggest replacing the Richardson iteration (7.2.39) with Chebyshev iteration, which is also considered by Zang et al. [422]. The idea is to scale the iterates in the smoother by using

$$\underline{u}^l = \underline{u}^{l-1} + \alpha_l B^{-1}(\underline{f} - K\underline{u}^{l-1}), \qquad l = 1, \ldots, m, \qquad (7.2.44)$$

so that the error polynomial (7.2.40) becomes

$$\hat{e}_k^m = p_m(\lambda_k)\hat{e}_k^0, \qquad p_m(\lambda) = \prod_{l=1}^{m}(1 - \alpha_l \lambda). \qquad (7.2.45)$$

One chooses the coefficients α_l to minimize

$$\max_{\lambda_{p+1} \leq \lambda \leq 1} |p_m(\lambda)|. \qquad (7.2.46)$$

The solution to (7.2.46) is well known (e.g., [44]) and is given by

$$p_m(\lambda) = \frac{T_m\left(1 - 2\frac{\lambda - 1}{\lambda_{p+1} - 1}\right)}{T_m\left(1 - 2\frac{-1}{\lambda_{p+1} - 1}\right)}, \qquad (7.2.47)$$

where T_m is the Chebyshev polynomial of degree m (Section B.1.3). Once an estimate of λ_{p+1} is available, one can find the coefficients α_l by equating $p_m(\lambda)$ in (7.2.45) and (7.2.47) at m different values of λ. This acceleration improves the convergence rate to $\rho \sim 1 - 2(C/N)^{\frac{1}{2}}$.

As we can see, the Chebyshev procedure requires an estimate of λ_{p+1}, which may be difficult to obtain in a general geometry application. An alternative suggested by many authors is to employ full projection (e.g., CG) in the smoothing steps, attacking the error over the entire interval $[\lambda_1, \lambda_n]$. In this case, Canuto et al. [64] suggest setting the number of smoothings to zero on the downward leg of the V-cycle, since the low-wavenumber components will be damped by the subsequent coarse-grid correction anyway. Working up from coarse to fine, the projection scheme will effectively reduce the remaining high-wavenumber components.

It is also possible to incorporate projection at the highest level of the V-cycle iteration by seeking the best fit in span$\{\underline{\hat{u}}^0, \underline{\hat{u}}^1, \ldots\}$ (see Figure 7.2.7). In this case, a single V-cycle will act as a preconditioner in an outer CG or GMRES iteration. If projection is also used in the inner sequence, the outer basis will

352   7. Domain Decomposition

not be a strict Krylov subspace in powers of $P^{-1}K$, because P will be changing from one iteration to the next. Because the polynomial dependencies of CG and GMRES will be compromised, it is probably best in this case to explicitly orthogonalize the outer basis by using a method similar to Algorithm 2.1 of Section 2.7.4, or to employ the flexible variant of GMRES [339].

Finally, given the spectral equivalence discussed in Section 2.7.3, the smoother B may also be derived from other discretizations. In early work on spectral multigrid, Zang et al. [422, 423] used incomplete factorizations of FD operators for B. A similar approach has been pursued by Heinrichs, who uses FD operators to precondition a number of spectral applications [188, 189, 190]. In an earlier work [187], Heinrichs also points out the advantages of line GS smoothing in spectral applications because of the high aspect ratio of the grid spacing near the borders.

7.3 The Mortar Element Method

The use of the conforming spectral elements presented so far frequently induces the propagation of strips or layers of densely packed elements throughout the mesh when refinement is needed close to solid objects in the computational domain, in boundary layers, or in the vicinity of geometrical singularities (e.g., reentrant corners), as shown in Figure 7.3.1. Such an arrangement incurs the penalty of obtaining elements with bad aspect ratios where they are unnecessary and of burdening the solver with an undue number of elements, particularly in far-field domains of external flow problems.

A cure for this difficulty is to relax the interface matching condition to allow for nonpropagating refinement. As illustrated in Figure 7.3.2, localized refinement can be achieved by varying the polynomial degree from one element to

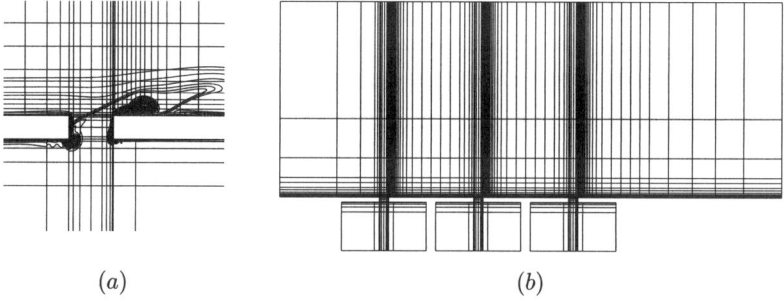

(a)   (b)

Figure 7.3.1. Close-up of mesh near a blowing or suction slot (*a*) reveals how propagating (conforming) refinement leads to undesirable (unnecessary, high-aspect-ratio) elements in the far field (*b*).

7.3. The Mortar Element Method

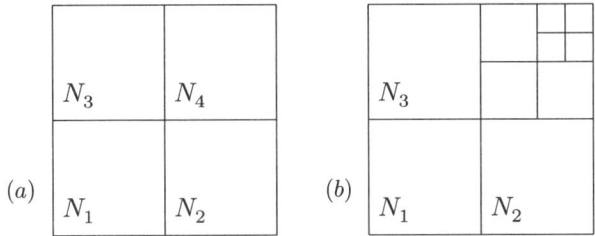

Figure 7.3.2. Functionally (a) and geometrically (b) varying refinement.

the next and/or by allowing multiple elements to share a single edge of an adjacent element. Such strategies improve the geometrical flexibility of the spectral discretization, allow for mesh adaptivity, and help to circumvent the loss of accuracy near singularities. The nonpropagating refinement can be accomplished either with conforming (interpolatory) bases or with nonconforming bases in which matching conditions are imposed through the use of L^2 projection operators defined on the *mortar* comprising the spectral-element interfaces. To illustrate the key points, we consider the *free-vertex* variant of the mortar spectral element (MSE) method in detail and illustrate how the conforming and constrained-vertex refinement strategies can be implemented in a similar framework.

The MSE method relaxes the H^1 continuity requirements of the conforming spectral-element method by considering each element individually and achieving matching or patching conditions through a variational process. The mortars play the role of binding the bricks of the spectral construction. Through the use of mortars, one can also couple domains discretized by spectral elements with others treated by finite elements. However, we will forgo that possibility and concentrate on nonconforming spectral methods.

The basic concept of the mortar element method can be sketched as follows. As in the conforming case, the entire domain Ω is subdivided into E nonoverlapping subdomains Ω^e, $e = 1, \ldots, E$, such that $\bar{\Omega} = \bigcup_{e=1}^{E} \bar{\Omega}^e$. The skeleton of the domain decomposition comprises all interfaces between subdomains and is also decomposed into mortars. These mortars are one-dimensional geometrical entities for 2D problems and surfaces for 3D problems. We require the mortars to coincide with a complete edge or face of one of the subdomains. The choice of mortars is not unique, as will be shown later in this section. Since full decoupling between subdomains is one of the goals, the choice of the local discretizations is left totally arbitrary.

A key issue is related to the choice of the discrete space for the mortar functions. In principle, we should ensure weak continuity between all subdomains; to this end, we impose the condition that the L^2 projection of the jump

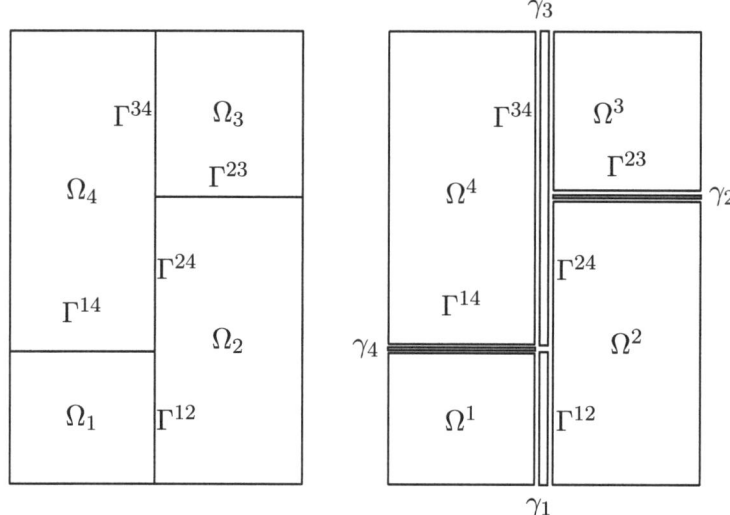

Figure 7.3.3. Example of nonconforming domain and mortar decomposition in $\mathbb{R}^2$.

across the interface vanishes. Early variants also required continuity wherever two spectral-element vertices (or edges in three dimensions) coincided. This approach was introduced in the theses by Mavriplis [267] and Anagnostou [5] for application to the Navier–Stokes equations. Numerical analysis results were presented in DD-methods conferences and by Bernardi, Maday, Patera, and coworkers [255, 6, 42, 43]. A more recent approach developed by Ben Belgacem and Maday [27, 28] relaxes the vertex matching condition. This simplifies the implementation for complex geometries and reduces the communication requirements in parallel computations.

To facilitate the presentation of the subject, we will restrict ourselves to 2D problems. Let us consider a rectangular domain divided into four subdomains, as shown in Figure 7.3.3. In each subdomain, the edges are designated by $\partial \Omega^{e,s}$, $s = 1, \ldots, 4$. For convenience, we attribute systematically $s = 1, \ldots, 4$ to the top, right, bottom, and left edges, respectively, when visited clockwise. For any $1 \le e < f \le E$, let Γ^{ef} denote the subdomain interface obtained by $\Gamma^{ef} = \bar{\Omega}^e \cap \bar{\Omega}^f$. In a nonconforming domain decomposition, since edges of subdomains and interfaces need not coincide, an edge may comprise several interfaces. The skeleton Γ is defined by $\Gamma = \bigcup_{1 \le e < f \le E} \Gamma^{ef}$ and corresponds to the complete interface. The skeleton itself is decomposed into disjoint mortars, which will be denoted by γ_m, $m = 1, \ldots, M$. By definition, a mortar is required to correspond to a complete edge of one of the subdomains. The intersection of the mortars is the empty space. To each mortar, we associate a unique index m

7.3. The Mortar Element Method

and an associated pair $e(m)$, $s(m)$ such that the mortar γ_m coincides with the edge $\partial\Omega^{e(m),s(m)}$ of $\Omega^{e(m)}$.

The geometrical mortars are the support of mortar functions chosen as polynomials of degree $N_{e(m)}$. The space of mortar functions φ for homogeneous Dirichlet conditions is defined by the relation

$$\mathcal{M}_N = \left\{ \varphi \in L^2(\Gamma); \varphi = \bigcup_{m=1}^{M} \varphi_m, \varphi_m = \varphi\big|_{\gamma_m} \in \mathbb{P}_{N_{e(m)}}(\gamma_m), \varphi\big|_{\partial\Omega} = 0 \right\}. \quad (7.3.1)$$

These mortar functions satisfy the essential boundary conditions of the problem; they are, however, not continuous. The spectral-element functional space is $\mathbb{P}_{N_e}(\Omega^e)$, $e = 1, \ldots, E$. One can define a trace operator on this space having as its image the space $\mathbb{P}_{N_{e(m)}}(\gamma_m)$ of polynomials with degree $\leq N_e$ on $\partial\Omega^{e(m),s(m)}$. Mortars and spectral elements are linked by matching conditions across the interfaces.

Let V_E denote the discrete space of functions

$$V_E = \left\{ v_E := (v_{N_1}, \ldots, v_{N_E}); v_{N_e} \in \mathbb{P}_{N_e}(\Omega^e); \text{ mortar conditions} \right\}. \quad (7.3.2)$$

We emphasize that v_E denotes the *set* of E polynomial functions in several space dimensions, with various polynomial degrees.

The mortar conditions are given as follows: For $1 \leq s \leq 4$ and $e \in \{1, \ldots, E\}$, $\exists \varphi \in \mathcal{M}_N$ such that:

Mortar Condition 1.

$$v_{N_e}\big|_{\partial\Omega^{e,s}} = \varphi_m\big|_{\partial\Omega^{e,s}} \quad \text{for} \quad (e, s) = (e(m), s(m)). \quad (7.3.3)$$

Mortar Condition 2.

$$\int_{\partial\Omega^{e,s}} (v_{N_e} - \varphi)\psi \, d\tau = 0 \quad \forall \psi \in \mathbb{P}_{N_e-2}(\partial\Omega^{e,s}) \quad \text{otherwise.} \quad (7.3.4)$$

The first condition imposes that the test and basis functions of the monitoring subdomain $\Omega^{e(m)}$ coincide with the mortar function φ_m on the geometrical mortar. The second integral condition (7.3.4) enforces weak continuity across the interface $\Gamma^{e(m)f}$. In contrast to earlier constrained-vertex variants [267, 5, 43], the mortar conditions do not explicitly enforce continuity at coincident spectral-element vertices unless they belong to the essential boundaries; only the $N_e - 2$ interior edge points are constrained by the test-function space, while the vertices are allowed to float. The degrees of freedom along a mortared

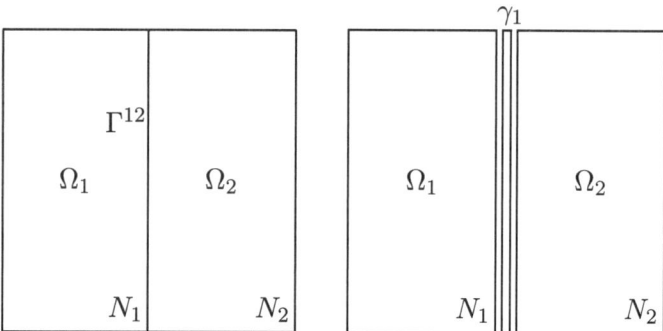

Figure 7.3.4. Example of geometrically conforming domain decomposition in $\mathbb{R}^2$.

edge are thus $N_{e(m)} + 1$ mortar coefficients and the two vertex coefficients on the dependent side. The integral matching allows for the computation of the interface nodes on the dependent domain and realizes the L^2 projection of the mortar values onto the interface.

We illustrate these concepts by a few examples. We begin with the simple case depicted in Figure 7.3.4, where a rectangular domain is split into two rectangular subdomains, Ω^1 and Ω^2, discretized by polynomials of degree N_1 and N_2, respectively. The decomposition is geometrically conforming. The interface Γ^{12} may produce two mortars: $\gamma_1 = \partial \Omega^{1,2}$ with $(e(1), s(1)) = (1, 2)$, or $\gamma_1 = \partial \Omega^{2,4}$ with $(e(1), s(1)) = (2, 4)$. Let $\tilde{u}^1$ and $\tilde{u}^2$ denote the restrictions of $u^1(\mathbf{x})$ and $u^2(\mathbf{x})$ to Γ^{12}, respectively. We assume without loss of generality that $\gamma_1 := \partial \Omega^{1,2}$, implying that $\tilde{u}^1 \equiv \varphi_1$ (Mortar Condition 1). The dependent variable is thus $\tilde{u}^2$, to be determined by the integral constraint

$$\int_{\Gamma^{12}} [\tilde{u}^2(s) - \varphi(s)] \psi(s)\, ds = 0 \qquad \forall \psi \in \mathbb{P}_{N_2-2}(\Gamma^{12}) \qquad (7.3.5)$$

(Mortar Condition 2). A matrix formulation of the integral constraint will be given after the introduction of elliptic problems in the next subsection. However, we note that a particularly simple form for $\tilde{u}^2$ results for the case $N_1 \leq N_2$:

$$\tilde{u}^2(s) = \tilde{u}^1(s) + \alpha L_{N_2}(s) + \beta L_{N_2-1}(s). \qquad (7.3.6)$$

The coefficients α and β can be expressed in terms of $\tilde{u}_0^2$ and $\tilde{u}_{N_2}^2$ by evaluating (7.3.6) at the vertices. We also note that the case $N_1 \leq N_2$ corresponds to a (variable-resolution) *conforming* case if vertex matching is imposed.

A second example is the nonconforming geometry of Figure 7.3.5. The three subdomains with discretizations N_1, N_2, N_3, respectively, are said to be *functionally conforming* if the SEM is built so that $u^2|_{\Gamma^{12}} \equiv u^1|_{\Gamma^{12}}$ and $u^3|_{\Gamma^{13}} \equiv u^1|_{\Gamma^{13}}$;

7.3. The Mortar Element Method

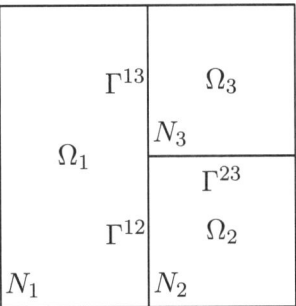

Figure 7.3.5. Example of nonconforming domain decomposition in $\mathbb{R}^2$.

otherwise they are *nonconforming*. Possible choices for mortars include $\gamma_m \in \{\Gamma^{12}, \Gamma^{13}, \Gamma^{23}\}$ and $\gamma_m \in \{\Gamma^{12} \cup \Gamma^{13}, \Gamma^{23}\}$.

7.3.1 Elliptic Problems

We are now ready to apply the discretization (7.3.2) to the continuous elliptic problem given by Equation (4.3.2) with Dirichlet boundary conditions. We will follow closely the line of reasoning drawn in Section 4.5, where conforming elements were introduced. For nonconforming (i.e. discontinuous) spectral element discretizations, the discrete operators for the scalar product, the bilinear form, and the source term are now defined as

$$(u_E, v_E)_{\text{nc}} := \sum_{e=1}^{E} \int_{\Omega^e} u_{N_e}(\mathbf{x}) v_{N_e}(\mathbf{x}) \, d\mathbf{x}, \tag{7.3.7}$$

$$\mathcal{A}_{\text{nc}}(u_E, v_E) := \sum_{e=1}^{E} \int_{\Omega^e} \left(p(\mathbf{x}) \nabla u_{N_e} \cdot \nabla v_{N_e} + q(\mathbf{x}) u_{N_e} v_{N_e} \right) d\mathbf{x}, \tag{7.3.8}$$

$$\mathcal{F}_{\text{nc}}(v_E) := \sum_{e=1}^{E} \int_{\Omega^e} f_{N_e}(\mathbf{x}) v_{N_e}(\mathbf{x}) \, d\mathbf{x}, \tag{7.3.9}$$

with quadratures over Ω replaced by a sum of quadratures over all nonconforming elements. The discrete problem becomes: Find $u_E \in V_E$ such that

$$\mathcal{A}_{\text{nc}}(u_E, v_E) := \mathcal{F}_{\text{nc}}(v_E) \quad \forall v_E \in V_E. \tag{7.3.10}$$

For error estimates, we introduce the broken norm

$$\|v_E\|_{H^1_*(\Omega)} = \left(\sum_{e=1}^{E} \|v_{N_e}\|^2_{H^1(\Omega^e)} \right)^{\frac{1}{2}}, \tag{7.3.11}$$

which corresponds to the rms of the elemental norms. Estimates bound the error by a term involving the approximation error per element and a consistency term resulting from the jumps of normal derivatives across the interfaces. Mortar Condition 2 (7.3.4) permits computing a bound of this consistency error. If we assume that $p(\mathbf{x}) = 1$, $q(\mathbf{x}) = 0$ in Equation (4.3.2) and if the source data f are such that $f_{N_e} \in H^{t_e}(\Omega^e)$ with $t_e > 1$, the final result for a 2D Poisson equation is given as

$$\|u - u_E\|_{H^1_*(\Omega)} \le C \sum_{e=1}^{E} N^{1-s_e} \|u\|_{H^{s_e}(\Omega^e)} + N^{-t_e} \|f\|_{H^{t_e}(\Omega^e)}. \quad (7.3.12)$$

The real number s_e indicates the regularity of the exact solution u in the subdomain Ω^e. From the error estimate (7.3.12), where the constant C is not a function of the local discretization, we conclude that the convergence of the nonconforming method depends essentially on the local regularity of the solution. Ben Belgacem and Maday [27] extend the previous methods to 3D problems and prove similar (almost optimal) results.

7.3.2 Implementation

The MSE formulation is implemented in the framework introduced in Section 4.5.2, in which matrix–vector products are effected through the action of connectivity matrices Q and Q^T, and of the block-diagonal matrix K_L comprising *local* stiffness matrices. The local stiffness matrices are unchanged from the forms given in Chapter 4. Imposition of the mortar constraints on the local degrees of freedom requires changes only to the connectivity matrices. The objective is to identify with Q a set of relationships between the global degrees of freedom, $\underline{u}$, and the local degrees of freedom, $\underline{u}_L$, such that $u(\mathbf{x})$ automatically satisfies the integral matching constraint (7.3.4). The methodology is illustrated by considering several nonconforming cases.

We begin with the simple two-element problem of Figure 4.5.2, with $N_1 = N_2 = N = 2$; here, however, we use the nonconforming discretization (7.3.10). In this case, the number of global degrees of freedom is increased from 15 to 17, to allow for the two additional vertex values that are allowed to float. This connectivity matrix is illustrated in Figure 7.3.6, in which the entries of the global vector $\underline{u}$ are denoted by their corresponding local notation. Q is no longer a Boolean matrix as it was in Figure 4.5.2. The five column vectors $\underline{a}, \ldots, \underline{e}$ reflect the influence of the global (mortar) degrees of freedom (here, $u^1_{2,0}, u^1_{2,1}, u^1_{2,2}, u^2_{0,0}$, and $u^2_{0,2}$) upon the dependent variables (here, $u^2_{0,1}$). By construction, $a_0 = b_0 = c_0 = e_0 = 0$, $d_0 = 1$, $a_2 = b_2 = c_2 = d_2 = 0$, and $e_2 = 1$, so that the obvious identities hold.

7.3. The Mortar Element Method

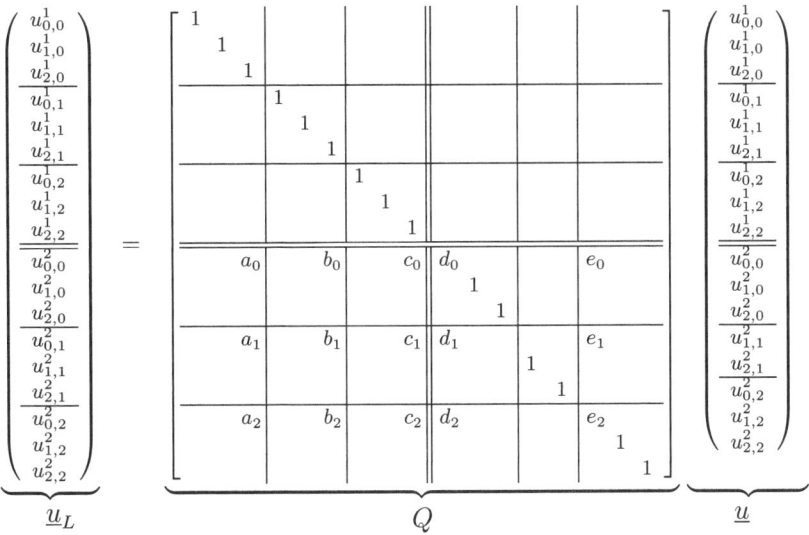

Figure 7.3.6. Nonconforming connectivity matrix for $(E, N) = (2, 2)$ spectral-element configuration of Figure 4.5.3.

To simplify the notation for Q, we reorder the local and global vectors with the interface nodes last. Because the bases are Lagrangian, the interior and interface degrees of freedom decouple, leading to the form

$$\begin{pmatrix} \underline{u}_{L,\,\text{interior}} \\ \underline{u}_{L,\,\text{interface}} \end{pmatrix} = \underbrace{\begin{bmatrix} I & \\ & \tilde{Q} \end{bmatrix}}_{Q} \begin{pmatrix} \underline{u}_{\text{interior}} \\ \underline{u}_{\text{interface}} \end{pmatrix}, \qquad (7.3.13)$$

where I is the identity and $\tilde{Q}$ is the connectivity matrix for the interface variables. In addition, because elements of the global vector $\underline{u}$ in Figure 7.3.6 are identified with their local counterparts, the interface relationship can be further reduced to

$$\begin{pmatrix} \underline{u}_{L,i} \\ \underline{u}_{L,d} \end{pmatrix} = \underbrace{\begin{bmatrix} I \\ Q_m \end{bmatrix}}_{\tilde{Q}} \underline{u}_{\text{interface}},$$

where Q_m reflects the required relationship between the independent (mortar) variables $\underline{u}_{L,i}$ and dependent variables $\underline{u}_{L,d}$. Implementation of the MSE formulation thus consists of (i) identifying the independent variables (the true degrees of freedom in the PDE approximation) and dependent variables (determined from Mortar Condition 2), and (ii) effecting the *action* of Q_m and Q_m^T through subroutine calls.

The matrix Q_m deserves careful inspection. In the two-element case of Figure 7.3.4, with, say, $\gamma_1 := \partial\Omega^{1,2}$, the dependent variable $\tilde{u}^2$ is determined by the integral constraint (7.3.5). To simplify the quadrature, we build the test function ψ on the polynomial basis

$$\hat{\pi}_i(r) := \frac{L'_{N_2}(r)}{L''_{N_2}(\xi_i^{N_2})(r - \xi_i^{N_2})} \in \mathbb{P}_{N_2-2}, \tag{7.3.14}$$

satisfying $\hat{\pi}_i(\xi_j^{N_2}) = \delta_{ij}$ at the *interior* GLL points $\xi_i^{N_2}$, $i \in [1, N_2 - 1]$.

Let us evaluate the first integral in (7.3.5). Substituting (7.3.14) for ψ and applying GLL quadrature, we obtain

$$\int_{\Gamma^{12}} \tilde{u}^2 \psi \, ds = \sum_{k=0}^{N_2} \rho_k^{N_2} \tilde{u}^2(\xi_k^{N_2}) \psi(\xi_k^{N_2}) = \sum_{k=0}^{N_2} \rho_k^{N_2} \sum_{j=0}^{N_2} \tilde{u}_j^2 \pi_j(\xi_k^{N_2}) \hat{\pi}_i(\xi_k^{N_2})$$

$$= \sum_{j=0}^{N_2} \rho_j^{N_2} \tilde{u}_j^2 \hat{\pi}_i(\xi_j^{N_2}) \quad \text{for } i \in [1, N_2 - 1]. \tag{7.3.15}$$

The other integral is evaluated similarly, except that we must use a higher-order quadrature rule if $N_1 > N_2 + 1$. Setting $N := \max(N_1, N_2)$ leads to

$$\int_{\Gamma^{12}} \varphi\psi \, ds = \sum_{k=0}^{N} \rho_k^N \varphi(\xi_k^N) \psi(\xi_k^N)$$

$$= \sum_{j=0}^{N_1} \varphi_j \sum_{k=0}^{N} \rho_k^N \pi_j^{N_1}(\xi_k^N) \hat{\pi}_i(\xi_k^N), \quad i \in [1, N_2 - 1]. \tag{7.3.16}$$

One may use (7.3.15)–(7.3.16) to recast the mortar constraint (7.3.5) in matrix notation:

$$B\,\underline{\tilde{u}}^2 = P\underline{\varphi}. \tag{7.3.17}$$

The matrix B has a special structure. Because $\hat{\pi}_i(\xi_j^{N_2}) = \delta_{ij}$ for $i \in [1, N_2 - 1]$, the interior part is diagonal. The first and last columns are full, but they are concerned only with degrees of freedom (namely, $\tilde{u}_0^2$ and $\tilde{u}_{N_2}^2$); they are not driven by the matching condition. We observe also that the matrix P is

7.3. The Mortar Element Method

full. The projection can be written in closed form as

$$\begin{pmatrix} \tilde{u}_1^2 \\ \vdots \\ \tilde{u}_{N_2-1}^2 \end{pmatrix} = B_{\text{int}}^{-1} \underbrace{\left[\begin{array}{ccc|cc} p_{1,0} & \cdots & p_{1N_1} & -b_{1,0} & -b_{1,N_2} \\ \vdots & \ddots & \vdots & \vdots & \vdots \\ p_{N_2-1,0} & \cdots & p_{N_2-1,N_1} & -b_{N_2-1,0} & -b_{N_2-1,N_2} \end{array} \right]}_{Q_m} \begin{pmatrix} \tilde{u}_0^1 \\ \vdots \\ \tilde{u}_{N_1}^1 \\ \tilde{u}_0^2 \\ \tilde{u}_{N_2}^2 \end{pmatrix},$$

with

$$p_{ij} := \sum_{k=0}^{N} \rho_k^N \pi_j^{N_1}(\xi_k^N) \hat{\pi}_i(\xi_k^N), \quad b_{ij} := \rho_i^{N_2} \hat{\pi}_i(\xi_j^{N_2}), \quad (7.3.18)$$

and $B_{\text{int}} := \text{diag}(\rho_i^{N_2})$, $i \in [1, N_2 - 1]$.

Other variable resolution schemes can be cast within the same framework. If vertex matching is enforced in addition to the integral constraint (e.g., as in [5, 267]), then $\tilde{u}_0^2$ and $\tilde{u}_{N_2}^2$ are no longer degrees of freedom. The continuity condition is given by $\underline{\tilde{u}}^2 = Q_m \underline{\tilde{u}}^1$, with $Q_m := B^{-1} P$, where P and B are given by (7.3.18) augmented with first and last rows $b_{0,j} = p_{0,j} = \delta_{0j}$ and $b_{N_2,j} = \delta_{N_2 j}$, $p_{N_2,j} = \delta_{N_1 j}$, respectively. If full continuity is enforced (i.e., the *conforming* case considered in [334]), then $\tilde{u}^1 \equiv \tilde{u}^2$, implying that the global variables must be identified with the local edge having *fewer* degrees of freedom. For $N_1 > N_2$, the connectivity relationship is thus $\underline{\tilde{u}}^1 = Q_m \underline{\tilde{u}}^2$, with matrix entries $(Q_m)_{ij} := \pi_j^{N_2}(\xi_i^{N_1})$, corresponding to the interpolant of $\tilde{u}^2(y)$ evaluated at the nodes of element 1.

As a second example, we consider the three-element geometry of Figure 7.3.7. Here, the integral constraint must be applied in physical coordinates to

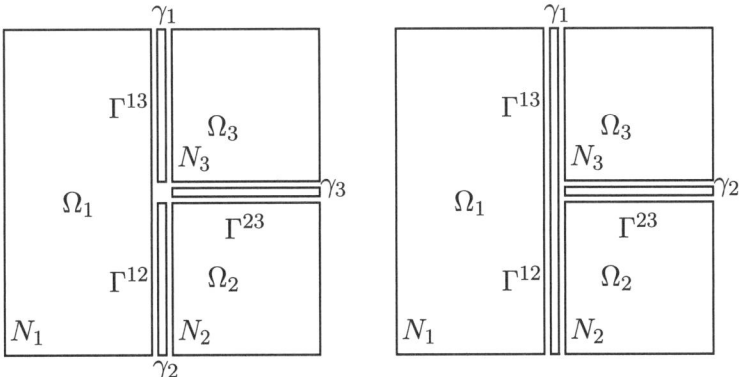

Figure 7.3.7. Mortar decompositions in a geometrically nonconforming case.

take account of the relative position and size of the interfaces. In addition, the long interface, $\partial\Omega^{1,2}$, can be decomposed into either one or two mortars, depending on which side is chosen to represent the global degrees of freedom.

We begin with the two-mortar case for the vertical interface shown on the left in Figure 7.3.7. Let $\tilde{u}^e$ denote the restriction of $u^e(\mathbf{x})$ to $\partial\Omega^{1,2}$, $e = 1, 2,$ or 3. The degrees of freedom are $\tilde{u}^2$ and $\tilde{u}^3$, plus the vertex values of $\tilde{u}^1$. Evaluation of the integrals required for the L^2 projection yields

$$\int_{\partial\Omega^{1,2}} \tilde{u}^1 \psi \, ds = B\underline{\tilde{u}}^1$$

$$\int_{\Gamma^{1,e}} \tilde{u}^e \psi \, ds = P^{1,e}\underline{\tilde{u}}^e, \qquad e = 2, 3.$$

If we let $|\cdot|$ denote the length of a segment, the matrix elements are

$$B_{ij} := \frac{|\partial\Omega^{1,2}|}{2} \sum_{k=0}^{N_1} \rho_k^{N_1} \hat{\pi}_i^{N_1}(\xi_k^{N_1}) \sum_{j=0}^{N_1} \pi_j^{N_1}(\xi_k^{N_1}),$$

$$P_{ij}^{1,e} := \frac{|\Gamma^{1,e}|}{2} \sum_{k=0}^{N} \rho_k^N \hat{\pi}_i^{N_1}(\zeta_k^e) \pi_j^{N_e}(\xi_k^N), \qquad N := \max(N_1, N_e),$$

with local coordinate transformations

$$\zeta_k^2 := (\xi_k^N + 1)\frac{|\partial\Omega^{2,4}|}{|\partial\Omega^{1,2}|} - 1, \qquad \zeta_k^3 := (\xi_k^N + 1)\frac{|\partial\Omega^{3,4}|}{|\partial\Omega^{1,2}|} - 1 + 2\frac{|\partial\Omega^{2,4}|}{|\partial\Omega^{1,2}|}. \tag{7.3.19}$$

The dependent side is computed as $\underline{\tilde{u}}^1_{\text{int}} = B^{-1}_{\text{int}}(P^{1,2}\underline{\tilde{u}}^2 + P^{1,3}\underline{\tilde{u}}^3 + B_v\underline{\tilde{u}}^1_v)$, where the subscripts int and v refer to interior and vertex coefficients for $\tilde{u}^1$. If vertex matching is imposed, one simply computes $\underline{\tilde{u}}^1 = B^{-1}P^{1,2}\underline{\tilde{u}}^2 + B^{-1}P^{1,3}\underline{\tilde{u}}^3$, with the first and last rows of B and $P^{1,e}$ modified accordingly. Here the situation is further complicated by the fact that $\tilde{u}^2_{N_2}$ and $\tilde{u}^3_0$ are no longer independent degrees of freedom, since they must also match.

The single-mortar case on the right in Figure 7.3.7 is implemented in a similar way. Since the degrees of freedom are associated with $\tilde{u}^1$, Mortar Condition 2 for $e = 2$ and 3 amounts to

$$\int_{\Gamma^{1e}} (\tilde{u}^e - \tilde{u}^1)\psi \, ds = 0 \quad \forall \psi \in \mathbb{P}_{N_e - 2}.$$

7.3. The Mortar Element Method

Applying quadrature in each case with the basis (7.3.14) for ψ yields

$$\int_{\Gamma^{1e}} \tilde{u}^e \psi \, ds = \sum_{k=0}^{N_e} \rho_k^{N_e} \sum_{j=0}^{N_e} \hat{\pi}_i^{N_e}(\xi_k^{N_e}) \pi_j^{N_e}(\xi_k^{N_e}) \tilde{u}_j^e$$

$$= \begin{cases} \rho_i^{N_e} \tilde{u}_i^e, & i = 1, \ldots, N_e - 1, \\ \sum_{j=0}^{N_e} \rho_j^{N_e} \hat{\pi}_i^{N_e}(\xi_j^{N_e}) \tilde{u}_j^e, & i = 0, N_e, \end{cases} \quad (7.3.20)$$

$$\int_{\Gamma^{1e}} \tilde{u}^1 \psi \, ds = \sum_{k=0}^{N} \rho_k^N \sum_{j=0}^{N_1} \hat{\pi}_i^{N_e}(\xi_k^N) \pi_j^{N_1}(\zeta_k^e) \tilde{u}_j^1, \quad (7.3.21)$$

where $N := \max(N_1, N_e)$ and ζ_k^e is given by (7.3.19). In this case, one obtains separate relationships for each of $\tilde{u}^2$ and $\tilde{u}^3$ by equating (7.3.20) and (7.3.21). Since there are fewer independent degrees of freedom ($\tilde{u}^1$) than dependent ones ($\tilde{u}^2$ and $\tilde{u}^3$), this configuration also lends itself to a (conforming) interpolation-based scheme, provided also that $N_1 \leq \min(N_2, N_3)$. The coefficients for $\tilde{u}^2$ and $\tilde{u}^3$ in this case would be computed as

$$\tilde{u}_i^e = \sum_{j=0}^{N_1} \pi_j^{N_1}(\zeta_i^e) \tilde{u}_j^1. \quad (7.3.22)$$

with N set equal to N_e in the definition (7.3.19).

Before leaving this subsection on implementations, we note that all of the preceding techniques carry through to the 3D case. A practical restriction in the geometrically nonconforming situation is that the refined interface should be a tensor-product decomposition of its parent. This allows the matrix operators (Q_m, B, P, etc.) to be expressed as tensor products of their one-dimensional counterparts, so that their application requires only $O(N^3)$ operations. Details of 3D implementations can be found in [114, 229].

7.3.3 Steady Stokes Problems

Let us now enlarge to the nonconforming case the $\mathbb{P}_N$–$\mathbb{P}_{N-2}$ spectral-element discretization of the Stokes problem (5.2.20)–(5.2.21). Here, the discrete space of functions is V_E^d, a d-dimensional extension to vector fields of the discrete space V_E defined by (7.3.2):

$$V_E^d = \{ \mathbf{v}_E := (\mathbf{v}_{N_1}, \ldots, \mathbf{v}_{N_E}); \, \mathbf{v}_{N_e} \in \mathbb{P}_{N_E}^d(\Omega^e); \, \text{mortar conditions} \}. \quad (7.3.23)$$

Again we emphasize that $\mathbf{v}_E$ denotes the set of E (polynomial) vector functions in d space dimensions with different polynomial degrees per element. With regard to the pressure, a space Z_E is built from $Z \in L_0^2(\Omega)$ such that

$$Z_E = \{q_E \in Z; \, q_E|_{\Omega^e} := q_{N_e} \in \mathbb{P}_{N_e-2}(\Omega^e), \, e = 1, \ldots, E\}. \quad (7.3.24)$$

The weak formulation (5.2.20)–(5.2.21) of the Stokes problem in the framework of nonconforming elements may now be stated: Find $(\mathbf{v}_E, p_E) \in (V_E^d, Z_E)$ such that $\forall (\mathbf{u}_E, q_E) \in (V_E^d, Z_E)$

$$\mathcal{A}_{\mathrm{nc}}(\mathbf{u}_E, \mathbf{v}_E) + \mathcal{B}_{\mathrm{nc,GL}}(\mathbf{u}_E, p_E) = \mathcal{F}_{\mathrm{nc}}(\mathbf{u}_E), \quad (7.3.25)$$

$$\mathcal{B}_{\mathrm{nc,GL}}(\mathbf{v}_E, q_E) = 0. \quad (7.3.26)$$

The bilinear forms appearing in (7.3.25)–(7.3.26) are given by the relationships

$$\mathcal{A}_{\mathrm{nc}}(\mathbf{u}_E, \mathbf{v}_E) = \sum_{e=1}^{E} \left(\nabla \mathbf{u}_{N_e}, \nabla \mathbf{v}_{N_e}\right)_{N_e}, \quad (7.3.27)$$

$$\mathcal{B}_{\mathrm{nc,GL}}(\mathbf{u}_E, q_E) = -\sum_{e=1}^{E} \left(\mathrm{div}\,\mathbf{u}_{N_e}, q_{N_e}\right)_{N_e}, \quad (7.3.28)$$

where the discrete scalar products in (7.3.27)–(7.3.28) are provided by the local GLL and GL rules, respectively. The mortar conditions apply to the d components of $\mathbf{v}$. The pressure being discontinuous in the $\mathbb{P}_N$–$\mathbb{P}_{N-2}$ context, the use of mortars does not require matching conditions for this variable. This is one of the benefits of this formulation.

Let us define the multiindex of polynomial degrees by $\epsilon = (N_1, \ldots, N_E)$. The inf–sup constant of the nonconforming approach β_ϵ is obtained by evaluating the relation

$$\beta_\epsilon = \inf_{q_E \in Z_E} \sup_{\mathbf{u}_E \in V_E^d} \frac{\mathcal{B}_{\mathrm{nc,GL}}(\mathbf{u}_E, q_E)}{\|\mathbf{u}_E\|_{H_*^1(\Omega)^d} \|q_E\|_{L_0^2(\Omega)}}. \quad (7.3.29)$$

Ben Belgacem et al. [26] proved that the discrete Stokes problem (7.3.25)–(7.3.26) is well posed. They obtained inf–sup conditions for a fixed-domain decomposition made of rectangles or parallelepipeds and for a hierarchical decomposition such as that in Figure 7.3.2.

Theorem 7.1 *If we assume that* $\forall e \in \{1, \ldots, E\}$

1. $N_e \geq 2$, *with the pressure p_E in Z_E,*
2. *the solution $(\mathbf{v}, p)|_{\Omega^e}$ of the weak continuous problem (5.2.12)–(5.2.13) is in $H^{s_e}(\Omega^e)^d \times H^{s_e-1}(\Omega^e)$ with $s_e \geq 1$, and the function $\mathbf{f}_{N_e}$ is in $H^{t_e}(\Omega^e)^d$ with $t_e > 1$,*

7.3. The Mortar Element Method

then the problem (7.3.25)–(7.3.26) possesses a unique solution $(\mathbf{v}_E, p_E)$ in $V_E^d \times Z_E$ with the following error bounds:

$$\|\mathbf{v} - \mathbf{v}_E\|_{H^1_*(\Omega)^d} + \beta_\epsilon \|p - p_E\|_{L^2(\Omega)}$$
$$\leq C_1 \left(1 + \frac{1}{\beta_\epsilon}\right) \sum_{e=1}^{N} N_e^{1-s_e} \left(\|\mathbf{v}\|_{H^{s_e}(\Omega^e)^d} + \|p\|_{H^{s_e-1}(\Omega^e)}\right)$$
$$+ C_2 \sum_{e=1}^{N} N_e^{-t_e} \|\mathbf{f}\|_{H^{t_e}(\Omega^e)^d}, \qquad (7.3.30)$$

with C_1 and C_2 constants independent of ϵ.

The mortar inf–sup constant is bounded from below so that

$$\beta_\epsilon \geq C(\mathrm{DD}) \inf_{1 \leq e \leq E} \beta_{N_e}. \qquad (7.3.31)$$

The "constant" $C(\mathrm{DD})$ depends on the domain decomposition itself, that is, the subdomain partitioning, going exponentially fast to zero with the refinement level. In practical calculations, however, the quad- or octtree is not very deep and C does not reach alarming values. For a fixed decomposition, $C(\mathrm{DD})$ becomes a true constant, again bringing optimal results for nonconforming subdomains.

Solvers for Stokes and Navier–Stokes rely on iterative methods like CG. Substructuring preconditioners taking advantage of the decoupling between interior unknowns and interface variables [see Equation (7.3.13)] are in order. This kind of algorithm was investigated for 2D elliptic problems by Achdou et al. [2], who proved that the condition number of the preconditioned matrix is bounded by $\max_{1 \leq e \leq E} (1 + \log N_e)^2$.

7.3.4 Applications

We demonstrate the flexibility of adaptive refinement schemes with a few examples.

Flow Around an Impeller

The first example, due to Anagnostou [5], consists of a rotating impeller of length $L/2$ centered at the origin inside a square domain, $\Omega := [-L/2, L/2]^2$. The Reynolds number, based on the rotation rate ω and kinematic viscosity ν, is defined as $Re = \omega L^2 / \nu$. The impeller has an aspect ratio of 5. Figure 7.3.8(a) shows the domain decomposed into two conforming meshes, joined by a circular interface that permits the inner domain to rotate relative to the outer domain. The mortar defined on this interface provides a mechanism to

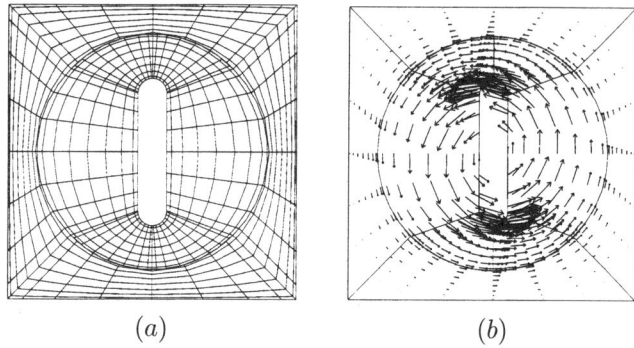

(a) (b)

Figure 7.3.8. Rotating-impeller problem.

impose the L^2 projection so as to achieve the desired continuity. Figure 7.3.8(b) shows the results at $Re = 1000$ after six impeller revolutions, at which point the solution has reached a steady periodic state.

Resonator Cavity Flow

Our second example, due to Kruse [229], shows the benefits of using nonpropagating refinement when simulating external flow problems. The 2D domain shown in Figure 7.3.9 consists of a flat plate over the interval $x \in [-12, 42]$, with unit-width slots centered at $x = 0, 10,$ and 20. The slots are connect to 9×4.5 chambers (*resonators*). Flow enters and exits the bottom of the chambers according to a time-oscillatory parabolic profile, $v = A \sin(2\pi f t)(1 - \tilde{x}^2)$, where $\tilde{x}$ is the local coordinate within each chamber such that the velocity vanishes at the left and right walls, $A = 0.05$, and $f = 0.2$. A Blasius profile is specified at

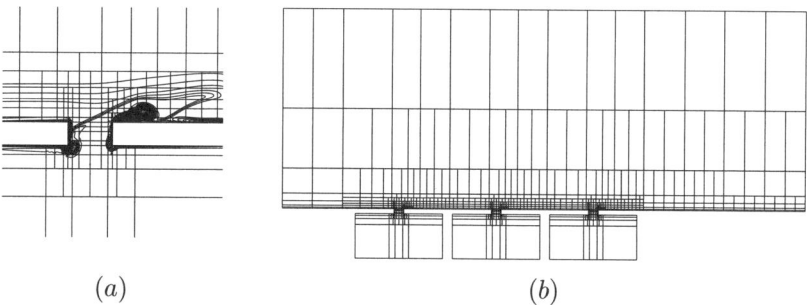

(a) (b)

Figure 7.3.9. Close-up of vorticity contours near a blowing or suction slot (a) for the resonator problem with nonpropagating mesh refinement [cf. Figure 7.3.1(a)]. The far-field view (b) reveals the absence of high-aspect-ratio elements that can degrade iterative solver performance.

$x = -12$ with unit velocity and $\delta_{99} = 1$. The Reynolds number based on inflow velocity and boundary-layer thickness is $Re = 200$.

The discretization is based on polynomial degree $N = 7$ in all elements and is conforming; nodal values along the daughter interfaces are computed as the interpolant of the solution along the parent edge. Figure 7.3.9 shows the computed vorticity at the same instant as in Figure 7.3.1. Clearly, both meshes obtain the same-quality solution. However, the present case, in addition to having fewer elements, requires five to ten times *fewer* pressure iterations because of the absence of the high-aspect-ratio elements that cause the iteration performance in the structured case to degrade.

Clearance-Gap Glow

Clearance-gap flow occurs at the tip of a blade in Kaplan hydraulic machines. This flow is highly turbulent and decreases dramatically the efficiency of the machine. Therefore, one must understand the physics of the phenomenon. The fluid near the blade tip is strongly accelerated in the gap and generates a vortical structure. A simpler situation to investigate is the internal flow in a channel partially obstructed by a blade. Figure 7.3.10 shows the computational domain and the boundary conditions. The geometrical parameters are the channel height $H = 1$; the channel length $L = 30$; the blade thickness $l = 0.2$; the gap clearance $h = 0.1$; the distance between the inlet section and the blade, $L_i = 4.0$; and the distance between the blade and the outflow section, $L_o = 25.8$. The Reynolds number of the simulation, based on the maximum velocity and the channel height, is equal to 288. Mortar elements have been used to obtain a better pressure field near the tip corners. A reference solution was computed with a polynomial order $N = 6$ in every element. The mortar elements are located around the blade tip as shown in Figure 7.3.11. The same polynomial degree applies to the elements recorded by the same letter according to the Table 7.3.1, which shows the various combinations used in these mortared elements.

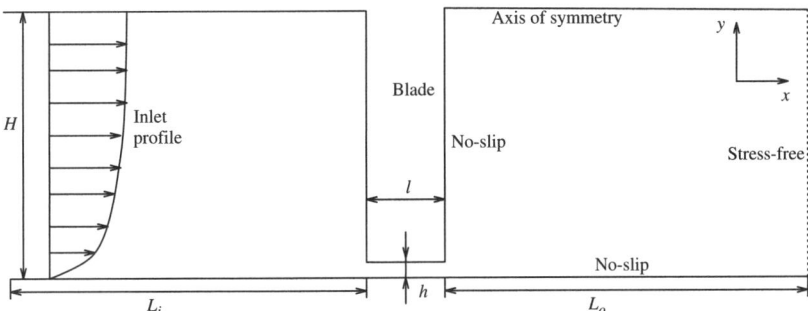

Figure 7.3.10. Clearance-gap problem: geometry and boundary conditions.

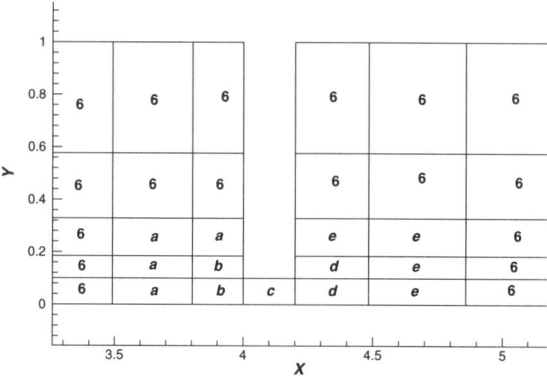

Figure 7.3.11. Clearance-gap problem: polynomial degree distribution in the mortar elements around the blade.

We will inspect the isocontour lines of the pressure field as a sensitive indicator of numerical instabilities. The reference case is case A in Figure 7.3.12. As soon as we increase the polynomial order by two units between adjacent elements as in case B, the pressure field displays wiggles around the blade tip. In case C, where the distribution is symmetric, the highest degree used is 8. Inspection of the corresponding plot leads to the conclusion that this degree is not yet high enough to remove all spurious oscillations. The final case, D, increases the polynomial order to 9 under the blade itself. The pressure field is smoother, and the isobars are almost vertical under the blade, indicating that the flow under the blade is Poiseuille-like. This may be considered as the best solution from the physical point of view.

7.4 Adaptivity and Singularity Treatment

In the case of spectral quadrilateral and hexahedral elements, adaptivity is achieved through the use of mortar elements. The mortar method allows the coupling without much difficulty between spectral- and finite-element

Table 7.3.1. *Polynomial degrees in the mortared elements of Figure 7.3.11.*

Case	a	b	c	d	e
A	6	6	6	6	6
B	8	10	10	10	8
C	7	8	8	8	7
D	7	8	9	8	7

7.4. Adaptivity and Singularity Treatment

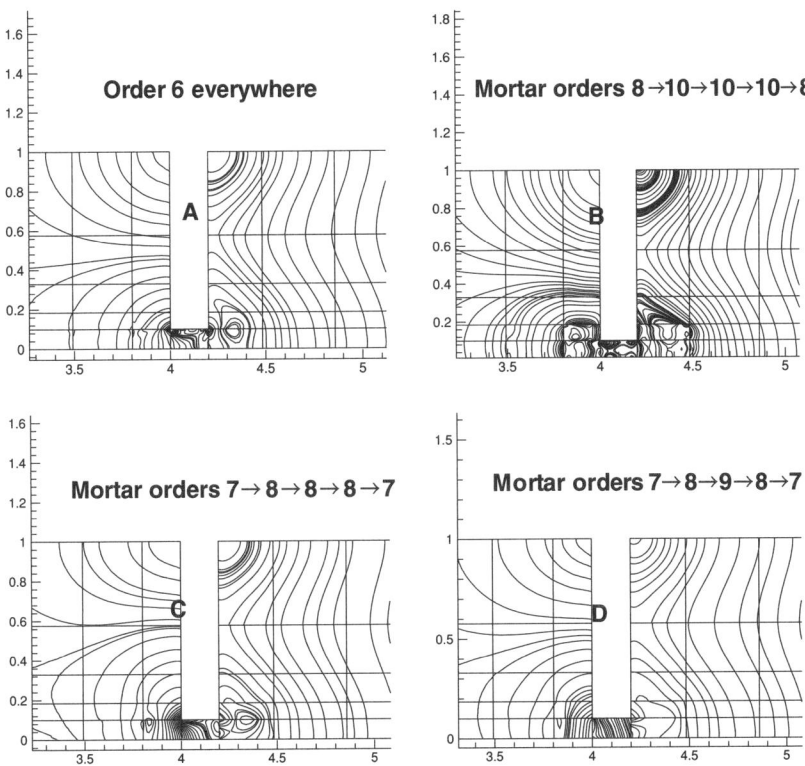

Figure 7.3.12. Clearance-gap problem: isobars.

discretizations. (This coupling will be addressed briefly in the next subsection.) However, the presence of singularities either in the geometry or in the function itself and its first derivatives, for example, affects the rate of convergence, which decreases to low order. To circumvent this difficulty, special treatments are possible and will be explained further.

During the 1990s, triangular and tetrahedral elements were developed. They will also be described in this section, because they bring much flexibility to cope with complicated geometries and, therefore, complex computational grids. Furthermore, the hierarchical property of the bases is well suited to local adaptivity.

7.4.1 Coupling between Finite and Spectral Elements

Coupling between finite and spectral elements was analyzed by Ben Belgacem and Maday [29] for the approximation of 3D second-order elliptic problems with a nonconforming domain decomposition. As an example, a Poisson equation is considered with Dirichlet homogeneous boundary conditions. Two matching procedures may be proposed. The FE matching corresponds to the case where the interface spectral-element values are obtained by a projection of the FE values. The symmetric case is the spectral-element matching where the interface FE values are produced via a projection of the spectral-element values. Ben Belgacem and Maday prove that the discretization error is bounded from above by three contributions: the quadrature error, the best approximation error, and the consistency error. This last contribution is in fact optimal for both matchings. From the practical point of view, it is most valuable to balance the spectral- and finite-element discretizations in such a way that the error is almost uniform over the complete domain.

7.4.2 Singularity Treatment

For 2D problems, most of the singularities are corner singularities. Let us suppose that the corner lies at the origin of the axes. The first side of the corner is located along the positive x-axis, and the other side makes an angle $\alpha\pi$, $0 < \alpha < 2$, in the counterclockwise direction to the first. Working with polar coordinates (r, θ) in the vicinity of the corner, one obtains the asymptotic form of the solution,

$$u(r, \theta) \sim Cr^\beta g(\theta) h(r, \theta), \qquad (7.4.1)$$

where C is a constant, $g(\theta)$ a smooth function, and $h(r, \theta)$ a smooth cutoff function. For a fixed spectral-element decomposition with polynomial degree N in each element, the error decreases like

$$\|u_N(r, \theta) - u(r, \theta)\|_{H^1(\Omega)} \leq C' N^{-2\beta - \epsilon}, \qquad (7.4.2)$$

where C' is a positive constant independent of N and ϵ a positive number. The exponent β depends both on the equation and on the angle α. Most of the problems are such that $\beta = 1/\alpha$, and the convergence rate is consequently between first and second order.

The method of auxiliary mapping was examined in the spectral-element context by Pathria and Karniadakis [299]. With the Schwarz–Christoffel conformal mapping used in complex variables, it is possible to send the domain containing the corner to a semicircle where the corner and the resulting singularities have been removed. Denoting by z the usual complex variable in polar form, the mapping $z = \xi^\alpha$ allows one to write the solution in terms of

$\xi = \zeta \exp(i\varphi)$:

$$u(\zeta, \varphi) = \sum_{p=0}^{\infty} a_p \zeta^p \sin(p\varphi). \tag{7.4.3}$$

This mapping may be used in conjunction with a domain decomposition procedure as presented in Section 7.2. Impressive spectral rates of convergence are recovered with this special treatment of the singularity.

Several elliptic models from Laplace to Helmholtz equations are examined in [299]. If this auxiliary mapping is not yet sufficient to weaken the singularity, a better rate of convergence can be obtained by augmenting the spectral-element basis with a few (one or two) singular basis functions. Augmentation has to be done *after* the mapping.

Extension of the asymptotic solution around 2D corners to the case of 3D corners or edges is lacking. Consequently, the only way to avoid the difficulty is to use a graded mesh with h-refinement; that is, we keep the polynomial degree constant, and we refine the size of the spectral elements through the 2D or 3D corner point. This strategy was adopted by Gerdes and Schötzau [154] in the hp-FEM context and used by Wilhelm [411] with the SEM. The refinement is characterized by a reduction factor $\sigma < 1$. One proceeds recursively, going to the corner and creating layers of elements such that the size ratio for each pair of adjacent elements is $\sigma \in (0, 1)$. The case $\sigma = 0.5$ is represented in Figure 7.3.2(*b*). In Wilhelm's thesis, the refinement operates quite well and yields excellent results for the flow of a Newtonian fluid over a forward-facing step at moderate Reynolds numbers around 300.

7.4.3 Triangular and Tetrahedral Elements

The first attempts at using triangular elements were carried out in the early nineties by Funaro [141] and Mavriplis and Van Rosendale [270]. The design of spectral methods on triangles and 3D geometrical objects like prisms, pyramids, and tetrahedra started with the seminal paper of Dubiner [113]. The book by Karniadakis and Sherwin [221] provides a complete view of the developments of the past decade for spectral/hp methods. Schwab [348, 349] describes the mathematical foundations of the p and hp FEM and in [349] focuses on fluid flow problems. Here, we will give only a short presentation of this new field, referring the reader to the cited books for full details.

We wish to keep the nice qualities of SEMs, namely, orthogonality, convergence, and computational efficiency. This last aspect was achieved through the tensor product of the bases and the sum factorization algorithm (see Chapter 4).

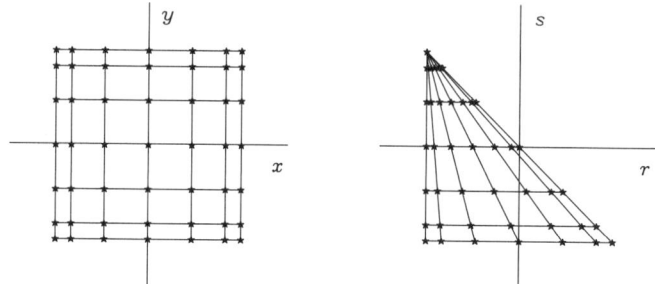

Figure 7.4.1. The reference square and triangle.

To work on the triangle of Figure 7.4.1 defined as the region

$$T^2 = \{(r, s); -1 \leq r, s, r+s \leq 0\}, \tag{7.4.4}$$

we need to specify the following mapping, which sends the reference square onto the triangle:

$$\begin{aligned} r &= \frac{(1+x)(1-y)}{2} - 1, \\ s &= y, \end{aligned} \tag{7.4.5}$$

and the inverse transformation

$$\begin{aligned} x &= 2\frac{1+r}{1-s} - 1, \\ y &= s. \end{aligned} \tag{7.4.6}$$

It is a simple exercise to check that the vertical lines in the reference square become a bundle of rays emanating from the vertex $(-1, 1)$ of the triangle. From (7.4.6), the ray coordinate at $(r = -1, s = 1)$ is multivalued. However, one can show that the behavior of x is bounded. This singularity comes from the fact that through the mapping, the top edge of the reference square degenerates into a single point. The 3D extension of the theory to construct tetrahedra from hexahedra requires three successive applications of mappings like Equation (7.4.5).

Modal Bases

Taking advantage of the orthogonality property of the Jacobi polynomials $P_n^{\alpha,\beta}$ [see Section B.1.1, Equation (B.1.6)], Dubiner [113] proposed the approximation

$$u_N(\mathbf{x}) = \sum_{m=0}^{N} \sum_{n=0}^{N-m} u_{mn} \psi_{mn}(r(\mathbf{x}), s(\mathbf{x})), \tag{7.4.7}$$

7.4. Adaptivity and Singularity Treatment

where the basis functions are given by

$$\psi_{mn}(r, s) = P_m^{0,0}\left(2\frac{1+r}{1-s} - 1\right)(1-s)^m P_n^{2m+1,0}(s). \tag{7.4.8}$$

This is a modal basis like that in the Galerkin framework (Section 2.2), and we cannot interpret the coefficients u_{mn} directly as nodal values. Moreover, the above relation is a polynomial in (r, s), since the factor $(1-s)^m$ combined with the first factor $P_m^{0,0}(2(1+r)/(1-s) - 1)$ yields a polynomial of mth order in r, s. By definition, $P_m^{0,0}$ is the Legendre polynomial L_m. The second polynomial $P_n^{2m+1,0}$ is introduced to maintain the orthogonality of the triangular basis. This property can be proven by evaluating

$$\int_{-1}^{+1}\int_{-1}^{-s} \psi_{mn}(r', s')\psi_{kl}(r', s')\, dr'\, ds' \tag{7.4.9}$$

and transforming this integral on the reference square (x, y). The Jacobian of the mapping is $(1-y)/2$.

The two-dimensional integral can be written as the product of two one-dimensional integrals. The first integral involves standard Legendre polynomials and is nonvanishing only when $m \neq k$. The second integral becomes

$$\int_{-1}^{+1} P_n^{2m+1,0} P_l^{2k+1,0}(1-y)^{m+k}\frac{1-y}{2}\, dy. \tag{7.4.10}$$

For $m = k$, (7.4.10) is zero when $n \neq l$. This is because the factor $(1-y)^{m+k+1}$ is the weight function for this integral.

The basis (7.4.7) may be cast as the product of two polynomials in the (x, y) coordinates

$$\psi_{mn}(r, s) = \psi_m^1(x)\psi_{mn}^2(y), \tag{7.4.11}$$

with the definitions

$$\psi_m^1(x) = P_m^{0,0}(x) = L_m(x), \qquad \psi_{mn}^2(y) = (1-y)^m P_n^{2m+1,0}(y). \tag{7.4.12}$$

The superscripts in (7.4.12) refer to the space direction. The factorization of the two-dimensional function as the product of two one-dimensional polynomials is called a *warped product* by Dubiner and may be considered merely as a generalized tensor product.

Keeping in mind that we want to use a weak formulation of the problem that necessitates C^0 continuity between elements, we note that this last property is not easily handled by the basis (7.4.11). Furthermore, the nice shapes (mostly diagonal) of the resulting matrices that were displayed in Figure 2.3.2 are lost on account of the constraints imposed by this continuity. To overcome the difficulty, as in the classical hp-method, Dubiner splits the modes into interior modes, which will be bubble functions, and boundary modes, which will be decomposed into vertex and edge modes. The vertex modes are approximated by Lagrangian linear shape functions, which will automatically enforce this C^0 continuity. The edge modes will be nonzero along one edge and will be vanishing at all other vertices and edges. Recalling the linear shape functions given by Equation (2.3.7), we have the following basis:

- Bubble modes ($2 \leq m; 1 \leq n, m < M; m + n < N$):

$$\psi_{mn} = \pi_0(x)\pi_1(x)P_{m-2}^{1,1}(x)\pi_0^m(y)\pi_1(y)P_{n-1}^{2m-1,1}(y). \quad (7.4.13)$$

- Edge modes ($2 \leq m; 1 \leq n, m < M; m + n < N$):

$$\psi_{m0}^{\text{side}(s=-1)} = \pi_0(x)\pi_1(x)P_{m-2}^{1,1}(x)\pi_0^m(y),$$
$$\psi_{1n}^{\text{side}(r+s=0)} = \pi_1(x)\pi_0(y)\pi_1(y)P_{n-1}^{1,1}(y), \quad (7.4.14)$$
$$\psi_{1n}^{\text{side}(r=-1)} = \pi_0(x)\pi_0(y)\pi_1(y)P_{n-1}^{1,1}(y),$$

- Vertex modes:

$$\psi_{00}(r = s = -1) = \pi_0(x)\pi_0(y),$$
$$\psi_{10}(r = 1, s = -1) = \pi_1(x)\pi_0(y), \quad (7.4.15)$$
$$\psi_{11}(r = -1, s = 1) = \pi_1(y).$$

The C^0 continuity in the discretization imposes $N = M$. The presence of the linear polynomials in the bubbles, namely, the product $\pi_0(x)\pi_1(x)$, ensures that they vanish on the triangle boundaries. The values of α, β used in the given basis are fixed in such a way that they achieve as much orthogonality as possible in the setup of the mass and stiffness matrices. This point is elaborated by Sherwin and Karniadakis [358].

In [359], Sherwin and Karniadakis investigate the tetrahedral bases; in [357], they investigate the 2D and 3D bases using area and volume coordinates, respectively. A unified description of high-order hp elements for hybrid 2D meshes composed of triangles and quadrilaterals is proposed by Warburton et al. [406].

Owens [290] has shown that a basis of the mentioned form can be derived from the eigenfunctions of a singular Sturm–Liouville problem solved on the triangle. These polynomials are orthogonal with a unit weight on this triangle.

7.4. Adaptivity and Singularity Treatment

Integration

Integration on the triangle is transferred via the mapping (7.4.6) to the rectangle. We obtain for the integration of u

$$\int_{T^2} u(r,s)\, dr\, ds = \int_{-1}^{1} \int_{-1}^{1} u(x,y) J\, dx\, dy, \qquad (7.4.16)$$

where $J = (1-y)/2$. We can still perform the quadrature of the integral of the r.h.s. by a Gauss–Legendre (GL) rule. Nonetheless, in this case, a more general quadrature rule helps to take the Jacobian into account automatically. This Gauss–Jacobi quadrature rule, introduced by Michelsen and Villadsen [400], was also used by Karniadakis and Sherwin [221]. The factor $(1-x)^\alpha (1+x)^\beta$ is incorporated in the integrand:

$$\int_{-1}^{1} (1-x)^\alpha (1+x)^\beta u(x)\, dx = \sum_{k=0}^{N} w^{\alpha,\beta} u\bigl(\zeta_k^{\alpha,\beta}\bigr), \qquad (7.4.17)$$

where $w^{\alpha,\beta}$ and $\zeta_k^{\alpha,\beta}$ are the weights and zeros corresponding to the choice of the indices α, β. For $\alpha = \beta = 0$, one recovers the well-known GL rule. Therefore, choosing $\alpha = 1, \beta = 0$, we include the Jacobian in the quadrature weights.

A last choice remains to be made: the quadrature nodes. On the triangle, although the vertex ($r = -1, s = 1$) presents no complication for integration, one is wise to avoid its use in differentiation procedures. A Gauss–Radau quadrature in the s-direction with $s = -1$ included solves the problem; a GLL rule in the r-direction is a standard option.

Differentiation

Differentiation can be carried out in the modal (spectral) or the physical space. The transformation tool is the matrix built with columns, where the coefficients are the discrete values of the spectral polynomials ψ_{mn} at the quadrature points. Assuming that we can go back and forth between spectral and physical space, we prefer to perform the computation of the derivative operators in physical space. On the triangular spectral element, the approximation can be written

$$u(r,s) = \sum_{i,j} u_{i,j} \pi_i(x) \pi_j^{1,0}(y), \qquad (7.4.18)$$

where $\pi_i(x)$ are given by Equation (2.4.3) and $\pi_j^{1,0}(y)$ denotes the Lagrange interpolant based on the zeros of the $(1, 0)$ Jacobi polynomial and the end points $y = \pm 1$. The partial derivative with respect to the first variable at the

m, n quadrature point gives

$$\frac{\partial u(r_m, s_n)}{\partial r} = \sum_{i,j} u_{i,j} \frac{2}{1-s_n} \frac{\partial \pi_i(x_m)}{\partial x} \pi_j^{1,0}(y_n)$$
$$= \sum_i u_{i,n} \frac{2}{1-s_n} \frac{\partial \pi_i(x_m)}{\partial x}. \qquad (7.4.19)$$

Besides the factor $2/(1-s_n)$, the last expression involves the derivative matrix defined by Equation (2.4.9). From this observation, we conclude that the operation cost to compute a derivative is the same as in the SEM. The condition numbers for the first-order derivative and the weak Laplacian scale as $O(N^2)$ and $O(N^4)$, respectively, as in the spectral-element case. Preconditioning is required to keep the computational cost reasonable.

One penalty for using triangular spectral elements is that the mass matrix is no longer diagonal, even for the interior modes (see Karniadakis and Sherwin [221]), imposing a system solution even for time-explicit problems. This difficulty has aroused much research in the spectral community. Wingate and Boyd [414] and Taylor et al. [372, 373] have attempted to obtain a diagonal mass matrix. However, the main motivation for undertaking this research was the design of Lagrange polynomial bases on the triangle and the tetrahedron.

Nodal Bases

Nodal bases offer the potential of a diagonal mass matrix, and so are of interest in time-dependent problems where inversion of such operators is repeatedly called for. Stability has been established for two sets of high-order interpolation points on triangles that have recently received much attention. These are the Fekete points, studied by Taylor et al. [373], and the electrostatic points, developed by Hesthaven [201].

The functional space of approximation on the triangle is given by

$$\mathbb{P}_N = \text{span}\{x^m y^n; \ m, n \geq 0, \ m + n \leq N\}. \qquad (7.4.20)$$

This space corresponds, for two space variables, to a triangular truncation of Pascal's triangle. Let us choose a basis $\pi_i = \psi_{mn}$, $i = 1, \ldots, N$, with i representing a unique pair (m, n) for the above space, and construct the generalized Vandermonde matrix $V(x_1, x_2, \ldots, x_N)$ defined at the points $\{x_i \in \Omega, i = 1, \ldots, N\}$. The matrix V has as coefficients the values $V_{ij} = \pi_j(x_i)$. The Fekete collocation points computed by Taylor et al. [373] are those that maximize the determinant of V. We note that Fekete points are independent of the choice for the basis π. For quadrilaterals, the Gauss–Lobatto nodes are Fekete points. For

7.4. Adaptivity and Singularity Treatment

triangles, the Fekete points are distributed along the edges as Gauss–Lobatto points of the polynomial of degree N. This property allows covering the 2D domain by triangles and quadrilaterals that are naturally conforming.

Another set of collocation points has been derived by Hesthaven [201] using an analogy between the zeros of Jacobi polynomials and the solution of an electrostatics problem. In this case, the collocation points correspond to the steady-state position of charges that minimizes the electrostatic energy. Extension to three-dimensional tetrahedral elements has been carried out by Hesthaven and Teng [203]. When these nodal bases are used to integrate advection problems or advection–diffusion problems, it has been proved that these spectral approximations produce energy-stable schemes [202, 203].

7.4.4 Error Estimates and Adaptivity

Two main streams have been used to deal with a posteriori error estimation. The first one works in spectral space and inspects the decay of the coefficient spectra. The second one operates in physical space and relies on residuals evaluation.

Spectral Error Estimator

Mavriplis [268, 269] and Henderson [196, 195] use an error estimate that is the sum of two contributions, namely, the truncation error and the quadrature error. Typically, if the numerical approximation to the spectral element e is written as

$$u_N^e(r) = \sum_{n=0}^{N} a_n^e L_n(r), \qquad (7.4.21)$$

where L_n is the Legendre polynomial of degree n and the a_n^e's are the spectral coefficients, we have

$$\epsilon_{\text{est}} = \left(\frac{a_N^2}{\frac{1}{2}(2N+1)} + \int_{N+1}^{\infty} \frac{[a(n)]^2}{\frac{1}{2}(2n+1)} \, dn \right)^{1/2}. \qquad (7.4.22)$$

The function $a(n)$ is constructed as a least-squares best fit of the last four points of the spectrum with an exponential decay $a(n) = C \exp(-\alpha n)$. The decay rate α indicates insufficient resolution if $\alpha < 1$ and good resolution if $\alpha > 1$. The refinement process uses the decay rate to decide whether it increases the number of elements and decreases accordingly the polynomial degree, or whether it has to move elements and reconstruct the grid. For that last purpose, Henderson [196] introduces the concept of a voxel database on geometric considerations.

Physical Error Indicators

Valenciano and Owens [394, 393] propose an adaptive modification strategy relying on h and p refinements. One of the ingredients of the numerical method uses the broken norm of the mortar elements [see Equation (7.3.11)]. In case of the steady Stokes problem, the error indicator is made of three terms: the elemental residual of the Stokes operator, the elemental residual of the divergence of the velocity field, and the jumps of the stress tensor across the interfaces. Both for spectral and for physical error estimates, the adaptivity strategies are described with full details in the references already given.

7.5 Further Reading

In order to obtain the collocation points inside the triangle, Heinrichs [193] proposes the solution of a Poisson equation by mapping the problem written on a sector of the unit disk onto a rectangle where Chebyshev collocation is applied. The singularity induced by the corner at the origin is removed by auxiliary mapping as proposed by Pathria and Karniadakis [299]. Numerical tests achieve spectral accuracy.

Henderson and Karniadakis [197, 198] have used mortar elements to compute the flow in complicated geometries such as the turbulent cylinder wake at $Re = 1000$ and the turbulent flow over riblets at $Re = 3280$. The spatial discretization has a homogeneous Fourier expansion in the crosswise direction. The simulation provides detailed results for mean-flow data and turbulence statistics. Hsu and Mavriplis [210] consider adaptive spectral elements for the driven-cavity problem. For convection-dominated flows, Rønquist [334] notes that special care is required when using mortar elements with variable polynomial degree from one domain to the next.

Warburton et al. [405] compute the 2D solution of the Neavier–Stokes equations on unstructured nodal triangles. The collocation points used in the simulation are Fekete points and the set derived by the electrostatics analogy. The Kovasznay flow, the shear-layer roll-up, and the flow behind a circular cylinder at $Re = 100$ are solved. The numerical results achieve spectral accuracy as expected and compare well with the presented hp and spectral-element results.

8
Vector and Parallel Implementations

The speed of a particular CFD code is the result of algorithm, hardware, and implementation. As discussed in Chapter 1, advances have proceeded on both algorithmic and architectural fronts. The theory and development of algorithms for high-order weighted residual techniques are the focus of the rest of this book. Here we consider their implementation and performance in serial and parallel architectures.

8.1 Introduction

Modern supercomputers employ multiple processors to work toward a common goal. The speed of such a computer, measured in millions of floating-point operations per second (Mflops), is expressed by the relationship

$$S_P = \eta P S_1,$$

where S_1 is the speed of a single processor, P is the number of processors, and η is the parallel efficiency. Somewhat surprisingly, the main challenge in achieving peak performance stems from the difficulty in increasing S_1, rather than in increasing η [168]. Fortunately, for locally structured, globally unstructured high-order methods, the operator evaluations that form the central kernel in most explicit or iteration-based implicit implementations can be naturally partitioned into separate serial and parallel components that can be optimized independently. For example, as shown in Section 4.5.2, spectral-element evaluation of the Poisson operator consists of local matrix–vector products

$$\underline{\tilde{v}}^e = K^e \underline{u}^e, \qquad e = 1, \ldots, E, \tag{8.1.1}$$

followed by direct-stiffness summation,

$$\underline{v}_L = QQ^T \underline{\tilde{v}}_L. \tag{8.1.2}$$

Good performance depends on (local) vectorization of (8.1.1) and efficient (interprocessor) communication for (8.1.2).

We begin with an overview of basic strategies for improving serial performance (S_1) (Section 8.2), followed by strategies specific to high-order methods (Section 8.3). We then introduce parallel architectures and performance (Section 8.4) and parallel algorithms for weighted residual methods (Section 8.5). Applications are presented in Section 8.6.

8.2 Serial Architectures

High-performance computers have clearly benefited over the past several decades from ever increasing clock cycle rates, which have roughly doubled every eighteen months, in accordance with Moore's law [278, 279]. In addition, features once found only on vector supercomputers, such as pipelined and multiple functional units (load, add, multiply, etc.), are becoming common in widely used reduced-instruction-set computer (RISC) architectures. These allow CPUs to produce one or more results per clock cycle, leading to higher performance potential.

Unfortunately, the peak performance of modern processors is rarely realized in practice. There are two basic reasons for this situation. The first arises from difficulties in scheduling operations to keep the pipelines full. The second, and more serious, results from the fact that improvements in memory access rates are not keeping pace with CPU cycle times, making it ever more difficult to keep the CPUs from being data-starved [200]. The following points should be considered when developing code for modern architectures:

- reducing vector dependences,
- eliminating loop clutter,
- increasing data reuse,
- using unit-stride data accesses.

The first pair of these addresses the issue of pipelining; the second pair focuses on memory subsystem performance. Although this list is by no means complete, these few basic concepts can easily affect performance by an order of magnitude.

8.2. Serial Architectures

Figure 8.2.1. Pipelined execution of $c_i = a_i + b_i$ for an s-stage pipe. Result c_1 is ready as a_{s+1} and b_{s+1} enter the pipe.

8.2.1 Pipelining

Floating-point operations such as addition or multiplication typically require $s = 2\text{--}5$ stages to produce a *single* result, with each stage requiring one clock cycle. As illustrated in Figure 8.2.1, pipelined functional units allow s operations to be in progress simultaneously, yielding one result per clock cycle. In order to realize this potential s-fold increase in performance, the operands entering the pipe must not depend on partially completed results still in the pipe. Such dependencies arise, for example, in the the backward substitution phase of a tridiagonal solve:

```
do i = n - 1, 1, -1
    x(i) = (b(i) - u(i)*x(i + 1))/d(i)
enddo
```

Here, the evaluation of $x(i)$ must wait until the result $x(i + 1)$ is complete.

More subtle examples arise from ambiguous references, such as the following:

```
(1) do i = 1, n                    (2) do i = 1, n
      x(i + k) = c*x(i) + a(i)           x(ind(i)) = x(ind(i)) + a(i)
    enddo                              enddo
```

In (1) the loop cannot be scheduled for pipelining unless the sign of k is known at compile time. The same is true of (2), because *ind(i)* can potentially point to a dependent address. Because they can inhibit pipelining, such constructs should be avoided whenever possible.

Other operations to be avoided in computation-intensive loops are subroutine calls, function evaluations, nested loops, *if* statements, and I/O – all of which can lead to ambiguous execution paths and can inhibit vectorization. In addition, unnecessary divides can significantly reduce performance on some RISC architectures.

8.2.2 Memory, Bandwidth, and Caches

By adhering to basic vectorization principles, one should be able to achieve a significant fraction of peak performance *provided there are enough data to keep the processor busy*. This caveat is significant because microprocessor and memory speeds are on divergent paths, with the former doubling every 18 months and the latter doubling only every 120 months [200].

To address this situation, manufacturers have developed architectures featuring hierarchies of data *caches*, that is, small amounts of fast memory between the processor and main memory. Most RISC processors today feature at least a level-one (L1) cache that, under ideal conditions, can provide data fast enough to keep the pipelines full. Some also feature larger L2 caches (and L3, and so on) that are somewhat slower than L1 but still faster than main memory. Typically, data can be delivered at speed to the CPU only if it is already in L1 as a result of an earlier call. Data not in L1 must be loaded from a higher-level cache or main memory, then passed to the CPU. This process, known as a *cache miss*, can take tens of clock cycles and severely degrade performance.

The following loops measure the time required to add a constant to a vector and are designed to illustrate some basic features of cache behavior:

$$\text{call } c\_add(a, c, n)$$
$$t_{start} = time()$$
$$\text{do } l = 1, loop$$
$$\quad \text{call } c\_add(a, c, n)$$
$$\text{enddo}$$
$$flops = n * loop / (time() - t_{start})$$

$$\text{subroutine } c\_add(a, c, n)$$
$$\text{real } a(n)$$
$$\text{do } i = 1, n$$
$$\quad a(i) = a(i) + c$$
$$\text{enddo}$$
$$\text{end}$$

The clock overhead is amortized by multiple calls to the c_add routine, and a single call to c_add outside the timing loop avoids timing the initial load of $\underline{a}$ into cache. Results of this test on a Sun Ultra-SPARC 30 are indicated by the unit-stride curve in Figure 8.2.2, which shows the speed S_1 in Mflops for 64-bit arithmetic. These results are typical of cached-based architectures. There is a rapid performance rise as n increases from 1 to 100, due to amortization of the subroutine call and loop overhead costs. A drop in performance beginning at $n = 2048$ indicates that the size of L1 is 16 Kbyte. As n exceeds 2048, the leading entries, $a_1, a_2, \ldots$, are pushed out of L1 by entries at the tail of $\underline{a}$. For $n \geq 4096$, no leading entry of $\underline{a}$ is in L1 by the time of the next c_add call, and, because each $a(i)$ must be transferred back into L1, performance is limited by the bandwidth between L2 and L1. This behavior is repeated for $n > 128,000$ (indicating that the size of L2 is 1 Mbyte), where performance

8.2. Serial Architectures

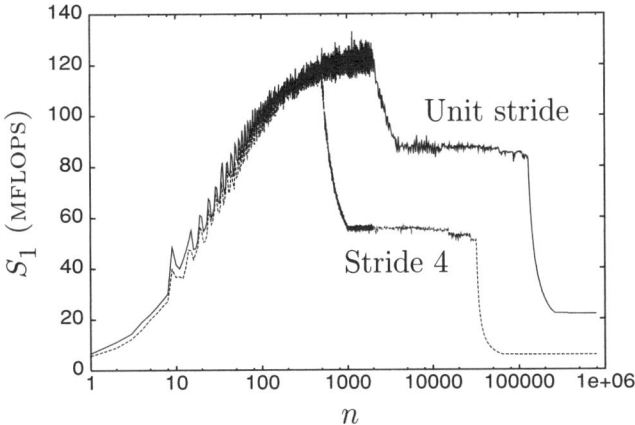

Figure 8.2.2. In-cache performance for $\underline{a} = \underline{a} + b$.

is limited by the speed of main memory. Note that the performance difference for the cached versus noncached data is a factor of six. If two *vectors* had been added, rather than a vector and a scalar, out-of-cache performance would drop further, because of the increase in memory traffic.

The preceding example illustrates the importance of reusing data once it is brought into cache. Another important design principle is to use *unit-stride* data accesses, that is, to operate on consecutive entries in memory. There are several reasons why this is a good idea, the principal one being that the computer retrieves an entire *cache line* when loading data into L1 (L2, L3, ...). A cache line is typically 128 to 256 bits long and contains adjacent entries in memory. For example, a load containing a_1 at the beginning of a 256-bit cache line will automatically load a_2–a_4, thus guaranteeing they will be in cache on any immediate succeeding reference. The effect of nonunit-stride referencing is illustrated by the stride-4 curve in Figure 8.2.2, which was obtained by replacing the *do* loop in the *c_add* procedure above by *do i = 1, 4*n, 4*. In this case, L1 has a 128-bit cache line, while L2 has a 256-bit line. As expected, the performance for intermediate (large) n drops by a factor of two (four), since only one out of two (four) values retrieved from main memory is actually used, while for small n the performance is unchanged from the unit-stride case. As long as data are in cache, good performance is attained. From the ratio of peak to bandwidth-limited performance one can conclude that a single noncached memory reference requires roughly 20–30 clock cycles on this particular computer. (In this example the effective cache size is also reduced because of the addressing strategy used to map data to cache. This effect would change under different strategies and with the number of vector operands being loaded. For reasonable performance, however, the subtleties of cache-mapping strategies are not of first-order concern. We refer to [110, 200] for more detail.)

The unit-stride issue most commonly arises when working with arrays. Consider the following simple loops for matrix addition:

(1) $do\ i = 1, n$
    $do\ j = 1, n$
    $a(i, j) = a(i, j) + b(i, j)$
    $enddo$
    $enddo$

(2) $do\ j = 1, n$
    $do\ i = 1, n$
    $a(i, j) = a(i, j) + b(i, j)$
    $enddo$
    $enddo$

The preferred order of the loops depends on the programming language. In case (1), one accesses a_{11}, followed by a_{12}, and so on, whereas in case (2), a_{11} is followed by a_{21}. Fortran uses *column major* ordering of arrays (see Figure 8.3.1), which implies that element a_{21} is adjacent to a_{11}, and so on. In this case, (2) will execute several times faster than (1) when out-of-cache conditions are encountered. C, on the other hand, uses *row major* ordering, implying that a_{12} is adjacent to a_{11}, and so on. Consequently, (1) should be used. Note that many modern compilers will actually reverse the loop order of this simple example. However, slight complications in the code usually defeat the compiler's analysis, so it is important to implement the correct ordering directly.

The issues discussed in this section are referred to in the high-performance literature as *locality of reference*. In the case of caches, it is data reuse that is important. In the case of cache lines, it is unit stride that one should strive for. These principles are important for other reasons (TLB misses, banked memories, etc.) that are considered in detail in [110, 200].

8.3 Tensor-Product Operator Evaluation

Efficient operator evaluation for high-order methods rests on the vectorization and data locality principles of the preceding sections. As an illustration, we consider evaluation of the spectral-element Laplacian. For each element, $e = 1, \ldots, E$, in $\mathbb{R}^d$, we have

$$\underline{v}^e = \sum_{j=1}^{d} \sum_{i=1}^{d} D_i^T G_{ij}^e D_j \underline{u}^e, \qquad (8.3.1)$$

where the G_{ij}^e's are the diagonal matrices of order $(N + 1)^d$ introduced in (4.4.7). We begin with the tensor-product forms ($D_i \underline{u}$, etc.) that account for

8.3. Tensor-Product Operator Evaluation

the leading-order work complexity of $O(N^{d+1})$. We then address the geometric terms involving G_{ij}^e, which have complexity $O(N^d)$.

8.3.1 Tensor-Product Evaluation

The majority of the work in (8.3.1) is associated with application of the derivatives D_j and their transposes. In $\mathbb{R}^3$, a typical form is

$$D_1 \underline{u}^e := (I \otimes I \otimes \hat{D})\underline{u}^e = \sum_{p=0}^{N} \hat{D}_{ip} u_{pjk}^e.$$

We noted in (4.2.20) that the 2D tensor-product evaluation, $\underline{v} = (A \otimes B)\underline{u}$, can be implemented as the matrix–matrix product sequence $V = BUA^T$. In view of data locality this sequence will perform well on any architecture. A similar transformation can be made for the 3D case by understanding how data are arranged in memory.

Consider the Fortran declaration for the vector $\underline{u}_L := \{\underline{u}^e\}_{e=1}^E$,

```
real u(0:N,0:N,0:N,1:E)
```

which implies that elements of $\underline{u}^e = u^e(i,j,k)$ are arranged with consecutive entries advancing with the leading (i) index, as indicated in Figure 8.3.1(a). We can reference $u^e(i,j,k)$ in whatever way is convenient for a given operation. For example, to compute $\underline{u} = \underline{u} + \underline{v}$, we can simply write

```
ntot = E*(N+1)**3
do i=0,ntot-1
   u(i,0,0,1) = u(i,0,0,1) + v(i,0,0,1)
enddo
```

which avoids (short) nested loops and ensures unit-stride access. Figure 8.3.1 shows other interpretations of the memory layout for $\underline{u}^e$ (with the superscript omitted). In (b), $\underline{u}^e$ is viewed as an $n \times n^2$ matrix; in (c), as a sequence of $n \times n$ matrices; and in (d), as an $n^2 \times n$ matrix.

The matrix interpretations of $\underline{u}$ allow tensor-product evaluations to be recast as *matrix–matrix* products. An example is illustrated by the following routine, which computes the gradient of u^e, $(\partial u/\partial r, \partial u/\partial s, \partial u/\partial t)$:

```
      subroutine local_grad(ur,us,ut,u,N,e,D,Dt)
c     Output: ur,us,ut           Input:u,N,e,D,Dt
```

386   8. *Vector and Parallel Implementations*

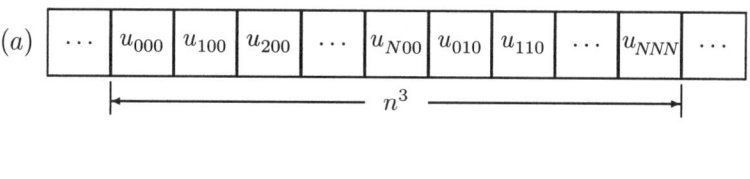
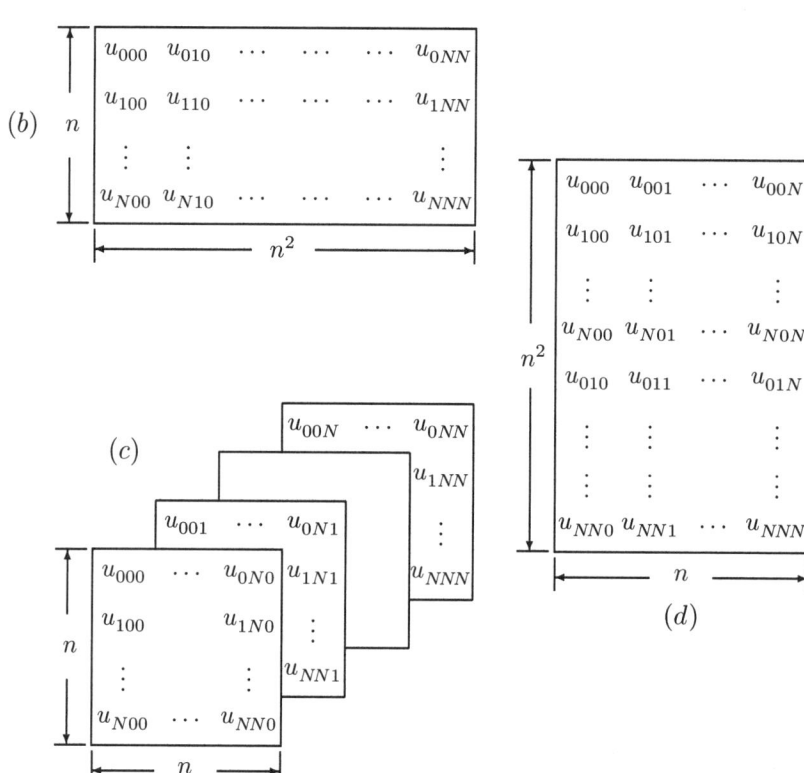

Figure 8.3.1. Interpretations of 3D data array as (*a*) consecutive elements in memory; (*b*) an $n \times n^2$ matrix ($n := N+1$); (*c*) a sequence of n $n \times n$ matrices; and (*d*) an $n^2 \times n$ matrix.

```
      real ur(0:N,0:N,0:N),us(0:N,0:N,0:N),ut(0:N,0:N,0:N)
      real u (0:N,0:N,0:N,1)
      real D (0:N,0:N),Dt(0:N,0:N)
c
      m1 = N+1
      m2 = m1*m1
c
      call mxm(D ,m1,u(0,0,0,e),m1,ur,m2)
```

8.3. Tensor-Product Operator Evaluation

```
      do k=0,N
        call mxm(u(0,0,k,e),m1,Dt,m1,us(0,0,k))
      enddo
      call mxm(u(0,0,0,e),m2,Dt,m1,ut,m1)
c
      return
      end
```

Here, D is assumed to be the one-dimensional derivative matrix $\hat{D}$ [see (2.4.9)], Dt its transpose, and mxm a matrix–matrix product routine described below. The first mxm call computes $(I \otimes I \otimes \hat{D})\underline{u} = \hat{D}U$, treating $\underline{u}$ as the $n \times n^2$ matrix of Figure 8.3.1(b). The second computes $(I \otimes \hat{D} \otimes I)\underline{u}$, treating $\underline{u}$ as the sequence of $n \times n$ matrices in Figure 8.3.1(c). The third call computes $(\hat{D} \otimes I \otimes I)\underline{u} = U\hat{D}^T$, treating $\underline{u}$ as the $n^2 \times n$ matrix of Figure 8.3.1(d). A similar sequence can be called for any desired tensor-product evaluation, such as interpolation, integration, or high-order differentiation. The advantage of recasting tensor-product forms as matrix–matrix products is that the majority of flops (>90% for typical 3D applications) is thereby concentrated into a single routine that can be highly optimized.

The mxm routine is designed to compute the product of (contiguously packed) matrices A and B, of sizes $n_1 \times n_2$ and $n_2 \times n_3$, respectively. A Fortran implementation of mxm is given below:

```
      subroutine mxm(A,n1,B,n2,C,n3)
      real A(n1,n2),B(n2,n3),C(n1,n3)
c
      do i=1,n1*n3
        C(i,1)=0.
      enddo
c
      do j=1,n3
      do k=1,n2
        tmp=B(k,j)
        do i=1,n1
          C(i,j)=C(i,j)+A(i,k)*tmp
        enddo
      enddo
      enddo
c
      return
      end
```

This routine requires $2n_1n_2n_3$ operations and only $(n_1 + n_3)n_2$ reads from memory, giving it a ratio of roughly $2(n_1n_3)/(n_1 + n_3)$ flops per data access. Because of this favorable ratio, multiplication of sufficiently large matrices can attain a significant fraction of peak performance, even on computers with severe memory bandwidth limitations.

Computer vendors and scientists have expended considerable effort on optimizing matrix–matrix products because of their widespread utility and importance to key performance benchmarks. Specific attention has been given to the BLAS3 routine dgemm (double-precision general matrix × matrix), which is more general than the mxm routine above. It supports the operation $C = C + AB$, will work with transposes, and does not require the array dimensions to match the matrix sizes. A dgemm-based implementation of mxm would read as follows:

```
      subroutine mxm(a,n1,b,n2,c,n3)
      real a(n1,n2),b(n2,n3),c(n1,n3)
c
      one = 1.
      zero = 0.
      call dgemm( 'N','N',n1,n3,n2,one,A,n1,B,n2,zero,C,n1)
c
      return
      end
```

A detailed example of performance tuning for dgemm can be found in [409].

While dgemm usually outperforms the triply nested loop in the mxm routine above, it is typically optimized only for large matrices important to the LINPACK benchmark. For multidomain spectral methods, one is often interested in small matrices. The following pair of routines illustrates an approach that has proven to be a good compromise between portability and high performance [390]:

```
      Subroutine mxm(a,n1,b,n2,c,n3)
      real a(n1,n2),b(n2,n3),c(n1,n3)
      if (n2.eq.1) then
         call mxm1(a,n1,b,n2,c,n3)
      elseif (n2.eq.2) then
         call mxm2(a,n1,b,n2,c,n3)
         :
      elseif (n2.eq.24) then
           call mxm24(a,n1,b,n2,c,n3)
      else
```

8.3. Tensor-Product Operator Evaluation

```
          one = 1.
          zero = 0.
          call dgemm( 'N','N',n1,n3,n2,one,A,n1,B,n2,zero,C,n1)
      endif
      return
      end
      Subroutine mxm8(a,n1,b,n2,c,n3)
      real a(n1,8),b(8,n3),c(n1,n3)
      do j=1,n3
         do i=1,n1
            c(i,j) = a(i,1)*b(1,j)
     $              + a(i,2)*b(2,j)
     $              + a(i,3)*b(3,j)
     $              + a(i,4)*b(4,j)
     $              + a(i,5)*b(5,j)
     $              + a(i,6)*b(6,j)
     $              + a(i,7)*b(7,j)
     $              + a(i,8)*b(8,j)
         enddo
      enddo
      return
      end
```

The main routine, mxm, has a branch that calls one of 25 different routines: mxmn_2, $n_2 = 1, \ldots, 24$, or dgemm. The *if*-statement overhead is minimal (less than 2% for $n_2 = 8$). The routines mxmn_2 have the inner-product (n_2) dimension completely unrolled into a single statement, as the second routine illustrates for $n_2 = 8$. (Note that the inner-product dimension always corresponds to $N + 1$ in spectral applications. See Figure 8.3.1.) The advantage of this approach is that there are now only two loops rather than three, with a short nested loop being eliminated in favor of one with more work per iteration. In addition, because the dimension n_2 is known prior to compilation, the compiler can hardcode address increments into memory read instructions.

Because mxm dominates the operation count, a small amount of testing to find an optimal variant is usually warranted when setting up simulations on a new platform. An example is presented in [390] for simulations on the Intel ASCI Red machine at Sandia National Laboratories. Careful analysis of a Navier–Stokes spectral-element code for $N = 15$ identified the specific calling patterns to mxm shown in Table 8.3.1. Performance was measured for several dgemm-based implementations of mxm (denoted lkm, ghm, and csm), the unrolled variant described above (denoted f2), and a similar unrolled version with the outer

Table 8.3.1. mxm *performance on* ASCI *Red.*

Dimension			Noncached Mflops				
n_1	n_2	n_3	lkm	ghm	csm	f3	f2
14	2	14	23	20	25	41	43
2	14	2	29	23	21	77	68
16	14	16	100	113	114	130	119
16	14	196	107	95	95	89	110
256	14	16	85	100	100	105	67
14	16	14	78	74	74	106	99
16	16	16	82	138	83	105	102
16	16	256	113	147	93	97	118
196	16	14	91	192	147	122	105
256	16	16	90	105	148	111	69

loops interchanged (denoted f3). The results show that no single version was fastest for all cases encountered in a given simulation. For production runs, a customized branching utility was used to select the fastest mxm variant as a function of the input arguments. With this approach, the program sustained 361 Gflops on 4096 333-MHz Pentium Pro processors – in excess of one-quarter of peak for this particular configuration.

8.3.2 Other Operations

Although tensor-product evaluations account for >90% of the *operations* in spectral methods, they typically account for a smaller fraction of *time* because of their high efficiency. Consequently, it is also important to address the implementation of other frequently called kernels. The following loop exploits the data locality issues discussed in Section 8.2 to evaluate the matrix–vector products $\underline{v}^e = K^e \underline{u}^e$ (4.4.7):

```
c body of Procedure_Ku()
  do e=1,E
    call local_grad(ur,us,ut,u,N,e,D,Dt)
    do i=1,nrst
      vr(i) = G(1,i,e)*ur(i) + G(2,i,e)*us(i) + G(3,i,e)*ut(i)
      vs(i) = G(2,i,e)*ur(i) + G(4,i,e)*us(i) + G(5,i,e)*ut(i)
      vt(i) = G(3,i,e)*ur(i) + G(5,i,e)*us(i) + G(6,i,e)*ut(i)
    enddo
    call local_gradt(v,vr,vs,vt,N,e,D,Dt)
  enddo
```

Here, it is assumed that nrst := $(N+1)^3$ and that local_gradt effects the transpose of local_grad. For moderate values of N, the work arrays ur, us,..., vt will stay in L2 for the duration of the outer loop. Note that the geometric array G places the greatest burden on memory bandwidth because each entry is used at most twice. By combining the evaluation of vr, vs, and vt into a single loop, the symmetric components of G are reused and thus brought into cache only once.

Other routines that merit careful inspection in Navier–Stokes implementations are the pressure operator, the advection operator, and any solve forming a local component of a preconditioner. If one is using the OIFS-based advection scheme of Section 4.9, the number of advection evaluations per step can be of the same order as the number of viscous and pressure operator evaluations, particularly if the temporal order is high.

8.4 Parallel Programming

Parallel computers provide a means for combining P individual processors to work in concert on a common task. A fundamental question early in the development of parallel computing was whether these should work in lockstep, each executing the same instruction on different data, corresponding to the single-instruction, multiple-data (SIMD) model, or whether each processor should be allowed to operate asynchronously, following its own instruction set, corresponding to the complementary multiple-instruction, multiple-data (MIMD) model [135]. Examples of the SIMD model include the very early ILLIAC-IV and the Thinking Machines CM-2. Although these machines were well suited to scientific computing, economics have favored the development of MIMD architectures based on commodity components. Examples include tightly coupled networks of high-performance microprocessors capable of running independently, networks of workstations, and networks of PCs.

Another central issue in the choice of architecture and programming model is whether the memory should be shared or distributed among the processors. In the shared-memory model, each processor has access to all the data (at nonuniform rates). In the distributed-memory model, each processor has direct (fast) access to its own local data only. Data from other processors must be obtained by extraordinary means. Such means can (in principle) be provided by the operating system, implementing a shared-memory programming model on a distributed-memory architecture, or by a user program that explicitly exchanges data packets, or messages, between processors. We will consider this distributed-memory message-passing programming model, following the approaches detailed in *Solving Problems on Concurrent Processors* by Fox et al.

[139], *Concurrent Scientific Computing* by van de Velde [395], and *Using MPI* by Gropp et al. [171].

The most commonly used programming model for numerical solution of PDEs on MIMD architectures is the single-program, multiple-data (SPMD) model in which each processor independently executes a copy of the same program on distinct subsets of data [186]. Coordinated data exchanges (messages) enforce a loose synchronization among the processors, thus emulating the SIMD programming model while preserving the flexibility of the MIMD approach.

The SPMD model is predicated on each processor having a private (or local) address space, implying that a given variable or array entry can have different values on different processors. Data transfers between address spaces (processors) are made through subroutine calls such as *send()*, used to send the contents of a variable or array to another processor, and *recv()*, used to direct incoming data to a specific variable or array location. One advantage of message passing is that synchronization is implicit in the message transfer; data are sent only when ready. In addition, because it is based on subroutine calls, it does not require language extensions and is therefore portable. This latter attribute is enhanced by the fact that there exists a well-developed standard interface, called the Message Passing Interface (MPI), that has found broad support among computer vendors and in the research community [171]. Other widely used message-passing libraries include PVM [152] and NXlib [366].

To provide a framework for assessing parallel performance and making software design decisions, we assume a model architecture consisting of a network of P processor/memory units (referred to as processors, or nodes) numbered from 0 to $P - 1$, as illustrated in Figure 8.4.1. The detailed topology of the network is not important in most domain decomposition applications, so we do not characterize it further at this point. We simply assume that data can be sent from one node to another, without interrupting processing on intermediate nodes. This

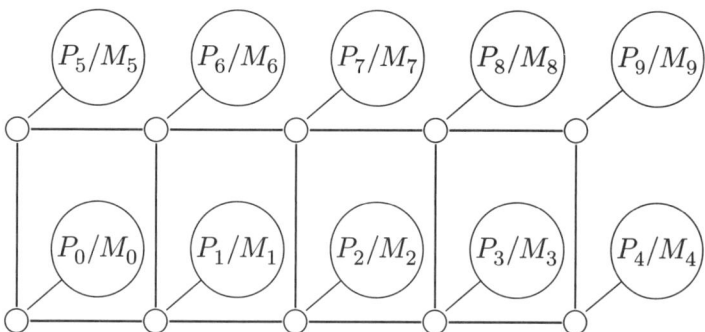

Figure 8.4.1. Network of $P = 10$ distributed processor/memory units.

model architecture encompasses loosely coupled networks of commodity components, shared-memory multiprocessors, and tightly coupled multicomputers consisting of thousands of nodes.

8.4.1 Communication Characteristics

This section introduces communication primitives and performance characteristics that are typical of distributed-memory programming models. While we do not wish to overemphasize low-level primitives, which distract from the higher-level objective of solving PDEs, it is important from a design standpoint to have some understanding of basic interprocessor interactions and their associated costs.

The basic elements of any message-passing implementation are exemplified by the following MPI calls:

```
call mpi_comm_rank(mpi_comm_world,MYID ,ierr)
call mpi_comm_size(mpi_comm_world,NPROCS,ierr)

call mpi_send(X,n,mpi_byte,id,mtag,mpi_comm_world,ierr)
call mpi_recv(Y,n,mpi_byte,id,mtag,mpi_comm_world,stat,ierr)
```

The first of these tells the calling processor its node identification (MYID := p). The second tells the number of processors participating in the current simulation (NPROCS := P). The third sends the contents of the vector X to processor id$\in$ $\{0, \ldots, P - 1\}$. In this case, the amount of data is n bytes, as denoted by the second and third arguments, and the data are identified with an integer tag (mtag) that allows the receiving processor to distinguish this message from others that it might receive. The fourth command receives incoming data from node id having tag mtag and places it in Y. Here, n is an upper bound on the expected number of bytes received. Also, the discriminator pair (id,mtag) can be loosened through the use of MPI's designated variables, mpi_any_source and mpi_any_tag. The mpi_comm_world argument is a handle that allows MPI to distinguish between calls made by the user's program and those made by library routines that might potentially send messages with the same integer tags. The ierr argument is an error condition that normally returns zero.

To estimate the costs of message passing, we assume, to leading order, a linear model of the form

$$t_c(m) = (\alpha + \beta m) t_a, \qquad (8.4.1)$$

where m is the length of the message, in 64-bit words, and t_a is the time required for a characteristic arithmetic operation (e.g., add or multiply). The

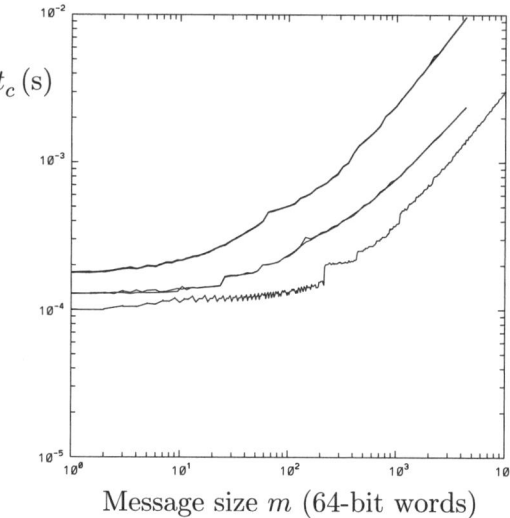

Message size m (64-bit words)

Figure 8.4.2. Measured communication times from ping-pong test for three distributed-memory computers.

constants α and β represent the nondimensional message startup cost (latency) and asymptotic transfer rate (inverse bandwidth), respectively, in the absence of other network traffic. This model assumes that the communication time is independent of the node addresses, which will be true if the network has sufficient connectivity and/or the size of the messages is sufficiently small, both of which tend to occur in most DD applications.

The values of α and β are important in making software design decisions and in assessing the parallel potential of a given algorithm–computer coupling. They are derived from the round-trip communication time, $2t_c(m)$, which is measured by a ping-pong test in which one processor sends a message of length m to another and then waits for a reply. Times for repeated cycles of this exchange are averaged to yield a graph of $t_c(m)$ versus m, as shown in Figure 8.4.2. The performance is relatively flat until $m \approx 20\text{--}200$. The left asymptote of this graph is used to deduce αt_a, while the right asymptote yields βt_a. Figure 8.4.3 shows measured values of α and β on a variety of platforms over the past decades, using a characteristic timescale (t_a) based on average performance for optimized noncached matrix–matrix products of order 10 and 16, which typify the computation in spectral-element applications. While the downward trend indicates significant improvement in the *absolute* times, we see from the ratios of the curves that the *relative* communication costs have exhibited less change.

8.4. Parallel Programming

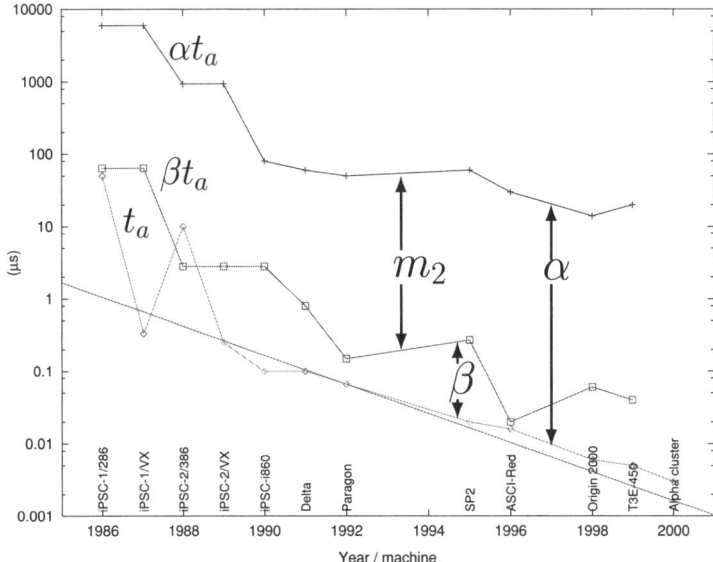

Figure 8.4.3. Performance parameter trends in distributed-memory computing. The straight line follows Moore's law.

Despite a couple of anomalies, the general trend has been toward increased latency, growing from $\alpha \sim 100$ to 4000, while the bandwidth term has stayed roughly in the range of $\beta \sim 1 - 10$.

As Figures 8.4.2 and 8.4.3 indicate, message-startup costs can be significant. Consequently, it is important to consider algorithm designs that use a few long messages rather than numerous short messages wherever possible. The ratio

$$m_2 := \alpha/\beta$$

provides a convenient demarkation between short and long messages (measured in 64-bit words). Any message of length $m = m_2$ has a communication time of just twice the startup cost αt_a. Messages shorter than m_2 are thus short, or latency-dominated, with a cost scaling as α. Messages longer than m_2 are bandwidth-limited, with costs scaling as βm.

Typical values of m_2 are illustrated by the ratios of the αt_a and βt_a curves in Figure 8.4.3. We can see that messages can be as long as 100 to 1500 words before a twofold increase in communication cost will be realized. In the long-message limit, one also needs to be concerned about the possibility of network contention if many processors are sending at once.

8.4.2 Vector Reductions

We examine vector reductions to illustrate an approach to the SPMD model that separates high-level objectives from detailed low-level source code and to illustrate the underlying costs associated with this commonly used operation. Vector reductions (also known as collective operations) take a (distributed) vector as input and return a scalar. Examples include $s = \sum_i x_i$, $\alpha = \underline{p}^T \underline{r}$, and $\|\underline{u}\|_\infty$. They may also be grouped together to return more than one scalar, for example, $(\alpha, \beta) = (\underline{u}, \underline{v})^T \underline{w}$, which in light of the preceding discussion can be expected to have the same communication overhead as an isolated vector reduction. Because of their global nature and frequent appearance in scientific applications, vector reductions are a common source of communication overhead in parallel codes.

The Fortran procedure below illustrates an MPI implementation of $s = \sum_i \underline{x}_i$, where the vector $\underline{x}$ is distributed across P processors:

```
real*8 function glsum(x,n)
real*8 x(n),s
include 'mpif.h'
s = 0.
do i=1,n
   s = s + x(i)
enddo
call mpi_allreduce(s,glsum,1,
$         mpi_double_precision,mpi_sum,mpi_comm_world,ierr)
return
end
```

According to the SPMD model, the contents of $\underline{x}$ are different on each processor, and the vector length n may also be different. The processor distribution can be viewed as providing an implicit index (p) such that $\underline{x} := \bigcup_{p=0}^{P-1} \underline{x}^{(p)}$, $\underline{x}^{(p)} \in \mathbb{R}^{n^{(p)}}$. The first part of the code is the standard loop for computing the sum $s^{(p)} = \sum_i x_i^{(p)}$. The mpi_allreduce call invokes the sum $s = \sum_p s^{(p)}$ across all processors and returns the result in glsum. Thus, all processors have the same global sum value at the end of the call. The third argument determines the length of the vector to be summed across processors. In this case, there is only one input (s) and output (glsum). The fourth argument tells MPI the size and type of data being transmitted, and the fifth determines which reduction operation to perform. The integers beginning with mpi_ are passed as part of the include 'mpif.h' statement.

8.4. Parallel Programming

Although the MPI call requires several arguments supplemental to the task of summing the elements of a vector, the use of the `glsum` function masks the details of *how* the vector reduction is implemented. Elsewhere in the code, it is sufficient simply to call an appropriate routine (such as `glsum`) written independently, wherever a vector reduction is encountered. The following code fragment illustrates this approach:

```
dmax = 0.
do i=1,n
   d=u(i)-v(i)
   w(i) = w(i) + d
   dmax = max(dmax,abs(d))
enddo
dmax = glmax(dmax,1)
```

The loop simultaneously computes $\underline{w} = \underline{w} + (\underline{u} - \underline{v})$ and $\|\underline{u} - \underline{v}\|_\infty$ and is typical of those found in many serial codes. The only change required to extend the code to parallel in the SPMD model is the addition of the `glmax` call, which would be implemented in the same fashion as the `glsum` routine above.

The communication cost of a vector reduction depends on the actual all_reduce implementation, which is most frequently cast as a fan-in–fan-out on a binary tree. Figure 8.4.4(*a*) illustrates a fan-in summation to node 0 on a one-dimensional network. Arrows indicate (contention-free) data movement. Upon arrival at each destination, data are summed with the existing partial

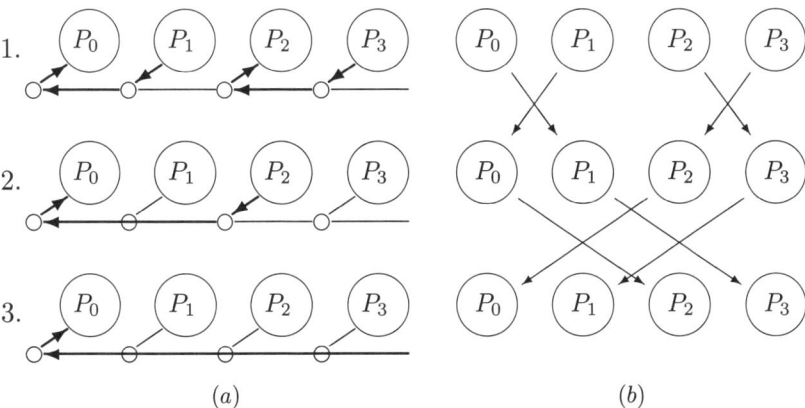

Figure 8.4.4. (*a*) First three stages of a contention-free binary-tree fan-in for vector reduction on a one-dimensional network. (*b*) Recursive doubling (butterfly) exchange for four-processor network.

result, which is then sent in the next phase. The number of communication cycles to produce the final result on node 0 is $\log_2 P$. An additional $\log_2 P$ cycles are required to spread the result to all processors, accomplished by reversing the tree traversal. The cost of `all_reduce` on an m-tuple is thus $2(\alpha + \beta m) \log_2 P$.

For short messages, $m < m_2$ (the most common case being $m = 1$), network contention is generally not an issue. Hence it is worthwhile to consider the recursive doubling scheme of Figure 8.4.4(b), which requires only $\log_2 P$ cycles to yield the correct result on each processor. If $P = 2^D$, a recursive doubling implementation of the `all_reduce` operation is given by

> *Procedure All_Reduce_Sum(v, w, m)*
> do $l = 1$ to D
>   send $v(1:m)$ to $mod\,(p + 2^{l-1}, P)$
>   recv $w(1:m)$ from $mod\,(P + p - 2^{l-1}, P)$
> $v(1:m) := v(1:m) + w(1:m)$
> enddo

The case of general P requires two additional send–recv pairs (see [395]).

We make a final important comment regarding send–recv commands. On some architectures, the procedure *All_Reduce_Sum* will deadlock, particularly for large values of m. The reason is that the send–recv construct above implicitly assumes that the system software provides a buffer (on an unspecified node) for the outbound contents of v. For example, the statement "send $v(1:m)$" to q is a *blocking* send, meaning that control is returned to the calling routine once the contents of v are safely stored and v can be safely reused. If no system buffer has sufficient space for v, then processing must stall until a corresponding "recv" is issued. However, if the receiving processor is also trying to send a large buffer, then deadlock will occur. To avoid deadlock, MPI provides a *nonblocking* receive, `mpi_irecv`, that allows one to specify a buffer for incoming data and that returns control to the calling routine before the receive request is fulfilled. The `mpi_wait` command allows one to pause at a later point for fulfillment of the `mpi_irecv`. The modified pseudocode is as follows:

> *Procedure All_Reduce_Sum(v, w, m)*
> do $l = 1$ to D
>   msg = irecv $w(1:m)$ from $mod\,(P + p - 2^{l-1}, P)$
>   send $v(1:m)$ to $mod\,(p + 2^{l-1}, P)$
>   wait (msg)
> $v(1:m) := v(1:m) + w(1:m)$
> enddo

For portability, we strongly recommend the nonblocking irecv approach in developing communication utilities. However, for ease of exposition, we will generally present the simpler send–recv variant of the algorithms.

8.5 Parallel Multidomain Methods

In this section, we develop data distribution, communication, and domain decomposition strategies for an SPMD implementation of high-order weighted residual methods.

8.5.1 Data Distribution and Operator Evaluation

The natural data distribution for high-order weighted residual methods is based on an elemental decomposition in which one partitions the domain Ω into P subdomains Ω^p, to be distributed to processors $p = 0, \ldots, P - 1$, with each subdomain comprising $E^{(p)}$ elements. Nodal values on subdomain interface boundaries, $\partial \Omega^p \setminus \partial \Omega$, are stored redundantly on each processor q for which $\bar{\Omega}^q \cap \bar{\Omega}^p \neq 0$. This approach is consistent with the element-based storage scheme, which lends itself naturally to element-by-element matrix–vector product evaluation.

Element-based data distribution allows loops without interelement dependencies to execute with P-fold concurrency. In the SPMD model, the element identifiers (e) are usually reassigned on each processor so that local loop indices range from 1 to $E^{(p)}$, as illustrated in Figure 8.5.1. The figure shows a five-element mesh partitioned into two subdomains with elements locally renumbered to $\{1, 2, 3\}$ on node 0 and to $\{1, 2\}$ on node 1. Because of the private address space, the loop index bound $E^{(p)}$ can be simply designated E on each processor. (For clarity, we therefore designate the number of elements in

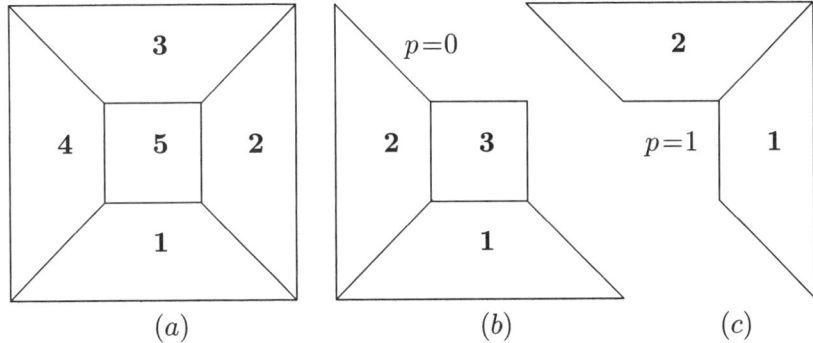

Figure 8.5.1. Decomposition of five-element mesh into two subdomains.

the global domain as E_g.) A typical element-based loop in an SPMD implementation thus is

do $e = 1$ to E
$\tilde{\underline{v}}^e = K^e \underline{u}^e$
enddo

where, for example, $K^e \underline{u}^e$ is implemented as in the body of *Procedure_Ku()* introduced in Section 8.3.2. Note that because of the reassignment of variables (performed in a preprocessing step) the above loop is *identical* to its serial counterpart. Provided the contents of all arrays involved are mapped according to the same distribution, this approach will always work. In addition, all *vector–vector* operations, such as $\underline{u} = \underline{u} + \underline{v}$, also remain unchanged from their serial counterparts.

Vector reductions, such as $\underline{u}^T \underline{v}$, will need to be modified to effect the necessary communication, as outlined in Section 8.4.2. In addition, when element-based storage is used (in serial or parallel), the inner products need to be weighted so that redundantly stored values are not counted twice. If w_{ijk}^e is the inverse of the multiplicity [Equation (4.5.17)], then the inner product $\alpha = \underline{u}^T \underline{v}$ is computed as

$$\alpha^{(p)} = \sum_{i=1}^{n} (\underline{u}_L)_i (\underline{v}_L)_i (\underline{w}_L)_i, \qquad \alpha = \sum_{p=0}^{P-1} \alpha^{(p)}, \qquad (8.5.1)$$

where $n = (N+1)^d E$ is the number of (local) grid points on processor p. Note that if contiguous storage is used, then (8.5.1) can be coded in a form similar to glsum given in Section 8.4.2. Note also that the multiplicity is not required when computing the continuous inner product, $(u, v) = \underline{u}_L^T M_L \underline{v}_L$ because the local Gauss–Lobatto quadrature weights take proper account of contributions from each element. The proposed routine for computing the algebraically weighted inner product (8.5.1) would equally serve to compute $\underline{u}_L^T M_L \underline{v}_L$ by simply passing the diagonal matrix M_L as the weight rather than the inverse multiplicity $\underline{w}_L$.

Steps requiring interprocessor exchanges, such as direct stiffness summation, and those involving global information, such as mesh input and solution output, typically need global references. These can be provided by establishing local-to-global and global-to-local integer mappings during problem setup. In the example of Figure 8.5.1 one might have arrays

$loc\_to\_glob\_el(:, 0) = \{1, 4, 5\}$,

$loc\_to\_glob\_el(:, 1) = \{2, 3\}$,

$glob\_to\_loc\_el(:) = \{1, 1, 2, 2, 3\}$,

$glob\_to\_loc\_proc(:) = \{0, 1, 1, 0, 0\}$,

where $loc\_to\_glob\_el(e, p)$ holds the global element number associated with the eth element on processor p, $glob\_to\_loc\_el(e_g)$ holds the local element number associated with element e_g, and $glob\_to\_loc\_proc(e_g)$ holds the processor number to which $e_g \in \{1, \ldots, E_g\}$ is mapped. Typically, these arrays are established during domain partitioning, which is the subject of Section 8.5.3.

We make a few comments regarding practical implementations of these methods. The first is that the code should be written so that the number of processors may be selected at run time. Provided that all array declarations scale as the number of elements *per processor* (E) and not as E_g, the simulation size should be able to increase linearly with the number of processors.

A related issue is how the mesh should be read from a file. Should one processor read the mesh and parcel out data to the others, or should each processor open the file and read only its portion of the mesh? On most systems, it is possible to have all processors open the same file for reading (not writing) provided $P \lesssim 100$. However, congestion can result for larger values of P, and it is more efficient to have just one or a few nodes read the mesh and distribute appropriate sections to the other processors. Fortunately, for high-order methods the number of mesh elements is relatively small, and congestion at the input stage is encountered only for large problems.

Similarly, for output, the most universally applicable approach is to have node 0 open a file and have every processor pass data to it for writing. Although parallel I/O is becoming more established, particularly with the development of MPI-2 [172], most machines will not support P processors simultaneously writing to the file system. To allow the flexibility of using parallel I/O for large P, one can generalize the "pass to node 0" scheme so that $P' \ll P$ nodes write to P' independent files, each handling the I/O for subgroups. For this reason, the I/O routines should support an unstructured file format with tags for each element, so that the order of the file contents is immaterial. This allows the P' output files simply to be concatenated at a later time for postprocessing without having to sort the data.

8.5.2 Direct Stiffness Summation

The principal communication utility in parallel weighted residual implementations is the gather–scatter operation, or direct stiffness summation, $\Sigma' := QQ^T$. As noted in Section 4.5.1, Σ' involves the summation of two or more *local* nodal values corresponding to the same *global* degree of freedom. As illustrated in Figure 8.5.2(a), nodal values on conforming element interfaces have a unique representation on the physical grid but are replicated two or more times in the element-based storage format in order that local matrix–vector products

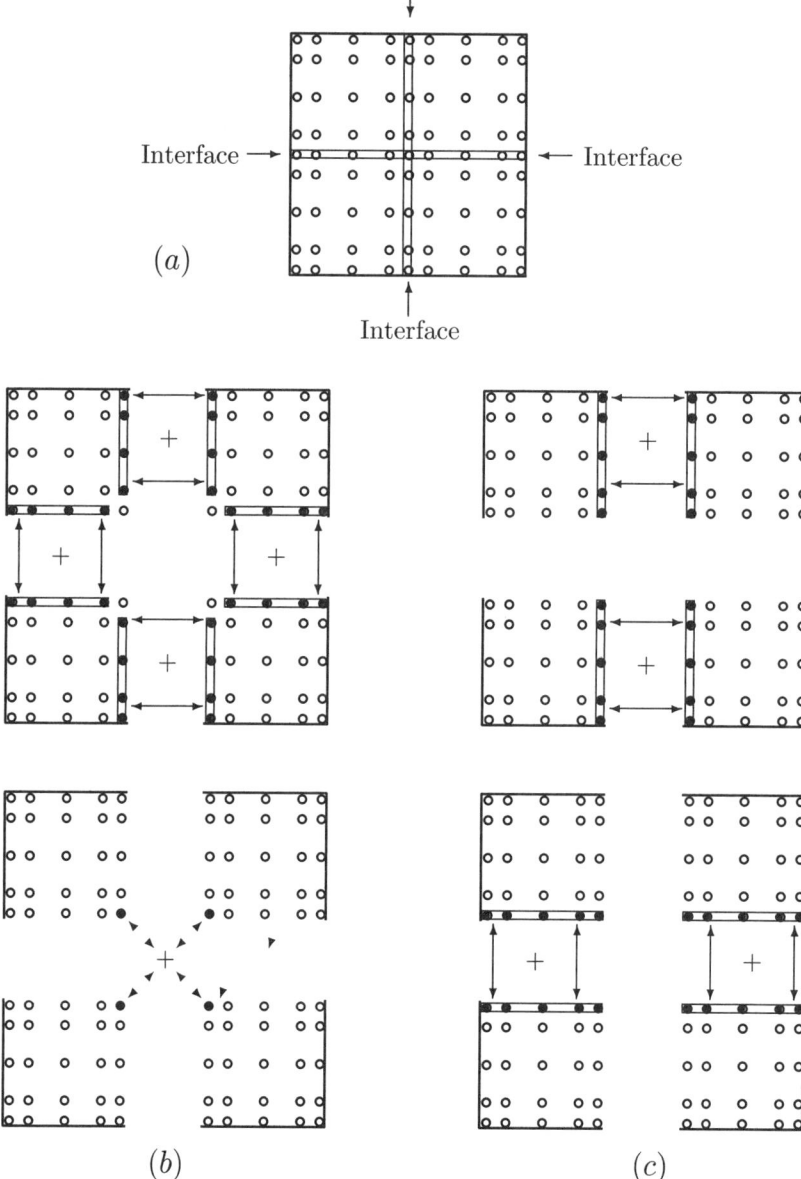

Figure 8.5.2. Direct stiffness communication patterns: (a) physical grid for $(E, N) = (4, 4)$ in $d = 2$ space dimensions; (b) edge–vertex exchange algorithm requiring three messages per element (processor); (c) synchronized edge–edge exchange algorithm requiring d messages per element.

$\tilde{\underline{v}}^e = K^e \underline{u}^e$ may be evaluated independently. Figure 8.5.2(b) and (c) illustrate two common approaches to implementing Σ' in parallel.

The first approach, illustrated in Figure 8.5.2(b), consists of separate edge exchanges of N words, followed by corner exchanges of one word each. In three dimensions, the edge exchange is preceded by a face exchange. Because individual edge, vertex, and face exchanges are all short ($m \ll m_2$), it is important to collect data destined for a given processor into a single longer message to reduce the latency costs. Given that a single element in $\mathbb{R}^3$ typically has 26 neighboring elements, the number of messages may still be quite large. For example, in the case of one element per processor with $N^2 \ll m_2$, the cost of Σ' will be roughly $26\alpha t_a$.

The second approach, suggested in [131], is based on the d-directional exchange of edge (face) data illustrated in Figure 8.5.2(c). By synchronizing the edge exchanges, the vertex values that are shared by 2^d processors are correctly updated without the need for expensive individual messages. This algorithm is efficient for tensor-product (globally structured) arrays of elements because it requires only $2d$ message cycles (two per spatial direction) – far less than the $3^d - 1$ of the vertex-edge algorithm. For a structured array of elements with $E = P$ and $N^2 \ll m_2$ the direct stiffness summation cost would be only $6\alpha t_a$. Couzy [83] found the d-directional exchange superior to the edge-vertex approach on the Cray T3D, but found no significant advantage on the Intel Paragon, which had relatively fast communication.

The directional exchange approach breaks down wherever the elements are not topologically aligned in a tensor-product array (e.g., in the five-element mesh of Figure 8.5.1). To overcome this difficulty, Fischer and Patera [131] suggested a two-phase strategy that uses the d-directional exchange to update the majority of the nodes, followed by a vertex-combine operation to update the remaining $\mathcal{N}_{\text{spec}}$ "special" nodes not correctly treated in the first phase. In the original implementation, the special-node combine was implemented as an all_reduce on an $\mathcal{N}_{\text{spec}}$-tuple. To reduce costs when $P > 128$, Fischer [125] developed a generalized combine based on the crystal router concept of Fox et al. [139]. This procedure begins by subdividing the processor group $\{0, \ldots, P-1\}$ into two consecutively ordered subgroups, each comprising $P' = P/2$ processors. If a nodal value resident on processor p is shared by *any* processor in the complementary subgroup, that value is added to a list to be sent to processor $q = p \pm P'$, where the sign is chosen such that q is in the complementary subgroup. If the node in question is also resident on q, the incoming value from p is summed to the current value on q. Otherwise, it is stored on q and labeled "resident on q." (Intermediate storage is therefore required on q.) Since the procedure is symmetric, reciprocal action is taken in the complementary subgroup (i.e., q sends to p). The process recurs $\log_2 P$ times by dividing each subgroup into new complementary pairs and repeating the exchange within each subgroup. Because each stage requires $P/2$ bidirectional exchanges (one send from each processor), the generalized combine can potentially suffer network contention. However, because data cross subgroup boundaries only where needed, the messages are much shorter than required for the all_reduce.

The potential of this approach was demonstrated in [125] for a spectral-element problem with $(E, N, d) = (512, 9, 3)$. Out of 375,000 grid points, 15,000 were located on the "wire frame" comprising the spectral element edges. For a $P = 512$ processor implementation, the synchronized exchange of faces correctly treated all but $\mathcal{N}_{spec} = 2000$ of the vertices. Treating these with the generalized combine required only $\log_2 P$ exchanges of less than 100 words each and required significantly less time than the all_reduce approach, which required $2 \log_2 P$ exchanges of length $\mathcal{N}_{spec}$.

Although potentially efficient, the implementation of the directional-exchange approach to Σ' can be quite tedious, error-prone, and inflexible in the face of new discretizations. A more general and straightforward approach to Σ' developed by Tufo [389] has proven successful in a number of high-order applications (see [119, 390]). Here, the geometry of the problem is discarded, and Σ' is viewed simply as an unstructured operation on a vector of data represented as the application of the Boolean matrix QQ^T to a set of reals (as in Figure 4.5.3). The code is implemented as a standalone utility with a simple interface requiring two calls, *handle=gs_init(global_index,n)*, and *ierr=gs_op(u,op,handle)*, where *global_index()* associates the $n := (N+1)^d E^{(p)}$ values of $\underline{u}_L$ on node p with their global counterparts, and *op* denotes the reduction operation (e.g., +, ∗, max) to be performed on shared elements of $\underline{u}_L$.

Suppose, for example, that one has the global orderings shown in Figure 8.5.3. If Ω^1 were assigned to node 0 and Ω^2 to node 1, then each processor would call *gs_init* with the respective lists:

node 0: *global_index* = {1 2 3 4 5 6 7 8 9}
node 1: *global_index* = {3 10 11 6 12 13 9 14 15}

The initialization phase sorts the lists, finds commonalities, and establishes the number and list of processors sharing each node. This is readily effected by using a bin sort in which the dynamic range (1 to 15 in the present case) is partitioned into bins B^p, $p = 0, \ldots, P-1$. Each processor p partitions its *global_index*

8.5. Parallel Multidomain Methods

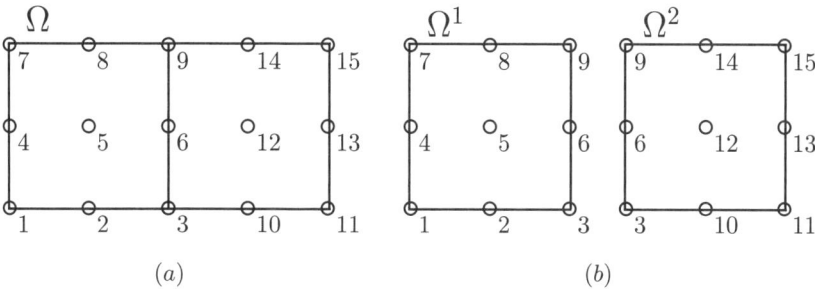

Figure 8.5.3. (*a*) Global numbering and (*b*) partition onto two processors.

array among the bins and sends the contents of B^q to processor q, where they are sorted with data received from other processors and then returned to the originating processor with the requisite pointers. The execution (*gs_op*) phase uses indirect addressing to extract shared elements from $\underline{u}_L$ ($\{u_3, u_6, u_9\}$ on node 0 and $\{u_1, u_4, u_7\}$ on node 1 in the present example) and combines these by using pairwise and/or tree exchanges, depending on the number of processors sharing a given vertex. Elements of $\underline{u}_L$ interior to an element are not touched during the update. To reduce communication, those that are shared within a processor are summed locally and then exchanged with other processors, if necessary.

Because it separates the geometry and PDE aspects of the problem from the communication and computer science issues, this "geometry-free" approach to implementing Σ' greatly simplifies parallel code development. It has proven quite successful in terascale spectral-element applications [124] and in tetrahedra-based computations [119]. It also has advantages in nonconforming methods where a geometric interpretation is not straightforward.

For example, in the interpolation-based case illustrated in Figure 8.5.4, Q_m^T is effectively factored into $\hat{Q}^T J^T$, where $\hat{Q}^T$ is the Boolean matrix accumulating the sum to the global unknowns (center), and J is a block-diagonal matrix comprising the local interpolation operators $J_{ee'}$, which map the global degrees of freedom onto element faces. Application of J and J^T is fully concurrent, whereas $\hat{Q}\hat{Q}^T$ can be implemented, for example, by using *gs_op* [390]. In the *gs_op* implementation, one simply lists the five highlighted vertices as corresponding to the same degrees of freedom (e.g., *global_index* = $\{1\,2\,3\,4\,5\ldots\}$). The entries of $J_{ee'}$ are determined by the geometry on the shared interface (see Section 7.3).

Finally, we note that much of the difficulty and communication overhead associated with Σ' is directly attributable to the need for vertex and edge

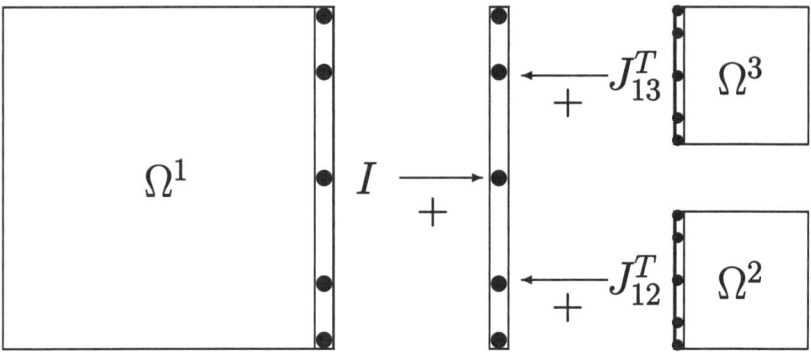

Figure 8.5.4. Schematic of $Q_m^T = \hat{Q}_m J$ applied to (interpolation-based) nonconforming interface variables.

updates. To alleviate this situation, Ben Belgacem and Maday [28] developed an interface condition that requires only face–face exchanges between elements and thus greatly reduces the connectivity requirements. This method appears to be particularly attractive in the context of nonconforming discretizations in $\mathbb{R}^3$.

8.5.3 Domain Partitioning

When partitioning the domain for distributed processing, one must consider the issues of load balance, that is, having a uniform distribution of work across processors, and of communication, which is generally proportional to the number of vertices on the boundaries of the processors subdomains. Although difficult to solve exactly, this optimization problem has a fairly broad minimum, so that heuristics-based approaches are successful. Of these, we mention three: the greedy algorithm, recursive coordinate bisection, and recursive spectral bisection. Each of these algorithms itself can be parallelized [413], a fact that may be important for large problems, or for problems that require adaptive mesh refinement [75, 356].

The greedy algorithm [121] is the simplest. Starting with any element, neighboring elements are added until E/P have been included; one then proceeds to a new seed element, and the process is repeated. Accumulation and seed heuristics can be added to yield reasonable subdomains. The recursive coordinate bisection method successively partitions subdomains along the longest axis (e.g., computed from the moment of inertia, which is readily computed in parallel). Element aspect-ratio effects can be mitigated by rescaling the coordinates so that the average aspect ratio is unity [356].

Recursive spectral bisection [314] successively orders elements within a subdomain according to their rank in the *Fiedler vector* – the vector associated with the second smallest eigenvalue of the dual graph Laplacian matrix. For a mesh containing E elements, the dual graph Laplacian G is an $E \times E$ matrix with G_{ii} equal to the number elements adjacent to Ω^i, $G_{ij} = -1$ for each adjacent element pair (Ω^i, Ω^j), and zeros elsewhere. The lowest eigenpairs can be readily found by modifying the CG algorithm as discussed in [160]. Mesh elements associated with the $E/2$ lowest Fiedler-vector entries are grouped into one subdomain, and the rest are put into the other. The process is repeated on each subdomain until P subdomains have been established. For example, for a one-dimensional array of consecutively ordered elements, G would correspond to the classic FD Laplacian with 2 along the main diagonal (save for the first and last entries, which would be 1), and -1 along the first upper and lower diagonals. The first eigenvector is the constant mode, and the second (the Fiedler vector) is the slowly varying wave, $\cos[\pi i/(E+1)]$, $i = 1, \ldots, E$. The midpoint of this wave corresponds to the dissection point that partitions the elements into two groups.

Several software packages are available to implement domain partitioning strategies, including Metis [222], Chaco [199], Jostle [403], and WGPP [178]. They offer many optimizations, including variable weighting, a feature that can be important if the work per element is not uniform.

8.5.4 Coarse-Grid Solves

An important consideration in developing scalable solvers for elliptic problems is parallelization of the coarse-grid solve. As noted in Section 7.2.3, this can be important for the convergence of iterative methods, particularly for the pressure computation, which is less well conditioned than the Helmholtz problems associated with implicit treatment of the viscous terms.

The coarse-grid problem, $K_0 \underline{x} = \underline{b}$, typically has $O(E) \approx P$ degrees of freedom, which makes it a relatively small parallel problem. Because K_0^{-1} is completely full, however, the solution $\underline{x}$ is dependent on every element of $\underline{b}$, implying that an all-to-all communication is required because both $\underline{b}$ and $\underline{x}$ are distributed vectors. Moreover, the coarse-grid problem is solved on every iteration, making a fast solution scheme imperative. Numerous articles have considered this problem in detail, including [169, 125, 120, 126, 391].

The simplest approach to the coarse-grid problem is to gather a copy of $\underline{b}$ onto each processor by using a concatenate variant of `all_reduce` and then to compute $\underline{x}$ redundantly on each processor. An optimal implementation would require $(\alpha \log_2 P + \beta n)t_a$ time for communication and $O(n \log n)$ or $O(n^{\frac{4}{3}})$

operations, assuming a nested-dissection ordering in $\mathbb{R}^2$ or $\mathbb{R}^3$, respectively. Note that the work does not decrease with increasing P and that the communication increases with P. Moreover, if $n \approx P$, then the work actually increases with P. As a result, the redundant solution approach is limited to approximately $P < 64$. For $P > 64$, a significant improvement can be obtained by explicitly computing K_0^{-1} in a preprocessing step and storing rows corresponding to local degrees of freedom on each processor [170, 125]. The solution is computed by gathering $\underline{b}$ onto each processor and computing $x_i = \underline{r}_i^T \underline{b}$, where $\underline{r}_i$ is the ith row of K_0^{-1}. The communication cost is identical to the redundant-solve approach. However, the work is $O(n^2/P)$: significant concurrency has been obtained.

To reduce the communication and work still further, Farhat and Chen [120], Fischer [126], and Tufo and Fischer [391] developed coarse-grid solvers based on projection. The solution is cast as the readily parallelized product $\underline{x} = XX^T \underline{b}$, where the columns of X are constructed so that $X^T K_0 X = I$ [see (2.7.36)]. Farhat and Chen based X on Krylov spaces generated during initial calls to their PCG solution of the coarse-grid problem. They found that accumulating $\approx n/4$ columns of X suffices to give an accurate projection without need for further PCG iterations.

A different approach was taken by Fischer and Tufo, who used projection onto a set of carefully ordered unit vectors to develop a quasi-sparse factorization of K_0^{-1}. They showed that, if K_0 is a sparse SPD matrix arising from a d-dimensional mesh, it is possible to find (via nested dissection) a factorization $K_0^{-1} = XX^T$, where X has $O(n^{(2d-1)/d})$ nonzeros, such that the parallel work to compute $\underline{x} = XX^T \underline{b}$ is $O(n^{(2d-1)/d}/P)$. Moreover, the communication cost is only $O(n^{(d-1)/d} \log_2 P)$. Examples in [391] show that this algorithm substantially outperforms the redundant solve and distributed K_0^{-1} approaches for large values of P and n.

8.6 Applications

We turn now to a number of applications that illustrate the combined potential of high-order methods and high-performance computing.

8.6.1 Hairpin Vortices

Since the early work of Theodorsen [378], hairpin vortices have been recognized as important dynamical structures in both transitional and turbulent boundary layers. Recently, a number of numerical and experimental studies have sought to understand the dynamics of multiple interacting hairpins (see [424] for a list of references). Among these, Acalar and Smith [1] undertook extensive

8.6. Applications

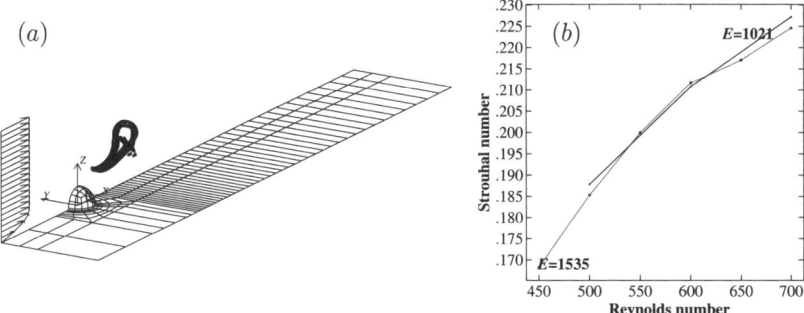

Figure 8.6.1. (*a*) Computational domain showing inlet velocity profile, flat plate, hemisphere, and isolated hairpin vortex. For clarity, the vortex has been reflected about the symmetry plane. (*b*) Strouhal number versus Reynolds number for the $E = 1021$ and $E = 1535$ meshes.

visualization studies of the growth and topology of hairpin-vortex chains induced by a hemispherical roughness element. Spectral-element simulations by Tufo et al. [392] presented detailed observations of hairpin-vortex evolution in the same configuration, using the computational domain shown in Figure 8.6.1(*a*), which consists of a unit-radius hemisphere centered at the origin, a Blasius inlet profile at $x = -10$, and a Neumann outflow condition at $x = 30$. Computational grids comprise either $E = 1021$ or 1535 elements, of order $N = 4$ to 15, depending on the Reynolds number. Scaling is based on the roughness height k and velocity U_k that would be observed at height k in the absence of the roughness element, giving rise to two independent parameters, namely, the Reynolds number $Re_k := kU_k/\nu$ and the nondimensional boundary layer thickness $\delta := \delta_{99}/k$. At sufficiently high Reynolds numbers, the steady inflow boundary layer destabilizes, leading to periodic shedding of hairpin vortices at nondimensional frequencies $St := fk/U_k$, as shown in Figure 8.6.1(*b*). Figure 8.6.2 shows a typical chain of of hairpin vortices at $Re_k = 922$, visualized by using the λ_2 criterion of Jeong and Hussain [213]. Stationary horseshoe vortices are visible upstream of the roughness element.

To study the scalability of the SEM, Tufo and Fischer [390] used an octree refinement of the $E = 1021$ mesh to obtain an $(E, N) = (8168, 15)$ mesh with 27.8 million velocity points, which was run on up to 4096 processors on the Intel ASCI Red machine at Sandia National Laboratories. Each node of ASCI Red is a dual-processor 333-MHz Pentium Pro, giving a peak theoretical performance of 666 Mflops/node. The results of this study, summarized in Table 8.6.1, reveal nearly linear speedup as P is increased from 512 to 2048. In addition, the importance of fast matrix–matrix routines is clearly evident. The Std. column represents the use of a single (optimized) variant of dgemm, and the Perf. column

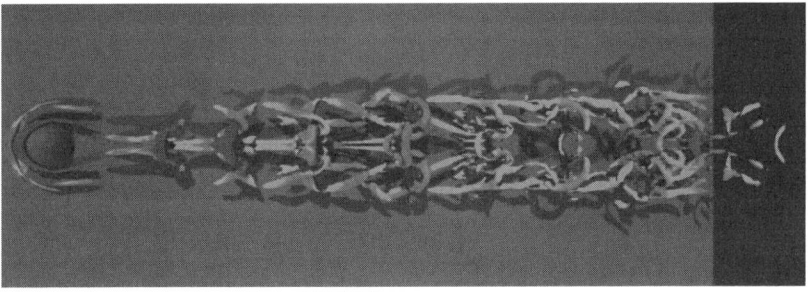

Figure 8.6.2. Profile (top) and planform (bottom) views of hairpin and secondary vortices generated in a boundary layer by a hemispherical roughness element for $Re_k = 922$. The spectral-element parameters are $(E, N) = (1021, 11)$.

represents the use of the best possible option in each of the configurations presented in Table 8.3.1. The coarse grid for this problem has 10,142 distributed degrees of freedom. The XX^T-based coarse-grid solver accounts for 1.0% of the total time in the worst-case scenario of 2048 nodes in dual-processor mode.

8.6.2 Driven Cavity

The DNS of the lid-driven cubic cavity flow was extensively investigated by Leriche and Gavrilakis [239, 240]. The basic algorithm is a Chebyshev collocation method with a direct solver based on the fast diagonalization method (4.2.15) [250, 181].

Table 8.6.1. ASCI *Red Performance (Gflops):* $E = 8168, N = 15$.

	Std. mxm		Perf. mxm	
P	Single	Dual	Single	Dual
512	47	67	65	94
1024	93	135	132	191
2048	183	267	257	361

8.6. Applications

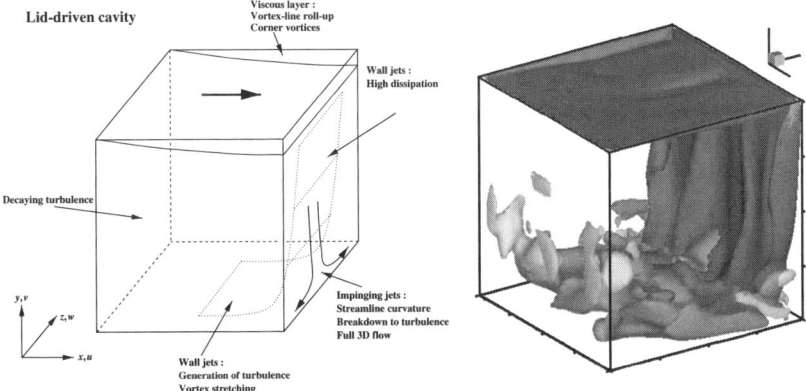

Figure 8.6.3. Three-dimensional driven cavity at $Re = 5000$ (due to Leriche [239]).

Figure 8.6.3 shows the problem definition. The flow field is contained within a cavity enclosure and is driven by imposing a velocity on the top wall. On the remaining five sides no-slip conditions are assumed. The top surface moves with a velocity component that is a high-degree polynomial to smooth out the edge singularities, still approximating the unit profile as closely as possible. The parallel version of the code is based on the parallel implementation of the BLAS routines developed by Nec Inc.

Table 8.6.2 gives the parallel performance of the code for a 129^3 mesh as a function of the number of processors on a Nec-SX4: the speedup factors, the Gflops sustained, and the CPU time in seconds per time step [241]. The degree of vectorization is 99%. The execution speed on 16 processors on a Nec-SX4 is 0.7 μs per collocation point per time step. For comparison, Table 8.6.3 shows

Table 8.6.2. *Parallel performance of the code with the final mesh 129^3, in terms of the speedup factors, the Gflops sustained, and the CPU time in seconds per time step, as a function of the number of processors on a Nec-SX4. The CPU time is the sum of the CPU time spent on each processor, divided by the number of processors.*

Processors	Speedup	Gflops	CPU (s/Δt)
1	1.00	1.937	18.955
2	1.97	3.803	9.653
4	3.85	7.452	4.926
8	7.36	14.249	2.577
10	9.22	17.852	2.057
12	10.70	20.723	1.772
16	13.40	25.962	1.415

Table 8.6.3. *Parallel performance of the real matrix–matrix multiplication of the NEC library (ASL/SX R13.1) in terms of the speedup factors and the Gflops sustained, as a function of the number of processors on a Nec-SX4. The matrix order is 5000.*

Processors	Speedup	Gflops
2	2.00	3.99
4	4.00	7.99
8	7.99	15.96
16	15.93	31.82

the speedup and the sustained Gflops for the parallel real matrix–matrix multiplication of the NEC library (ASL/SX R13.1) as a function of the number of processors. The results are similar up to 8 processors. For 16 processors, the parallel matrix–matrix multiplication continues to exhibit perfect linear scaling with the number of processors, whereas the performance of the parallel Chebyshev code slows because of the small size of submatrices on each processor, which leads to communication between processors being more important than the local work on a single processor.

At $Re \geq 10^4$, the DNS of the cubical cavity flow shows both transient and steady-in-the-mean states that possess long time scales requiring long integration times. A large fraction of the total kinetic energy and dissipation is concentrated in the near-lid mean flow. The flow over most of the domain is laminar, with distinct wall-jet profiles found near three of the walls. The high-momentum fluid near the lid transmits its energy into a downflowing nonparallel wall jet, which separates ahead of the bottom wall. From the collision of this separated layer against the bottom wall, two wall jets emerge. This process is clearly illustrated in right part of Figure 8.6.3, where isosurfaces of the u velocity component are shown near the the lid and the bottom of the cavity, while isosurfaces of the v velocity components are shown near the two vertical lateral walls. In this process the energy lost to turbulence by the impingement is partly recovered by the emerging wall jets.

8.6.3 Backward-Facing Step

Spectral-element calculations of the 3D flow over a backward-facing step were undertaken by Couzy [83]. The geometry is given in Figure 8.6.4. This geometry corresponds to the experimental setup of Armaly et al. [7]. The length of the channel behind the step is 19, which is long enough at the Reynolds number considered ($Re = 343$) to accommodate a fully developed velocity profile at

8.6. Applications 413

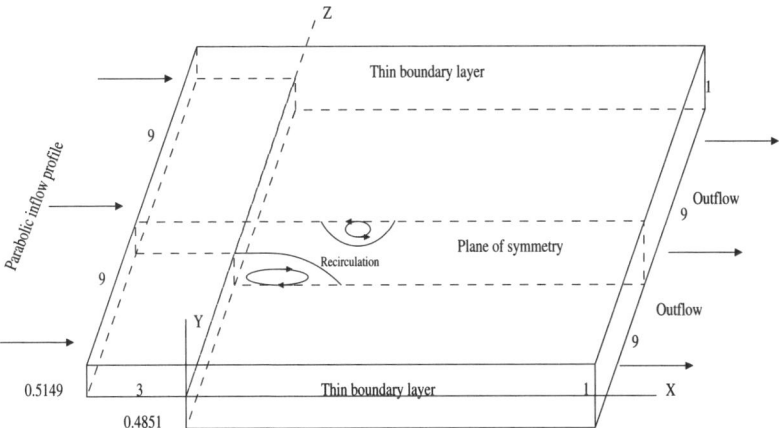

Figure 8.6.4. Geometry of the backward-facing step with a 1 : 1.94 expansion ratio. For the Reynolds number considered here, two recirculation zones are of interest. Their locations are indicated in the symmetry plane.

the exit. The Reynolds number is defined with a reference velocity based on the average inlet velocity, and the reference length is twice the inlet channel height. Figure 8.6.5 shows the mesh, which comprises 128 spectral elements. The numerical results will be presented for two polynomial degrees, namely, $N = 9$ and $N = 11$. They produce almost coincident results, showing the good resolution of both discretizations. As soon as all spurious numerical transients associated with the start of the computation are over, a third-order projection scheme (6.2.31) is used together with a subcycling method based on RK4.

The length of the recirculation zone behind the step increases linearly with the Reynolds number, as is seen in Figure 8.6.6, and agrees well with the experimental data and the DNS results of Kim and Moin [223]. A 3D flow develops with a growing boundary layer along the side wall, interacting with the recirculation zone.

In Figures 8.6.7 and 8.6.8, the streamwise velocity component is shown at various locations. Note that the origin of the axes is now at the bottom of the step. In particular, the observation line at $x = 3.22$, $y = 0.75$ shows the vertical recirculation at the side wall.

Figure 8.6.9 indicates clearly that 3D effects are not limited to the boundary layer. Moreover, the spanwise velocity component also changes sign near the

Figure 8.6.5. Spectral-element distribution. True aspect ratio.

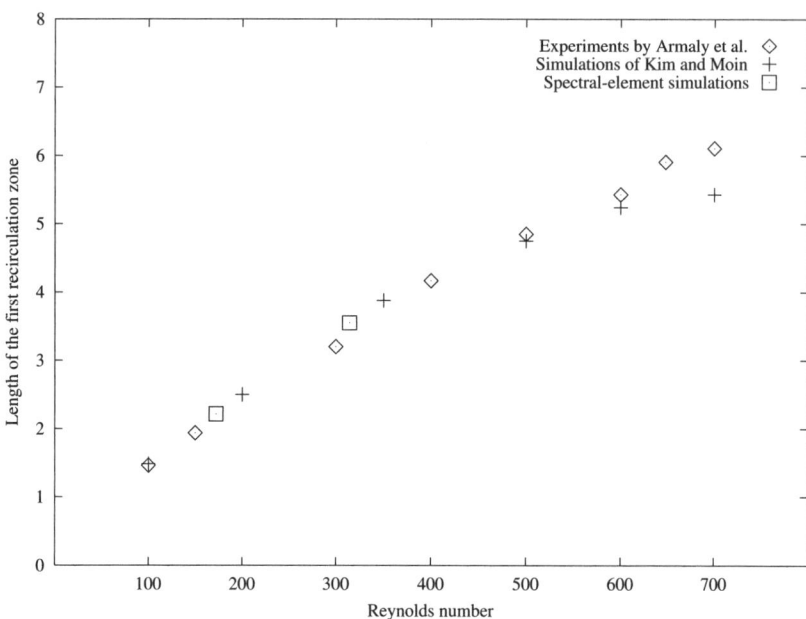

Figure 8.6.6. Comparison of the length of the first recirculation zone as a function of the Reynolds number.

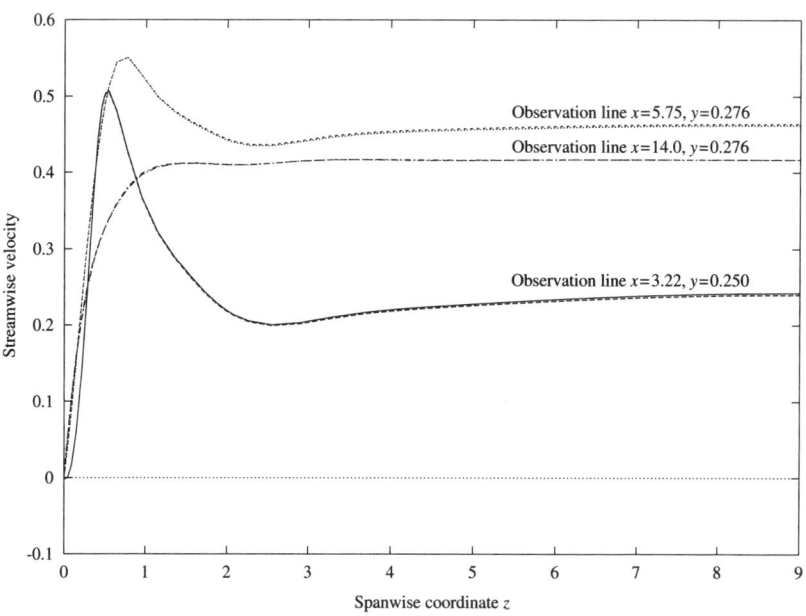

Figure 8.6.7. $Re = 343$. Streamwise velocity component along the three observation lines in the lower part of the geometry, downstream of the recirculation zone. Results for $N = 9$ and $N = 11$ are shown but are virtually indistinguishable.

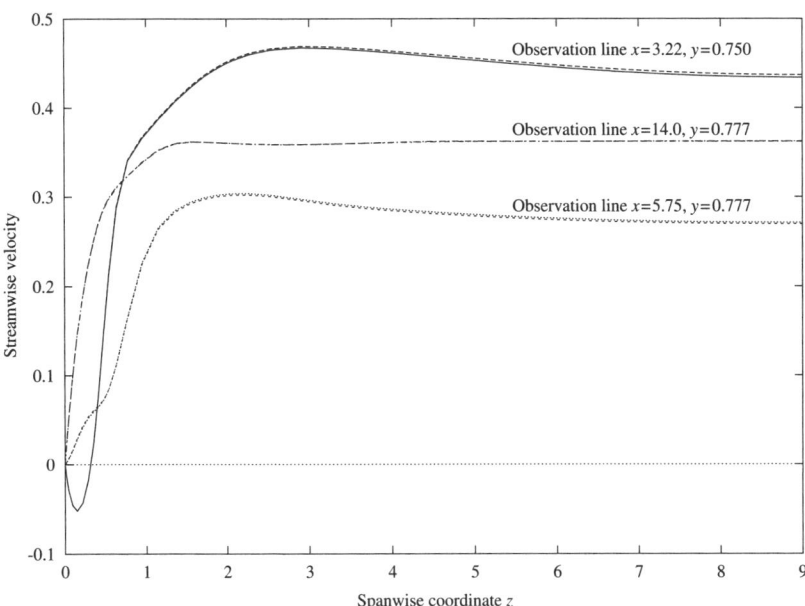

Figure 8.6.8. $Re = 343$. Streamwise velocity component along the three observation lines in the upper part of the geometry, downstream of the recirculation zone. The results for $N = 9$ and $N = 11$ are given, but are difficult to distinguish.

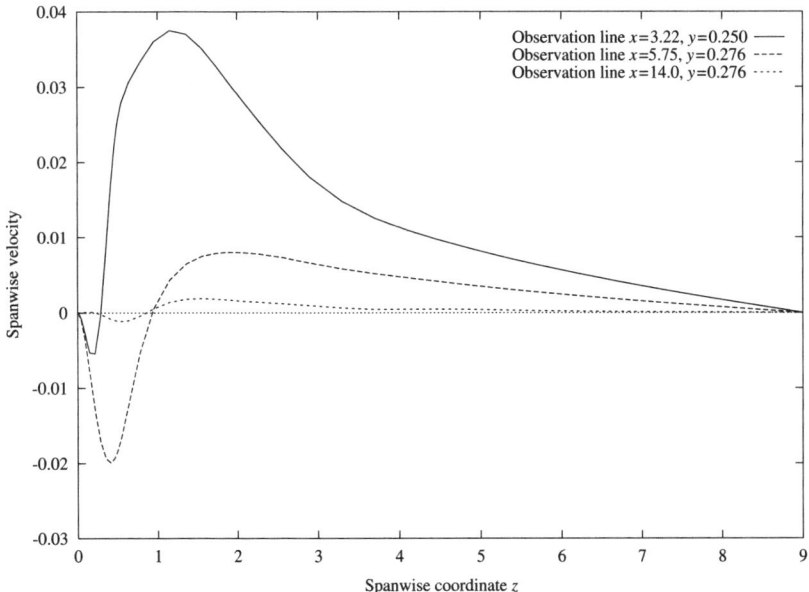

Figure 8.6.9. $Re = 343$. Spanwise velocity component along the three observation lines in the lower part of the geometry, downstream of the recirculation zone.

side wall. A test for spatial convergence is obtained by increasing the polynomial degree from $N = 9$ to $N = 11$. No noticeable change can be observed.

8.7 Further Reading

Parallel-performance analysis of high-order multidomain methods can be found in Fischer and Patera [131], Fischer [125], and Evangelinos et al. [119]. For analyses of global spectral methods, see for example Pelz [304] and Wray and Rogallo [415]. For a discussion of object-oriented spectral-element implementations, see the papers by Dubois-Pèlerin and coworkers [116, 114, 115].

A significant challenge in parallel PDE solution that has not been addressed in this chapter is the issue of dynamic load balancing in the presence of adaptive refinement. For low-order FEs there have been many significant efforts in this area. The works of Shephard et al. [356], Castaños and Savage [76], and the references therein give an overview of the fundamental issues. Parallel adaptivity has also been addressed in the high-order context by Henderson [195].

Appendix A
Preliminary Mathematical Concepts

This appendix deals with a few mathematical concepts pervading the book, which are related to spaces of functions and systematically use the language of geometry.

A.1 Metric Spaces

A metric space is the simplest mathematical structure where the evaluation of some "distance" between mathematical objects is made possible.

A.1.1 Definition

Let X denote a nonempty set of elements $x, y, z, \ldots$, the nature of which is unspecified, and let $d : X \times X \to \mathbb{R}$ denote a real-valued mapping such that for every couple $(x, y) \in X \times X$ one has

$$d(x, y) = d(y, x), \tag{A.1.1}$$

$$d(x, y) = 0 \iff x = y, \tag{A.1.2}$$

$$d(x, y) \leq d(x, z) + d(z, y). \tag{A.1.3}$$

It is easy to show that these properties imply that $d(x, y)$ is finite and nonnegative. The pair (X, d) is a *metric space*, and $d(x, y)$ is a measure of the distance between the elements x and y. For obvious reasons, the axiom (A.1.3) is known as the *triangle inequality*.

A metric space $(S, \hat{d})$ is a subspace of (X, d) if S is a subset of X and if the metric $\hat{d}$ is a restriction of d to the elements of S:

$$\hat{d} := d|_{S \times S}. \tag{A.1.4}$$

The distance $\hat{d}$ is the metric induced by d, and most of the time one represents the subspace by (S, d).

Examples of Metric Spaces

Example 1: Space $\mathbb{R}^n$
For any two elements of $\mathbb{R}^n$, $x = (\xi_1, \ldots, \xi_n)$ and $y = (\eta_1, \ldots, \eta_n)$, the distance $d(x, y)$ may be measured by

$$d(x, y) := [(\xi_1 - \eta_1)^2 + \cdots + (\xi_n - \eta_n)^2]^{1/2}. \tag{A.1.5}$$

For $n = 2$, the expression (A.1.5) is the Euclidean metric of plane geometry.

A more abstract example, useful for understanding some of the subtleties of the theory, is the following.

Example 2: Discrete metric space
Let X denote any collection of elements. The distance $d(x, y)$ between two elements $x, y \in X$ is defined by

$$d(x, x) := 0, \qquad d(x, y) := 1 \quad \text{if } x \neq y. \tag{A.1.6}$$

The relation (A.1.6) satisfies the axioms (A.1.1)–(A.1.3). Therefore (X, d) is a metric space, usually called the *discrete metric space*.

As the latter example shows, a metric space is not necessarily a vector space.

A.1.2 Open Set, Closed Set, Neighborhood

Given a metric space (X, d) and an element $x \in X$, one defines the *open ball* $B(x; r)$ and the *closed ball* $\bar{B}(x; r)$ of center x and radius r as, respectively,

$$B(x; r) := \{y \in X \mid d(x, y) < r\}, \tag{A.1.7}$$

$$\bar{B}(x; r) := \{y \in X \mid d(x, y) \leq r\}. \tag{A.1.8}$$

Let (S, d) denote a subspace of (X, d). The set S is said to be *open* if it includes a ball about each of its elements. The set is said to be *closed* if its complement S^C ($:= X - S$) is open.

Examples taken from geometry usually suggest that subsets of metric spaces are either open or closed. The reader must be aware that this is not so in general. Applying the definitions to the discrete metric space given in Example 2, one verifies easily that for this particular case, all subsets are both open and closed.

A.1. Metric Spaces

In limiting processes one often uses the concept of ε-neighborhood. An ε-*neighborhood* of $x \in X$ is an open ball of radius ε. By extension, a *neighborhood* of $x \in X$ is any subset containing an ε-neighborhood of x.

A.1.3 Cauchy Sequence, Limit Points, Dense Sets

Let (X, d) denote a metric space. A sequence $\{x_n\}_{n=1}^{\infty}$ of elements of X is a *Cauchy sequence* if, for every $\varepsilon > 0$, there is an index $N = N(\varepsilon)$ such that

$$d(x_m - x_n) < \varepsilon \qquad \forall m, n > N. \tag{A.1.9}$$

Assume S is a subset of X. Any element $x \in X$, not necessarily in S, is called a *limit point* (or an accumulation point) of S if every ε-neighborhood of x contains at least one element $y \in S$ distinct from x. The set comprising the points of S and the limit points of S is called the *closure* of S and is denoted by $\bar{S}$.

Let (X, d) be a metric space and S a subset of X. The subset S is said to be *dense in* X if its closure $\bar{S}$ is identical to X, that is if $\bar{S} = X$. A metric space (X, d) is said to be *separable* if X has a countable subset (i.e., a subset where the elements may be put in a one-to-one correspondence with the natural numbers) that is dense in X.

A.1.4 Mapping, Domain, Range, Continuity

Let (X, d_X) and (Y, d_Y) denote two metric spaces and M be a subset of X. A *mapping* T from M into Y is an operation that associates with each $x \in M$ a unique $y \in Y$. One usually writes $y = Tx$. The element y is the *image* of x with respect to the mapping. The subset M is the *domain* $\mathcal{D}(T)$ of T, and the set of elements $\mathcal{R}(T) \subseteq Y$ such that

$$\mathcal{R}(T) := \{y \in Y;\ y = Tx,\ x \in \mathcal{D}(T)\}$$

is called the *range* of T.

The set of all elements $x \in \mathcal{D}(T)$ such that $Tx = y_0$ for a given y_0 is called the *inverse image* of y_0.

A mapping T is called a *surjection* if $\mathcal{R}(T)$ covers the whole metric space, that is, if $\mathcal{R}(T) = Y$. It is called an *injection* if different elements in $\mathcal{D}(T)$ have different images: $x_1 \neq x_2 \Rightarrow Tx_1 \neq Tx_2$. Finally, a mapping that is both surjective and injective is called a *bijection*.

If T is a bijection of $M \subset X$ into Y, then there exists an *inverse mapping* T^{-1} such that for any $y \in Y$, a unique element $x \in M$ exists such that $T^{-1}y = x$.

420   Appendix A. Preliminary Mathematical Concepts

Let $T : X \to Y$ denote a mapping of X onto Y. The mapping is said to be *continuous at the element* $x_0 \in X$ if for every $\varepsilon > 0$ there exists a $\delta > 0$ such that

$$d_Y(Tx, Tx_0) < \varepsilon \quad \forall x \in X \text{ with } d_X(x, x_0) < \delta.$$

A mapping T is *continuous* if it is continuous at every point in the domain $\mathcal{D}(T)$. The following important theorem relates continuous mappings to open (or closed) sets.

Theorem A.1 (Continuous mapping) *A mapping T of a metric space X onto a metric space Y is continuous if and only if the inverse image of any open (closed) subset of Y is an open (closed) subset of X.*

A.1.5 Convergence, Completeness, Completion Process

Consider $S := \{x_n\}_{n=1}^{\infty}$, a sequence of elements of X. Clearly, S is a subset of X. A sequence S is said to be *convergent* if there exists a (unique) element $x \in X$ that is a limit point of S.

Every convergent sequence in a metric space (X, d) is a Cauchy sequence, but the converse is not true. Metric spaces in which every Cauchy sequence is a convergent sequence are said to be *complete*. If this condition is not fulfilled, the metric space is *incomplete*.

Completeness is the most desirable property of spaces of functions, because it warrants the existence of solutions obtained by using limiting processes. A fundamental theorem of functional analysis states that every incomplete metric space can be completed.

Theorem A.2 (Completion process) *Let $X := (X, d)$ be an incomplete metric space. There exists a complete metric space $\hat{X} := (\hat{X}, \hat{d})$ having a subspace V that is dense in $\hat{X}$ and such that V and X may be related by a one-to-one transformation T preserving the metric, in other words, such that*

$$\hat{d}(Tx, Ty) = d(x, y), \quad x, y \in X, \quad Tx, Ty \in \hat{X}.$$

The space $\hat{X}$ is unique except for the transformation T.

Most of the time, the subspace V is identified with the incomplete space X itself, and $\hat{X}$ is the closure of X. The construction process for $\hat{X}$ is the following. One identifies the set of all equivalent Cauchy sequences in X. Two Cauchy sequences $\{x_n\}_{n=1}^{\infty}$ and $\{x_n'\}_{n=1}^{\infty}$ are equivalent ($\{x_n\} \sim \{x_n'\}$), or belong to the same equivalence class $\hat{x}$, if

$$\lim_{n \to \infty} d(x_n, x_n') = 0.$$

The set $\hat{X}$ is made of all equivalence classes $\hat{x}, \hat{y}, \hat{z}, \ldots$, and one defines the distance $\hat{d}$ between $\hat{x}$ and $\hat{y}$ by

$$\hat{d}(\hat{x}, \hat{y}) := \lim_{n \to \infty} d(x_n, y_n), \qquad (A.1.10)$$

where the Cauchy sequences $\{x_n\}$ and $\{y_n\}$ are representatives of their equivalence classes. One may easily show that the limit in (A.1.10) is independent of the selection of the representatives and that $\hat{d}$ satisfies the axioms (A.1.1)–(A.1.3) (see [228]).

A.2 Normed Spaces

The real interest for equation solution lies in real (or complex) *abstract* vector spaces where linear combinations of elements (e.g. functions, or distributions) from a given set also belong to it. Normed spaces and inner-product spaces are mathematical structures that allow the generalization of concepts such as "length," "distance," and "angle" in abstract vector spaces. We begin with the concepts of length and distance.

A.2.1 Definition

Let X denote a real vector space. A *norm* on X is a nonnegative real function $\|\cdot\|: X \to \mathbb{R}$ such that for any $x \in X$ one has

$$\|cx\| = |c| \|x\|, \qquad (A.2.1)$$

$$\|x + y\| \leq \|x\| + \|y\|, \qquad (A.2.2)$$

$$\|x\| = 0 \iff x = O, \qquad (A.2.3)$$

where O is the null vector of X.

In some cases only the properties (A.2.1) and (A.2.2) are satisfied. If there exist elements $x \neq O$ such that $\|x\| = 0$, then the application $X \to \mathbb{R}$ is called a *seminorm*. A seminorm is usually denoted by $|\cdot|$.

A *normed space* is a vector space equipped with a norm. The norm of a vector is a measure of its "length." For any two vectors $x, y \in X$, therefore,

$$d(x, y) := \|x - y\| \qquad (A.2.4)$$

is a measure of their distance. Normed spaces are metric spaces with a metric *induced* by the norm as shown by (A.2.4), the property (A.2.2) being another expression of the triangle inequality (A.1.3). As a consequence, all concepts such as Cauchy sequences, limit points, and dense subsets introduced in Section A.1 for metric spaces are readily transposable to normed spaces.

Before giving two examples of normed spaces, we point out that different normed spaces may be built on the same vector space, depending on the norm.

Examples of Normed Spaces

Example 1: Space $\mathbb{R}^n$
Let $x = (\xi_1, \ldots, \xi_n) \in \mathbb{R}^n$ and $p \geq 1$. The expression

$$\|x\| := (|\xi_1|^p + \cdots + |\xi_n|^p)^{1/p}, \tag{A.2.5}$$

satisfies the axioms (A.2.1)–(A.2.3) and therefore defines a norm over $\mathbb{R}^n$, called the *p-norm*. The case $p = 2$ is important in practice and corresponds to the familiar Euclidean (or square) norm. The property (A.2.2) applied to the p-norm (A.2.5) is also known as Minkowski's inequality:

$$\left(\sum_{i=1}^n |\xi_i + \eta_i|^p\right)^{1/p} \leq \left(\sum_{i=1}^n |\xi_i|^p\right)^{1/p} + \left(\sum_{i=1}^n |\eta_i|^p\right)^{1/p}.$$

Example 2: Space $C[a, b]$
To every element $x \in C[a, b]$ where $C[a, b]$ is the set of continuous real functions in $[a, b]$, we associate the real quantity

$$\|x\| := \max_{t \in [a,b]} |x(t)|. \tag{A.2.6}$$

$C[a, b]$ is a vector space because any combination of continuous functions is continuous. The quantity (A.2.6) is called the *maximum norm*, and $C[a, b]$ is a normed space. One verifies easily that the expression

$$\|x\| := \left(\int_a^b x^2(t)\, dt\right)^{1/2}, \tag{A.2.7}$$

called the L^2 norm, also satisfies the axioms for a norm and therefore leads to another normed space.

A.2.2 Banach Spaces

Normed spaces in which every Cauchy sequence has a limit are complete. Complete normed spaces are called *Banach spaces*. Since normed spaces are

A.3. Linear Operators and Functionals in Normed Spaces

also metric spaces, incomplete normed spaces may be completed by using the completion process described in Section A.1.5 with the metric (A.2.4).

One may show that the vector space $C[a, b]$ with the maximum norm (A.2.6) is complete, whereas Cauchy sequences of elements of $C[a, b]$ in the metric induced by (A.2.7) may tend toward discontinuous functions.

A.3 Linear Operators and Functionals in Normed Spaces

In Section A.1.4, we introduced mappings between metric spaces. When mappings apply between vector spaces, they are called *operators*. Linear operators form an important class of mappings, for which we give a few basic results.

A.3.1 Linear Operator, Domain, Range, Nullspace

Let $(X, \|\cdot\|_X)$ and $(Y, \|\cdot\|_Y)$ denote two real normed spaces. A *linear operator* is a mapping $T : \mathcal{D}(T) \to \mathcal{R}(T)$ between subspaces of X and Y that satisfies the following property:

$$T(\alpha x_1 + \beta x_2) = \alpha T x_1 + \beta T x_2 \quad \forall x_1, x_2 \in \mathcal{D}(T) \subseteq X,$$

where α and β are any real quantities. The element $y = Tx$ is called the *image* of x by T.

The *domain* $\mathcal{D}(T)$ of T is the set of elements $x \in X$ belonging to X on which the operator T acts. The image of $\mathcal{D}(T)$ is called the *range* $\mathcal{R}(T) \subseteq Y$ of T.

The *nullspace* (or kernel) $\mathcal{N}(T) \subseteq X$ of T is the set of elements $x \in X$ such that $Tx = O$.

The domain, range, and nullspace of a linear operator T are vector spaces. Furthermore, one may show that if the domain $\mathcal{D}(T)$ is finite-dimensional with $\dim \mathcal{D}(T) = n$, then $\dim \mathcal{R}(T) \leq n$.

A linear operator is called a *surjection* if $\mathcal{R}(T)$ covers the whole space Y. It is called an *injection* if distinct elements in $\mathcal{D}(T)$ have distinct images in $\mathcal{R}(T)$ (one-to-one correspondence). In this case, it is easy to show that the nullspace $\mathcal{N}(T)$ reduces to the zero element. Injective operators are especially important in that they possess an *inverse*. Operators that are both injections and surjections are called *bijections*.

A.3.2 The Inverse Operator

Let $(X, \|\cdot\|_X)$ and $(Y, \|\cdot\|_Y)$ denote two real normed spaces with T, a mapping between X and Y with domain $\mathcal{D}(T)$ and range $\mathcal{R}(T)$. The *inverse operator*

T^{-1}, if it exists, is the mapping $T^{-1}: \mathcal{R}(T) \to \mathcal{D}(T)$ such that

$$T^{-1}Tx = x \quad \forall x \in \mathcal{D}(T), \quad \text{or} \quad TT^{-1}y = y \quad \forall y \in \mathcal{R}(T).$$

The inverse of a linear operator T exists and is unique iff its nullspace $\mathcal{N}(T)$ reduces to the zero element. Furthermore, in the finite-dimensional case, for the inverse to exist, the condition $\dim \mathcal{D}(T) = \dim \mathcal{R}(T)$ must be satisfied.

Linear algebra offers a simple example to illustrate these properties.

Example: Matrices

Let $A = [a_{i,j}]$ denote an $m \times n$ real matrix that maps $\mathcal{D}(A) = \mathbb{R}^n$ onto $\mathcal{R}(A) \subseteq \mathbb{R}^m$:

$$\begin{bmatrix} y_1 \\ y_2 \\ \vdots \\ y_m \end{bmatrix} = \begin{bmatrix} a_{11} & a_{12} & \cdots & a_{1n} \\ a_{21} & a_{22} & \cdots & a_{2n} \\ \vdots & \vdots & & \vdots \\ a_{m1} & a_{m2} & \cdots & a_{mn} \end{bmatrix} \begin{bmatrix} x_1 \\ x_2 \\ \vdots \\ x_n \end{bmatrix}.$$

The range $\mathcal{R}(A)$ is the column space of A, and the nullspace $\mathcal{N}(A)$ is the subspace of $\mathbb{R}^n$ the elements of which are solution of the system $Ax = 0$.

A basic result of linear algebra states that the dimensions of $\mathcal{R}(A)$ and $\mathcal{N}(A)$ are equal respectively to r and $n - r$, where $r \leq \min(m, n)$, the *rank* of A, is the number of nonzero pivots in the Gaussian elimination process applied to Ax. The inverse operator A^{-1} exists iff $m = n$ and the rank r is equal to n (see [367]).

A.3.3 Bounded Operators, Compact Operators

Assume $(X, \|\cdot\|_X)$ and $(Y, \|\cdot\|_Y)$ are two real normed spaces. A linear operator $T: \mathcal{D}(T) \to Y$ is said to be *bounded* if there exists a real constant $C > 0$ such that for any element $x \in \mathcal{D}(T) \subseteq X$ one has

$$\|Tx\| \leq C \|x\| \quad \forall x \in \mathcal{D}(T). \tag{A.3.1}$$

By definition the smallest constant satisfying (A.3.1) is the norm $\|T\|$ of the operator T. Using a suitable transformation, one may show that

$$\|T\| = \sup_{\substack{x \in \mathcal{D}(T) \\ \|x\|=1}} \|Tx\|. \tag{A.3.2}$$

One easily verifies that $\|T\|$ given by (A.3.2) satisfies the axioms (A.2.1)–(A.2.3).

A.3. Linear Operators and Functionals in Normed Spaces

Following Theorem A.1, an operator $T : \mathcal{D}(T) \subseteq X \to Y$, not necessarily linear, between normed spaces is *continuous* at $x_0 \in \mathcal{D}(T)$ if the inverse image of an ε-neighborhood of Tx_0 is an ε-neighborhood of x_0. The operator T is continuous in $\mathcal{D}(T)$ if it is continuous at every $x \in \mathcal{D}(T)$. For linear operators, one has the following important theorem that links continuity and boundedness.

Theorem A.3 (Continuity and boundedness) *Let $T : \mathcal{D}(T) \subseteq X \to Y$ be a linear operator between normed spaces. Then*

1. *T is continuous if and only if T is bounded.*
2. *If T is continuous at a single point, it is continuous.*

A linear operator $T : X \to Y$ between normed spaces is *compact* if for every bounded sequence $\{x_n \in X, n = 1, 2, \ldots\}$, the corresponding image $\{Tx_n \in Y, n = 1, 2, \ldots\}$ has a convergent subsequence $\{Tx_{n_k} \in Y, k = 1, 2, \ldots\}$. In finite-dimensional normed spaces, every linear operator is compact.

A.3.4 Bounded Linear Functionals, Dual Spaces

Linear functionals are a special category of linear operators that map normed spaces onto the real axis $\mathbb{R}$ or into the complex plane $\mathbb{C}$, depending on the scalar field associated to the vector space.

A linear functional $l : X \to \mathbb{R}$ is said to be *bounded* if there exists a constant $C > 0$ such that

$$|l(x)| \leq C \|x\| \qquad \forall x \in X. \tag{A.3.3}$$

As a consequence of Theorem A.3, a bounded linear functional is continuous.

Since every combination of bounded linear functionals is also a bounded linear functional, the set of bounded linear functionals on a normed space is itself a vector space. This space is called the *dual space* of X and is denoted by X'. Endowed with the norm defined by (A.3.2), the dual space X' is a Banach space. The mapping from $X' \times X$ into $\mathbb{R}$ defined by

$$\langle l, x \rangle = l(x)$$

is known as the *duality pairing* between the spaces X and X'.

A.3.5 The Fréchet Derivative of an Operator

Let X and Y be two normed linear spaces. Let T be an operator from X into Y. This operator is said to be Fréchet differentiable at a point u if there exists a continuous linear operator $L(u): X \to Y$ such that

$$T(u + w) - T(u) = Lw + \varepsilon(u, w), \tag{A.3.4}$$

where

$$\lim_{\substack{w \in X \\ \|w\| \to 0}} \frac{\|\varepsilon(u, w)\|_Y}{\|w\|_X} = 0. \tag{A.3.5}$$

If both relations hold, the linear operator L is unique; it is called the Fréchet derivative of T at point u and is denoted as $T'(u)$.

Examples of Fréchet Derivatives

Example 1
Consider the nonlinear function

$$T(u) = \Delta u - u^n,$$

where $n \geq 2$, $n \in \mathbb{N}$, and Δ is the Laplacian operator. If $u \in C^2(\Omega)$, then $\Delta u \in C^0(\Omega)$. We choose the range $\mathcal{R}(T)$ as $Y = \{y; y \in C^0(\Omega), \|y\|_Y = \sup_{\underline{x} \in \Omega} |y(\underline{x})|\}$. Then Equation (A.3.4) yields

$$T(u + w) - T(u) = \Delta w - nwu^{n-1} - \left(\frac{n(n-1)}{2} w^2 u^{n-2} + \cdots\right).$$

We set

$$\varepsilon(u, w) = -\left(\frac{n(n-1)}{2} w^2 u^{n-2} + \cdots\right).$$

It is easy to show that $\|\varepsilon(u, w)\|_Y / \|w\|_X$ goes to zero when $\|w\|_X \to 0$. One therefore concludes that

$$T' = \Delta - nu^{n-1},$$

or

$$T'(u, w) = \Delta w - nu^{n-1} w.$$

Example 2
Consider now the nonlinear Burgers equation

$$T(u) = \frac{\partial u}{\partial t} + u \frac{\partial u}{\partial x} - \nu \frac{\partial^2 u}{\partial x^2} = 0, \tag{A.3.6}$$

with $u \in C^2(\Omega)$. Following the same steps, we obtain

$$T(u+w) - T(u) = \frac{\partial w}{\partial t} + u\frac{\partial w}{\partial x} + w\frac{\partial u}{\partial x} - \nu\frac{\partial^2 w}{\partial x^2} + w\frac{\partial w}{\partial x}.$$

In this case, the term $\varepsilon(u, w)$ is $w\, \partial w/\partial x$. Choosing again Y as the space defined in the previous example and assuming that $\sup_{x \in \Omega} |\partial y/\partial x|$ is bounded, we check that $\|\varepsilon\|_Y/\|w\|_X$ goes to zero. The Fréchet derivative of the Burgers operator is consequently

$$T'(u) = \frac{\partial}{\partial t}(\cdot) + \left(\frac{\partial u}{\partial x}\right)(\cdot) + u\frac{\partial}{\partial x}(\cdot) - \nu\frac{\partial^2}{\partial x^2}(\cdot), \qquad (A.3.7)$$

where the symbol $(\cdot)$ denotes the variable of the problem at hand.

A.4 Inner-Product Spaces

Normed spaces are unsuitable for defining the "orientation" of an element in a vector space with regard to a given reference element. This is precisely the purpose of inner-product spaces, defined as follows.

A.4.1 Definition

Given a real vector space X, an *inner product* on X is a real-valued mapping $(\cdot, \cdot) : X \times X \to \mathbb{R}$ such that, for any elements $x, y, z \in X$ and any real number c, the following properties hold:

$$(x + y, z) = (x, z) + (y, z), \qquad (A.4.1)$$

$$(cx, y) = c\,(x, y), \qquad (A.4.2)$$

$$(x, y) = (y, x), \qquad (A.4.3)$$

$$(x, x) \geq 0, \quad \text{and} \quad (x, x) = 0 \Leftrightarrow x = O. \qquad (A.4.4)$$

A real *inner-product space* (also called a pre-Hilbert space) is a real vector space equipped with an inner product. Inner-product spaces may be endowed with a metric, because from the inner product $(\cdot, \cdot)$ one induces a nonnegative quantity

$$\|x\| := (x, x)^{1/2} \qquad (A.4.5)$$

that satisfies the axioms (A.2.1)–(A.2.3) for a norm. The distance between two elements $x, y \in X$ is then given by

$$d(x, y) := (x - y, x - y)^{1/2}.$$

This shows that inner-product spaces are also normed spaces and metric spaces. The converse, however, is not true. Given a real-valued function $\|\cdot\|$ satisfying the axioms of a norm, most of the time there is no associated inner product $(\cdot, \cdot)$.

Examples of Inner-Product Spaces

Example 1: Space $\mathbb{R}^n$
The n-dimensional Euclidean space $\mathbb{R}^n$ is an inner-product space where the inner product (also called *scalar* or *dot product*) of two vectors $x = (\xi_1, \ldots, \xi_n)$ and $y = (\eta_1, \ldots, \eta_n)$ is defined by the symmetric bilinear form

$$(x, y) := \xi_1 \eta_1 + \cdots + \xi_n \eta_n.$$

One easily verifies that all axioms (A.4.1)–(A.4.4) are satisfied.

Example 2: Space $C[a, b]$
The space of continuous real functions $C[a, b]$, with the inner product

$$(x, y) := \int_a^b x(t) y(t) \, dt, \qquad (A.4.6)$$

is an inner-product space with induced norm identical to (A.2.7).

A.4.2 Hilbert Spaces

Cauchy sequences and convergent sequences in inner-product spaces have the same definition as in normed spaces, except that one uses the norm (A.4.5) induced by the inner product instead of (A.1.9).

Inner-product spaces in which every Cauchy sequence has a limit in the space are complete. Complete inner-product spaces are called *Hilbert spaces*. Hilbert spaces are especially important in the numerical approximation of PDEs.

Examples of Hilbert Spaces

Example 1: Space $L^2(a, b)$
The space $C[a, b]$ of continuous real functions in $[a, b]$ equipped with the inner product (A.4.6) and induced norm is an incomplete inner-product space, because, as mentioned earlier, Cauchy sequences in the metric associated to (A.2.7) may converge toward discontinuous functions. Completing this inner-product space yields the Hilbert space of (classes of) Lebesgue-square-integrable functions $L^2(a, b)$:

$$L^2(a, b) := \{v; \ (a, b) \to \mathbb{R}, v \text{ defined a.e.}, \|v\| < \infty\}, \qquad (A.4.7)$$

with

$$\|v\| := \left(\int_a^b |v(t)|^2\, dt \right)^{1/2}. \tag{A.4.8}$$

A function $v \in L^2(a, b)$ is said to be defined *almost everywhere* (a.e.) in the interval $[a, b]$ if it is continuous except at a finite or infinite countable set of points in $[a, b]$, where its value is (usually) unspecified. Such a function, though not integrable in the Riemann sense, has a Lebesgue integral. Elements $v_1, v_2 \in L^2(a, b)$ are said to belong to the same class if $v_1 - v_2$ is equal to zero a.e., in which case the Lebesgue integral $\int_a^a (v_1 - v_2)^2\, dt = 0$. Discontinuous piecewise functions are a simple example of functions defined a.e.

Example 2: Space $L_w^2(a, b)$
Let $w(x)$ denote a nonnegative and integrable function in the interval (a, b). This function may be used as a *weight* function in the definition of an inner product. Consider, for instance, the bilinear form $(\cdot, \cdot) : C[a, b] \times C[a, b] \to \mathbb{R}$:

$$(x, y) := \int_a^b w(t)\, x(t)\, y(t)\, dt, \tag{A.4.9}$$

which verifies all the axioms (A.4.1)–(A.4.4). From (A.4.9), $C[a, b]$ is a (weighted) inner-product space with associated norm equal to

$$\|x\| := \left(\int_a^b w(t)\, |x(t)|^2\, dt \right)^{1/2} \qquad \forall x \in C[a, b]. \tag{A.4.10}$$

The completion of this inner-product space yields the Hilbert space of Lebesgue-weighted-square-integrable functions $L_w^2(a, b)$.

Remark. Inner products and norms on Hilbert spaces will include the Hilbert space in a subscript. For instance, the extensions of (A.4.9) and (A.4.10) to $L_w^2(a, b)$ are denoted $(x, y)_{L_w^2(a,b)}$ and $\|x\|_{L_w^2(a,b)}$, respectively.

Whenever the subscript is absent, we assume that the underlying vector space is either a generic inner-product space or the Hilbert space L^2.

A.4.3 Cauchy–Schwarz Inequality

Let x, y denote two arbitrary elements of an inner-product space (or Hilbert space) X with inner product $(\cdot, \cdot)$ and induced norm $\|\cdot\|$. For *every* real constant c, one has $\|x + cy\| \geq 0$. This condition entails the fundamental inequality

$$|(x, y)| \leq \|x\|\, \|y\|, \tag{A.4.11}$$

which is known as the *Cauchy–Schwarz inequality*. The equality in (A.4.11) holds iff the two vectors x, y are linearly dependent.

Further important relationships involving the vector norm may be obtained by using the Cauchy–Schwarz inequality, among them the triangle inequality (A.2.2), the *parallelogram equality*

$$\|x+y\|^2 + \|x-y\|^2 = 2(\|x\|^2 + \|y\|^2), \quad (A.4.12)$$

and the *polarization equality*

$$(x, y) = \frac{1}{4}(\|x+y\|^2 - \|x-y\|^2). \quad (A.4.13)$$

The relation (A.4.12) is often used to check whether a given norm has a related inner product.

A.4.4 The Riesz Representation

The definitions and basic properties of linear operators and functionals have been given in Section A.3.1, in the context of normed spaces. Since inner-product (or Hilbert) spaces are also normed spaces, the results given in Section A.3.1 apply to Hilbert spaces as well.

Let X denote a Hilbert space with inner product $(\cdot, \cdot)$ and associated norm $\|\cdot\|$. An important example of bounded linear functional is given by the inner product. Consider a *fixed* element $x_0 \in X$. The application $l : X \to \mathbb{R}$ given by

$$l(x) := (x_0, x) \quad \forall x \in X \quad (A.4.14)$$

is a linear functional. By application of the Cauchy–Schwarz inequality (A.4.11), one has

$$|l(x)| \leq \|x_0\| \|x\|, \quad (A.4.15)$$

showing that l is bounded with a norm equal to $\|x_0\|$ [see (A.3.2)].

The following fundamental theorem states that the dual space X' can be identified with X.

Theorem A.4 (Riesz representation) *Given a bounded linear functional $l^* : X \to \mathbb{R}$ on a Hilbert space X, there exists a unique element $\hat{x} \in X$ depending on l^* such that*

$$l^*(y) := (\hat{x}, y) \quad \forall y \in X \quad (A.4.16)$$

with, in addition, $\|l^*\| = \|\hat{x}\|$.

A.4. Inner-Product Spaces

The element $\hat{x}$ is known as the *Riesz representation* of the functional l^*.

A.4.5 Orthogonality, Orthogonal Projection

Assume X is an inner-product space (not necessarily complete) with inner product $(\cdot, \cdot)$ and induced norm $\|\cdot\|$, and let V and W denote subspaces of X.

Two vectors $x, y \in X$ are *orthogonal* if their inner product $(x, y) = 0$. This property is sometimes denoted by $x \perp y$. Further, the vector x is *orthogonal to V* ($x \perp V$) if for all $v \in V$ one has $(x, v) = 0$. Finally, V and W are *orthogonal subspaces of X* if for all $v \in V$ and all $w \in W$ the property $(v, w) = 0$ holds.

For a given vector $x \in X$, the set $x^\perp := \{y \in X; (x, y) = 0\}$ of all elements of X orthogonal to x forms a subspace called the *orthogonal complement* of x. Since the inner product is a bounded linear functional [see (A.4.15)] and the element $\{0\}$ is a closed subset of $\mathbb{R}$, from Theorem A.1 we see that $x^\perp$ is a closed subspace of X.

One defines the orthogonal complement of subspace V in a similar way, as the set of elements $V^\perp := \{x \in X; \ x \perp V\}$. The subspace $V^\perp$ is closed for the same reasons as above.

With these definitions, the essential "geometric" properties of Hilbert spaces are contained in the following theorem.

Theorem A.5 (Orthogonal projection) *Let V denote a closed subspace of a Hilbert space X, and $x \in X$ an arbitrary vector. There exists a unique element $\hat{v} \in V$ such that*

$$\|x - \hat{v}\| = \inf_{v \in V} \|x - v\|, \tag{A.4.17}$$

$$\|x - \hat{v}\| \in V^\perp. \tag{A.4.18}$$

The vector $\hat{v}$ is the orthogonal projection of $x \in X$ onto V.

The operator P_V associating to any $x \in X$ an element $\hat{v} \in V$ is, according to (A.4.17), the orthogonal projection operator of X onto V. It is characterized by the *idempotence* property

$$P_V^2 = P_V. \tag{A.4.19}$$

The Hilbert space X may then be decomposed into the direct sum of the closed subspace V and its orthogonal complement $V^\perp$:

$$X = V \oplus V^\perp. \tag{A.4.20}$$

For every $x \in X$, elements v and w exist such that

$$x = v + w, \qquad v \in V, \qquad w \in V^\perp. \tag{A.4.21}$$

A.4.6 Separable Hilbert Spaces, Basis

An *orthogonal set* $V \subset X$ in an inner-product space X is a collection of elements $\{v_\alpha\}_{\alpha \in I}$ depending on an index α such that for any two values of the index, one has

$$(v_\mu, v_\nu) = \begin{cases} 0, & \mu \neq \nu, \\ \gamma_\nu, & \mu = \nu. \end{cases} \tag{A.4.22}$$

If, moreover, $\gamma_\nu = 1$ for all values of ν, then V is an *orthonormal set*.

An orthogonal set $V \subset X$ in the inner-product space X is said to be *total* if the closure of the vector space spanned by the elements of V is dense in X:

$$\overline{\text{span } V} = X. \tag{A.4.23}$$

Particularly simple and important in practice are the *countable* sets that are closely related to separable Hilbert spaces (see Section A.1.3), as stated by the following theorem.

Theorem A.6 (Separable Hilbert spaces) *Let X denote a Hilbert space. Then:*

1. *If X is separable, every orthogonal set in X is countable.*
2. *If X contains a countable orthogonal set that is total in X, then X is separable.*

In every separable Hilbert space X, there exists a total orthogonal set V that is countable. In the simplest case, $V = \{v_i\}_{i=1}^n$ has a *finite* number n of elements. Then, as a consequence of (A.4.23), any vector $x \in X$ may be decomposed into a unique combination of the elements of V:

$$x = \sum_{k=1}^n \mu_k v_k, \tag{A.4.24}$$

where the coefficients μ_l may be obtained by taking the inner product of (A.4.24) with v_l, yielding $\mu_l = (x, v_l)/\gamma_l$. The elements of V form a *basis* for the Hilbert space X, which has dimension n.

The case where V, the total orthogonal set of the Hilbert space X, is countable and *infinite* is a little more delicate. Given an infinite countable set of elements,

A.4. Inner-Product Spaces

$W = \{w_i\}_{i=1}^{\infty}$, it is always possible to extract an infinite countable subsequence $\{w_{i_k}\}_{k=1}^{\infty}$. What really matters is the relation (A.4.23). Therefore, to form a basis for X, any infinite countable set of elements $\{v_i\}_{i=1}^{\infty}$ must be complete, that is, must ensure the existence for each $x \in X$ of a unique expansion

$$x = \sum_{k=1}^{\infty} \mu_k v_k. \tag{A.4.25}$$

A.4.7 Gram–Schmidt Orthonormalization Process

Because it is more convenient to work with an orthonormal basis rather than with an arbitrary basis of vectors, one needs an algorithm to transform a given basis into a basis satisfying (A.4.22) with $\gamma_\nu = 1$ for all values of ν. This algorithm is known as the *Gram–Schmidt orthonormalization process*.

Suppose $\{x_i\}_{i=1}^{n}$ is a given set of n linearly independent vectors forming a basis of X_n, an n-dimensional inner-product (or Hilbert) space with inner product $(\cdot, \cdot)$ and associated norm $\|\cdot\|$. This basis may be transformed into an orthonormal basis $\{e_i\}_{i=1}^{n}$, by using the following computational scheme:

- Step 1. $e_1 := x_1/\|x_1\|$.
- Step 2. Assume the vectors $e_1, \ldots, e_k$ to be known. Set

$$y_{k+1} := x_{k+1} - \sum_{i=0}^{k} (x_{k+1}, e_i) e_i.$$

- Step 3. The vector e_{k+1} is given by

$$e_{k+1} := y_{k+1}/\|y_{k+1}\|.$$

Apply steps 2 and 3 repeatedly until $k = n$. The Gram–Schmidt process is then completed, and one may verify that $\{e_i\}_{i=1}^{n}$ forms an orthonormal set.

The adaptation of the Gram–Schmidt algorithm to weighted inner-product spaces is straightforward.

Example: Orthogonal Polynomials

Let $L_w^2(\Omega)$ denote the Hilbert space of Lebesgue-weighted-square-integrable functions on $\Omega := (-1, +1)$. The inner product is given by (A.4.9), where $w(x)$ is a nonnegative integrable function over Ω. The elementary polynomials $\{x^k, k = 0, 1, \ldots\}$ are linearly independent, and one may show that they form a basis for $L_w^2(\Omega)$. These polynomials, however, do not form a w-orthonormal set. For

$w(x) = 1$ for instance, one has

$$(x^p, x^q) = \begin{cases} \frac{2}{p+q+1}, & p+q \text{ even,} \\ 0 & \text{otherwise.} \end{cases}$$

Applying the Gram–Schmidt algorithm with weight w yields the family of orthonormal polynomials in Ω with respect to w. Two important families are the *Legendre polynomials* corresponding to the weight $w(x) = 1$ and the *Chebyshev polynomials* corresponding to $w(x) = (1 - x^2)^{-1/2}$ (see Appendix B).

A.5 Distributions

The theory of partial differential equations has made significant progress with the understanding that the solutions of the equations of physics are distributions (or generalized functions), not classical functions. The concept of distribution is so natural in physics (where one deals with charged particles, localized forces, dipoles, multipoles, etc.) that Dirac and Heaviside used it well before its formal definition by Schwartz [351]. Since then, the theory of distributions has been considerably extended. For obvious reasons we will only deal with elementary concepts of distributions in the field of real numbers for one-dimensional problems, and focus on their basic properties. All definitions and properties given in the next subsections may be extended to the multidimensional case and to distributions in the complex field. Readers interested in a thorough discussion of these topics related to the solution of partial differential equations should consult Richtmyer [326], Oden and Reddy [286], or Showalter [361].

A.5.1 Definitions

As the definition of distributions involves a vector space of *test functions* with a particular concept of convergence, we first deal with that concept.

Test Functions

Let $C_0^\infty(\mathbb{R})$ denote the set of infinitely differentiable functions having compact support in $\mathbb{R}$. The support of a function $\varphi(x)$ on $\mathbb{R}$ (usually denoted supp φ) is the closure of the set of points $I \subset \mathbb{R}$ on which $\varphi(x) \neq 0$. Any element $\varphi(x) \in C_0^\infty(\mathbb{R})$ must therefore identically vanish outside a bounded interval on the real axis.

Consider a sequence of elements $\{\varphi_n(x)\}_{n=1}^\infty \in C_0^\infty(\mathbb{R})$. We say that this sequence *converges* to $\varphi(x) \in C_0^\infty(\mathbb{R})$ if the following conditions are met:

A.5. Distributions

- the supports of $\{\varphi_n(x)\}_{n=1}^{\infty}$ and $\varphi(x)$ are all contained in a fixed compact set $K \subset \mathbb{R}$;
- the derivatives of any order r, $\varphi_n^{(r)}(x)$ converge uniformly to $\varphi^{(r)}(x)$ as n tends to infinity.

This stringent convergence condition is sometimes called the $\mathcal{D}$-convergence condition. The set $C_0^{\infty}(\mathbb{R})$ with the $\mathcal{D}$-convergence condition is usually denoted $\mathcal{D}(\mathbb{R})$ and called the class of *test functions*.

Example of Test Functions

A classical example of a function belonging to $\mathcal{D}(\mathbb{R})$ is

$$\varphi(x) = \begin{cases} 0, & |x| \geq 1, \\ \exp\left(-\frac{1}{1-x^2}\right), & -1 < x < 1. \end{cases} \quad (A.5.1)$$

This function has a finite support $|x| \leq 1$, is infinitely differentiable in $\mathbb{R}$, and vanishes identically together with its derivatives of every order at $|x| = 1$. The reader will easily verify that any (finite) translation and homothetic transformation of the real axis on (A.5.1) produces another test function

$$\phi(x) = \varphi\left(\frac{x-a}{\lambda}\right), \quad (A.5.2)$$

with nonzero values in the interval $a - \lambda < x < a + \lambda$.

Distribution

Let q denote a linear functional, and $\langle q, \phi \rangle$ represent the value of q when applied to the function $\phi(x)$.

Any linear functional q that is defined on all elements $\varphi(x) \in C_0^{\infty}(\mathbb{R})$ and satisfies a continuity condition with regard to the $\mathcal{D}$-convergence,

$$\varphi_n(x) \xrightarrow{\mathcal{D}} \varphi(x) \quad \Longrightarrow \quad \langle q, \varphi_n(x) \rangle \to \langle q, \varphi(x) \rangle, \quad (A.5.3)$$

is a *Schwartz distribution*. In (A.5.3), the infinite sequence $\{\varphi_n\}_{n=1}^{\infty}$ converges to φ in the sense defined above.

Obviously, any linear combination of distributions is itself a distribution. Therefore, the set of distributions based on $\mathcal{D}(\mathbb{R})$ is a linear space. The space of continuous linear functionals on $\mathcal{D}(\mathbb{R})$ is called the *dual space* of $\mathcal{D}(\mathbb{R})$ and is denoted $\mathcal{D}'(\mathbb{R})$.

The value of a functional $q \in \mathcal{D}'(\mathbb{R})$ applied to $\varphi \in \mathcal{D}(\mathbb{R})$ is $\langle q, \varphi \rangle$ where the symbol $\langle \cdot, \cdot \rangle$ may now be interpreted as a duality pairing, that is, a bilinear map of $\mathcal{D}'(\mathbb{R}) \times \mathcal{D}(\mathbb{R})$ into $\mathbb{R}$.

The dual space $\mathcal{D}'(\mathbb{R})$ is endowed with the so-called weak* topology, according to which an infinite set of functionals $\{q_n\}_{n=1}^{\infty} \in \mathcal{D}'(\mathbb{R})$ is said to converge distributionally to $q \in \mathcal{D}'(\mathbb{R})$ if

$$\forall \varphi(x) \in \mathcal{D}(\mathbb{R}), \qquad \langle q_n, \varphi(x) \rangle \to \langle q, \varphi(x) \rangle. \qquad \text{(A.5.4)}$$

Examples of Distributions

Example 1
Consider a function $f(x) \in C(\mathbb{R})$ such as, for instance, the function shown on Figure A.5.1(a). Clearly, the linear functional $\int_{-\infty}^{+\infty} f(x)\varphi(x)\,dx$ with $\varphi(x) \in \mathcal{D}(\mathbb{R})$ satisfies the requirement (A.5.3); for we have

$$\lim_{n \to \infty} \int_{-\infty}^{+\infty} f(\varphi_n - \varphi)\,dx \leq \lim_{n \to \infty} \sup_{x \in K} |\varphi_n - \varphi| \int_K f\,dx = 0, \qquad \text{(A.5.5)}$$

because of the uniform convergence of the sequence of test functions to $\varphi(x) \in \mathcal{D}(\mathbb{R})$. In this relationship, $K \subset \mathbb{R}$ is the compact set containing the support of the test functions. Thus, the linear functional $\int_{-\infty}^{+\infty} f(x)\varphi(x)\,dx$ is a distribution on $\mathcal{D}(\mathbb{R})$. For the sake of convenience one often denotes this distribution by f. However, we emphasize that by definition, the distribution is the linear mapping $\langle f, \cdot \rangle$ of $\mathcal{D}(\mathbb{R})$ into $\mathbb{R}$ given by

$$f(x): C_0^{\infty}(\mathbb{R}) \to \mathbb{R}: \qquad \varphi(x) \to \langle f, \varphi \rangle := \int_{-\infty}^{+\infty} f(x)\varphi(x)\,dx. \qquad \text{(A.5.6)}$$

The reader should notice a particular feature of the function displayed on the figure: it is piecewise continuous with undefined first derivative at a finite number of points on the real axis. Functions of this type are at the heart of the FE theory (see [369, 79, 370]).

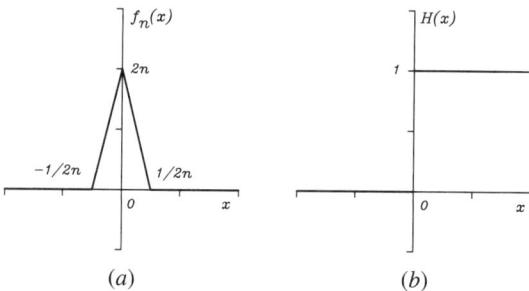

Figure A.5.1. Two examples of distributions: (*a*) a piecewise linear function; (*b*) the Heaviside step function.

A.5. Distributions

This example shows that continuous functions defined pointwise may also be viewed as distributions. Actually, the two standpoints (i.e. pointwise defined functions and linear functionals) are equivalent according to the following theorem [326].

Theorem A.7 *A continuous function $f(x)$, defined for $-\infty < x < +\infty$, is completely determined if the value of the integral $\int_{-\infty}^{+\infty} f(x)\varphi(x)\,dx$ is known for every continuous function $\varphi(x)$ that vanishes outside a finite interval, different intervals being allowed for different functions $\varphi(x)$.*

Finally, we emphasize that no problems arise with the behavior of $f(x)$ at infinity, since by definition the support of φ is bounded.

Example 2
Figure A.5.1(b) shows the Heaviside step function

$$H(x) := \begin{cases} 0, & x < 0, \\ 1/2, & x = 0, \\ +1, & x > 0, \end{cases} \tag{A.5.7}$$

which is discontinuous at $x = 0$. This function is locally Riemann-integrable, that is,

$$\int_a^b |H(x)|\,dx \tag{A.5.8}$$

exists for any compact interval $[a, b]$. Using again the argument (A.5.5), one easily verifies that the Heaviside step function also generates a distribution.

Incidentally we remark that the value of the step function at $x = 0$ has no influence whatsoever on the distribution $\langle H, \cdot \rangle$, since $x = 0$ is an isolated point. We then conclude that two locally integrable functions $f_1(x)$ and $f_2(x)$ that differ on a (countable) set of isolated points generate the same distribution

$$\langle f_1, \varphi \rangle = \langle f_2, \varphi \rangle \qquad \forall \varphi \in \mathcal{D}(\mathbb{R}). \tag{A.5.9}$$

The functions $f_1(x)$ and $f_2(x)$ are said to be equal *almost everywhere*. Distributions generated by locally integrable functions are said to be *regular*. Distributions that are not regular are called *singular distributions*.

Example 3
We come back to Figure A.5.1(a) and go one step further by considering an

infinite sequence of distributions $\{\langle f_n, \cdot \rangle\}_{n=1}^{\infty}$, where the integer n is indicated on the figure, and

$$f_n(x) = \begin{cases} 0, & |x| \geq 1/2n, \\ 4n^2 x + 2n, & -1/2n < x \leq 0, \\ -4n^2 x + 2n, & 0 \leq x < 1/2n. \end{cases} \quad (A.5.10)$$

As $h := 1/n$ becomes small, one gets

$$\forall \varphi(x) \in \mathcal{D}(\mathbb{R}): \quad \int_{-\infty}^{+\infty} f_n(x)\varphi(x)\,dx = \varphi(0) + O(h^2), \quad (A.5.11)$$

and

$$\lim_{n \to \infty} \langle f_n, \varphi \rangle = \varphi(0). \quad (A.5.12)$$

The result of the limiting process is another distribution, since the continuity condition (A.5.3) is satisfied for all $\varphi(x) \in \mathcal{D}(\mathbb{R})$. This distribution is the *Dirac distribution* $\langle \delta, \cdot \rangle$, often (improperly) called the Dirac "function" $\delta(x)$ in the literature:

$$\forall \varphi \in \mathcal{D}(\mathbb{R}): \quad \langle \delta, \varphi \rangle := \int_{-\infty}^{+\infty} \delta(x)\varphi(x)\,dx = \varphi(0). \quad (A.5.13)$$

The $\delta(x)$ notation is misleading. Unlike for the functions $f_n(x)$, there is no pointwise definition for $\delta(x)$, and the relation (A.5.13) must be regarded as the *definition* of the distribution $\langle \delta, \cdot \rangle$.

One may easily show that no locally integrable function $f(x)$ exists such that $\int_{-\infty}^{-\infty} f(x)\varphi(x)\,dx = \varphi(0)$. Hence, the Dirac distribution is a singular distribution.

A.5.2 Basic Properties of Distributions

We make a short review of some elementary properties of distributions.

Translation and Change in Scale

Let $f(x)$ denote a locally integrable function generating the distribution $\langle f, \cdot \rangle$, and $\alpha \neq 0$ a given real number. The translation of $f(x)$ by a finite value a generates a new distribution, and one has

$$\langle f(x - a), \varphi(x) \rangle = \langle f(x), \varphi(x + a) \rangle \quad \forall \varphi(x) \in \mathcal{D}(\mathbb{R}). \quad (A.5.14)$$

A.5. Distributions

Similarly, the change in scale by α into $f(x)$ generates a distribution such that

$$\langle f(\alpha x), \varphi(x) \rangle = \frac{1}{|\alpha|} \left\langle f(x), \varphi\left(\frac{x}{\alpha}\right) \right\rangle \qquad \forall \varphi(x) \in \mathcal{D}(\mathbb{R}). \tag{A.5.15}$$

Multiplication by a Continuous Function

Consider any distribution $\langle f, \cdot \rangle$ on $\mathcal{D}(\mathbb{R})$ and a continuous function $g(x)$. One has

$$\langle gf, \varphi \rangle = \langle f, g\varphi \rangle \qquad \forall \varphi(x) \in \mathcal{D}(\mathbb{R}). \tag{A.5.16}$$

So, if $g \in C^\infty(\mathbb{R})$, then $g\varphi \in \mathcal{D}(\mathbb{R})$ and the product gf is a distribution. If $g \notin C^\infty(\mathbb{R})$, then $g\varphi$ is not necessarily a test function and a close examination must be performed, case by case. In general, even though taken separately $\langle f, \cdot \rangle$ and $\langle g, \cdot \rangle$ are distributions, their product is not necessarily a distribution. For instance, the product of the Dirac distributions $\delta^2(x) = \delta(x)\delta(x)$ is not defined.

Derivatives of Distributions

Assume that the function $f(x)$ has a first derivative $f'(x)$ that is a locally integrable function. We have seen with Example 2 that $f'(x)$ generates a distribution $\langle f', \cdot \rangle$. Using integration by parts and the properties of the test functions, one gets the relationship

$$\langle f', \varphi \rangle = -\langle f, \varphi' \rangle \qquad \forall \varphi(x) \in \mathcal{D}(\mathbb{R}). \tag{A.5.17}$$

The value of the distribution $\langle f', \cdot \rangle$ applied to $\varphi(x) \in \mathcal{D}(\mathbb{R})$ is related to the value of the distribution $\langle f, \cdot \rangle$ applied to $\varphi'(x) \in \mathcal{D}(\mathbb{R})$.

Because, for any convergent sequence of test functions $\{\varphi_n(x)\}_{n=1}^\infty \in \mathcal{D}(\mathbb{R})$ one has

$$\varphi_n(x) \xrightarrow{\mathcal{D}} \varphi(x) \quad \Longrightarrow \quad \varphi_n'(x) \xrightarrow{\mathcal{D}} \varphi'(x), \tag{A.5.18}$$

one may use (A.5.17) as a *definition* for the derivative of *any* distribution $\langle f, \cdot \rangle$. The quantity $\langle f', \cdot \rangle$ defined by (A.5.17) is then known as the *distributional derivative* of $\langle f, \cdot \rangle$. This procedure is very useful for generalizing the concept of derivative to functions having no ordinary derivative.

Successive applications of integration by parts and convergence arguments analogous to (A.5.18) allow the introduction of higher distributional derivatives. For instance, the derivative of order k ($k > 0$) of any distribution $\langle f, \cdot \rangle$, denoted $\langle f^{(k)}, \cdot \rangle$, again is defined by

$$\langle f^{(k)}, \varphi \rangle := (-1)^k \langle f, \varphi^{(k)} \rangle \qquad \forall \varphi(x) \in \mathcal{D}(\mathbb{R}). \tag{A.5.19}$$

440   Appendix A. Preliminary Mathematical Concepts

To end, we illustrate the concept of distributional derivative, going back to the examples introduced above.

Examples of Distributional Derivatives

Example 4
We first evaluate the distributional derivative of the functional $\langle f_n, \cdot \rangle$ with pointwise definition given by (A.5.10). Applying the definition (A.5.17), one gets

$$\langle f_n', \varphi \rangle = \int_{-1/2n}^{0} f_n(x) \varphi'(x)\, dx + \int_{0}^{1/2n} f_n(x) \varphi'(x)\, dx, \qquad (A.5.20)$$

or, with the help of integration by parts,

$$\langle f_n', \varphi \rangle = \int_{-1/2n}^{1/2n} f_n'(x) \varphi(x)\, dx. \qquad (A.5.21)$$

The functions $f_n(x)$ being continuous everywhere (although with only left and right derivatives at some isolated points), the distributional derivative in this case is identical to the ordinary derivative *almost everywhere*:

$$f_n'(x) = \begin{cases} 0, & |x| > 1/2n, \\ 4n^2, & -1/2n < x < 0, \\ -4n^2, & 0 < x < 1/2n. \end{cases} \qquad (A.5.22)$$

The absence of the (ordinary) first derivative $f_n'(x)$ at some isolated points is of no concern for the evaluation of functionals.

Example 5
A more interesting example comes with the discontinuous Heaviside step function. Its distributional derivative $\langle H', \cdot \rangle$ again may be defined using (A.5.17). One gets

$$\langle H', \varphi \rangle = -\int_{0}^{\infty} \varphi'(x)\, dx = \varphi(0) \qquad \forall \varphi(x) \in \mathcal{D}(\mathbb{R}), \qquad (A.5.23)$$

showing that functional $\langle H', \cdot \rangle$ is the Dirac distribution $\langle \delta, \cdot \rangle$.

Example 6
Finally, one may define the first derivative of the functional $\langle \delta, \cdot \rangle$. This distribution denoted $\langle \delta', \cdot \rangle$ is equal to

$$\langle \delta', \varphi \rangle := -\langle \delta, \varphi' \rangle = -\varphi'(0) \qquad \forall \varphi(x) \in \mathcal{D}(\mathbb{R}). \qquad (A.5.24)$$

A.5. Distributions

The reader should observe that the same distribution obtains as for the limit case of the functionals $\langle f'_n, \cdot \rangle$ given by (A.5.22). Using a Maclaurin expansion of $\varphi(x)$, it is easy to show that for all $\varphi(x) \in \mathcal{D}(\mathbb{R})$

$$\langle f'_n, \varphi \rangle = \int_{-1/2n}^{1/2n} f'_n(x)\varphi(x)\,dx \approx -\varphi'(0) - \frac{1}{48n^2}\varphi'''(0). \quad (\text{A.5.25})$$

As a conclusion and to summarize this short review on distributions, let us stress that their main interest is for an unambiguous definition of the derivatives, of any order, of functions that may lack derivatives in the ordinary sense.

Appendix B
Orthogonal Polynomials and Discrete Transforms

In Section A.4.2 of Appendix A, we introduced the Hilbert space of Lebesgue-weighted-square-integrable functions L_w^2. Unless otherwise stated, this space will be the framework of most developments made here. To conform with our notation, we will designate continuous inner products and norms in L_w^2 by $(\cdot, \cdot)_{L_w^2}$ and $\|\cdot\|_{L_w^2}$, but the discrete inner products and norms introduced in this appendix will be designated by $(\cdot, \cdot)_N$ and $\|\cdot\|_N$, respectively.

B.1 Systems of Orthogonal Polynomials

Let Ω denote the open reference interval $(-1, +1)$, with closure $\hat{\Omega}$. Let also w denote any nonnegative integrable weight function in $\hat{\Omega}$. The sequence of polynomial functions $\{p_k\}_{k=0}^{\infty}$, where the degree of p_k is equal to k, forms a *system of orthogonal polynomials* with respect to w if

$$(p_k, p_l)_{L_w^2(\Omega)} := \int_{-1}^{1} w(x)\, p_k(x)\, p_l(x)\, dx = \gamma_k\, \delta_{kl}, \qquad (B.1.1)$$

where δ_{kl} denotes the Kronecker symbol and $\gamma_k := \|p_k\|_{L_w^2(\Omega)}^2$ depends on the polynomial degree.

For convenience, the corresponding *orthonormal* sequence will be labeled $\{p_k^*\}_{k=0}^{\infty}$, with

$$p_k^*(x) := \gamma_k^{-1/2} p_k(x).$$

Systems of orthogonal polynomials have the following properties, details and proofs of which can be found in the literature (see, for instance, [92]).

B.1. Systems of Orthogonal Polynomials

Property B.1 Let $\{p_k\}_{k=0}^{\infty}$ denote the family of orthogonal polynomials in $\hat{\Omega}$ with respect to the weight function w. The zeros of the polynomial p_k are real, simple, and located in Ω.

The fact that orthogonal polynomials possess *real* zeros is crucial for Gaussian quadrature rules (see Section B.2).

Property B.2 (*Three-term recursion formula*) Let $\{p_k\}_{k=0}^{\infty}$ be the family of orthogonal polynomials in $\hat{\Omega}$ with respect to the weight function w. There exists a three-term recursion relationship between the elements of the family such that p_{k+1} is related to its predecessors p_k and p_{k-1} by the expression

$$p_{k+1}(x) = (a_k + b_k x) \, p_k(x) - c_k \, p_{k-1}(x), \qquad k \geq 1, \qquad \text{(B.1.2)}$$

where the coefficients a_k, b_k, and c_k depend on the degree k and on the orthogonal family.

Starting from $p_0(x)$ and $p_1(x)$, (B.1.2) allows a determination of the polynomials $p_k(x)$ in increasing order.

Property B.3 (*Christoffel–Darboux relation*) Let $\{p_k^*\}_{k=0}^{\infty}$ be the set of orthonormal polynomials in $\hat{\Omega}$ with respect to the weight function w. Assume the coefficient of x^n in p_n^* to be equal to s_n^*. Then one has the following expression:

$$\sum_{k=0}^{n} p_k^*(x) \, p_k^*(y) = \frac{s_n}{s_{n+1}} \frac{p_{n+1}^*(x) \, p_n^*(y) - p_n^*(x) p_{n+1}^*(y)}{x - y}, \qquad \text{(B.1.3)}$$

known as the Christoffel–Darboux relation.

Finally, one has the following property.

Property B.4 Let $\{p_k\}_{k=0}^{\infty}$ denote the family of orthogonal polynomials with respect to the weight function w. One can easily show that the infinite set $\{\frac{d}{dx} p_k\}_{k=0}^{\infty}$ is also a family of orthogonal polynomials.

The interest in orthogonal polynomials in numerical methods for PDEs originates from two reasons: a theoretical one and a practical one. The theoretical reason is that, as a consequence of the Weierstrass approximation theorem (see [142, 326, 418]), the infinite sequence of polynomials $\{p_k^*\}_{k=0}^{\infty}$ is *complete* and forms a basis of $L_w^2(\Omega)$. The practical reason lies in the fact that polynomials are easy to generate and easy to use.

444   Appendix B. Orthogonal Polynomials and Discrete Transforms

Although, in principle, series expansions could be carried with *any* orthogonal system, in actual computations only those orthogonal systems are used that ensure the fastest convergence in the approximation of smooth functions. Such convergence, called *spectral convergence*, is obtained with orthogonal systems of functions related to *singular* Sturm–Liouville problems.

B.1.1 Eigensolutions of Sturm–Liouville Problems

A Sturm–Liouville problem is an eigenvalue problem of the form

$$-[p(x)u'(x)]' + q(x)u(x) = \lambda w(x)u(x), \qquad x \in \Omega, \qquad (\text{B.1.4})$$

with suitable boundary conditions. In (B.1.4), $w(x)$ is a nonnegative integrable weight function, and $p(x)$ and $q(x)$ are real-valued functions satisfying the following requirements: p is differentiable and positive in Ω and continuous at $x = \pm 1$, whereas q is continuous, nonnegative, and bounded in Ω (see [326]). When these conditions are satisfied, the Sturm–Liouville problem is *regular*. If, additionally, p vanishes at one boundary at least, the Sturm–Liouville problem is *singular*.

Singular Sturm–Liouville problems occupy a key position in spectral methods. Gottlieb and Orszag have shown that, if any infinitely smooth function u is expanded in series of eigensolutions of a singular problem (B.1.4), then the coefficients of the expansion decay faster than algebraically in k (see [163]). This property is known as *spectral accuracy*. Most of the time, only a small number of terms are needed to approximate u with high precision. Singular Sturm–Liouville eigensolutions are the only functions that satisfy this property. Spectral accuracy cannot be reached with eigensolutions of regular problems unless the smooth function to be approximated satisfies additional constraints at the boundary.

Focusing on polynomials, one can show that the only polynomial eigensolutions of a singular Sturm–Liouville problem are the Jacobi polynomials $P_n^{(\alpha,\beta)}(x)$ of degree n ($n = 0, 1, \ldots$), that depend on two real parameters α and β (with the constraint $\alpha, \beta > -1$) and that satisfy the differential problem

$$-\frac{d}{dx}\left((1-x^2) w^{(\alpha,\beta)} \frac{d}{dx} P_n^{\alpha,\beta}(x) \right) = \lambda_n^{(\alpha,\beta)} w^{(\alpha,\beta)} P_n^{\alpha,\beta}(x). \qquad (\text{B.1.5})$$

The weight function $w^{(\alpha,\beta)}$ and the eigenvalues $\lambda_n^{(\alpha,\beta)}$ are given by

$$\begin{aligned} w^{(\alpha,\beta)} &= (1-x)^\alpha (1+x)^\beta, \\ \lambda_n^{(\alpha,\beta)} &= n(n+\alpha+\beta+1). \end{aligned} \qquad (\text{B.1.6})$$

The eigensolutions of (B.1.5) are defined up to a normalization constant. Setting the normalization constant to a value such that

$$P_n^{\alpha,\beta}(1) := \binom{n+\alpha}{n} = \frac{\Gamma(n+\alpha+1)}{\Gamma(n+1)\Gamma(\alpha+1)}, \quad \text{(B.1.7)}$$

one may show that the Jacobi polynomials $P_n^{\alpha,\beta}(x)$ satisfy a relationship known as *Rodrigues's formula*:

$$P_n^{\alpha,\beta}(x) = \frac{(-1)^n}{2^n n!} (1-x)^{-\alpha}(1+x)^{-\beta} \frac{d^n}{dx^n}[(1-x)^{n+\alpha}(1+x)^{n+\beta}]. \quad \text{(B.1.8)}$$

Explicitly, $P_n^{(\alpha,\beta)}(x)$ has the following expansion:

$$P_n^{\alpha,\beta}(x) = 2^{-n} \sum_{k=0}^{n} \binom{n+\alpha}{k}\binom{n+\beta}{n-k}(x-1)^{n-k}(x+1)^k. \quad \text{(B.1.9)}$$

Multiplying both sides of (B.1.5) by $P_m^{\alpha,\beta}(x)$ (with $m \neq n$), integrating over the interval Ω, and taking (B.1.1) into account, one gets

$$\int_{-1}^{1} (1-x^2) w^{(\alpha,\beta)} \frac{dP_n^{\alpha,\beta}}{dx} \frac{dP_m^{\alpha,\beta}}{dx} dx = 0, \quad \text{(B.1.10)}$$

which is nothing else than Property B.4.

By carefully selecting the values of the parameters α and β, one obtains well-known families of orthogonal polynomials such as the Legendre polynomials $L_k(x)$ (with $\alpha = \beta = 0$), the Chebyshev polynomials of the first kind $T_k(x)$ (with $\alpha = \beta = -1/2$), or the Chebyshev polynomials of the second kind $U_k(x)$ (with $\alpha = \beta = 1/2$). Since spectral methods rely almost exclusively on Chebyshev polynomials of the first kind and on Legendre polynomials, our analysis will deal with these functions only, and we summarize their main properties in Sections B.1.2 and B.1.3. For further details, the reader should refer to the literature on orthogonal polynomials, in particular the books by Askey [12] and Szegö [371]. We will adhere to the common usage in calling $T_k(x)$ simply the "Chebyshev polynomials."

B.1.2 The Legendre Polynomials

The Legendre polynomial of degree k, $L_k(x)$, is the eigensolution of the differential problem (B.1.5) where α and β have been set to zero:

$$-\frac{d}{dx}\left((1-x^2)\frac{d}{dx}L_k(x)\right) = k(k+1)L_k(x), \quad k = 0, 1, \ldots. \quad \text{(B.1.11)}$$

The set of Legendre polynomials form an orthogonal family with respect to the weight function $w(x) = 1$, and one has

$$\int_{-1}^{1} L_k(x) L_l(x) \, dx = \gamma_k \, \delta_{kl}, \tag{B.1.12}$$

where the normalization constant γ_k is equal to

$$\gamma_k := \int_{-1}^{1} L_k^2(x) \, dx = \frac{2}{2k+1}. \tag{B.1.13}$$

The elements of the orthonormal family $\{L_k^*(x)\}_{k=0}^{\infty}$ are therefore related to the standard polynomials by

$$L_k^*(x) := \left(\frac{2k+1}{2}\right)^{1/2} L_k(x). \tag{B.1.14}$$

No compact analytic expression exists for the Legendre polynomial $L_k(x)$. To evaluate a polynomial explicitly, one may use either Rodrigues's formula (B.1.8) or the expansion (B.1.9), with the appropriate values of the parameters α and β. However, by far, the most efficient method to determine the Legendre polynomials is the three-term recursion relationship (B.1.2) which, in this case, is written as

$$L_0(x) = 1 \quad \text{and} \quad L_1(x) = x,$$
$$(k+1) L_{k+1}(x) = (2k+1) x \, L_k(x) - k \, L_{k-1}(x), \quad k \geq 1. \tag{B.1.15}$$

Further useful properties of the Legendre polynomial $L_k(x)$ are the following:

$$|L_k(x)| \leq 1, \quad -1 \leq x \leq 1, \tag{B.1.16}$$

$$L_k(\pm 1) = (\pm 1)^k, \tag{B.1.17}$$

$$|L_k'(x)| \leq \frac{1}{2} k(k+1), \quad -1 \leq x \leq 1, \tag{B.1.18}$$

$$L_k'(\pm 1) = \frac{1}{2}(\pm 1)^{k+1} k(k+1). \tag{B.1.19}$$

Finally, we mention an important relationship between Legendre polynomials and their derivatives:

$$(2k+1) L_k(x) = L_{k+1}'(x) - L_{k-1}'(x), \quad k \geq 1, \tag{B.1.20}$$

with $L_0 = L_1'$. This relationship is at the basis of the Legendre polynomial expansion of derivatives of functions (see Section B.3.3). The proof of (B.1.20) is obtained from the integral version of this relationship and may be found in Bernardi and Maday (see Chapter 1, Lemma 3.8, of [38]).

B.1.3 The Chebyshev Polynomials

The Chebyshev polynomial of degree k, $T_k(x)$, is given explicitly in terms of trigonometric functions by

$$T_k(x) = \cos k\theta, \qquad x = \cos\theta. \tag{B.1.21}$$

One easily verifies that $T_k(x)$ is indeed a polynomial function of x and that it satisfies the differential equation

$$-\frac{d}{dx}\left(\sqrt{1-x^2}\,\frac{d}{dx}T_k(x)\right) = \frac{k^2}{\sqrt{1-x^2}}\,T_k(x). \tag{B.1.22}$$

The set of Chebyshev polynomials $\{T_k(x)\}_{k=0}^{\infty}$ forms an orthogonal family with respect to the weight function $w(x) = (1-x^2)^{-1/2}$, and one has

$$\int_{-1}^{1} \frac{T_k(x)\,T_l(x)}{\sqrt{1-x^2}}\,dx = \gamma_k\,\delta_{kl}, \tag{B.1.23}$$

where the constant γ_k is equal to $c_k\pi/2$, with

$$c_k = \begin{cases} 2 & \text{if } k = 0, \\ 1 & \text{if } k \geq 1. \end{cases} \tag{B.1.24}$$

The relationships (B.1.23) and (B.1.24) give the elements of the orthonormal family $\{T_k^*(x)\}_{k=0}^{\infty}$ with

$$T_k^*(x) = \left(\frac{2}{\pi c_k}\right)^{1/2} T_k(x). \tag{B.1.25}$$

The Chebyshev polynomials $T_k(x)$ obey the three-term recursion relation

$$\begin{aligned} &T_0(x) = 1 \quad \text{and} \quad T_1(x) = x, \\ &T_{k+1}(x) = 2x\,T_k(x) - T_{k-1}(x), \qquad k \geq 1, \end{aligned} \tag{B.1.26}$$

used to evaluate the polynomials explicitly.

Some further useful properties of the polynomial $T_k(x)$ are as follows:

$$\frac{dT_k(x)}{dx} = k\,U_{k-1}(x), \tag{B.1.27}$$

$$|T_k(x)| \leq 1, \qquad -1 \leq x \leq 1, \tag{B.1.28}$$

$$T_k(\pm 1) = (\pm 1)^k, \tag{B.1.29}$$

$$|T_k'(x)| \leq k^2, \qquad -1 \leq x \leq 1, \tag{B.1.30}$$

$$T_k'(\pm 1) = (\pm 1)^{k+1} k^2, \tag{B.1.31}$$

where $U_k(x)$ is the Chebyshev polynomial of second kind and degree k.

Finally, the relationship between the Chebyshev polynomials and their derivatives, analogous to (B.1.20), is

$$2 T_k(x) = \frac{1}{k+1} T'_{k+1}(x) - \frac{1}{k-1} T'_{k-1}(x), \qquad k \geq 1. \qquad \text{(B.1.32)}$$

B.2 Gaussian-Type Quadratures

We limit ourselves to the case of a bounded interval, and we assume that this interval has been mapped onto the reference interval $\hat{\Omega}$.

B.2.1 Fundamental Theorems

Let $w(x)$ denote a nonnegative integrable weight function in $\hat{\Omega}$ to which corresponds a family of orthogonal polynomials $\{p_k(x)\}_{k=0}^{\infty}$. Then we have the following theorem.

Theorem B.5 (Gauss integration) *Let $\{\xi_k\}_{k=0}^{N}$ denote the $N+1$ zeros of the polynomial $p_{N+1}(x)$ belonging to the orthogonal family, located in the open interval $(-1, 1)$ with $\xi_0 < \cdots < \xi_N$. One can find $N+1$ positive constants $\rho_0, \ldots, \rho_N$ such that the following relationship holds for all polynomials through degree $2N+1$:*

$$\int_{-1}^{1} w(x) f(x) \, dx = \sum_{k=0}^{N} \rho_k f(\xi_k) \qquad \forall f(x) \in \mathbb{P}_{2N+1}. \qquad \text{(B.2.1)}$$

For polynomials with degree larger than $2N+1$, or nonpolynomial functions in $L_w^2(\Omega)$, the r.h.s. of Equation (B.2.1), known as the Gauss–Jacobi integration scheme with $N+1$ nodes, yields an approximate value of $\int_{-1}^{1} w(x) f(x) \, dx$ with an error, the leading term of which is proportional to the derivative of order $2N+2$, $f^{(2N+2)}(\xi)$, with $-1 < \xi < 1$. The quantities $\{\xi_k\}_{k=0}^{N}$ and $\{\rho_k\}_{k=0}^{N}$ are known as the Gauss integration *nodes* and *weights*, respectively.

In some circumstances, it may be desirable to have one or both end points included among the quadrature knots. With one end point included (e.g. $x = -1$), one has the following theorem.

Theorem B.6 (Gauss–Radau integration) *Let $\{\xi_k\}_{k=0}^{N}$ denote the zeros of $q(x) = p_{N+1}(x) + a\, p_N(x)$ with $-1 = \xi_0 < \cdots < \xi_N$. The parameter a is chosen such that*

$$a = -p_{N+1}(-1)/p_N(-1).$$

In this case, one can find $N+1$ positive constants $\rho_0, \ldots, \rho_N$ such that

$$\int_{-1}^{1} w(x) f(x)\, dx = \sum_{k=0}^{N} \rho_k\, f(\xi_k) \qquad \forall f(x) \in \mathbb{P}_{2N}. \tag{B.2.2}$$

Arbitrarily putting a quadrature node at $x = -1$ represents a constraint. As a consequence, one unit is lost in the degree of the polynomials that are integrated exactly by the Gauss–Radau scheme with $N+1$ nodes. If, instead of choosing the left end point, one had imposed a quadrature node at $x = 1$, the value of the parameter a would have changed, but the conclusions would have remained the same.

Finally, with the two end points included among the $N+1$ quadrature nodes, one has the following theorem.

Theorem B.7 (Gauss–Lobatto integration) *Let $\{\xi_k\}_{k=0}^{N}$ denote the zeros of $q(x) = p_{N+1}(x) + a\, p_N(x) + b\, p_{N-1}(x)$ with $-1 = \xi_0 < \cdots < \xi_N = 1$. The parameters a and b are chosen such that $q(-1) = q(1) = 0$. Then one can find $N+1$ positive constants $\rho_0, \ldots, \rho_N$ such that*

$$\int_{-1}^{1} w(x) f(x)\, dx = \sum_{k=0}^{N} \rho_k\, f(\xi_k) \qquad \forall f(x) \in \mathbb{P}_{2N-1}. \tag{B.2.3}$$

With quadrature nodes at both end points, two units are lost in the degree of the polynomials that are integrated exactly. The Gauss–Lobatto scheme with $N+1$ nodes has an error term proportional to $f^{(2N)}(\xi)$, $-1 < \xi < 1$, which vanishes for all polynomials through degree $2N-1$. On the other hand, the inclusion of the end points in the quadrature nodes makes the Gauss–Lobatto rule extremely useful in spectral methods, as the end points will take care of the boundary conditions in a BVP.

Few computationally useful results about the zeros of orthogonal polynomials exist, except of course that they are real, as stated in Property B.1. However, for Jacobi polynomials with parameters α and β restricted to $-1/2 \leq \alpha \leq 1/2$ and $-1/2 \leq \beta \leq 1/2$, the following estimation about the location of the zeros of p_N may be found in [371]:

$$-1 \leq -\cos\left(\frac{k + \frac{\alpha+\beta-1}{2}}{N + \frac{\alpha+\beta+1}{2}}\pi\right) \leq \xi_k^{(N)} \leq -\cos\left(\frac{k}{N + \frac{\alpha+\beta+1}{2}}\pi\right) \leq 1, \tag{B.2.4}$$

with $p_N(\xi_k^{(N)}) = 0$, $1 \leq k \leq N$. The result (B.2.4) explains why, for large values of N, the zeros of p_N have a tendency to accumulate at the end points of Ω. Expanding the cosine function, one can easily show that the distance between two consecutive zeros is proportional to $1/N^2$ near the end points, whereas it is proportional to $1/N$ in the middle of the interval. While concentrating more nodes near the end points is generally regarded as favorable from the computational point of view (for an accurate description of boundary layers, etc.), it has also a drawback, namely, the increase of the condition number of the algebraic system to be solved.

Finally, we note that with the Jacobi orthogonal family, a more convenient characterization of Gauss–Lobatto quadrature nodes exists, which is given by the following theorem.

Theorem B.8 (Gauss–Lobatto–Jacobi quadrature nodes) *The quadrature nodes of the Gauss–Lobatto–Jacobi (GLJ) integration rule with $N + 1$ points are solutions of the algebraic equation*

$$\chi_N^{(\alpha,\beta)}(x) := (1 - x^2) \frac{d}{dx} P_N^{\alpha,\beta}(x) = 0. \tag{B.2.5}$$

Analytic expressions for the quadrature weights in terms of the quadrature nodes can be found, for instance, in [92]. Below, we summarize the basic parameters for the two most important families of quadrature schemes for spectral methods, corresponding to Legendre and Chebyshev polynomials.

B.2.2 Gaussian Rules Based on Legendre Polynomials

In the Legendre case, the weight function is $w(x) = 1$. Given any function $u \in L^2(\Omega)$, one has the approximate quadrature schemes

$$\int_{-1}^{1} u(x)\, dx \approx \sum_{k=0}^{N} \rho_k\, u(\xi_k) \tag{B.2.6}$$

with the following parameters:

Gauss–Legendre (GL) *integration scheme:*

$$\begin{aligned}\xi_k &: \text{zeros of } L_{N+1}(x), & 0 \leq k \leq N, \\ \rho_k &= \frac{2}{(1 - \xi_k^2)[L'_{N+1}(\xi_k)]^2}, & 0 \leq k \leq N.\end{aligned} \tag{B.2.7}$$

Gauss–Radau–Legendre (GRL) integration scheme:

$$\xi_k : \text{zeros of } L_N(x) + L_{N+1}(x), \quad 0 \le k \le N,$$

$$\rho_0 = \frac{2}{(N+1)^2},$$

$$\rho_k = \frac{1}{(N+1)^2} \frac{1 - \xi_k}{[L_N(\xi_k)]^2}, \quad 1 \le k \le N. \tag{B.2.8}$$

Gauss–Lobatto–Legendre (GLL) integration scheme:

$$\xi_k : \xi_0 = -1, \xi_N = 1, \text{ zeros of } L'_N(x), \quad 1 \le k \le N-1,$$

$$\rho_k = \frac{2}{N(N+1)} \frac{1}{[L_N(\xi_k)]^2}, \quad 0 \le k \le N. \tag{B.2.9}$$

B.2.3 Gaussian Rules Based on Chebyshev Polynomials

In the Chebyshev case, the weight function $w(x) = (1 - x^2)^{-1/2}$. Given any function $u \in L^2_w(\Omega)$, one has the approximate quadrature schemes

$$\int_{-1}^{1} \frac{u(x)}{\sqrt{1-x^2}} dx \approx \sum_{k=0}^{N} \rho_k^c u(\xi_k^c), \tag{B.2.10}$$

with the following parameters:

Gauss–Chebyshev (GC) integration scheme:

$$\xi_k^c = \cos \frac{(2k+1)\pi}{2N+2}, \quad 0 \le k \le N,$$

$$\rho_k^c = \frac{\pi}{N+1}, \quad 0 \le k \le N. \tag{B.2.11}$$

Gauss–Radau–Chebyshev (GRC) integration scheme:

$$\xi_k^c = -\cos \frac{2\pi k}{2N+1}, \quad 0 \le k \le N,$$

$$\rho_0^c = \frac{\pi}{2N+1},$$

$$\rho_k^c = \frac{2\pi}{2N+1}, \quad 1 \le k \le N. \tag{B.2.12}$$

452   Appendix B. Orthogonal Polynomials and Discrete Transforms

Gauss–Lobatto–Chebyshev (GLC) integration scheme:

$$\xi_k^c = \cos \frac{\pi k}{N}, \qquad 0 \leq k \leq N,$$
$$\rho_k^c = \frac{\pi}{2N}, \qquad k = 0, N, \qquad \text{(B.2.13)}$$
$$\rho_k^c = \frac{\pi}{N}, \qquad 1 \leq k \leq N-1.$$

B.2.4 Discrete Inner Products and Norms

The use of Gaussian quadrature rules as a convenient numerical approximation of definite integrals leads to the replacement of the (conventional) inner products and norms in spaces of functions by discrete inner products and norms.

Let u, v denote any two elements in $L_w^2(\Omega)$; their inner product and norm are given respectively by (A.4.9) and (A.4.10):

$$(u, v)_{L_w^2(\Omega)} := \int_{-1}^{1} w(x) u(x) v(x) \, dx, \qquad u, v \in L_w^2(\Omega),$$

$$\|u\|_{L_w^2(\Omega)} := \left(\int_{-1}^{1} w(x) u^2(x) \, dx \right)^{1/2}, \qquad u \in L_w^2(\Omega).$$

Discretizing both right-hand sides with a Gaussian quadrature with $N+1$ nodes yields the following quantities:

$$(u, v)_N := \sum_{k=0}^{N} \rho_k \, u(\xi_k) \, v(\xi_k), \qquad u, v \in L_w^2(\Omega), \qquad \text{(B.2.14)}$$

$$\|u\|_N := \left(\sum_{k=0}^{N} \rho_k \, u^2(\xi_k) \right)^{1/2}, \qquad u \in L_w^2(\Omega). \qquad \text{(B.2.15)}$$

One easily verifies that, given the positivity of the coefficients ρ_k, the quantities (B.2.14) and (B.2.15) satisfy all the axioms for an inner product and norm and may therefore be considered as such in their own right (see Sections A.2.1 and A.4.1). The quantities related to a Gaussian quadrature are the *discrete inner product* $(\cdot, \cdot)_N$ and the *discrete norm* $\|\cdot\|_N$ defined on $L_w^2(\Omega)$.

Remark B.1 Applying one of Theorems B.5–B.7, depending on the underlying Gaussian quadrature, gives the following *exact* results for polynomial functions:

$$(u, v)_{L_w^2(\Omega)} = (u, v)_N \qquad \forall \, u(x), v(x) \in \mathbb{P}_{2N+\varepsilon}, \qquad \text{(B.2.16)}$$

B.2. Gaussian-Type Quadratures

with $\varepsilon = 1, 0, -1$ corresponding to Gauss (G), Gauss–Radau (GR) and Gauss–Lobatto (GL) quadratures. Furthermore, one has

$$\|u\|_{L^2_w(\Omega)} = \|u\|_N, \quad \begin{cases} \forall\ u(x) \in \mathbb{P}_N & \text{(G and GR)}, \\ \forall\ u(x) \in \mathbb{P}_{N-1} & \text{(GL)}. \end{cases} \quad \text{(B.2.17)}$$

Remark B.2 Denoting by $\gamma_k^{(N)} := \|p_k\|_N$ the discrete norm associated to a Gaussian quadrature with $N+1$ nodes of the orthogonal polynomial p_k, one may show that

$$(p_k, p_l)_N = \gamma_k^{(N)} \delta_{kl}, \quad 0 \le k, l \le N. \quad \text{(B.2.18)}$$

In other words, orthogonal polynomials with respect to the L^2_w inner product remain orthogonal with respect to $(\cdot, \cdot)_N$. Further evaluation of the quantities $\gamma_k^{(N)}$ gives, for the Legendre polynomials,

$$\gamma_k^{(N)} := \sum_{j=0}^{N} \rho_j L_k^2(\xi_j) = \frac{2}{2k+1}, \quad 0 \le k \le N-1,$$

$$\gamma_N^{(N)} := \sum_{j=0}^{N} \rho_j L_N^2(\xi_j) = \begin{cases} \dfrac{2}{2N+1} & \text{(G and GR)}, \\ \dfrac{2}{N} & \text{(GL)}, \end{cases} \quad \text{(B.2.19)}$$

and, for the Chebyshev polynomials,

$$\gamma_k^{(N)} := \sum_{j=0}^{N} \rho_j^c T_k^2(\xi_j^c) = c_k \frac{\pi}{2}, \quad c_0 = 2, \quad c_k = 1, \ 1 \le k \le N-1,$$

$$\gamma_N^{(N)} := \sum_{j=0}^{N} \rho_j^c T_N^2(\xi_j^c) = \begin{cases} \dfrac{\pi}{2} & \text{(G and GR)}, \\ \pi & \text{(GL)}. \end{cases} \quad \text{(B.2.20)}$$

In accordance with (B.2.17), all discrete norms $\gamma_k^{(N)}$ are *exact*, except $\gamma_N^{(N)}$ for the Gauss–Lobatto quadrature scheme.

Remark B.3 For the Gauss–Lobatto integration scheme where the discrete norm $\|p_N\|_N$ is only approximate, one may show that the norms $\|\cdot\|_{L^2_w(\Omega)}$ and $\|\cdot\|_N$ are *uniformly equivalent*, which means that constants C_1 and C_2 exist, independent of N, such that

$$C_1 \|u\|_{L^2_w(\Omega)} \le \|u\|_N \le C_2 \|u\|_{L^2_w(\Omega)} \quad \forall u \in \mathbb{P}_N. \quad \text{(B.2.21)}$$

Using the results (B.2.19) and (B.2.20), one easily verifies that the constants C_1 and C_2 are equal to 1 and $\sqrt{3}$ for the Legendre polynomials and equal to 1 and $\sqrt{2}$ for the Chebyshev polynomials. This result is due to Canuto and Quarteroni [68]. Inequalities such as (B.2.21) are used in the error analysis of spectral methods.

B.3 Spectral Approximation and Interpolation

Having recalled the main properties of orthogonal polynomials and Gaussian quadrature, we may now summarize the essentials on spectral approximation and interpolation of functions.

B.3.1 Preliminaries

Consider the family of orthogonal polynomials $\{p_k\}_{k=0}^{\infty}$ with respect to the weight function $w(x)$, as a basis for $L_w^2(\Omega)$. Because of the completeness property, any function $u \in L_w^2(\Omega)$ may be decomposed into a formal series

$$u(x) := \sum_{k=0}^{\infty} \hat{u}_k \, p_k(x), \qquad (B.3.1)$$

where the coefficients $\hat{u}_k$ are defined uniquely. For the sake of convenience we denote the r.h.s. of (B.3.1) by $Su(x)$.

Multiplying both sides of (B.3.1) by $w(x) \, p_m(x)$, integrating over Ω, and applying the orthogonality relationship (B.1.1), one gets

$$\hat{u}_m = \frac{1}{\gamma_m} \int_{-1}^{1} w(x) u(x) \, p_m(x) \, dx, \qquad m = 0, \ldots, \qquad (B.3.2)$$

with $\gamma_m := \|p_m\|_{L_w^2(\Omega)}^2$. The series (B.3.1) is the *spectral expansion* of u with respect to the orthogonal family $\{p_k\}_{k=0}^{\infty}$.

By squaring (B.3.1) and integrating, one gets

$$\|u(x)\|_{L_w^2(\Omega)}^2 = \sum_{k=0}^{\infty} \|p_k\|_{L_w^2(\Omega)}^2 \, \hat{u}_k^2, \qquad (B.3.3)$$

a property known as *Parseval's identity*.

Two fundamental questions arise: What is the nature of the relation between the function $u(x)$ and its spectral expansion $Su(x)$; and what is, for a converging series, the decay rate of the coefficients $\hat{u}_k$ as a function of k, that is, the rate of convergence? We immediately deal with the first question, using the concepts developed in Appendix A. The second question will be answered in Section B.3.4.

B.3. Spectral Approximation and Interpolation

Let $P_N u$ represent the formal series (B.3.1) truncated at the degree N:

$$P_N u(x) = \sum_{k=0}^{N} \hat{u}_k \, p_k(x), \qquad u \in L^2_w(\Omega). \tag{B.3.4}$$

Completeness of the polynomial basis in $L^2_w(\Omega)$ implies that for any $u \in L^2_w(\Omega)$, the expansion (B.3.1) converges to u in the *mean*, namely,

$$\lim_{N \to \infty} \|u - P_N u\|_{L^2_w(\Omega)} = 0. \tag{B.3.5}$$

Locally, however, $Su(x)$ may differ from $u(x)$. One may show that, at any abscissa where u is discontinuous, one has

$$Su(x) = \frac{1}{2}[u(x^-) + u(x^+)], \tag{B.3.6}$$

provided that u has bounded variations. The quantities in the r.h.s. of (B.3.6) are the left and right limits of u at the discontinuity. This property is a classical property of Fourier series for periodic functions [64].

B.3.2 Discrete Spectral Transforms

Let S_N denote the linear space spanned by the elements $\{p_k\}_{k=0}^{N}$.[1] One can show that the element $P_N u \in S_N$ is the *best approximation* of $u(x) \in L^2_w(\Omega)$ in S_N in the norm $\|\cdot\|_{L^2_w(\Omega)}$, or, in other words, that among all elements $\Phi \in S_N$, $P_N u$ has the shortest distance to $u \in L^2_w(\Omega)$ in the norm $\|\cdot\|_{L^2_w(\Omega)}$:

$$\inf_{\Phi \in S_N} \|u - \Phi\|_{L^2_w(\Omega)} = \|u - P_N u\|_{L^2_w(\Omega)}, \qquad u \in L^2_w(\Omega). \tag{B.3.7}$$

Another characterization of $P_N u$ is that, from (B.1.1), this element is the *orthogonal projection* of u onto S_N, with respect to the inner product $(\cdot, \cdot)_{L^2_w(\Omega)}$. The idempotence property of projectors (A.4.19) is satisfied, and for any $u \in L^2_w(\Omega)$ one has

$$(u - P_N u, p)_{L^2_w(\Omega)} = 0 \qquad \forall \, p \in S_N. \tag{B.3.8}$$

Let $I_N u \in S_N$ be the *Lagrange interpolation polynomial* of $u(x) \in L^2_w(\Omega)$ at the $N+1$ nodes of one of the Gaussian quadrature rules associated to the weight function w:

$$I_N u(x) := \sum_{k=0}^{N} \tilde{u}_k \, p_k(x), \qquad u \in L^2_w(\Omega). \tag{B.3.9}$$

---

[1] S_N is, in fact, identical to $\mathbb{P}_N$ except for the basis.

The coefficients $\tilde{u}_k$ in the expansion (B.3.9) are determined by the interpolation conditions

$$(I_N u)(\xi_k) = u(\xi_k), \qquad 0 \le k \le N, \tag{B.3.10}$$

where $\{\xi_k\}_{k=0}^N$ are the Gaussian quadrature nodes. For trivial reasons, one also has

$$(u, v)_N = (I_N u, v)_N \qquad \forall v \in L_w^2(\Omega). \tag{B.3.11}$$

Using (B.3.10), (B.3.11), and the orthogonality property (B.2.18), one can easily show that the nodal values of $\{u(\xi_l)\}_{l=0}^N$ and the discrete coefficients $\tilde{u}_k$ are related by

$$u(\xi_k) = \sum_{\ell=0}^{N} \tilde{u}_\ell \, p_\ell(\xi_k), \qquad u \in L_w^2(\Omega), \tag{B.3.12}$$

and

$$\tilde{u}_k = \frac{1}{\gamma_k^{(N)}} \sum_{j=0}^{N} \rho_j \, u(\xi_j) \, p_k(\xi_j), \qquad u \in L_w^2(\Omega). \tag{B.3.13}$$

The expansions (B.3.12) and (B.3.13) are known as the *inverse discrete spectral transform* (IDST) and the *discrete spectral transform* (DST) of u, respectively.

Combining (B.3.9) and (B.3.13) yields the canonical form of the interpolation function

$$I_N u(x) = \sum_{j=0}^{N} u(\xi_j) \, \psi_j(x), \tag{B.3.14}$$

where the basis elements $\psi_j(x)$, explicitly given by

$$\psi_j(x) := \rho_j \sum_{\ell=0}^{N} \frac{1}{\gamma_\ell^{(N)}} p_\ell(\xi_j) \, p_\ell(x), \qquad 0 \le j \le N, \tag{B.3.15}$$

satisfy the bi-orthogonality property

$$\psi_j(\xi_k) = \delta_{jk}, \qquad 0 \le j, k \le N. \tag{B.3.16}$$

Using $\chi_N^{(\alpha,\beta)}(x)$, generating function of the GLJ nodes $\{\xi_j^J\}_{j=0}^N$ defined by (B.2.5), the interpolation polynomial on the GLJ grid takes the form (B.3.14) with basis elements

$$\psi_j^J(x) = C_j \frac{1-x^2}{x-\xi_j^J} \frac{d}{dx} P_N^{\alpha,\beta}(x). \tag{B.3.17}$$

B.3. Spectral Approximation and Interpolation

The constant C_j is a normalization constant such that the condition (B.3.16) is satisfied. An elementary calculation based on the differential equation (B.1.5) satisfied by the Jacobi polynomials gives

$$C_j = -\frac{1}{\bar{c}_j \lambda_N^{(\alpha,\beta)}} \frac{1}{P_N^{\alpha,\beta}(\xi_j^J)}, \tag{B.3.18}$$

where $\lambda_N^{(\alpha,\beta)} := N(N + \alpha + \beta + 1)$ is the eigenvalue (B.1.6) of the Sturm–Liouville problem corresponding to N, and

$$\bar{c}_j = \begin{cases} 1/(\beta+1), & j = 0, \\ 1, & 1 \le j \le N-1, \\ 1/(\alpha+1), & j = N. \end{cases} \tag{B.3.19}$$

In particular, for the Legendre ($\alpha = \beta = 0$) and Chebyshev ($\alpha = \beta = -1/2$) interpolation expansions of degree N on the Gauss–Lobatto quadrature grid, the canonical basis elements $\{\psi_j^J(x)\}_{j=0}^N$ are given by

$$\psi_j(x) = -\frac{1}{N(N+1)} \frac{1}{L_N(\xi_j)} \frac{(1-x^2) L_N'(x)}{x - \xi_j} \quad \text{(Legendre)},$$

$$\psi_j^c(x) = \frac{(-1)^{j+1}}{\bar{c}_j N^2} \frac{(1-x^2) T_N'(x)}{x - \xi_j^c} \quad \text{(Chebyshev)}, \tag{B.3.20}$$

with $\bar{c}_0 = \bar{c}_N = 2$ and $\bar{c}_j = 1$ for $1 \le j \le N - 1$.

The reader should notice that (for similar reasons to those for P_N), $I_N u$ is the *orthogonal projection* of u onto S_N with respect to the discrete inner product $(\cdot, \cdot)_N$. One has $I_N^2 = I_N$, and the property (B.3.11) may be cast into

$$(u - I_N u, v)_N = 0 \quad \forall v \in S_N. \tag{B.3.21}$$

Introducing the spectral expansion (B.3.1) in the r.h.s. of (B.3.13) yields

$$\tilde{u}_k = \frac{1}{\gamma_k^{(N)}} \sum_{\ell=0}^{\infty} \hat{u}_\ell \left(\sum_{j=0}^{N} \rho_j \, p_\ell(\xi_j) \, p_k(\xi_j) \right). \tag{B.3.22}$$

Taking into account the result (B.2.16) for the Gauss–Lobatto quadrature of polynomials up to the degree $2N - 1$, one gets

$$\tilde{u}_k = \hat{u}_k + \frac{1}{\gamma_k^{(N)}} \sum_{\ell=2N-k}^{\infty} \hat{u}_\ell \, (p_k, p_\ell)_N. \tag{B.3.23}$$

Finally, multiplication of (B.3.23) by p_k, followed by summation, gives the important relationship

$$I_N u(x) = P_N u(x) + R_N u(x), \tag{B.3.24}$$

where $R_N u \in S_N$ is given by

$$R_N u = \sum_{k=0}^{N} \left(\frac{1}{\gamma_k^{(N)}} \sum_{\ell=2N-k}^{\infty} \hat{u}_\ell \, (p_k, p_\ell)_N \right) p_k. \qquad (B.3.25)$$

This quantity is known as the *aliasing error* due to interpolation. Clearly, $R_N u$ is a low-order contribution to the error, since it involves polynomials of degree less than or equal to N. However, it depends exclusively on the high-order terms of the spectral expansion of u (i.e. the terms with $\ell > N$). In $L_w^2(\Omega)$ norm, the interpolation error $u - I_N u$ is written as

$$\|u - I_N u\|_{L_w^2(\Omega)}^2 = \|u - P_N u\|_{L_w^2(\Omega)}^2 + \|R_N u\|_{L_w^2(\Omega)}^2. \qquad (B.3.26)$$

This expression shows that, for any element $u \in L_w^2(\Omega)$, the error norm $\|u - I_N u\|_{L_w^2(\Omega)}$ is an upper bound of the truncation error $\|u - P_N u\|_{L_w^2(\Omega)}$ and that the faster the coefficients $\hat{u}_k$ decay with respect to k, the closer the interpolation error gets to the truncation error.

In the case of GLC integration, one can show that the aliasing contribution has a simple form

$$\tilde{u}_k = \hat{u}_k + \sum_{m=1}^{\infty} (\hat{u}_{2mN+k} + \hat{u}_{2mN-k}), \qquad 0 \le k \le N. \qquad (B.3.27)$$

B.3.3 Approximate Evaluation of Derivatives

The most delicate (and most important) point in any approximate method for partial differential equations lies in the treatment of derivative terms of the unknown. We first show on finite Legendre and Chebyshev expansions that the problem of evaluating derivatives of functions may be tackled from different points of view. Then, using results established in the previous sections, we indicate a very convenient technique to evaluate the first and second derivatives of functions, based on interpolation polynomials on Gaussian quadrature grids. Compact expressions for these *collocation derivatives* are given in the general framework of Jacobi polynomials (see also [100]).

Finite Legendre and Chebyshev Expansions

Assume, for the sake of simplicity, that u is an infinitely smooth function having a uniformly convergent Legendre expansion

$$u = \sum_{k=0}^{\infty} \hat{u}_k \, L_k(x), \qquad u \in C^\infty(\Omega), \qquad (B.3.28)$$

B.3. Spectral Approximation and Interpolation

where the $\hat{u}_k$'s are given by (B.3.2) and decay exponentially in k. The successive derivatives of u are infinitely smooth and therefore may also be expanded into Legendre series. For instance, u' is equal to

$$u' = \sum_{k=0}^{\infty} \hat{u}_k^{(1)} L_k(x), \qquad (B.3.29)$$

where the coefficients $\hat{u}_k^{(1)}$ are evaluated by using the same process as $\hat{u}_k$. The r.h.s. of (B.3.29) is the *spectral derivative* of u.

Taking the derivative of the expansion (B.3.28) term by term and using (B.1.20), one gets a relationship between the coefficients u_k and $\hat{u}_k^{(1)}$, which is written as

$$\hat{u}_k^{(1)} = (2k+1) \sum_{\substack{p=k+1 \\ p+k \text{ odd}}}^{\infty} \hat{u}_p, \qquad k = 0, \ldots, \qquad (B.3.30)$$

showing that low-order modes of the derivative depend on the high-order modes of the function.

Now, let $P_N u$ denote the orthogonal $L^2(\Omega)$ projection of u onto S_N:

$$P_N u = \sum_{k=0}^{N} \hat{u}_k L_k(x). \qquad (B.3.31)$$

Its derivative $(P_N u)'$ belongs to $\mathbb{P}_{N-1}$. On the other hand, projecting u' orthogonally into S_{N-1} gives

$$P_{N-1} u' = \sum_{k=0}^{N-1} \hat{u}_k^{(1)} L_k(x). \qquad (B.3.32)$$

In light of the relation (B.3.30) it is easily seen that, in the most general case,

$$(P_N u)' \neq P_{N-1} u'. \qquad (B.3.33)$$

The difference, $(P_N u)' - P_{N-1} u'$, decays exponentially as a function of N for infinitely smooth functions. For less smooth functions, however, one may show that the error $\|u' - P_{N-1} u'\|_{L^2(\Omega)}$ is smaller than $\|u' - (P_N u)'\|_{L^2(\Omega)}$ (see Section B.3.4). The derivative $(P_N u)'$ is known as the *Legendre Galerkin derivative* of u. Its evaluation rests essentially upon the knowledge of the Legendre coefficients of u. Once the calculation has been performed for the first derivative, it can readily be extended to higher derivatives.

Another, more practical way exists to evaluate the first derivative of a smooth function u, starting from its projection $I_N u$ onto S_N with respect to the discrete inner product $(\cdot, \cdot)_N$ associated to a Gaussian quadrature rule:

$$\mathcal{D}_N u = (I_N u)'. \tag{B.3.34}$$

The derivative $\mathcal{D}_N u$ is called the (Legendre) *collocation derivative* of u. Before we examine this derivation process in more detail, we perform the same finite-expansion analysis with Chebyshev polynomials.

Once again, assume u is an infinitely smooth function expanded in Chebyshev series,

$$u = \sum_{k=0}^{\infty} \hat{u}_k T_k(x), \qquad u \in C^{\infty}(\Omega). \tag{B.3.35}$$

The spectral expansion of its first derivative is

$$u' = \sum_{k=0}^{\infty} \hat{u}_k^{(1)} T_k(x), \tag{B.3.36}$$

where, as before, the spectral coefficients $\hat{u}_k^{(1)}$ are related to those of u. With the help of (B.1.32), one gets

$$\hat{u}_k^{(1)} = \frac{2}{c_k} \sum_{\substack{p=k+1 \\ p+k \text{ odd}}}^{\infty} p \, \hat{u}_p, \tag{B.3.37}$$

for which similar conclusions hold as for (B.3.30).

Chebyshev Galerkin and Chebyshev collocation derivatives are defined in the same way as they are for Legendre expansions. For instance the Chebyshev Galerkin derivative writes

$$(P_N u)' = \sum_{k=0}^{N} \hat{u}_k \frac{d}{dx} T_k(x). \tag{B.3.38}$$

The Gauss–Lobatto–Jacobi Collocation Derivatives

We go back to the collocation derivative $\mathcal{D}_N u$ of a smooth function. Two different techniques can be used to evaluate the collocation derivatives from the values of u on the quadrature nodes. The first technique uses the discrete spectral coefficients $\tilde{u}_k$ obtained from the nodal values by (B.3.13). This is followed by (B.3.30), which gives the discrete spectral coefficients of the derivative.

B.3. Spectral Approximation and Interpolation

The second technique only rests upon the nodal values of u. From the canonical expansion (B.3.14) of the interpolation polynomial on the GLJ grid, one gets for the first two derivatives

$$(I_N u)^{(\ell)}(x) = \sum_{i=0}^{N} u(\xi_i^J) \frac{d^\ell}{dx^\ell} \psi_j^J(x), \qquad \ell = 1, 2. \quad (B.3.39)$$

The collocation derivatives at the quadrature nodes $\{\xi_k^J\}_{k=0}^N$ can then be obtained very simply from the derivatives of the basis functions (B.3.17)–(B.3.19) at these nodes. To get the ℓth collocation derivative on the GLJ grid, one multiplies the vector $\underline{u} := \{u(\xi_0^J), u(\xi_1^J), \ldots, u(\xi_N^J)\}^T$ of nodal values by the matrix $D_N^{J(\ell)}$:

$$(I_N u)^{(\ell)}(\xi_k^J) = \sum_{i=0}^{N} D_{N,ki}^{J(\ell)} u(\xi_i^J), \qquad \ell = 1, 2, \quad (B.3.40)$$

with

$$D_N^{J(\ell)} = \left\{ D_{N,ki}^{J(\ell)} := \frac{d^\ell}{dx^\ell} \psi_i^J(\xi_k^J) \right\}_{k,i=0}^N, \qquad \ell = 1, 2.$$

Straightforward calculations on (B.3.17)–(B.3.19) give the first derivatives of the interpolation basis functions $\psi_j^J(x)$ on the GLJ grid, which are the components of $D_N^{J(1)}$. One gets

$$\left.\frac{d\psi_i^J(x)}{dx}\right|_{\xi_k^J} = \frac{\bar{c}_k}{\bar{c}_i} \frac{1}{\xi_k^J - \xi_i^J} \frac{P_N^{\alpha,\beta}(\xi_k^J)}{P_N^{\alpha,\beta}(\xi_i^J)}, \quad \xi_k^J \neq \xi_i^J, \quad 0 \leq i, k \leq N, \quad (B.3.41)$$

$$\left.\frac{d\psi_i^J(x)}{dx}\right|_{\xi_i^J} = -\frac{1}{2} \frac{\beta - \alpha - (\alpha + \beta)\xi_i^J}{1 - \xi_i^{J2}}, \quad 1 \leq i \leq N-1, \quad (B.3.42)$$

$$\left.\frac{d\psi_0^J(x)}{dx}\right|_{-1} = -\frac{1}{2} \frac{\lambda_N^{(\alpha,\beta)} - \alpha}{\beta + 2}, \quad (B.3.43)$$

$$\left.\frac{d\psi_N^J(x)}{dx}\right|_{1} = \frac{1}{2} \frac{\lambda_N^{(\alpha,\beta)} - \beta}{\alpha + 2}, \quad (B.3.44)$$

Similarly, the values of the second derivatives of the basis functions on the

GLJ grid are

$$\left.\frac{d^2\psi_i^J(x)}{dx^2}\right|_{\xi_k^J} = -\frac{1}{\bar{c}_i} \frac{(\beta - \alpha - (\alpha+\beta)\xi_k^J)(\xi_k^J - \xi_i^J) + 2(1 - \xi_k^{J2})}{(\xi_k^J - \xi_i^J)^2 (1 - \xi_k^{J2})} \frac{P_N^{\alpha,\beta}(\xi_k^J)}{P_N^{\alpha,\beta}(\xi_i^J)},$$
$$\xi_k^J \neq \xi_i^J, \quad 0 \leq i \leq N, \quad 1 \leq k \leq N-1, \quad \text{(B.3.45)}$$

$$\left.\frac{d^2\psi_i^J(x)}{dx^2}\right|_{\xi_i^J} = -\frac{1}{3(1-\xi_i^{J2})}\left(\lambda_N^{(\alpha,\beta)} - 2(\alpha+\beta)\right.$$
$$\left. -[\beta - \alpha - (\alpha+\beta)\xi_i^J]\frac{\beta - \alpha - (\alpha+\beta+4)\xi_i^J}{1 - \xi_i^{J2}}\right),$$
$$1 \leq i \leq N-1, \quad \text{(B.3.46)}$$

$$\left.\frac{d^2\psi_i^J(x)}{dx^2}\right|_{-1} = -\frac{1}{2}\frac{\bar{c}_0}{\bar{c}_i}\left(\frac{4}{(1+\xi_i^J)^2} - \frac{2}{\beta+2}\frac{\lambda_N^{(\alpha,\beta)} - \alpha}{1+\xi_i^J}\right)\frac{P_N^{\alpha,\beta}(-1)}{P_N^{\alpha,\beta}(\xi_i^J)},$$
$$1 \leq i \leq N, \quad \text{(B.3.47)}$$

$$\left.\frac{d^2\psi_i^J(x)}{dx^2}\right|_1 = -\frac{1}{2}\frac{\bar{c}_N}{\bar{c}_i}\left(\frac{4}{(1-\xi_i^J)^2} - \frac{2}{\alpha+2}\frac{\lambda_N^{(\alpha,\beta)} - \beta}{1-\xi_i^J}\right)\frac{P_N^{\alpha,\beta}(1)}{P_N^{\alpha,\beta}(\xi_i^J)},$$
$$0 \leq i \leq N-1, \quad \text{(B.3.48)}$$

$$\left.\frac{d^2\psi_0^J(x)}{dx^2}\right|_{-1} = \frac{1}{4}\frac{(\lambda_N^{(\alpha,\beta)} - 2\alpha)(\lambda_N^{(\alpha,\beta)} - (\alpha+\beta+2))}{(\beta+2)(\beta+3)}, \quad \text{(B.3.49)}$$

$$\left.\frac{d^2\psi_N^J(x)}{dx^2}\right|_1 = \frac{1}{4}\frac{(\lambda_N^{(\alpha,\beta)} - 2\beta)(\lambda_N^{(\alpha,\beta)} - (\alpha+\beta+2))}{(\alpha+2)(\alpha+3)}, \quad \text{(B.3.50)}$$

First-Order Legendre and Chebyshev Collocation Derivatives

On the widely used GLL and GLC quadrature grids, the elements of the differentiation matrices $D_N^{(1)}$ and $D_N^{c(1)}$ are given by

$$D_{N,ki}^{(1)} = \begin{cases} \dfrac{L_N(\xi_k)}{L_N(\xi_i)}\dfrac{1}{\xi_k - \xi_i}, & k \neq i, \\[6pt] -\dfrac{N(N+1)}{4}, & k = i = 0, \\[6pt] \dfrac{N(N+1)}{4}, & k = i = N, \\[6pt] 0 & \text{otherwise} \end{cases} \quad \text{(B.3.51)}$$

B.3. Spectral Approximation and Interpolation

and

$$D^{c(1)}_{N,ki} := \begin{cases} \dfrac{\bar{c}_k}{\bar{c}_i} \dfrac{(-1)^{k+i}}{\xi^c_k - \xi^c_i}, & k \neq i, \\[2mm] \dfrac{2N^2 + 1}{6}, & k = i = 0, \\[2mm] -\dfrac{2N^2 + 1}{6}, & k = i = N, \\[2mm] \dfrac{-\xi^c_i}{2(1 - \xi^{c2}_i)}, & 1 \leq k = i \leq N - 1. \end{cases} \quad (B.3.52)$$

In the case of finite Legendre expansions we emphasize that the faster evaluation of the collocation derivative (of the two choices mentioned above) is obtained by using the matrix multiplication alternative [$O(N^2)$ operations]. This is not always the best solution, however. Indeed, for Chebyshev expansions it is computationally more efficient (on a scalar computer) to evaluate collocation derivatives in the transformed space. This is because GLC quadrature nodes are evenly distributed in $\theta = \arccos x$ on $[0, \pi]$ [see Equation (B.2.13)]. Therefore, the inverse and direct transforms (B.3.12) and (B.3.13) can be evaluated in $O(5N \log_2 N)$ arithmetic operations using the FFT. This is done as follows: starting from the nodal values of u, the DST (B.3.13) produces the spectral coefficients $\tilde{u}_k$. Then we use (B.3.37) to determine the $\tilde{u}^{(1)}_k$, and finally, going back to physical space, the IDST (B.3.12) gives the nodal values of the collocation derivative $\mathcal{D}^c_N u$.

Most of the time, instead of using (B.3.37) to obtain the discrete spectral coefficients of the derivative, one uses a numerically stable recurrence relation in decreasing order:

$$c_k \hat{u}^{(1)}_k = \hat{u}^{(1)}_{k+2} + 2(k+1)\hat{u}_{k+1}, \quad 0 \leq k \leq N-1, \quad (B.3.53)$$

with c_k given by (B.1.24). The recurrence is started with $\hat{u}^{(1)}_N \equiv 0$. Efficient Fortran routines for the evaluation of the Chebyshev FFT can be found in the literature, namely in the studies by Canuto et al. [64] and Deville and Labrosse [97].

As a final remark, we notice that the elements of the second-order differentiation matrices $D^{(2)}_N$ and $D^{c(2)}_N$ can be calculated from the expressions (B.3.45)–(B.3.50). They can also be evaluated numerically by multiplying the first-order differentiation matrices $D^{(1)}_N$ [given by (B.3.51)] and $D^{c(1)}_N$ [given by (B.3.52)], by themselves.

B.3.4 Estimates for Truncation and Interpolation Errors

To conclude this review of discrete transforms, we go back to the question raised in Section B.3, about the decay rate of the coefficients of (B.3.1). We summarize the most important results for truncation and interpolation errors in Legendre and Chebyshev expansions, without proof. We also show their implications in practice. Details and proofs may be found in the literature, in particular in [64] and in [142]. We start by dealing with truncation errors; interpolation errors come next.

A fundamental result for truncation errors is the following: For any element $u \in L_w^2(\Omega)$ having a weighted-square-integrable mth derivative in the standard interval, the truncated spectral expansion $P_N u := \sum_{k=0}^{N} \hat{u}_k p_k$ converges as

$$\|u - P_N u\|_{L_w^2(\Omega)} \leq C N^{-m} \|u\|_{H_w^m(\Omega)}. \tag{B.3.54}$$

The polynomials p_k are either the Legendre polynomials ($w = 1$) or the Chebyshev polynomials $[w = (1 - x^2)^{-1/2}]$, C is a generic constant independent of N, and $\|u\|_{H_w^m(\Omega)}$ denotes the Sobolev norm[2] $H_w^m(\Omega)$ of u:

$$\|u\|_{H_w^m(\Omega)} := \left(\sum_{i=0}^{m} \int_{-1}^{1} w(x) \left(u^{(i)} \right)^2 dx \right)^{1/2}, \quad u \in H_w^m(\Omega). \tag{B.3.55}$$

The result (B.3.54) shows that functions belonging to $H_w^m(\Omega)$ have spectral coefficients $\hat{u}_k$ with asymptotic decay rates of at least $O(k^{-m})$. For infinitely smooth functions, clearly the convergence of P_N is *faster than algebraic*.

Another important result involves derivatives of the truncation error. One may show that for $u \in H_w^m(\Omega)$ the truncation error in the Sobolev norm H_w^l satisfies the inequality

$$\|u - P_N u\|_{H_w^l(\Omega)} \leq C N^{-1/2} N^{2l-m} \|u\|_{H_w^m(\Omega)}. \tag{B.3.56}$$

This implies, in particular, that if u belongs to $H_w^m(\Omega)$, then obviously $u' \in H_w^{m-1}(\Omega)$, and because of (B.3.54),

$$\|u' - P_N u'\|_{L_w^2(\Omega)} \approx C N^{1-m}. \tag{B.3.57}$$

On the other hand, because of (B.3.56) and the definition of the Sobolev norm, we find

$$\|u' - (P_N u)'\|_{L_w^2(\Omega)} \approx C N^{3/2-m}. \tag{B.3.58}$$

---

[2] Sobolev spaces are treated in Section 2.1.3.

B.3. Spectral Approximation and Interpolation

This result shows that the Galerkin derivative $(P_N u)'$ has a slightly lower convergence rate than the orthogonal projection of u'. Furthermore, if the smoothness of u is such that it belongs to $H_w^1(\Omega)$ [but not to $H_w^2(\Omega)$], then clearly $P_N u$ converges to u but $(P_N u)'$ does not converge to u'.

Similar results occur for the interpolation error $u - I_N u$. In L^2 norm, the result corresponding to (B.3.54) is written as

$$\|u - I_N u\|_{L^2(\Omega)} \leq C N^{1/2} N^{-m} \|u\|_{H^m(\Omega)}, \tag{B.3.59}$$

whereas in the Sobolev norm $H^m(\Omega)$, one may show that

$$\|u - I_N u\|_{H^l(\Omega)} \leq C N^{1/2} N^{2l-m} \|u\|_{H^m(\Omega)}. \tag{B.3.60}$$

Bibliography

[1] M.S. Acalar and C.R. Smith. A study of hairpin vortices in a laminar boundary layer: Part 1, hairpin vortices generated by a hemisphere protuberance. *J. Fluid Mech.*, 175:1–41, 1987.

[2] Y. Achdou, Y. Maday, and O.B. Widlund. Iterative substructuring preconditioners for mortar element methods in two dimensions. *SIAM J. Numer. Anal.*, 36:551–580, 1999.

[3] R.A. Adams. *Sobolev Spaces*. Academic Press, New York, 1975.

[4] G.S. Almasi and A. Gottlieb. *Highly Parallel Computing*. Benjamin/Cummings, Redwood City, CA, 1994.

[5] G. Anagnostou. *Nonconforming Sliding Spectral Element Methods for Unsteady Incompressible Navier–Stokes Equations*. PhD thesis, Massachusetts Institute of Technology, Cambridge, MA, 1991.

[6] G. Anagnostou, Y. Maday, C. Mavriplis, and A.T. Patera. On the mortar element method: Generalizations and implementation. In T.F. Chan, R. Glowinski, J. Périaux, and O.B. Widlund, editors, *Third International Symposium on Domain Decomposition Methods for Partial Differential Equations*, pages 157–173. SIAM, 1990.

[7] B.F. Armaly, F. Durst, J.C.F. Pereira, and B. Schönung. Experimental and theoretical investigation of backward-facing step flow. *J. Fluid Mech.*, 127:473–496, 1983.

[8] D.N. Arnold, F. Brezzi, and M. Fortin. A stable finite element for the Stokes equations. *Calcolo*, 21:337–344, 1984.

[9] K. Arrow, L. Hurwicz, and H. Uzawa. *Studies in Nonlinear Programming*. Stanford University Press, Stanford, 1958.

[10] N. Arslan. *Experimental Characterization of Transitional Unsteady Flow inside a Graft-to-Vein Junction*. PhD thesis, Dept. of Mech. Eng., University of Illinois, Chicago, 1999.

[11] U.M. Ascher, S.J. Ruuth, and B.T.R. Wetton. Implicit–explicit methods for time-dependent partial differential equations. *SIAM J. Numer. Anal.*, 32:797–823, 1995.

[12] R Askey. *Orthogonal Polynomials and Special Functions*. SIAM, Philadelphia, 1975.

[13] O. Axelsson. *Iterative Solution Methods*. Cambridge University Press, Cambridge, 1994.

[14] O. Axelsson and V.A. Barker. *Finite Element Solution of Boundary Value Problems. Theory and Computation*. Computer Science and Applied Mathematics. Academic Press, New York, 1984.

[15] M. Azaïez, F. Ben Belgacem, M. Grundmann, and H. Khallouf. Staggered grids hybrid-dual spectral element method for second-order elliptic problems. Application to high-order time splitting for Navier–Stokes equations. *Comput. Methods Appl. Mech. Engrg.*, 166:183–199, 1998.

[16] M. Azaïez, C. Bernardi, and M. Grundmann. Spectral methods applied to porous media equations. *East-West J. Numer. Math.*, 2:91–105, 1994.

[17] M. Azaïez and G. Copoletta. Calcul de la pression dans le problème de Stokes par une méthode spectrale de quasi-collocation à grille unique. *Ann. Maghrébines Ingénieur*, 6:41–61, 1992.

[18] I. Babuška. The finite element method with Lagrangian multipliers. *Numer. Math.*, 20:179–192, 1973.

[19] C. Baiocchi, F. Brezzi, and L.P. Franca. Virtual bubbles and Galerkin least-squares type methods. *Comput. Methods Appl. Mech. Engrg.*, 105:125–141, 1993.

[20] A.K. Bangia, P.F. Batcho, I.G. Kevrekidis, and G.E. Karniadakis. Unsteady two-dimensional flows in complex geometries: Comparative bifurcation studies with global eigenfunction expansions. *SIAM J. Sci. Comput.*, 18:775–805, 1997.

[21] R.E. Bank and C.C. Douglas. Sharp estimates for multigrid rates of convergence with general smoothing and acceleration. *SIAM J. Numer. Anal.*, 22:617–643, 1985.

[22] C. Basdevant, M. Deville, P. Haldenwang, J.M. Lacroix, J. Ouazzani, R. Peyret, P. Orlandi, and A.T. Patera. Spectral and finite difference solutions of the Burgers equation. *Comput. Fluids*, 14:23–41, 1986.

[23] G.K. Batchelor. *An Introduction to Fluid Dynamics*. Cambridge University Press, Cambridge, 1967.

[24] P.F. Batcho and G.E. Karniadakis. Generalized Stokes eigenfunctions: A new trial basis for the solution of incompressible Navier–Stokes equations. *J. Comput. Phys.*, 115:121–146, 1994.

[25] J.B. Bell, P. Collela, and H.M. Glaz. A second-order projection method for the incompressible Navier–Stokes equations. *J. Comput. Phys.*, 85:257–283, 1989.

[26] F. Ben Belgacem, C. Bernardi, N. Chorfi, and Y. Maday. Inf–sup conditions for the mortar spectral element method discretization of the Stokes problem. *Numer. Math.*, 85:257–281, 2000.

[27] F. Ben Belgacem and Y. Maday. La méthode des éléments avec joints en 3D: Mise en oeuvre du couplage spectral-éléments finis pour le problème de Poisson. Technical Report HI–72/8065-B, Electricité de France, 1993.

[28] F. Ben Belgacem and Y. Maday. A spectral element methodology tuned to parallel implementation. *Comput. Methods Appl. Mech. Engrg.*, 116:59–67, 1994.

[29] F. Ben Belgacem and Y. Maday. Coupling spectral and finite elements for second order elliptic three-dimensional equations. *SIAM J. Numer. Anal.*, 36:1234–1263, 1999.

[30] C.M. Bender and S.A. Orszag. *Advanced Mathematical Methods for Scientists and Engineers*. International Series in Pure and Applied Mathematics. McGraw-Hill, New York, 1978.

[31] E.R. Benton and G.N. Platzmann. A table of solutions of one-dimensional Burgers equation. *Quart. Appl. Math.*, 29:195–212, 1972.

[32] P. Bergé, Y. Pomeau, and Ch. Vidal. *L' ordre dans le chaos*. Hermann, Paris, 1988.

[33] G. Berkooz, P. Holmes, and J.L. Lumley. The proper orthogonal decomposition in the analysis of turbulent flows. *Annu. Rev. Fluid Mech.*, 25:539–575, 1993.

[34] C. Bernardi, M. Dauge, and Y. Maday. *Spectral Methods for Axisymmetric Domains*. Gauthier-Villars, Paris, 1999.
[35] C. Bernardi, C. Canuto, and Y. Maday. Generalized inf–sup conditions for Chebyshev spectral approximation of the Stokes problem. *SIAM J. Numer. Anal.*, 25:1237–1271, 1988.
[36] C. Bernardi, C. Canuto, Y. Maday, and B. Métivet. Single-grid spectral collocation for the Navier–Stokes equations. *IMA J. Numer. Anal.*, 10:253–297, 1990.
[37] C. Bernardi and Y. Maday. A collocation method over staggered grids for the Stokes problem. *Int. J. Numer. Meth. Fluids*, 8:537–557, 1988.
[38] C. Bernardi and Y. Maday. *Approximations spectrales de problèmes aux limites elliptiques*. Springer, Paris, 1992.
[39] C. Bernardi and Y. Maday. Spectral methods. In P.G. Ciarlet and J.L. Lions, editors, *Handbook of Numerical Analysis*, Volume V of *Techniques of Scientific Computing*, pages 209–486. North Holland, Amsterdam, 1997.
[40] C. Bernardi and Y. Maday. Uniform inf–sup conditions for the spectral discretization of the Stokes problem. *Math. Models Meth. Appl. Sci.*, 9:395–414, 1999.
[41] C. Bernardi, Y. Maday, and B. Métivet. Calcul de la pression dans la résolution spectrale du problème de Stokes. *La Recherche Aérospatiale*, 1:1–21, 1987.
[42] C. Bernardi, Y. Maday, and A.T. Patera. Domain decomposition by the mortar element method. In H.G. Kaper and M. Garbey, editors, *Asymptotic and Numerical Methods for Partial Differential Equations with Critical Parameters*, Volume C 384 of *NATO ASI Series*, pages 269–286. 1993.
[43] C. Bernardi, Y. Maday, and A.T. Patera. A new nonconforming approach to domain decomposition: The mortar element method. In H. Brezis and J.L. Lions, editors, *Collège de France Seminar*, volume XI, pages 13–51. Pitman, Boston, 1994.
[44] G. Birkhoff and R.E. Lynch. *Numerical Solution of Elliptic Problems*. SIAM Studies in Applied Mathematics. SIAM, Philadelphia, 1984.
[45] J. Boussinesq. Théorie de l'écoulement tourbillonnant. *Mém. Prés. Div. Savants Acad. Sci. Paris*, 23:46–50, 1877.
[46] J.P. Boyd. The erfc–log filter and the asymptotics of the Euler and Vandeven sequence accelerations. In A.V. Ilin and L.R. Scott, editors, *Third International Conference on Spectral and High Order Methods*, pages 267–275. Houston Journal of Mathematics, 1996.
[47] J.P. Boyd. Two comments on filtering for Chebyshev and Legendre spectral and spectral element methods. *J. Comput. Phys.*, 143:283–288, 1998.
[48] J.P. Boyd. *Chebyshev and Fourier Spectral Methods*. Dover, Mineola, New York, 2nd edition, 2001.
[49] J.U. Brackbill, D.B. Kothe, and C. Zemach. A continuum method for modeling surface tension. *J. Comput. Phys.*, 100:335–354, 1992.
[50] A. Brandt. Multi-level adaptive solution to boundary-value problems. *Math. Comput.*, 31:333–390, 1977.
[51] F. Brezzi. On the existence, uniqueness and approximation of saddle point problems arising from Lagrangian multipliers. *R.A.I.R.O. Anal. Numer.*, 8:129–151, 1974.
[52] F. Brezzi, M-O. Bristeau, L. Franca, M. Mallet, and G. Rogé. A relationship between stabilized finite element methods and the Galerkin method with bubble functions. *Comput. Methods Appl. Mech. Engrg.*, 96:117–129, 1992.
[53] F. Brezzi and M. Fortin. *Mixed and Hybrid Finite Element Methods*. Springer Series in Computational Mathematics 15. Springer, Berlin, 1991.
[54] F. Brezzi and J. Pitkäranta. On the stabilization of finite element approximations

of the Stokes equations. In W. Hackbusch, editor, *Efficient Solutions of Elliptic Systems*, pages 11–19. Vieweg, Braunschweig, 1984.

[55] A.N. Brooks and T.J.R. Hughes. Streamline upwind/Petrov–Galerkin formulation for convection dominated flow with particular emphasis on the incompressible Navier–Stokes equations. *Comput. Methods Appl. Mech. Engrg.*, 32:199–259, 1982.

[56] D.L. Brown and M.L. Minion. Performance of under-resolved two-dimensional incompressible flow simulations. *J. Comput. Phys.*, 122:165–183, 1995.

[57] J.C. Butcher. Implicit Runge–Kutta processes. *Math. Comput.*, 18:50–64, 1964.

[58] J.C. Butcher. *The Numerical Analysis of Ordinary Differential Equations. Runge–Kutta and General Linear Methods*. Wiley, New York, 1987.

[59] B.L. Buzbee, G.H. Golub, and C.W. Nielson. On direct methods for solving Poisson's equations. *SIAM J. Numer. Anal.*, 7:627–656, 1970.

[60] J. Cahouet and J.P. Chabard. Some fast 3D finite element solvers for the generalized Stokes problem. *Int. J. Numer. Methods Fluids*, 8:869–895, 1986.

[61] A. Campion-Renson and M.J. Crochet. On the stream function–vorticity finite element solutions of Navier–Stokes equations. *Int. J. Numer. Methods Engrg.*, 12:1809–1818, 1978.

[62] C. Canuto. Spectral methods and the maximum principle. *Math. Comput.*, 51:615–629, 1988.

[63] C. Canuto. Stabilization of spectral methods by finite element bubble functions. *Comput. Methods Appl. Mech. Engrg.*, 116:1–19, 1994.

[64] C. Canuto, Y. Hussaini, A. Quarteroni, and T.A. Zang. *Spectral Methods in Fluid Dynamics*. Springer Series in Computational Physics. Springer, Berlin, 1988.

[65] C. Canuto and V. Van Kemenade. Bubble-stabilized spectral methods for the incompressible Navier–Stokes equations. *Comput. Methods Appl. Mech. Engrg.*, 135:35–61, 1996.

[66] C. Canuto and P. Pietra. Boundary and interface conditions within a finite element preconditioner for spectral methods. *J. Comput. Phys.*, 91:310–343, 1990.

[67] C. Canuto and G. Puppo. Bubble stabilization of spectral Legendre methods for the advection–diffusion equation. *Comput. Methods Appl. Mech. Engrg.*, 118:239–263, 1994.

[68] C. Canuto and A. Quarteroni. Approximation results for orthogonal polynomials in Sobolev spaces. *Math. Comput.*, 38:67–86, 1982.

[69] C. Canuto and A. Quarteroni. Preconditioned minimal residual methods for Chebyshev spectral calculations. *J. Comput. Phys.*, 60:315–337, 1985.

[70] C. Canuto, A. Russo, and V. Van Kemenade. Stabilized spectral methods for the Navier–Stokes equations: Residual-free bubbles and preconditioning. *Comput. Methods Appl. Mech. Engrg.*, 166:65–83, 1998.

[71] G.F. Carey and J.T. Oden. *Finite Elements: A Second Course*. The Texas Finite Element Series, Vol. II. Prentice-Hall, Englewood Cliffs, NJ, 1983.

[72] M.A. Casarin. Diagonal edge preconditioners in p-version and spectral element methods. *SIAM J. Sci. Comput.*, 18:610–620, 1997.

[73] M.A. Casarin. Quasi-optimal Schwarz methods for the conforming spectral element discretization. *SIAM J. Numer. Anal.*, 34:2482–2502, 1997.

[74] M.A. Casarin. Schwarz preconditioners for the spectral element Stokes and Navier–Stokes discretizations. In P. Bjørstad, M. Espedal, and D. Keyes, editors, *Domain Decomposition 9 Proceedings*, pages 72–79. Wiley, New York, 1998.

[75] J. Castaños. *Parallel Adaptive Refinement and Derefinement*. PhD thesis, Division of Computer Science, Brown University, 2000.

[76] J. Castaños and J. Savage. Repartitioning unstructured adaptive meshes. In

Proceedings of the International Parallel and Distributed Processing Symposium, 2000. 2000.
[77] C. Cercignani. *The Boltzmann Equation and Its Applications.* Applied Mathematical Sciences No. 67. Springer, New York, 1988.
[78] A.J. Chorin. Numerical solution of the Navier–Stokes equations. *Math. Comput.*, 22:745–762, 1968.
[79] P.G. Ciarlet. *The Finite Element Method for Elliptic Problems.* Studies in Mathematics and its Applications. North-Holland, Amsterdam, 1978.
[80] B. Cockburn, C. Johnson, C.-W. Shu, and E. Tadmor. *Advanced Numerical Approximation of Nonlinear Hyperbolic Equations.* Lecture Notes in Mathematics 1697. Springer, Berlin, 1998.
[81] J.D. Cole. On a quasi-linear parabolic equation occurring in aerodynamics. *Quart. Appl. Math.*, 9:225–236, 1951.
[82] P. Constantin and C. Foias. *Navier–Stokes Equations.* Chicago Lectures in Mathematics. University of Chicago Press, Chicago, 1988.
[83] W. Couzy. *Spectral Element Discretization of the Unsteady Navier–Stokes Equations and its Iterative Solution on Parallel Computers.* PhD thesis, Swiss Federal Institute of Technology, Lausanne, 1995.
[84] W. Couzy. Iterative solutions of the 3D Navier–Stokes equations on parallel computers. In A.V. Ilin and L.R. Scott, editors, *Third International Conference on Spectral and High Order Methods*, pages 585–594. Houston Journal of Mathematics, 1996.
[85] W. Couzy and M.O. Deville. Spectral-element preconditioners for the Uzawa pressure operator applied to incompressible flows. *J. Sci. Comput.*, 9:107–112, 1994.
[86] W. Couzy and M.O. Deville. A fast Schur complement method for the spectral element discretization of the incompressible Navier–Stokes equations. *J. Comput. Phys.*, 116:135–142, 1995.
[87] M. Crouzeix. Sur les méthodes de Runge–Kutta pour l'approximation des problèmes d'évolution. In R. Glowinski and J.L. Lions, editors, *Computing Methods in Applied Sciences and Engineering: Second International Symposium, 1975*, Lecture Notes in Econom. and Math. Systems 134, pages 206–223. Springer, Berlin, 1976.
[88] M. Crouzeix and P.A. Raviart. Conforming and nonconforming finite element methods for solving the stationary Stokes equations I. *R.A.I.R.O. Anal. Numér.*, 7:33–75, 1973.
[89] G.G. Dahlquist. A special stability problem for linear multistep methods. *BIT*, 3:27–43, 1963.
[90] O. Daube. Resolution of the 2D Navier–Stokes equations in velocity–vorticity form by means of an influence matrix technique. *J. Comput. Phys.*, 103:402–414, 1992.
[91] P.J. Davis. *Interpolation and Approximation.* Dover, New York, 4th edition, 1975.
[92] P.J. Davis and P. Rabinowitz. *Methods of Numerical Integration.* Academic Press, New York, 2nd edition, 1984.
[93] G. de Vahl Davis. Natural convection of air in a square cavity: A bench mark numerical solution. *Int. J. Numer. Methods Fluids*, 3:249–267, 1983.
[94] P. Demaret and M.O. Deville. Chebyshev pseudo-spectral solution of the Stokes equations using finite element preconditioning. *J. Comput. Phys.*, 83:463–484, 1989.
[95] P. Demaret and M.O. Deville. Chebyshev collocation solutions of the Navier–Stokes equations using multi-domain decomposition and finite element preconditioning. *J. Comput. Phys.*, 95:359–386, 1991.

[96] P. Demaret, M.O. Deville, and C. Schneidesch. Thermal convection solutions by Chebyshev pseudospectral multi-domain decomposition and finite element preconditioning. *Appl. Numer. Math.*, 6:107–121, 1989.

[97] M. Deville and G. Labrosse. An algorithm for the evaluation of multidimensional (direct and inverse) discrete Chebyshev transform. *J. Comput. Appl. Math.*, 8:293–304, 1982.

[98] M. Deville, T.-H. Lê, and Y. Morchoisne (Eds.). *Numerical Simulation of 3-D Incompressible Unsteady Viscous Laminar Flows*, Volume 36 of *Notes on Numerical Fluid Mechanics*. Vieweg, Braunschweig, 1992.

[99] M.O. Deville and E.H. Mund. Chebyshev pseudospectral solution of second-order elliptic equations with finite element preconditioning. *J. Comput. Phys.*, 60:517–533, 1985.

[100] M.O. Deville and E.H. Mund. Finite element preconditioning for pseudospectral solutions of elliptic problems. *SIAM J. Sci. Stat. Comput.*, 11:311–342, 1990.

[101] M.O. Deville and E.H. Mund. Finite element preconditioning of collocation schemes for advection–diffusion equations. In P. de Groen and R. Beauwens, editors, *Proceedings of the IMACS International Symposium on Iterative Methods in Linear Algebra, Brussels, 1991*, pages 181–189. Elsevier, Amsterdam, 1992.

[102] M.O. Deville and E.H. Mund. Fourier analysis of finite element preconditioned collocation schemes. *SIAM J. Sci. Stat. Comput.*, 13:596–610, 1992.

[103] M.O. Deville, E.H. Mund, and V. Van Kemenade. Preconditioned Chebyshev collocation methods and triangular finite elements. *Comput. Methods Appl. Mech. Engrg.*, 116:193–200, 1994.

[104] M.O. Deville, E.H. Mund, and A.T. Patera. Iterative solution of isoparametric spectral element equations by low-order finite element preconditioning. *J. Comput. Appl. Math.*, 20:189–197, 1987.

[105] J. Devooght and E.H. Mund. Numerical solution of neutron kinetics equations using A-stable algorithms. *Progr. Nucl. Energy*, 16:97–126, 1985.

[106] W.S. Don and D.I. Gottlieb. Spectral simulation of supersonic reactive flows. *SIAM J. Numer. Anal.*, 35:2370–2384, 1998.

[107] J. Donea. A Taylor–Galerkin method for convective transport problems. *Int. J. Numer. Meth. Engrg.*, 20:101–119, 1984.

[108] J. Donea, S. Giuliani, and J.P. Halleux. An arbitrary Lagrangian–Eulerian finite element method for transient dynamic fluid–structure interactions. *Comput. Methods Appl. Mech. Engrg.*, 33:689–723, 1982.

[109] J. Donea and L. Quartapelle. An introduction to finite element methods for transient advection problems. *Comput. Methods Appl. Mech. Engrg.*, 95:169–203, 1992.

[110] K. Dowd and C. Severance. *High Performance Computing*, O'Reilly, Sebastopol, CA, 2nd edition, 1998.

[111] M. Dryja and O.B. Widlund. An additive variant of the Schwarz alternating method for the case of many subregions. Technical Report TR 339, Dept. of Comput. Sci., Courant Institute, NYU, 1987.

[112] M. Dryja and O.B. Widlund. Domain decomposition algorithms with small overlap. *SIAM J. Sci. Comput.*, 15:604–620, 1994.

[113] M. Dubiner. Spectral methods on triangles and other domains. *J. Sci. Comput.*, 6:345–390, 1991.

[114] Y. Dubois-Pèlerin, V. Van Kemenade, and M.O. Deville. An object-oriented toolbox for spectral element analysis. *J. Sci. Comput.*, 14:1–29, 1999.

[115] Y. Dubois-Pèlerin and P. Pegon. Object-oriented programming in nonlinear finite element analysis. *Comput. Structures*, 67:225–241, 1998.
[116] Y. Dubois-Pèlerin, T. Zimmermann, and P. Bomme. Object-oriented finite element programming concepts. In *New Advances in Computational Structural Mechanics*, P. Ladevèze and O. Zienkiewiczs, eds., pages 457–466. Elsevier, 1992.
[117] I. Duff, A.M. Erisman, and J.K. Reid. *Direct Methods for Sparse Matrices*. Clarendon Press, Oxford, 1989.
[118] B.L. Ehle. On Padé approximations to the exponential function and A-stable methods for the numerical solution of initial value problems. Technical Report CSRR 2010, Dept. of Applied Analysis and Computer Science, University of Waterloo, 1969.
[119] C. Evangelinos, S.J. Sherwin, and G.E Karniadakis. Parallel DNS algorithms on unstructured grids. *Comput. Methods Appl. Mech. Engrg.*, 184:401–425, 2000.
[120] C. Farhat and P.S. Chen. Tailoring domain decomposition methods for efficient parallel coarse grid solution and for systems with many right hand sides. *Contemp. Math.*, 180:401–406, 1994.
[121] C. Farhat and M. Lesoinne. Automatic partitioning of unstructured meshes for the parallel solution of problems in computational mechanics. *Int. J. Numer. Meth. Engrg.*, 36:745–764, 1993.
[122] J.H. Ferziger. Simulation of incompressible turbulent flows. *J. Comput. Phys.*, 69:1–48, 1987.
[123] B.A. Finlayson. *The Method of Weighted Residuals and Variational Principles*. Academic Press, New York, 1972.
[124] P.F. Fischer and H.M. Tufo. High performance spectral element algorithms and implementations. In D.E. Keyes, A. Ecer, N. Satofuka, P. Fox, and J. Périaux, editors, *Parallel Computational Fluid Dynamics. Towards Teraflops, Optimization and Novel Formulations*, pages 17–26. North Holland, Amsterdam, 2000.
[125] P.F. Fischer. Parallel domain decomposition for incompressible fluid dynamics. *Contemp. Math.*, 157:313–322, 1994.
[126] P.F. Fischer. Parallel multi-level solvers for spectral element methods. In A.V. Ilin and L.R. Scott, editors, *Third International Conference on Spectral and High Order Methods*, pages 595–604. Houston Journal of Mathematics, 1996.
[127] P.F. Fischer. An overlapping Schwarz method for spectral element solution of the incompressible Navier–Stokes equations. *J. Comput. Phys.*, 133:84–101, 1997.
[128] P.F. Fischer. Projection techniques for iterative solution of $A\underline{x} = \underline{b}$ with successive right-hand sides. *Comput. Methods Appl. Mech. Engrg.*, 163:193–204, 1998.
[129] P.F. Fischer, N.I. Miller, and H.M. Tufo. An overlapping Schwarz method for spectral element simulation of three-dimensional incompressible flows. In P. Bjørstad and M. Luskin, editors, *Parallel Solution of Partial Differential Equations*, pages 158–180. Springer, Berlin, 2000.
[130] P.F. Fischer and J.S. Mullen. Filter-based stabilization of spectral element methods. *C. R. Acad. Sci. Sér. I – Anal. Numér.*, 332:265–270, 2001.
[131] P.F. Fischer and A.T. Patera. Parallel spectral element of the Stokes problem. *J. Comput. Phys.*, 92:380–421, 1991.
[132] P.F. Fischer and E.M. Rønquist. Spectral element methods for large scale parallel Navier–Stokes calculations. *Comput. Methods Appl. Mech. Engrg.*, 116:69–76, 1994.
[133] C.A.J. Fletcher. *Computational Galerkin Methods*. Springer Series in Computational Physics. Springer, New York, 1984.

[134] C.A.J. Fletcher. *Computational Techniques for Fluid Dynamics I. Fundamental and General Techniques*. Springer Series in Computational Physics. Springer, New York, 1990.
[135] M.J. Flynn. Some computer organizations and their effectiveness. *IEEE Trans. Computers*, C-21:948–960, 1972.
[136] L. Formaggia and F. Nobile. A stability analysis for the arbitrary Lagrangian Eulerian formulation with finite elements. *East-West J. Numer. Math.*, 7:105–132, 1999.
[137] B. Fornberg. *A Practical Guide to Pseudospectral Methods*. Cambridge University Press, Cambridge, 1996.
[138] M. Fortin. Finite element solution of the Navier–Stokes equations. In A. Iserles, editor, *Acta Numerica 1993*, pages 239–284. Cambridge University Press, Cambridge, 1993.
[139] G.C. Fox, M.A. Johnson, G.A. Lyzenga, S.W. Otto, J.K. Salmon, and D.W. Walker. *Solving Problems on Concurrent Processors*. Prentice-Hall, Englewood Cliffs, NJ, 1988.
[140] P. Francken, M.O. Deville, and E.H. Mund. On the spectrum of the iteration operator associated to the finite element preconditioning of Chebyshev collocation calculations. *Comput. Methods Appl. Mech. Engrg.*, 80:295–304, 1990.
[141] D. Funaro. Pseudospectral approximation of a p.d.e. defined on a triangle. *Appl. Math. Comput.*, 42:121–138, 1991.
[142] D. Funaro. *Polynomial Approximation of Differential Equations*. Lecture Notes in Physics. Springer, Berlin, 1992.
[143] D. Funaro. A new scheme for the approximation of advection–diffusion equations by collocation. *SIAM J. Numer. Anal.*, 30:1664–1676, 1993.
[144] D. Funaro. *Spectral Elements for Transport-Dominated Equations*. Lecture Notes in Computational Science and Engineering. Springer, Berlin, 1997.
[145] G.P. Galdi. *An Introduction to the Mathematical Theory of the Navier–Stokes Equations. Volume I. Linearized Steady Problems*. Springer, New York, 1994.
[146] G.P. Galdi. *An Introduction to the Mathematical Theory of the Navier–Stokes Equations. Volume II. Nonlinear Steady Problems*. Springer, New York, 1994.
[147] G.P. Galdi. *An Introduction to the Mathematical Theory of the Navier–Stokes Equations. Volume III. Linearized Unsteady Problems*. Springer, New York, 1995.
[148] G.P. Galdi. *An Introduction to the Mathematical Theory of the Navier–Stokes Equations. Volume IV. Nonlinear Unsteady Problems*. Springer, New York, 1995.
[149] T.B. Gatski, C.E. Grosch, and M.E. Rose. The numerical solution of the Navier–Stokes equations for three-dimensional, unsteady, incompressible flows by compact schemes. *J. Comput. Phys.*, 82:298–329, 1989.
[150] S. Gavrilakis. Numerical simulation of low-Reynolds-number turbulent flow through a straight square duct. *J. Fluid Mech.*, 244:101–129, 1992.
[151] C.W. Gear. *Numerical Initial Value Problems in Ordinary Differential Equations*. Prentice-Hall, Englewood Cliffs, NJ, 1971.
[152] A. Geist, A. Beguelin, J. Dongarra, W. Jiang, R. Manchek, and V.S. Sunderam. *PVM: Parallel Virtual Machine: A Users' Guide and Tutorial for Networked Parallel Computing*. MIT Press, Cambridge, MA, 1994.
[153] A. George and J.W.H. Liu. *Computer Solution of Large Sparse Positive Definite Systems*. Prentice Hall, Englewood Cliffs, NJ, 1981.
[154] K. Gerdes and D. Schötzau. hp–finite element simulations for Stokes flow – stable and stabilized. *Finite Elem. Anal. Des.*, 33:143–165, 1999.
[155] M.I. Gerritsma and T.N. Phillips. Compatible spectral approximations for the velocity–pressure–stress formulation of the Stokes problem. *SIAM J. Sci. Comput.*, 20:1530–1550, 1998.

Bibliography

[156] M.I. Gerritsma and T.N. Phillips. Discontinuous spectral element approximations for the velocity–pressure–stress formulation of the Stokes problem. *Int. J. Numer. Meth. Engng.*, 43:1401–1419, 1998.

[157] V. Girault and P.A. Raviart. *Finite Element Methods for Navier–Stokes Equations.* Springer, Berlin, 1986.

[158] R. Glowinski and O. Pironneau. On mixed finite element approximation of the Stokes problem (I). *Numer. Math.*, 33:397–424, 1979.

[159] K. Goda. A multistep technique with implicit difference schemes for calculating two- or three-dimensional cavity flows. *J. Comput. Phys.*, 30:76–95, 1979.

[160] G. Golub and C.F. Van Loan. *Matrix Computations.* Johns Hopkins University Press, Baltimore, 1996.

[161] G. Golub and J.M. Ortega. *Scientific Computing: An Introduction with Parallel Computing.* Academic Press, San Diego, 1993.

[162] W.J. Gordon and C.A. Hall. Transfinite element methods: Blending-function interpolation over arbitrary curved element domains. *Numer. Math.*, 21:109–129, 1973.

[163] D. Gottlieb and S.A. Orszag. *Numerical Analysis of Spectral Methods: Theory and Applications.* SIAM-CBMS, Philadelphia, 1977.

[164] S. Gottlieb, C.-W. Shu, and E. Tadmor. Strong stability-preserving high-order time discretization methods. *SIAM Rev.*, 74:89–112, 2001.

[165] D.D. Gray and A. Giorgini. The validity of the Boussinesq approximation for liquids and gases. *Int. J. Heat Mass Transfer*, 19:545–551, 1976.

[166] P.M. Gresho. Incompressible fluid dynamics: Some fundamental formulation issues. *Annu. Rev. Fluid Mech.*, 23:413–453, 1991.

[167] P.M. Gresho and R.L. Sani. On pressure boundary conditions for the incompressible Navier–Stokes equations. *Int. J. Numer. Meth. Fluids*, 7:1111–1145, 1987.

[168] W.D. Gropp, D.K. Kaushik, D.E. Keyes, and B.F. Smith. Towards realistic performance bounds for implicit CFD codes. In D.E. Keyes, A. Ecer, N. Satofuka, P. Fox, and J. Périaux, editors, *Parallel Computational Fluid Dynamics. Towards Teraflops, Optimization and Novel Formulations*, pages 241–248. North Holland, Amsterdam, 2000.

[169] W.D. Gropp. Parallel computing and domain decomposition. In D.E Keyes, T.F. Chan, G. Meurant, J.S. Scroggs, and R.G. Voigt, editors, *Fifth International Symposium on Domain Decomposition Methods for Partial Differential Equations*, pages 349–361. SIAM, 1992.

[170] W.D. Gropp and D.E. Keyes. Domain decomposition with local mesh refinement. *SIAM J. Sci. Stat. Comput.*, 13:967–993, 1992.

[171] W.D. Gropp, E. Lusk, and A. Skjellum. *Using MPI: Portable Parallel Programming with the Message-Passing Interface*, MIT Press, Cambridge, MA, 2nd edition, 1999.

[172] W.D. Gropp, E. Lusk, and R. Thakur. *Using MPI-2: Advanced Features of the Message-Passing Interface.* MIT Press, Cambridge, MA, 1999.

[173] W. Gui and I. Babuška. The h-, p- and hp-versions of the finite element method in one dimension. Part I: The error analysis of the p-version. *Numer. Math.*, 49:577–612, 1986.

[174] W. Gui and I. Babuška. The h-, p- and hp-versions of the finite element method in one dimension. Part II: The error analysis of the h- and hp-versions. *Numer. Math.*, 49:613–657, 1986.

[175] W. Gui and I. Babuška. The h-, p- and hp-versions of the finite element method in one dimension. Part III: The adaptive hp-version. *Numer. Math.*, 49:659–683, 1986.

[176] M. Gunzburger. *Finite Element Methods for Viscous Incompressible Flows: A Guide to Theory, Practice and Algorithms.* Academic Press, Boston, 1989.
[177] B.-Y. Guo. *Spectral Methods and Their Applications.* World Scientific, London, 1998.
[178] A. Gupta. WGPP: Watson graph partitioning (and sparse matrix ordering) package. Technical Report RC 20453, IBM, T.J. Watson Research Center, Yorktown Heights, NY, 1996.
[179] M.E. Gurtin. *An Introduction to Continuum Mechanics.* Academic Press, San Diego, 1981.
[180] D.R. Gwynllyw, A.R. Davies, and T.N. Phillips. A moving spectral element approach to the dynamically loaded journal bearing problem. *J. Comput. Phys.*, 123:476–494, 1996.
[181] D. Haidvogel and T.A. Zang. The accurate solution of Poisson's equation by expansion in Chebyshev polynomials. *J. Comput. Phys.*, 30:167–180, 1979.
[182] P. Haldenwang, G. Labrosse, S. Abboudi, and M.O. Deville. Chebyshev 3-D spectral and 2-D pseudospectral solvers for the Helmholtz equation. *J. Comput. Phys.*, 55:115–128, 1984.
[183] G. Hamel. Spiralförmige bewegung zäher flüssigkeiten. *Jahresber. Dt. Math. Verein.*, 25:34–60, 1916.
[184] R.W. Hamming. *Numerical Methods for Scientists and Engineers.* McGraw Hill, New York, 1962.
[185] F.H. Harlow and J.E. Welsh. Numerical calculation of time-dependent viscous incompressible flow. *Phys. Fluids*, 8:2182–2189, 1965.
[186] M. Heath. The hypercube: A tutorial overview. In *Hypercube Multiprocessors 1986*, pages 1–7. SIAM, Philadelphia, 1986.
[187] W. Heinrichs. Line relaxation for spectral multigrid methods. *J. Comput. Phys.*, 77:166–182, 1988.
[188] W. Heinrichs. Multigrid methods for combined finite difference and Fourier problems. *J. Comput. Phys.*, 78:424–436, 1988.
[189] W. Heinrichs. A spectral multigrid method for the Stokes problems in streamfunction formulation. *J. Comput. Phys.*, 102:310–318, 1992.
[190] W. Heinrichs. Spectral multigrid methods for the reformulated Stokes equations. *J. Comput. Phys.*, 107:213–224, 1993.
[191] W. Heinrichs. Splitting techniques for the pseudospectral approximation of the unsteady Stokes equations. *SIAM J. Numer. Anal.*, 30:19–39, 1993.
[192] W. Heinrichs. High order time splitting for the Stokes equations. *J. Sci. Comput.*, 11:397–410, 1996.
[193] W. Heinrichs. Spectral collocation on triangular elements. *J. Comput. Phys.*, 145:743–757, 1998.
[194] W. Heinrichs. Splitting techniques for the unsteady Stokes equations. *SIAM J. Numer. Anal.*, 35:1646–1662, 1998.
[195] R. Henderson. Adaptive spectral element methods for turbulence and transition. In T.J. Barth and H. Deconinck, editors, *High-Order Methods for Computational Physics*, pages 225–324. Springer, Berlin, 1999.
[196] R.D. Henderson. Dynamic refinement algorithms for spectral element methods. *Comput. Methods Appl. Mech. Engrg.*, 175:395–411, 1999.
[197] R.D. Henderson and G.E. Karniadakis. Hybrid spectral-element–low-order methods for incompressible flows. *J. Sci. Comput.*, 6:79–100, 1991.
[198] R.D. Henderson and G.E. Karniadakis. Unstructured spectral element methods for simulation of turbulent flows. *J. Comput. Phys.*, 122:191–217, 1995.
[199] B. Hendrickson and R. Leland. The Chaco user's guide, version 1.0. Technical

Report TR SAND93-2339, Sandia National Laboratories, Albuquerque, NM, 1993.
[200] J.H. Hennesy and D.A. Patterson. *Computer Architecture: A Quantitative Approach*. Morgan Kaufmann, San Francisco, 1996.
[201] J. Hesthaven. From electrostatics to almost optimal nodal sets for polynomial interpolation in a simplex. *SIAM J. Numer. Anal.*, 35:655–676, 1998.
[202] J. Hesthaven and D. Gottlieb. Stable spectral methods for conservation laws on triangles with unstructured grids. *Comput. Methods Appl. Mech. Engrg.*, 175:361–381, 1999.
[203] J. Hesthaven and C.H. Teng. Stable spectral methods on tetrahedral elements. *SIAM J. Sci. Comput.*, 21:2352–2380, 2000.
[204] R.S. Hirsh. Higher order accurate difference solutions of fluid mechanics problems by a compact differencing scheme. *J. Comput. Phys.*, 19:90–109, 1975.
[205] C.W. Hirt, A.A. Amsden, and J.L. Cook. An arbitrary Lagrangian–Eulerian computing method for all flow speeds. *J. Comput. Phys.*, 14:227–253, 1974.
[206] L.W. Ho, Y. Maday, A.T. Patera, and E. Rønquist. A high-order Lagrangian-decoupling method for the incompressible Navier–Stokes equations. *Comput. Methods Appl. Mech. Engrg.*, 80:65–90, 1990.
[207] L.W. Ho and A.T. Patera. A Legendre spectral element method for simulation of unsteady incompressible viscous free-surface flows. *Comput. Methods Appl. Mech. Engrg.*, 80:355–366, 1990.
[208] L.W. Ho and A.T. Patera. Variational formulation of three-dimensional viscous free-surface flows: Natural imposition of surface tension boundary conditions. *Int. J. Numer. Methods Fluids*, 13:691–698, 1991.
[209] R.W. Hockney and C.R. Jesshope. *Parallel Computers: Architecture, Programming and Algorithms*. A. Hilger, Bristol, 1981.
[210] L.C. Hsu and C. Mavriplis. Adaptive meshes for the spectral element method. In P. Bjørstad, M. Espedal, and D. Keyes, editors, *Domain Decomposition 9 Proceedings*, pages 374–381. Wiley, New York, 1998.
[211] T.J.R. Hughes, W.K. Liu, and T.K. Zimmermann. Lagrangian–Eulerian finite element formulation for incompressible viscous flows. *Comput. Methods Appl. Mech. Engrg.*, 29:329–349, 1981.
[212] S. Jensen and S. Zhang. The p and h-p versions of some finite element methods for Stokes problem. *Comput. Methods Appl. Mech. Engrg.*, 116:147–155, 1994.
[213] J. Jeong and F. Hussain. On the identification of a vortex. *J. Fluid Mech.*, 285:69–94, 1995.
[214] C. Johnson. *Numerical Solution of Partial Differential Equations by the Finite Element Method*. Cambridge University Press, Cambridge, 1992.
[215] C. Johnson and J. Saranen. Streamline diffusion methods for the incompressible Euler and Navier–Stokes equations. *Math. Comput.*, 47:1–18, 1986.
[216] D.D. Joseph. *Fluid Dynamics of Viscoelastic Liquids*. Springer, New York, 1990.
[217] G.-S. Karamanos and S.J. Sherwin. A high-order splitting scheme for the Navier–Stokes equations with variable viscosity. *Appl. Numer. Math.*, 33:455–462, 2000.
[218] G.E. Karniadakis. Toward a numerical error bar in CFD. *J. Fluids Eng.*, 117:7–9, 1995.
[219] G.E. Karniadakis, M. Israeli, and S.A. Orszag. High-order splitting methods for the incompressible Navier–Stokes equations. *J. Comput. Phys.*, 97:414–443, 1991.
[220] G.E. Karniadakis and S.A. Orszag. Nodes, modes and flow codes. *Phys. Today*, 46:34–42, 1993.

[221] G.E. Karniadakis and S.J. Sherwin. *Spectral/hp Element Methods for CFD.* Oxford University Press, Oxford, 1999.
[222] G. Karypis and V. Kumar. Metis: Unstructured graph partitioning and sparse matrix ordering system, version 2.0. Technical Report, Dept. of Computer Science, University of Minnesota, 1995.
[223] J. Kim and P. Moin. Applications of a fractional step method to incompressible Navier–Stokes equations. *J. Comput. Phys.*, 59:308–323, 1985.
[224] S.D. Kim and S.V. Parter. Preconditioning Chebyshev spectral collocation method for elliptic partial differential equations. *SIAM J. Numer. Anal.*, 33:2375–2400, 1996.
[225] S.D. Kim and S.V. Parter. Preconditioning Chebyshev spectral collocation by finite-difference operators. *SIAM J. Numer. Anal.*, 34:939–958, 1997.
[226] L. Kleiser and U. Schumann. Treatment of incompressibility and boundary conditions in 3-D numerical spectral simulations of plane channel flows. In E.H. Hirschel, editor, *Proceedings of the Third GAMM Conference on Numerical Methods in Fluid Mechanics*, pages 165–173. Vieweg, Braunschweig, 1980.
[227] L.I.G. Kovasznay. Laminar flow behind a two-dimensional grid. *Proc. Cambridge Philos. Soc.*, 44:58–62, 1948.
[228] E. Kreyszig. *Introductory Functional Analysis with Applications.* Wiley, New York, 1978.
[229] G. Kruse. *Parallel Nonconforming Spectral Element Solution of the Incompressible Navier–Stokes Equations in Three Dimensions.* PhD thesis, Division of Applied Mathematics, Brown University, 1997.
[230] H.C. Ku. Solution of flow in complex geometries by the pseudospectral element method. *J. Comput. Phys.*, 117:215–227, 1995.
[231] H.C. Ku, R.S. Hirsh, and T.D. Taylor. A pseudospectral method for solution of three-dimensional incompressible Navier–Stokes equations. *J. Comput. Phys.*, 70:439–462, 1987.
[232] H.C. Ku, R.S. Hirsh, T.D. Taylor, and A. Rosenberg. A pseudospectral matrix element method for solution of three-dimensional incompressible flows and its parallel implementation. *J. Comput. Phys.*, 83:260–291, 1989.
[233] H.C. Ku, T.D. Taylor, and R.S. Hirsh. Pseudospectral matrix element methods for flow in complex geometries. *Comput. Methods Appl. Mech. Engrg.*, 80:381–388, 1990.
[234] G. Labrosse. Compatibility conditions for the Stokes system discretized in 2-D Cartesian domains. *Comput. Methods Appl. Mech. Engrg.*, 106:353–365, 1993.
[235] B. Lafaurie, C. Nardone, R. Scardovelli, S. Zaleski, and G. Zanetti. Modelling, merging and fragmentation in multiphase flows with SURFER. *J. Comput. Phys.*, 113:134–147, 1994.
[236] B.E. Launder. Second-moment closure: present... and future? *Int. J. Heat and Fluid Flow*, 10:282–300, 1989.
[237] P. Le Quéré. Mono and multidomain Chebyshev algorithm on a staggered grid. In T.J. Chung and G.R. Karr, editors, *Proceedings of the 7th International Conference on Finite Element Methods in Flow Problems.* UAH Press, Huntsville, AL, 1989.
[238] P. Le Quéré and T. Alziary de Roquefort. Computation of natural convection in two-dimensional cavities with Chebyshev polynomials. *J. Comput. Phys.*, 57:210–228, 1985.
[239] E. Leriche. *Direct Numerical Simulation of a Lid-Driven Cavity Flow by a Chebyshev Spectral Method.* PhD thesis, Swiss Federal Institute of Technology, Lausanne, 1998.

[240] E. Leriche and S. Gavrilakis. Direct numerical simulation of the flow in a lid-driven cubical cavity. *Phys. Fluids*, 12:1363–1376, 2000.
[241] E. Leriche, S. Gavrilakis, and G. Labrosse. Direct numerical simulation of lid-driven cavity flow within a 3D inhomogeneous domain on an NEC-SX4 supercomputer. In M. Ingber, H. Power, and C.A. Brebbia, editors, *Proceedings of the 6th International Conference on Applications of High Performance Computing in Engineering 2000 (HPC 2000)*, pages 359–369. Wit Press, Ashurst, Southampton, 2000.
[242] E. Leriche and G. Labrosse. High-order direct Stokes solvers with or without temporal splitting: Numerical investigations of their comparative properties. *SIAM J. Sci. Comput.*, 22:1386–1410, 2001.
[243] M. Lesieur. *Turbulence in Fluids*. Kluwer, Dordrecht, 1990.
[244] R.J. LeVeque. *Numerical Methods for Conservation Laws*. Lectures in Mathematics 1, ETH Zürich. Birkhäuser, Basel, 1992.
[245] J.G. Levin, M. Iskandarani, and D.B. Haidvogel. A spectral filtering procedure for eddy-resolving simulations with a spectral element ocean model. *J. Comput. Phys.*, 137:130–154, 1997.
[246] X.K. Li, D.R.H. Gwynllyw, A.R. Davies, and T.N. Phillips. Three-dimensional effects in dynamically loaded journal bearings. *Int. J. Numer. Meth. Fluids*, 29:311–341, 1999.
[247] F. Loth, N. Arslan, P.F. Fischer, C.D. Bertram, S.E. Lee, T.J. Royston, R.H. Song, W.E. Shaalan, and H.S. Bassiouny. Transitional flow at the venous anastomosis of an arteriovenous graft: Potential relationship with activation of the ERK 1/2 mechanotransduction pathway. *ASME J. Biomech Engr.*, (accepted).
[248] D.B. Loveman. High performance Fortran. *IEEE Parallel and Distributed Technology*, 1:25–42, 1993.
[249] J.L. Lumley. Transition and turbulence. In R. Meyer, editor, *Coherent Structures in Turbulence*, pages 215–242. Academic Press, New York, 1981.
[250] R.E. Lynch, J.R. Rice, and D.H. Thomas. Direct solution of partial difference equations by tensor product methods. *Numer. Math.*, 6:185–199, 1964.
[251] L. Machiels. *Simulation and Theory of Randomly Forced Turbulence*. PhD thesis, Swiss Federal Institute of Technology-Lausanne, 1997.
[252] L. Machiels and M.O. Deville. Fortran 90: An entry to object-oriented programming for solution of partial differential equation. *ACM Trans. Math. Softw.*, 23:32–49, 1997.
[253] L. Machiels and M.O. Deville. Numerical simulation of randomly forced turbulent flows. *J. Comput. Phys.*, 145:246–279, 1998.
[254] Y. Maday, S. Ould Kaber, and E. Tadmor. Legendre pseudospectral viscosity method for nonlinear conservation laws. *SIAM J. Numer. Anal.*, 30:321–342, 1993.
[255] Y. Maday, C. Mavriplis, and A.T. Patera. Nonconforming mortar element methods: Application to spectral discretizations. In T.F. Chan, R. Glowinski, J. Périaux, and O.B. Widlund, editors, *Domain Decomposition Methods*, Chapter 31, pages 392–418. SIAM, 1989.
[256] Y. Maday, D. Meiron, A.T. Patera, and E.M. Rønquist. Analysis of iterative methods for the steady and unsteady Stokes problem: Application to spectral element discretizations. *SIAM J. Sci. Comput.*, 14:310–337, 1993.
[257] Y. Maday and R. Muñoz. Spectral element multigrid: Numerical analysis. *J. Sci. Comput.*, 3:323–354, 1988.
[258] Y. Maday and R. Muñoz. Numerical analysis of a multigrid method for spectral approximations. In M. Y. Hussaini, D. L. Dwoyer, and R. G. Voigt, editors, *Lecture*

Notes in Physics, Volume 323: Proceedings of the 11th International Conference on Numerical Methods in Fluid Dynamics, pages 389–394. Springer, 1989.

[259] Y. Maday, R. Muñoz, A.T. Patera, and E.M. Rønquist. Spectral element multigrid methods. In P. de Groen and R. Beauwens, editors, *Proceedings of the IMACS International Symposium on Iterative Methods in Linear Algebra, Brussels, 1991*, pages 191–201. Elsevier, Amsterdam, 1992.

[260] Y. Maday, A.T. Patera, and E.M. Rønquist. A well-posed optimal spectral element approximation for the Stokes problem. Technical Report 87–48, ICASE, Hampton, VA, 1987.

[261] Y. Maday and A.T. Patera. Spectral element methods for the Navier–Stokes equations. In A.K. Noor and J.T. Oden, editors, *State-of-the-Art Surveys in Computational Mechanics*, pages 71–143. ASME, New York, 1989.

[262] Y. Maday, A.T. Patera, and E.M. Rønquist. An operator-integration-factor splitting method for time-dependent problems: Application to incompressible fluid flow. *J. Sci. Comput.*, 5:263–292, 1990.

[263] Y. Maday, A.T. Patera, and E.M. Rønquist. The $\mathbb{P}_N \times \mathbb{P}_{N-2}$ method for the approximation of the Stokes problem. Technical Report 92009, Department of Mechanical Engineering, MIT, Cambridge, MA, 1992.

[264] Y. Maday and E.M. Rønquist. Optimal error analysis of spectral methods with emphasis on non-constant coefficients and deformed geometries. *Comput. Methods Appl. Mech. Engrg.*, 80:91–115, 1990.

[265] J.F. Maitre and O. Pourquier. Condition number and diagonal preconditioning: Comparison of the p-version and the spectral element methods. *Numer. Math.*, 74:69–84, 1996.

[266] P.S. Marcus. Jupiter's great red spot and other vortices. *Annu. Rev. Astron. Astrophys.*, 31:523–573, 1993.

[267] C. Mavriplis. *Nonconforming Discretizations and a posteriori Error Estimations for Adaptive Spectral Element Techniques*. PhD thesis, Massachusetts Institute of Technology, Cambridge, MA, 1989.

[268] C. Mavriplis. A posteriori error estimates for adaptive spectral element techniques. In P. Wesseling, editor, *Notes on Numerical Methods in Fluid Mechanics*, Volume 29, pages 333–342, Vieweg, Braunschweig, 1990.

[269] C. Mavriplis. Adaptive mesh strategies for the spectral element method. *Comput. Methods Appl. Mech. Engrg.*, 116:77–86, 1994.

[270] C. Mavriplis and J. Van Rosendale. Triangular spectral elements for incompressible fluid flow. Technical Report 3346-CP, AIAA, 1993.

[271] O.A. McBrian, P.O. Frederickson, J. Linden, A. Schüller, K. Solchenbach, K. Stüben, C.-A. Thole, and U. Trottenberg. Multigrid methods on parallel computers – a survey of recent developments. *Impact Comput. Sci. Engrg.*, 3:1–75, 1991.

[272] M.R. Mehrabi and R.A. Brown. Parallel implementation of finite-element/Newton method for solution of steady-state and transient nonlinear partial differential equations. *J. Sci. Comput.*, 10:93–137, 1995.

[273] B. Mercier. *An Introduction to the Numerical Analysis of Spectral Methods*. Springer, Berlin, 1989.

[274] P. Mercier and M.O. Deville. A multidimensional compact higher order scheme for 3-D Poisson's equation. *J. Comput. Phys.*, 39:443–455, 1981.

[275] B. Métivet. *Résolution spectrale des équations de Navier–Stokes par une méthode de sous-domaines courbes*. PhD thesis, Université Pierre et Marie Curie, 1987.

[276] S.G. Mikhlin. *Mathematical Physics, an Advanced Course*. North-Holland, Amsterdam, 1970.

[277] H.K. Moffatt. Viscous and resistive eddies near a sharp corner. *J. Fluid Mech.*, 18:1–18, 1964.

[278] G. Moore. Cramming more components onto integrated circuits. *Electron. Mag.*, 38:114–117, 1965.
[279] G. Moore. Lithography and the future of Moore's law. In *Optical/Laser Microlithography VIII: Proceedings of the SPIE*, Volume 2240, pages 2–17, 1995.
[280] Y. Morchoisne. Resolution of Navier–Stokes equations by a space–time pseudospectral method. *Rech. Aérospat.*, 5:293–306, 1979.
[281] R.D. Moser, P. Moin, and A. Leonard. A spectral numerical method for the Navier–Stokes equations with applications to Taylor–Couette flow. *J. Comput. Phys.*, 52:524–544, 1983.
[282] E.H. Mund. A short survey on preconditioning techniques in spectral calculations. *Appl. Numer. Math.*, 33:61–70, 2000.
[283] R. Muñoz. Theoretical analysis of some spectral multigrid methods. *Comput. Methods Appl. Mech. Engrg.*, 80:287–294, 1990.
[284] R.A. Nicolaides. Deflation of conjugate gradients with application to boundary value problems. *SIAM J. Numer. Anal.*, 24:355–365, 1987.
[285] J. Nitsche. Ein Kriterium für die Quasi-optimalität des Ritzschen Verfahrens. *Numer. Math.*, 11:346–348, 1968.
[286] J.T. Oden and J.N. Reddy. *An Introduction to the Mathematical Theory of Finite Elements*. Wiley, New York, 1976.
[287] J.T. Oden, W.H. Wu, and V. Legat. An *hp* adaptive strategy for finite element approximation of the Navier–Stokes equations. *Int. J. Numer. Meth. Fluids*, 20:831–851, 1995.
[288] S.A. Orszag. Spectral methods for problems in complex geometry. *J. Comput. Phys.*, 37:70–92, 1980.
[289] S.A. Orszag, M. Israeli, and M.O. Deville. Boundary conditions for incompressible flows. *J. Sci. Comput.*, 1:75–111, 1986.
[290] R.G. Owens. Spectral approximations on the triangle. *Proc. R. Soc. London A*, 454:857–872, 1998.
[291] R.G. Owens and T.N. Phillips. Mass- and momentum-conserving spectral methods for Stokes flow. *J. Comput. Appl. Math.*, 53:185–206, 1994.
[292] S.S. Pahl. *Schwarz Type Domain Decomposition Methods for Spectral Element Discretizations*. Univ. of Witwatersrand, Johannesburg, South Africa, 1993. Master's Thesis, Dept. of Computational and Applied Math.
[293] R.L. Panton. *Incompressible Flow*. Wiley, New York, 1984.
[294] S.V. Parter and E.E. Rothman. Preconditioning Legendre spectral collocation approximations to elliptic problems. *SIAM J. Numer. Anal.*, 32:333–385, 1995.
[295] F. Pasquarelli. Domain decomposition for spectral approximation to Stokes equations via divergence-free functions. *Appl. Numer. Math.*, 8:493–514, 1991.
[296] F. Pasquarelli, A. Quarteroni, and G. Sacchi-Landriani. Spectral approximations of the Stokes problem by divergence-free functions. *J. Sci. Comput.*, 2:195–226, 1987.
[297] A.T. Patera. A spectral element method for fluid dynamics: Laminar flow in a channel expansion. *J. Comput. Phys.*, 54:468–488, 1984.
[298] A.T. Patera. Fast direct (Poisson) solvers for high order finite element discretizations in rectangularly decomposable domains. *J. Comput. Phys.*, 65:474–480, 1986.
[299] D. Pathria and G.E. Karniadakis. Spectral element methods for elliptic problems in nonsmooth domains. *J. Comput. Phys.*, 122:83–95, 1995.
[300] L.F. Pavarino and T. Warburton. Overlapping Schwarz methods for unstructured spectral elements. *J. Comput. Phys.*, 160:298–317, 2000.
[301] L.F. Pavarino and O.B. Widlund. A polylogarithmic bound for an iterative

substructuring method for spectral elements in three dimensions. *SIAM J. Numer. Anal.*, 33:1303–1335, 1996.
[302] L.F. Pavarino and O.B. Widlund. Balancing Neumann–Neumann methods for incompressible Stokes equations. Technical Report TR2001-813, Computer Science Dept., Courant Institute of Mathematics, 2001.
[303] D.W. Peaceman and H.H. Rachford. The numerical solution of parabolic and elliptic differential equations. *J. Soc. Indust. Appl. Math.*, 3:28–41, 1955.
[304] R.B. Pelz. The parallel Fourier pseudospectral method. *J. Comput. Phys.*, 92:296–312, 1991.
[305] J.B. Perot. An analysis of the fractional step method. *J. Comput. Phys.*, 108:51–58, 1993.
[306] J.B. Perot. Comments on the fractional step method. *J. Comput. Phys.*, 121:190–191, 1995.
[307] T.N. Phillips and G.W. Roberts. The treatment of spurious pressure modes in spectral incompressible flow calculations. *J. Comput. Phys.*, 105:150–164, 1993.
[308] A. Pinelli. *Preconditioned Parallel Algorithms and Spectral Solutions for the Incompressible Navier–Stokes Equations*. PhD thesis, Swiss Federal Institute of Technology, Lausanne, 1995.
[309] A. Pinelli, C. Benocci, and M.O. Deville. A Chebyshev collocation algorithm for the solution of advection–diffusion equations. *Comput. Methods Appl. Mech. Engrg.*, 116:201–210, 1994.
[310] A. Pinelli, C. Benocci, and M.O. Deville. Chebyshev pseudo-spectral solution of advection–diffusion equations with mapped finite difference preconditioning. *J. Comput. Phys.*, 112:1–11, 1994.
[311] A. Pinelli, W. Couzy, M.O. Deville, and C. Benocci. An efficient iterative solution method for the Chebyshev collocation of advection-dominated transport problems. *SIAM J. Sci. Comput.*, 17:647–657, 1996.
[312] U. Piomelli. Applications of large eddy simulation in engineering: An overview. In B. Galperin and S.A. Orszag, editors, *Large Eddy Simulation of Complex Engineering and Physical Flows*. Cambridge University Press, Cambridge, 1993.
[313] S. Pissanetsky. *Sparse Matrix Technology*. Academic Press, New York, 1984.
[314] A. Pothen, H.D. Simon, and K.P. Liou. Partitioning sparse matrices with eigenvectors of graphs. *SIAM J. Matrix Anal. Appl.*, 11:430–452, 1990.
[315] W.E. Pracht. Calculating three-dimensional fluid flows at all speeds with an Eulerian–Lagrangian computing mesh. *J. Comput. Phys.*, 17:132–159, 1975.
[316] W.H. Press, S.A. Teukolsky, W.T. Vetterling, and B.P. Flannery. *Numerical Recipes – The Art of Scientific Computing*. Cambridge University Press, Cambridge, New York, 1992.
[317] G. Puppo. Bubble stabilization of spectral methods: The multidimensional case. *J. Sci. Comput.*, 13:115–149, 1998.
[318] A. Quarteroni, F. Saleri, and A. Veneziani. Factorization methods for the numerical approximation of Navier–Stokes equations. *Comput. Methods Appl. Mech. Engrg.*, 188:505–526, 2000.
[319] A. Quarteroni and A. Valli. *Numerical Approximation of Partial Differential Equations*. Springer Series in Computational Mathematics. Springer, Berlin, 1994.
[320] A. Quarteroni and A. Valli. *Domain Decomposition Methods for Partial Differential Equations*. Oxford University Press, Oxford, 1999.
[321] A. Quarteroni and E. Zampieri. Finite element preconditioning for Legendre spectral collocation approximations to elliptic equations and systems. *SIAM J. Numer. Anal.*, 29:917–936, 1992.

[322] B. Ramaswamy and M. Kawahara. Arbitrary Lagrangian–Eulerian finite element method for unsteady, convective, incompressible viscous free surface fluid flow. In R.H. Gallagher, R. Glowinski, P.M. Gresho, J.T. Oden, and O.C. Zienkiewicz, editors, *Finite Elements in Fluids*, Volume 7, Chapter 4, pages 65–87. Wiley, 1987.

[323] S.C. Reddy and L.N. Trefethen. Pseudospectra of the convection–diffusion operator. *SIAM J. Appl. Math.*, 54:1634–1649, 1994.

[324] W.C. Reynolds. Fundamentals of turbulence for turbulence modelling and simulation. In *Lecture Notes for von Karman Institute*, AGARD Lect. Ser. No. 86, pages 1–66. NATO, New York, 1987.

[325] W.C. Reynolds. The potential and limitations of direct and large eddy simulations. In J.L. Lumley, editor, *Proceedings of Whither Turbulence Workshop, Ithaca, N.Y.*, pages 313–342. Springer, Berlin, 1990.

[326] R.D. Richtmyer. *Principles of Advanced Mathematical Physics*. Texts and Monographs in Physics, Volume 1. Springer, Berlin, 1978.

[327] R.D. Richtmyer and K.W. Morton. *Difference Methods for Initial-Value Problems*. Interscience, New York, 1967.

[328] P.J. Roache. *Computational Fluid Dynamics*. Hermosa, Albuquerque, NM, 1982.

[329] I. Robertson and S. Sherwin. Free-surface flow simulation using hp/spectral elements. *J. Comput. Phys.*, 155:26–53, 1999.

[330] R.S. Rogallo and P. Moin. Numerical simulation of turbulent flows. *Annu. Rev. Fluid Mech.*, 16:99–137, 1984.

[331] E. Rønquist. *Optimal Spectral Element Methods for the Unsteady Three-Dimensional Incompressible Navier–Stokes Equations*. PhD thesis, Massachusetts Institute of Technology, Cambridge, MA, 1988.

[332] E.M. Rønquist. Spectral element methods for the unsteady Navier–Stokes equations. In *Computational Fluid Dynamics Lecture Series 1991-01*. von Karman Institute for Fluid Dynamics, Rhode St-Genèse, Belgium, 1991.

[333] E.M. Rønquist. A domain decomposition method for elliptic boundary value problems: Application to unsteady incompressible fluid flow. In D.E. Keyes, T.F. Chan, G. Meurant, J.S. Scroggs, and R.G.Voigt, editors, *Fifth Conference on Domain Decomposition Methods for Partial Equations*, pages 545–557. SIAM, 1992.

[334] E.M. Rønquist. Convection treatment using spectral elements of different order. *Int. J. Numer. Meth. Fluids*, 22:241–264, 1996.

[335] E.M. Rønquist. A domain decomposition solver for three-dimensional steady free surface flows. In P. Bjørstad, M. Espedal, and D. Keyes, editors, *Domain Decomposition 9 Proceedings*, pages 792–801. Wiley, New York, 1998.

[336] E.M. Rønquist and A.T. Patera. A Legendre spectral element method for the Stefan problem. *Int. J. Numer. Meth. Engrg.*, 24:2273–2299, 1987.

[337] E.M. Rønquist and A.T. Patera. Spectral element multigrid. I. Formulation and numerical results. *J. Sci. Comput.*, 2:389–406, 1987.

[338] H.H. Rosenbrock. Some general implicit processes for the numerical solution of differential equations. *Comput. J.*, 5:329–330, 1963.

[339] Y. Saad. *Iterative Methods for Sparse Linear Systems*. The PWS series in Computer Science. PWS, Boston, 1996.

[340] Y. Saad and M.H. Schultz. GMRES: A generalized minimal residual algorithm for solving nonsymmetric linear systems. *SIAM J. Sci. Stat. Comput.*, 7:856–869, 1986.

[341] P. Sagaut. *Large Eddy Simulation for Incompressible Flows: An Introduction*. Springer, Berlin, 2001.

[342] R. Sani, P. Gresho, R. Lee, D. Griffiths, and M. Engleman. The cause and cure (?) of the spurious pressures generated by certain FEM solutions of the incompressible Navier–Stokes equations. *Int. J. Numer. Meth. Fluids*, 1:17–43 and 171–204, 1981.

[343] R. Scardovelli and S. Zaleski. Direct numerical simulation of free-surface and interfacial flow. *Annu. Rev. Fluid Mech.*, 31:567–603, 1999.

[344] H. Schlichting. *Boundary Layer Theory*. McGraw-Hill, New York, 1950.

[345] C. Schneidesch and M. Deville. Chebyshev collocation method and multi-domain decomposition for Navier–Stokes equations in complex curved geometries. *J. Comput. Phys.*, 106:234–257, 1993.

[346] C. Schneidesch and M. Deville. Multidomain decomposition of curved geometries in the Chebyshev collocation method for thermal problems. *Comput. Methods Appl. Mech. Engrg.*, 116:87–94, 1994.

[347] M.R. Schumack, W.W. Schultz, and J.P. Boyd. Spectral method solution of the Stokes equations on nonstaggered grids. *J. Comput. Phys.*, 94:30–58, 1991.

[348] C. Schwab. *p and hp-FEM*. Oxford University Press, Oxford, 1998.

[349] C. Schwab. HP-FEM for fluid flow simulation. In T.J. Barth and H. Deconinck, editors, *High-Order Methods for Computational Physics*, pages 325–438. Springer, Berlin, 1999.

[350] C. Schwab and M. Suri. Mixed HP finite element methods for Stokes and non-Newtonian flow. *Comput. Methods Appl. Mech. Engrg.*, 175:217–241, 1999.

[351] L. Schwartz. *Théorie des Distributions*. Hermann, Paris, 1950.

[352] L.E. Scriven and C.V. Sternling. On cellular convection driven by surface-tension gradients: Effects of mean surface tension and surface viscosity. *J. Fluid Mech.*, 19:321–340, 1964.

[353] J.A. Sethian. *Level Set Methods and Fast Marching Methods*. Cambridge University Press, Cambridge, 1999.

[354] S. Shaanan and J.H. Ferziger. A direct method of solution of Poisson equation accurate to fourth order. In U. Schumann, editor, *Fast Elliptic Solvers*, Advance Publication, London, U.K., 1978.

[355] J.E. Shen. On fast direct Poisson solver, inf–sup constant and iterative Stokes solver by Legendre–Galerkin method. *J. Comput. Phys.*, 116:184–188, 1995.

[356] M.S. Shephard, J.E. Flaherty, H. L. DeCougny, C. Ozturan, C. L. Bottasso, and M. Beall. Parallel automated adaptive procedures for unstructured meshes. Technical Report 11, AGARD, 1995.

[357] S.J. Sherwin and G.E Karniadakis. A new triangular and tetrahedral basis for high-order (HP) finite element methods. *Int. J. Numer. Meth. Engrg.*, 38:3775–3802, 1995.

[358] S.J. Sherwin and G.E Karniadakis. A triangular spectral element method; applications to the incompressible Navier–Stokes equations. *Comput. Methods Appl. Mech. Engrg.*, 123:189–229, 1995.

[359] S.J. Sherwin and G.E. Karniadakis. Tetrahedral hp finite elements: Algorithms and flow simulations. *J. Comput. Phys.*, 124:14–45, 1996.

[360] S.J. Sherwin, O. Shah, D.J. Doorly, J. Peiró, Y. Papapharilaou, N. Watkins, C.G. Caro, and C.L. Demoulin. The influence of out-of-plane geometry on the flow within a distal end-to-side anastomosis. *ASME J. Biomech. Engrg.*, 122:1–10, 2000.

[361] R.E. Showalter. *Hilbert Space Methods for Partial Differential Equations*. Pitman, San Francisco, 1979.

[362] J.S. Smagorinsky. General circulation experiments with the primitive equations. *Mon. Weather Rev.*, 91:99–164, 1963.

[363] B. Smith, P. Bjørstad, and W. Gropp. *Domain Decomposition: Parallel Multilevel Methods for Elliptic PDEs.* Cambridge University Press, Cambridge, 1996.
[364] C.G. Speziale. On nonlinear $k-l$ and $k-\varepsilon$ models of turbulence. *J. Fluid Mech.*, 178:459–475, 1987.
[365] C.G. Speziale. Analytical methods for the development of Reynolds-stress closures in turbulence. *Annu. Rev. Fluid Mech.*, 23:107–157, 1991.
[366] G. Stellner, A. Bode, S. Lamberts, and T. Ludwig. NXLib – a parallel programming environment for workstation clusters. In C. Halatsis, D.G. Maritsas, G. Philokyprou, and S. Theodoridis, editors, *Lecture Notes in Computer Science, Volume 817: Proceedings of the 6th International Parallel Architectures and Languages Europe (PARLE) Conference*, pages 745–748. Springer, 1994.
[367] G. Strang. *Linear Algebra and Its Applications.* Saunders, 3rd edition, 1988.
[368] G. Strang. *Introduction to Applied Mathematics.* Wellesley Cambridge Press, Wellesley, MA, 1986.
[369] G. Strang and G.J. Fix. *An Analysis of the Finite Element Method.* Prentice-Hall Series in Automatic Computation. Prentice-Hall, Englewood Cliffs, NJ, 1973.
[370] B. Szabó and I. Babuška. *Finite Element Analysis.* Wiley, New York, 1991.
[371] G. Szegö. *Orthogonal Polynomials.* AMS, Providence, RI, 3rd edition, 1967.
[372] M.A. Taylor and B.A. Wingate. A generalized diagonal mass matrix spectral element method for non-quadrilateral elements. *Appl. Numer. Math.*, 33:259–265, 2000.
[373] M.A. Taylor, B.A. Wingate, and R.E. Vincent. An algorithm for computing Fekete points in the triangle. *SIAM J. Numer. Anal.*, 38:1707–1720, 2000.
[374] R. Témam. Sur l'approximation de la solution de Navier–Stokes par la méthode des pas fractionnaires. I,II. *Arch. Rat. Mech. Anal.*, 32:135–153, 377–385, 1969.
[375] R. Témam. *Navier–Stokes Equations. Theory and Numerical Analysis.* North-Holland, Amsterdam, 1977.
[376] R. Témam. *Navier–Stokes Equations and Nonlinear Functional Analysis.* SIAM CBMS-NSF Regional Conference Series in Applied Mathematics. Philadelphia, 1983.
[377] H. Tennekes and J.L. Lumley. *A First Course in Turbulence.* MIT Press, Cambridge, MA, 1972.
[378] T. Theodorsen. Mechanism of turbulence. In *Proceedings of the 2nd Midwestern Conference on Fluid Mechanics, Ohio State University*, page 1. Columbus, OH, 1952.
[379] L. Timmermans. *Analysis of Spectral Element Methods with Application to Incompressible Flow.* PhD thesis, Eindhoven University of Technology, Eindhoven, 1994.
[380] L.J.P. Timmermans, F.N. Van de Vosse, and P.D. Minev. Taylor–Galerkin based spectral element methods for convection–diffusion problems. *Int. J. Numer. Meth. Fluids*, 18:853–870, 1994.
[381] L.J.P. Timmermans, P.D. Minev, and F.N. Van de Vosse. An approximate projection scheme for incompressible flow using spectral elements. *Int. J. Numer. Meth. Fluids*, 22:673–688, 1996.
[382] A.G. Tomboulides, M. Israeli, and G.E. Karniadakis. Efficient removal of boundary-divergence errors in time-splitting methods. *J. Sci. Comput.*, 4:291–308, 1989.
[383] L.N. Trefethen. *Spectral Methods in Matlab.* SIAM, Philadelphia, 2000.
[384] L.N. Trefethen and D. Bau III. *Numerical Linear Algebra.* SIAM, Philadelphia, 1997.
[385] C. Truesdell. *Rational Thermodynamics.* McGraw-Hill, New York, 1969.

[386] C. Truesdell and K.R. Rajagopal. *An Introduction to the Mechanics of Fluids.* Birkhäuser, Boston, 2000.
[387] C. Truesdell and R. Toupin. The classical field theories. In S. Fluegge, editor, *Handbuch der Physik, III–1*, pages 226–858. Springer, Berlin, 1960.
[388] C. Tu, M.O. Deville, L. Dheur, and L. Vanderschuren. Finite element simulation of pulsatile flow through arterial stenosis. *J. Biomech.*, 25:1141–1152, 1992.
[389] H.M. Tufo. *Algorithms for Large-Scale Parallel Simulation of Unsteady Incompressible Flows in Three-Dimensional Complex Geometries.* PhD thesis, Division of Applied Mathematics, Brown University, 1998.
[390] H.M. Tufo and P.F. Fischer. Terascale spectral element algorithms and implementations. In *Proceedings of the ACM/IEEE SC99 Conference on High Performance Networking and Computing*, CD-ROM, IEEE Computer Soc., 1999.
[391] H.M. Tufo and P.F. Fischer. Fast parallel direct solvers for coarse-grid problems. *J. Parallel Distrib. Comput.*, 61:151–177, 2001.
[392] H.M. Tufo, P.F. Fischer, M.E. Papka, and K. Blom. Numerical simulation and immersive visualization of hairpin vortices. In *Proceedings of the ACM/IEEE SC99 Conference on High Performance Networking and Computing*, CD-ROM, IEEE Computer Society, 1999.
[393] J. Valenciano and R.G. Owens. An h-p adaptive spectral element method for Stokes flow. *Appl. Numer. Math.*, 33:365–371, 2000.
[394] J. Valenciano and R.G. Owens. A new adaptive modification strategy for numerical solutions to elliptic boundary value problems. *Appl. Numer. Math.*, 32:305–329, 2000.
[395] E.F. Van de Velde. *Concurrent Scientific Computing.* Texts in Applied Mathematics 16. Springer, Berlin, 1994.
[396] J. Van Kan. A second-order accurate pressure-correction scheme for viscous incompressible flow. *SIAM J. Sci. Stat. Comput.*, 7:870–891, 1986.
[397] H. Vandeven. Compatibilité des espaces discrets pour l'approximation spectrale du problème de Stokes périodique–non périodique. *Model. Math. Anal. Numer.*, 23:649–688, 1989.
[398] H. Vandeven. On the eigenvalues of second-order spectral differentiation operators. *Comput. Methods Appl. Mech. Engrg.*, 80:313–318, 1990.
[399] H. Vandeven. Family of spectral filters for discontinous problems. *J. Sci. Comput.*, 6:159–192, 1991.
[400] J. Villadsen and M.L. Michelsen. *Solution of Differential Equation Models by Polynomial Approximation.* Prentice-Hall, Englewood Cliffs, NJ, 1978.
[401] E.L. Wachspress. *Iterative Solution of Elliptic Systems and Applications to the Neutron Diffusion Equations of Reactor Physics.* Prentice-Hall, Englewood Cliffs, NJ, 1966.
[402] E.L. Wachspress. *The ADI Model Problem.* E.L. Wachspress, Windsor, CA, 1995.
[403] C. Walshaw, M. Cross, and M. G. Everett. Parallel dynamic graph partitioning for adaptive unstructured meshes. *J. Parallel Dist. Comput.*, 47:102–108, 1997. (Originally published as Univ. Greenwich Tech. Rep. 97/IM/20.)
[404] G.H. Wannier. A contribution to the hydrodynamics of lubrication. *Quart. Appl. Math.*, 8:1–32, 1950.
[405] T.C. Warburton, L.F. Pavarino, and J.S. Hesthaven. A pseudo-spectral scheme for the incompressible Navier–Stokes equations using unstructured nodal elements. *J. Comput. Phys.*, 164:1–21, 2000.
[406] T.C. Warburton, S.J. Sherwin, and G.E Karniadakis. Basis functions for triangular and quadrilateral high-order elements. *SIAM J. Sci. Comput.*, 20:1671–1695, 1999.

[407] V. Warichet and V. Legat. Adaptive hp–finite element viscoelastic flow calculations. *Comput. Methods Appl. Mech. Engrg.*, 136:93–110, 1996.
[408] J.A.C. Weideman and L.N. Trefethen. The eigenvalues of second-order spectral differentiation matrices. *SIAM J. Numer. Anal.*, 25:1279–1298, 1988.
[409] R.C. Whaley and J.J. Dongarra. Automatically tuned linear algebra software. In *Proceedings of the ACM/IEEE SC98 Conference on High Performance Networking and Computing*, IEEE Computer Society, Orlando, FL, 1998, http://www.cs.utk.edu/rwhaley/ATL/INDEX.HTM.
[410] O.B. Widlund. Iterative substructuring methods: Algorithms and theory for elliptic problems in the plane. In R. Glowinski, G.H. Golub, G. Meurant, and J. Périaux, editors, *First International Symposium on Domain Decomposition Methods for Partial Differential Equations*, pages 113–128. SIAM, 1988.
[411] D. Wilhelm. *Numerical Investigation of Three-Dimensional Separation in a Forward-Facing Step Flow Using Spectral Element Method*. PhD thesis, Swiss Federal Institute of Technology, Zürich, 2000.
[412] D. Wilhelm and L. Kleiser. Stable and unstable formulations of the convection operator in spectral element simulations. *Appl. Numer. Math.*, 33:275–280, 2000.
[413] R.D. Williams. Performance of dynamic load balancing for unstructured mesh calculations. *Concurrency: Practice and Experience*, 3:457–481, 1991.
[414] B.A. Wingate and J.P. Boyd. Triangular spectral element methods for geophysical fluid dynamics applications. In A.V. Ilin and L.R. Scott, editors, *Third International Conference on Spectral and High Order Methods*, pages 305–314. Houston Journal of Mathematics, 1996.
[415] A.A. Wray and R.S. Rogallo. Simulation of turbulence on the Intel Gamma and Delta. Technical report, NASA Ames Research Center, Mountain View, CA, 1992.
[416] V. Yakhot and S.A. Orszag. Renormalization group analysis of turbulence. I. Basic theory. *J. Sci. Comput.*, 1:3–51, 1986.
[417] N.N. Yanenko. *The Method of Fractional Steps*. Springer, New York, 1971.
[418] K. Yosida. *Functional Analysis*. Springer, Berlin, 4th edition, 1974.
[419] D.M. Young. Iterative methods for solving partial difference equations of elliptic type. *Trans. Amer. Math. Soc.*, 76:92–111, 1954.
[420] D.M. Young. *Iterative Solutions of Large Linear Systems*. Academic Press, New York, 1971.
[421] E. Zampieri. On the condition number of some spectral collocation operators and their finite element preconditioning. *J. Sci. Comput.*, 9:419–443, 1994.
[422] T.A. Zang, Y.S. Wong, and M.Y. Hussaini. Spectral multigrid methods for elliptic equations. *J. Comput. Phys.*, 48:485–501, 1982.
[423] T.A. Zang, Y.S. Wong, and M.Y. Hussaini. Spectral multigrid methods for elliptic equations II. *J. Comput. Phys.*, 54:489–507, 1984.
[424] J. Zhou, R.J. Adrian, S. Balachandar, and T. M. Kendall. Mechanisms for generating coherent packets of hairpin vortices in near-wall turbulence. *J. Fluid Mech.*, 387:353–396, 1999.

Index

H^1 space, 41, 44, 56, 92
L^2 norm, 41–43, 72, 95, 97
$\mathbb{P}_N$–$\mathbb{P}_N$ method, 235, 239, 240, 245, 257, 282
$\mathbb{P}_N$–$\mathbb{P}_{N-2}$ method, 241–244, 253, 268, 272, 289

A-inner product, 76, 87
A-norm, 76, 89
A-stable scheme
  IRK schemes, 118
  consistency less than 2, 109
  forward step methods, 105
  implicit, 108
accuracy
  of splitting method, 296
  for Stokes analytical problem, 258
  global third-order, 307
  loss of, with pressure discretization, 253
  loss of, with standard Stokes eigenfunctions, 280
  of split scheme, 299
  of Taylor–Galerkin method, 315
  spatial, 279
  time, 295, 332
  with spectral methods, 289
action of a matrix, 167, 168, 173, 176, 193, 360
action of an operator, 198
Adams–Bashforth (AB) method, 105, 107
Adams–Bashforth/Crank-Nicolson (ABCN) method, 119, 120
Adams–Moulton (AM) method, 105, 106
advection-diffusion equation, 17, 50, 145, 147, 317
  steady, 216
  unsteady, 230
algorithm robustness, 208

aliasing error, 458
angular momentum conservation, 6
approximation methods, 92
arbitrary Lagrangian–Eulerian (ALE) formulation, 293, 319
Arnoldi iteration, 91
ASCI project, 23
assembled Neumann operator, 195, 198
autonomous form, 99

backward differentiation (BDF) method, 109
backward-facing step, 267, 412
Banach space, 422
bandwidth
  in distributed memory architecture, 394
  of matrix, 69, 73, 205
basis functions
  Chebyshev, 457
  collocation, 52, 68, 69, 142, 152, 203, 461
  divergence-free, 265
  finite element, 54–56, 58, 61, 272
  Lagrangian, 54, 56, 57, 62, 63, 456
  Legendre, 457
  modal on triangles, 373
  mortar conditions for, 355
  nodal on triangles, 376, 377
  overlapping Schwarz, 345
  singular, 371
  spectral element, 62, 63, 138, 174, 191
  Stokes eigenfunctions, 266, 281
Bernoulli, 34
bilinear form, 37, 46, 47, 49, 51, 79, 170, 175, 191, 194, 195, 237, 238, 240, 252, 357, 364
block-Jacobi preconditioner, 338, 343, 348
blocking send, 398
Boolean sum, 183

489

490 Index

boundary condition
  Dirichlet, 39, 43, 51, 52, 57, 92, 125, 129
  essential, 36, 40, 43, 195, 323
  homogeneous, 39, 126, 137, 176
  inflow/outflow, 137, 147
  inhomogeneous, 39, 176, 196, 200, 209, 222, 310
  natural, 36, 39
  Neumann, 125
  periodic, 130
  Robin, 40, 169, 186, 200
  variational, 34
boundary value problem
  effect of collocation grid on, 67
  variational, 34
  weak form, 47
  with Dirichlet conditions, 51
boundedness condition, 42
Boussinesq approximation, 1, 8, 9
Boussinesq equations, 9, 12
brachistochrone problem, 34
Brinkman number, 13
broken norm, 357, 378
bubble functions, 61, 269, 282, 374
  definition of, 269
  interaction with polynomial functions, 270
  interior nodes, 272
  residual-free, 140
  stabilization with, 139, 140, 282
  with finite elements, 273
bulk viscosity, 7
buoyancy force, 9
Burgers equation, 17, 151, 426
  and time derivative, 159
  spectral element discretization of, 152, 155
  with Taylor–Galerkin time integration, 158, 161
Bézier curve, 25

cache, 382–384, 391
capillary tensor, 322
Cauchy principle, 5
Cauchy sequence, 420, 422, 428
  and coercivity condition, 42
  convergence of, 41, 43, 420
  definition, 419
Cauchy–Schwarz inequality, 429, 430
CFL condition, 17, 131, 133, 135, 136, 228
  scaling parameter, 133
chaos, 17
characteristic length, 22
characteristic polynomial, 104
characteristics, 231, 232
Chebyshev polynomials, 447, 451, 453
  fast transforms with, 62
  preconditioning, 84
  spectral case, 79

Cholesky factorization, 77
Clausius–Duhem inequality, 7, 8
clearance gap flow, 367
coarse-grid operator, 302, 344–348, 407, 410
coefficient of thermal conductivity, 8
coercivity condition, 42, 46, 48, 51
collocation
  approximation, 275
  Chebyshev, 248, 257, 283, 288, 290, 313, 410
  Chebyshev finite element, 332
  Chebyshev single-grid, 247
  derivative, 131, 153, 460, 462
  discretization, 202
  equations, 203, 210, 213
  Legendre single-grid, 245
  motivation for, 245
  multidomain, 212, 214, 215
  one-dimensional, 69
  orthogonal, 67, 69, 70, 94, 202
  points in triangle, 376–378
  preconditioned, 258, 289
  single-grid, 252
  spectral element compared with, 279
  spurious pressure modes eliminated with, 257
  staggered Legendre, 250
collocative spectral element, 242
column-major ordering, 384
commodity technology, 5, 23, 24, 391, 393
complete set, 41
compression algorithm, 25
computational fluid dynamics (CFD)
  hardware history and solution speed, 23
  high-order methods with, 2
  present status, 22
condition number, 29, 77, 78, 80, 83–85, 90, 95, 97, 205
  for triangular elements, 376
  with additive Schwarz, 345
  with block-Jacobi, 338
conditioning, 74, 257
cone problem, 229
conforming case, 353, 356, 361
conjugate gradient method, 201
  lowest eigenpairs found with, 407
  preconditioned, 29, 90, 256, 302
  relationship with variational method, 86
  semiiterative methods, 73
  spectral element implementation, 197
connectivity matrix, 193, 221, 358
conservation
  form, 227, 228
  momentum, 288
consistency error, 358, 370
consistency property, 102, 104, 107

Index

constitutive equations
  for stress tensor and heat conduction flux, 7
  Newtonian, 37
contact angles, 324
contact forces, 5, 235
convection operator, 100, 132, 133, 228, 318
convergence
  algebraic, 72, 73
  and compliance with inf-sup condition, 237
  and maximum principle, 339
  consistency and stability, 104
  dependence on Green's functions, 340
  effect of singularities on, 369
  enhancement with Laplacian operator, 266
  exponential, 72, 80, 273
  for nonsmooth functions, 267
  for single-grid collocation, 247
  initial, 348
  of Navier–Stokes problem, 277
  of nonconforming method, 358
  of sequence, 420
  property, 206
  spectral, 444
  test for spatial, 416
  with coarse-grid solve, 344–346, 348, 407
  with collocation method, 235
  with generalization of finite element method, 244
  with iterative techniques, 277
  with Newton method, 278
  with Richardson method, 258
  with special treatment of singularity, 371
corner singularity, 370
Courant–Friedrichs–Lewy (CFL) condition, 17, 133
Crank–Nicolson (CN) method, 106, 108, 128, 221, 228
creeping flow, 2, 12
Crouzeix DIRK scheme, 118, 119
crystal router, 403
cubic Hermite element, 55

deflation, 302, 347, 348
deformation tensor, 324
deformed geometry, 178
derivative operator, 181, 225
developed turbulence, 18
dgemm, 388, 389, 410
diagonal preconditioning, 198, 201
diagonalization method, 26, 28–30, 256, 303
  fast, 167, 177, 178, 410
differentiation matrix, 64, 461, 463
dimensional analysis
  guidelines for model selection, 14
  Pi theorem, 11
dimensionless numbers, 11–13, 16

direct numerical simulation (DNS)
  limits of, 2
  of turbulent flow, 23
  with low Reynolds number, 17
direct stiffness summation, 57, 65, 194, 196, 400, 401, 403
  geometry-free approach, 405
direction derivative, 35
Dirichlet boundary condition, 39, 43, 51, 52, 57, 92
  inhomogeneous, 176, 196, 210
discrete transform, 33, 456
discrete variable method
  basic requirements for, 102
  convergence with, 103
  forward step method, 101
dispersion error, 135, 149, 230
dissipation
  hyperbolic, 129
  parabolic, 124
dissipation function, 8
distributed memory, 391, 393
distribution, 43, 46, 434
divergence theorem, 5–7
driven cavity, 410
DuFort–Frankel method, 219
dyadic product, 8
dynamic pressure, 11

eddy-promoter domain, 267, 330
edge mode, 374
eigenvalue
  hyperbolic problem, 107, 129, 131, 139
  in fluid mechanics, 100
  model problem, 100
  of skew-symmetric operator, 228
  parabolic problem, 125
  preconditioned collocation, 208
  pseudo-spectra, 147
  Sturm–Liouville, 444
  tensor-product form, 165–167
  unsteady advection-diffusion, 145–147
Einstein summation convention, 5
elasticity problem, 324
electrostatic points, 376–378
element mass matrix, 58
element stiffness matrix, 57, 61
elliptic problem, 33, 162
  d-dimensional, 169
ellipticity condition, 42
ellipticity constant, 51
energy functional, 236
energy inner product, 38, 76, 174, 177, 179, 181
energy norm, 38, 41–43, 46, 50, 76
error estimation, 49, 71, 377, 464
Euclidean norm, 75, 77

Euler equations, 12, 259
Euler splitting, 225
Euler–Lagrange equation, 36, 38, 46
Eulerian representation, 4, 318
explicit Euler method (EM), 107, 127
explicit scheme
   produced with backward formula, 160
   spectral methods coupled with, 128
   stability control of, 119
   time step selection for, 128
exponential convergence, 80, 273, 285
external flow, 2, 17

fan-in/fan-out, 397
fast diagonalization method
   and Chebyshev collocation, 410
   as preconditioner, 177, 346
   for domain decomposition preconditioners, 167
   for inverse action, 177
   for solution of global spectral methods, 167
   limitations of, 177, 178
Fekete points, 376, 378
Fick's law, 37
Fiedler vector, 407
filter
   for stabilizing spectral methods, 150, 328
   for stabilizing unsteady flow, 293, 315
   higher Reynolds number flows with, 326
   with high Reynolds numbers, 21
   with moderate to high Reynolds numbers, 18
fine-grid operator, 302, 303
finite difference
   as a filter, 21
   boundary conditions determined by, 15
   centered, 26, 79
   five-point stencil, 73
   preconditioning, 84, 143
   upwind, 139
   vs. weighted residuals, 128
finite element
   as preconditioner, 345, 346
   automation of equation generation, 60
   cubic Hermite, 55
   ease of subspace generation with, 44
   error estimates in approximations, 49
   Galerkin approximations implemented with, 53
   h-type, 60
   h-version, 54
   in Sobolev space, 55
   interpolation, 57–59
   Lagrangian, 92, 95
   Lagrangian basis functions, 56
   linear, 72, 129, 132, 135
   linear systems induced by, 69
   mixed model, 20
   p- vs. h-version, 61
   p-version, 60
   partition of integration domain, 53
   preconditioning, 29, 85
   quadratic, 31
   second-order isoparametric, 56
   structure compared with spectral element, 65
   tensorized Lagrangian, 55
   vs. spectral element, 62
   weak formulation, 15
finite volume, 21
flow, external, 2, 17
fluid flow equations, 8
forward step method
   applied to initial value problem, 103
   discrete variable method, 101
   stability region of, 105
Fourier's law, 37
fractional step method, 296, 298, 299
free surface, 10, 15, 293, 318, 320–327
Froude number, 12
Fréchet derivative, 153, 274–276, 278, 425–427

Galerkin approximation, 330
   finite elements with, 53, 54
   numerical performance poor, 50
   of variational problem, 51
   principle of, 47
   projection method with, 47
   stability of, 49
   theorem for, 48
   with quadrature, 65
Galerkin method
   and spectral elements, 62
   detailed description of, 47
   discontinuous, 233
   generalized, 51
   implementation, 61
   implementation with spectral elements, 64
   matrix evaluation with, 57
   modified, 270, 271
   Petrov, 50
   vs. Ritz method, 50
gather-scatter operation, 193, 401
Gauss–Chebyshev, 142
Gauss–Jacobi quadrature, 375
Gauss–Legendre quadrature, 144
Gauss–Lobatto quadrature, 62
Gauss–Lobatto–Chebyshev (GLC) quadrature, 68, 70
Gauss–Lobatto–Chebyshev points, 79
Gauss–Lobatto–Jacobi grid, 85
Gauss–Lobatto–Jacobi quadrature, 131, 142, 152

Gauss–Lobatto–Legendre quadrature, 163, 190
Gauss–Seidel iteration, 73, 82
Gaussian elimination, 117
  comparison with multigrid, 27
  variants detailed, 74
Gaussian factorization, 73
Gaussian quadrature rule, 51, 448, 460
generalized conjugate residual (GCR) method, 89, 91
generalized derivative, 43
generalized Galerkin method, 51
generalized method of residuals (GMRES), 73, 86, 89, 91, 342, 352
geometric factor, 180, 181
glass flow, 3, 9, 13, 14
Gordon–Hall method, 183–185
Gram–Schmidt orthogonalization, 222, 434
  classical, 89
  definition, 433
  modified, 88
Grashof number, 13
greedy algorithm, 406
Green's function, 260, 262, 263, 340
grooved channel flow, 267, 280, 285

h-version of finite elements
  description, 54
  error decay rate, 72
  vs. Galerkin method with spectral elements, 64
  vs. p-version, 61
  vs. spectral elements, 62
  with homogeneous domain partition, 71
hairpin vortices, 408
Hamel flow, 187
hardware, 22, 23
heat capacity, 8
heat flux, 6, 7, 11, 125
Helmholtz equation, 16, 264
Helmholtz operator, 175, 176, 181, 222
Hermitian interpolation, 54
Hessenberg matrix, 92
hierarchical structure, 61, 382
high-order methods
  boundary value problems, 33
  conditioning, 77
  limitations, 2
  on massively parallel computers, 32
  orthogonal collocation, 67
  preconditioning, 84
  sophisticated data structures needed with, 25
  spectral convergence of, 31
  spectral elements, 62
  stabilization, 140
  time evolution of, 29
  vs. Lagrangian finite elements, 95
  vs. linear finite elements, 129

Hilbert space, 43, 45, 46, 428–433
hp method, 235, 272, 273
hydrostatic pressure, 13
hyperbolic problems, 107, 129, 233

implementation
  decomposition strategies for SPMD, 399
  for 2D problem, 263
  for 3D problem, 264
  for Chebyshev collocation, 257
  for complex geometry, 354
  in spectral frame, 265
  iteration-based implicit, 379
  Navier–Stokes, 391
  of algorithms for high-order weighted residual methods, 379
  of frequently called kernels, 390
  of object-oriented spectral element, 416
  of parallel BLAS, 411
  of reduce operation, 397
  weighted residual, 401
  with collocation, 253, 289
implicit Euler (IE) method, 108, 127
implicit scheme
  A-stable, 108, 124
  Crank–Nicolson scheme similar to, 120
  explicit counterpart to, 128
  for viscous term, 152
  leading order cost, 168
  produced with forward formula, 160
  stability regions of, 107
incompressibility constraint, 5, 260, 266, 288, 300, 314, 332
incompressible fluid, 124
  bottleneck for simulation, 26
  heat capacity in, 8
  Navier–Stokes equations for, 9
  Newtonian, 8
  stress tensor in, 7
  volumic mass as constant in, 4
incremental loading, 277
indirect addressing, 193, 405
inertial forces, 2
inertial time, 11, 156
inexact factorization, 299, 300
inf–sup problem, 237
inf–sup constant, 364, 365
influence matrix technique, 260, 263
initial conditions, 292, 299, 308, 326, 327
inner product
  A, 76, 87
  discrete, 452, 460
  energy, 38, 50, 76
  of elements in V, 37
  of spaces, 44
  Sobolev, 41, 43–44

inner product (*contd.*)
  space, 427, 429–433
  with basis functions, 48
internal energy, 6
internal flow, 2, 17
interpolation error, 71, 317
interpolation operator, 52, 85, 183, 211, 214, 343, 350
inviscid fluid, 10
irrotational field, 261
isoparametric element, 56
isoparametric mapping, 183, 190, 191
iteration matrix, 258
iterative method
  approximation of solutions with, 86
  finite termination property lacking in, 73
  for three-dimensional problems, 74
  general, 74
  Richardson fixed point iteration, 81

Jacobi iterative scheme, 82
Jacobian, 180
Jacobian matrix, 101
Jacobian, surface, 186, 187, 201

k–ϵ model, 19–21
k–l model, 20
kinematic viscosity, 12
kinematical constraint, 322
Kovasznay flow, 283, 378
Kronecker delta function, 182
Kronecker product, 164
Kronecker symbol, 52
Krylov method, 89
Krylov subspace, 89–91, 408

L-stable method, 109, 119, 124
Lagrange multiplier, 236
Lagrangian formulation, 319
Lagrangian interpolation, 54, 56, 62, 67, 68, 70, 455
Lagrangian polynomial, 55
laminar flow, 2
Laplace equation, 16, 58, 84, 185
Laplace relation, 321
large eddy simulation (LES), 2, 17, 224
latency, 394, 395, 403
Lax equivalence theorem, 49, 104
Lax–Milgram theorem, 33, 45, 48
Legendre polynomials, 60, 63, 445, 450, 453, 464
lexicographical ordering, 26, 73, 164, 203, 205, 219, 220
linear momentum conservation, 5
linear multistep method
  A-stable schemes, 109

  examples, 105
  explicit, 111
  formula for, 101
  implicit, 110
  numerical solution with, 104
  stability of, 113
  vs. IRK schemes, 117
low-order preconditioning, 205
LU decomposition, 26, 60

Marangoni effect, 324
mask array, 195, 221
mass conservation
  local equation, 4
  principle, 3
mass lumping, 78, 128, 132, 133, 174
mass matrix
  diagonal, 174, 255, 264, 376
  diagonal pressure, 255, 256
  element, 58
  finite element, 61, 97
  global, 58, 78
  integration of basis function with, 78
  local, 58
  lumped, 79, 128, 132, 133, 174
  two dimensional, 173
  with triangular spectral elements, 376
matching condition, 335, 352–355, 364
material time derivative, 4, 10, 19
matrix bandwidth, 69, 73, 205
matrix norm, 74, 75
matrix-matrix product
  dgemm, 388
  code, 387
  importance for speedup, 409
  in NEC library, 412
  optimized for benchmarks, 388
  performance, 394
  tensor-product evaluation recast as, 169, 385
mesh velocity, 319, 324, 326
message passing, 391, 392
  costs, 393
  implementation, 393, 396
Message Passing Interface (MPI), 22, 23, 25, 392, 393, 396, 398, 401
method of characteristics, 231, 232
metric space, 417, 418
metrics, 180, 181, 185
mixing length theory, 19, 20
modified Adams–Bashforth/Crank–Nicolson (MABCN) method, 120
modified Galerkin method, 270, 271
Moore's law, 380
mortar conditions, 355, 364
mortar spectral element (MSE), 25, 353
multidomain collocation, 67, 212, 214, 215

multigrid method, 288, 330
  as preconditioner, 29
  comparison with Gaussian elimination, 28
  convergence factor, 349
  introduction of, 26
  on parallel computers, 27
  spectral element, 348
multiplicity, 198, 223, 400
multistep
  consistent, 103
  cost of implicit, 102
  examples of linear, 105
  implicit linear, 117
  linear, 101, 104, 119
  stable, 103
  starting procedures needed with, 102

natural boundary conditions, 39
natural convection, 9, 13
natural ordering, 167, 173, 180, 305
Navier–Stokes equations
  choice of discrete space for mortar
    functions, 354
  diffusion and advection contributions, 119
  fractional-step scheme with, 346
  on unstructured nodal triangles, 378
  space discretization, 99
  splitting with, 120, 124
  stabilization, 140
  vs. Burgers equation, 151
Navier–Stokes equations
  as singular perturbation problem, 12
  derivation of, 2
  dimensionless, 12
  for compressible fluids, 9
  for incompressible fluids, 9
  for moderate to high Reynolds numbers, 18
  for very high Reynolds numbers, 18
  fourth-order compact schemes for, 28
  isothermal incompressible, 11
  Reynolds averaged, 2, 18, 23
  steady-state, 234
  subsets used for testing numerical methods, 16
  three dimensional, 31
  transient, 10
  unsteady, 291
  vs. Burgers equation, 17
Neumann boundary condition
  enforced asymptotically, 66
  homogeneous, 39, 57
  inhomogeneous, 40, 200, 209
Neumann operator
  assembled, 195, 198
  definition, 176
Newton method, 235, 277, 278

Newton–Cotes quadrature scheme, 105
Newtonian fluid, 7
Nitsche's trick, 71
no-slip wall condition, 10
non-Newtonian fluid, 20, 32
nonconforming
  3D case, 363
  convergence, 358
  discretization, 358
  domain decomposition, 354, 369
  elements, 357, 364
  implementation, 358
  spectral elements, 353, 357
  steady Stokes case, 363
nonlinear term, 152, 153, 161, 274, 276
nonlinear term evaluation, 305
nonuniform rational B-splines (NURBS), 25
norm
  A-norm, 89
  L^2, 43, 72, 95, 97
  bounded solution of, 49
  broken, 357, 378
  conditions of, 41
  decay rate of, 93
  energy, 38, 41–43, 50, 76
  energy vs. Sobolev, 42, 46
  equivalence of, 82, 453
  error, 94
  Euclidean, 77
  matrix, 74, 75
  maximum, 422
  notation of, 35
  on vector space, 421
  semi-, 421
  Sobolev, 41
  vector, 76
normal pressure gradient, 259
null space, 176, 195, 196, 423
NUMA, 22, 23

object-oriented concepts, 25
Oldroyd-B model, 20
one-equation model, 20˙
Operator-Integration-Factor Splitting (OIFS)
  method, 121, 123, 155–157, 161, 230
operator–splitting scheme, 100
orthogonal collocation method, 67, 69, 70, 94
orthogonal polynomial, 62, 433, 442
  three-term recursion, 443
overlapping Schwarz method, 338, 345

p-version of finite elements, 61, 62, 72
parabolic problems, 220, 224
parallel computation
  effects of hardware cost, 23
  high-order methods, 25, 32

parallel computation (*contd.*)
  improvements in, 28
  introduction of parallel machines, 24
  message passing libraries, 25
  multigrid method, 26
  new languages for, 25
  programming difficulty, 24
  RISC chip revolution, 22
parallel processing
  visualization needs, 26
  with MPPs, 24
Parallel Virtual Machine (PVM), 25, 392
parent element, 53, 162, 239, 269, 272
Parseval's identity, 454
Petrov–Galerkin method, 50
Picard iterative method, 278
pipelined execution, 381
plane channel, 267
plastics processing, 9
Poincaré–Friedrichs inequality, 42
Poisson equation
  3D, 26, 28
  collocation schemes for, 29
  continuous presentation, 334
  diagonal scaling with, 201
  for pressure, 259
  from simplified Helmholtz equation, 16
  inversion of, 29
  mapping with, 378
  overlapping Schwarz approach with, 338
  pressure, 223, 292, 301, 310, 311
  pressure correction, 315
  pseudo, 264
  Schwarz iteration with, 339
  velocity field obtained from, 15
  with Dirichlet homogeneous boundary
    conditions, 370
  with pressure boundary condition, 260
polymer processing, 9
Prandtl number, 13
preconditioner
  diagonal, 198
  fast diagonalization method, 177
  finite difference, 84–86, 207–209, 215
  finite element, 29, 85, 95, 208, 209,
    213–215, 345, 346
  for Chebyshev collocation, 257
  for pressure matrix, 258
  for Uzawa method, 264
  Funaro's finite difference, 217
  of conjugate gradient method, 256
  of linear systems, 288
  of Schur complement, 338
  with collocation, 289
preconditioning
  diagonal, 201, 222
  domain decomposition, 334
  low order, 205
  Schwarz, 342, 343
  two-stage, 303
preconditioning matrix, 81, 83, 218
predictor-corrector (PC) method, 111, 112, 313
pressure condition, inhomogeneous, 262
pressure correction, 296, 299, 313–315, 329
pressure gradient, normal, 259
pressure method, 293, 296
pressure modes
  weakly spurious, 245
pressure modes, spurious, 240, 241, 245–247,
    250, 252, 256–258, 279
pressure operator, 296, 303
primitive variables, 15, 240, 291
principle of thermodynamics, 6, 7
projection, 223, 224, 254, 257, 270, 286, 342,
    408, 413
  in Schwarz method, 344
  method, 297, 299
  onto previous solutions, 222
  orthogonal, 50, 87, 249, 349, 431, 432, 455,
    457
projection-diffusion method, 331
prolongation, 176, 341, 342, 345–347, 349
pseudo–Laplacian matrix, 300, 301
pseudo–Poisson equation, 264
pseudo–spectra, 147
Prandtl number, 13

quadrature rule
  Gauss–Lobatto–Jacobi, 69
  Gaussian, 33, 51, 448, 460

rate of deformation tensor, 7, 21, 289
rate of rotation tensor, 8
Rayleigh number, 13
recursive bisection, 406
reduced equations, 16
reduced instruction set computer (RISC), 22,
    380
reduced instruction set computer (RISC), 381,
    382
reference pressure level, 235, 253, 292,
    323
relaxation factor, 81, 83, 85, 97, 206, 208, 258
renormalization group (RNG), 21
residual-free bubbles, 140
restriction operator, 176, 186, 198, 342, 346,
    349
Reynolds number
  and computational grids, 409
  and recirculation zone, 413
  and LES, 2, 21
  based on inflow velocity, 367
  based on rotation rate, 365
  choice of numerical scheme, 18

convergence with collocation, 275
critical, 17
definition of, 11
dimensionless, 13
divergence of solutions, 277
effects when high, 10
high, 137
in advection-dominated flows, 12
in fully developed turbulence, 2
influence of nonlinear advection, 307
ratio of two time scales, 156
stability conditions with, 295
with filtering, 326
Reynolds transport theorem, 4, 7
Reynolds–averaged Navier–Stokes (RANS), 18, 23
Richardson extrapolation, 28
Richardson iteration, 258, 348, 351
Richardson iterative scheme
　as Krylov method, 89
　convergence rate, 82, 83
　evaluation of right-hand sides in, 84
　fixed point, 81
　preconditioned, 82, 86, 97, 206
Richardson number, 13
Ritz method, 50
Robin boundary condition, 40, 86, 186
　　implementation, 200
　　in heat transfer, 186
　　mixed, 169
robustness of algorithm, 208
Rosenbrock DIRK scheme, 118
row-major ordering, 384
Runge–Kutta (RK) method, 113, 308
　diagonally implicit (DIRK), 118
　explicit (ERK), 114–116, 123, 157
　implicit (IRK), 117, 118
　nonlinear essence of, 114
　splitting schemes with, 121

saddle-point problem, 237
Schur complement, 337, 338, 342
Schwarz method
　additive, 343, 345
　alternating, 339, 341
　multiplicative, 341, 343
　overlapping, 338, 345
Schwarz preconditioner
　additive, 343, 347
　multiplicative, 343
semi-discrete problem, 126, 294
semi-implicit scheme, 100, 149
seminorm, 421
　Sobolev, 41
shared-memory model, 391
shared-memory multiprocessor, 22, 393
shear flows, 20

shear viscosity, 7
similarity transform, 165, 166
single domain, 67, 194
single-step method, 102
　AB1, 106
　BDF1, 109
　numerical scheme of, 101
　with AM method, 106
singular basis function, 371
singular perturbation, 138, 147
skew-symmetric form, 227, 228, 304
smoothing step, 351
SMP machine, 22, 24, 393
Sobolev inner product, 41, 43
Sobolev norm, 41, 464
Sobolev seminorm, 41
Sobolev space, 43, 138
　H^1, 44, 56
　constrained, 264
　definition, 41, 171
　embedding properties, 171
　in d-dimension, 170
　negative, 45
　notation, 43
　weighted, 44
software engineering, 25
solenoidal field, 261
spectral accuracy, 444
spectral approximation, 33, 62, 139, 140
spectral collocation, 25
spectral condition number, 77, 83–85, 95, 205
spectral convergence, 444
spectral derivative, 459
spectral element
　2D, 229
　accuracy, 30
　and AB3, 135
　approximation space, 190
　collocative, 242
　convergence, 31
　discretization of Burgers equation, 152
　flexibility of, 190
　for solving elliptic equation, 25
　grid points, 143
　inner nodes, 144
　inner-product definition, 174
　Lagrangian basis functions with, 177
　Laplacian, 384
　Legendre, 29, 138
　mass matrix, 195
　Navier–Stokes, 389
　nodes, 144
　nonconforming, 353, 357
　numerical values of S, 133
　object-oriented, 416
　operators, 194
　preconditioned, 30

spectral element (*contd.*)
　resolvable spectrum, 129
　scalability, 409
　scaling, 124
　staggered, 240, 242, 331
　steps, 238
　stiffness matrix, 178, 181
　vs. linear finite element, 149
spectral element method
　basis functions, 63
　convergence, 72
　free vertex variant, 353
　Legendre, 65, 67
　mass lumping with, 79
　mortar, 353
　Schwarz preconditioners for, 345
spectral radius, 75, 77, 83, 133, 206, 221, 258
splitting error, 293, 296, 297, 308, 311
splitting technique, 98
SPMD, 392, 399, 400
spurious pressure modes, 240, 241, 245–247, 250, 252, 256–258, 279
stability
　CFL condition, 133
　with explicit time-marching schemes, 295
stability condition, 49, 156, 157, 228, 230, 304
stability property, 102, 103
　governed by eigenproblem, 146
　governed by eigenvalue problem, 131
　of $P(EC)^m E$ methods, 112
　of IRK schemes, 117
　problem dependency of, 103
stability region
　of A-stable scheme, 110
　of IRK schemes, 118
　Adams–Bashforth, 107
　Adams–Moulton, 107
　definition, 105
　for BDFk, 110
　Runge–Kutta, 113
　splitting schemes (ABCN), 120
stabilization
　bubble function, 139, 140, 269, 282
　by upwind grids, 140, 144, 145, 216, 217, 220
　filter-based, 316, 328
　upwind, 139
stabilizing term, 271, 318
staggered grid, 86, 240, 242, 250, 278, 305
　example, 144
　for Chebyshev case, 216
　generation of, 143
　Legendre, 145
　orthogonal collocation solutions of, 141
　stabilization techniques based on, 143
　upwind grid, 217
　with advection-diffusion, 219

staggered spectral element, 240, 242
Steklov–Poincaré operator, 334–337
stiff system, 19, 105, 109, 120, 149
stiffness matrix
　coarse-grid, 350
　derivation, 174
　element, 57, 61
　evaluation of, 59
　global, 58, 65
　in deformed geometry, 178
　local, 57, 194, 358
　masks applied to, 195
　orthogonality in, 374
　quadratures for evaluation, 62
　spectral element, 178, 181, 194
　storage requirement, 181
　structure of, 65
　unacceptable fill in, 177
　unassembled, 194
　viscous term with, 153
stiffness summation, 65
Stokes equations, 12, 264, 273, 276, 278, 288
stream function, 15, 16
streamline-upwind Petrov–Galerkin (SUPG) method, 139
stress tensor, 5–7, 19, 21
stress vector, 5
strong formulation, 38
Sturm–Liouville problem, 444
subcycling, 155, 156, 230–232, 413
subdomain, 47, 69
　load balancing with, 406
　motivation for choosing, 333
　nonconforming, 365
　with high-order methods, 399
　with recursive bisection, 406
subgrid-scale approximation, 2
subgrid-scale stress tensor, 21
substructuring, 333, 334, 338, 342
successive overrelaxation (SOR) method, 26
surface Jacobian, 186, 187, 201
symmetric positive definite (SPD) matrix, 76, 77, 149

Taylor–Galerkin method
　explicit, 160, 161
　three-step explicit, 314
　time integration, 158
　two-step, 161
　with spectral elements, 158, 159
tensor field, 5
tensor product, 26, 29, 117, 164, 305
　diagonalization, 165
　domain, 162
　evaluation, 167, 385
　factorization, 166

generalized, 373
grids, 180, 202
in 2D case, 170
in 3D case, 168
interpolation, 211
inverse, 165
matrix, 163
of 1D finite elements, 346
of 1D operators, 363
of GLL quadrature, 163
of Lagragian interpolants, 191
polynomial, 163
symmetric use of, 202
test function, 33, 38–40, 146, 171, 186, 210, 211, 213, 220, 225, 227, 434, 435, 439
thermal conduction, 13, 14
thermal convection, 1, 3
thermal diffusivity, 13, 14
thermodynamics
  first law, 6
  second law, 7
third-order pressure correction scheme, 315
three-field approach, 288, 289
trace operator, 6
transport operator, 227
trapezoidal rule, 106
trial function, 54, 55, 57, 61
triangle inequality, 344, 417
triangular element, 371
trilinear form, 274
turbulence
  as flow property, 19
  fully developed, 2
  memory effects, 20
  models in industrial applications, 23
  simulation important for large-scale applications, 32
  stages, 2, 17
  two-layer model of, 21
  with varying Reynolds number, 18
turbulent viscosity, 19, 21

unit-stride data access, 380, 383
unrolling, 389
unstable mode, 134
upwind grid, 140, 144, 217, 218, 220
Uzawa algorithm, 253, 256, 257, 268, 295, 296, 331
Uzawa operator, 255, 264

variational crime, 182
variational formulation, 38
variational functional, 58, 59
variational method, 34
  for multidimensional problems, 162
variational principle, 33, 46, 49, 57
vector potential, 15

vector reduction, 396
velocity-pressure formulation, 15
vertex modes, 374
viscosity
  artificial, 139
  bulk, 7
  effects, 10
  for modeling creeping flow, 12
  kinematic, 12
  shear, 7
  small values of, 153
  turbulent, 19, 21
  vanishing, 159
  variable, 224, 225
viscous dissipation, 13, 17
viscous effects, 1, 17
viscous forces, 2, 11
volumetric mass, 9
vortex sheet roll-up, 326, 327
vorticity, 283
vorticity equation, 15
vorticity-stream function formulation, 16

Wannier–Stokes flow, 282, 284
warped product, 373
wave equation, 16
weak formulation, 15, 39–41, 170, 186, 226, 230
  abstract form, 237
  definition of, 236
  in parabolic case, 130
  local mass and stiffness matrices with, 204
  modified with constrained Sobolev space, 264
  of Green's function, 263
  of Poisson equation, 340
  of Steklov–Poincaré operator, 337
  of Stokes problem, 364
  quadrature rules with, 239
  recasting with, 138
  vs. collocation, 278
weakly spurious pressure (WSP) modes, 245
Weber number, 321
weighted residual, 379, 399
weighted residual method, 34, 49, 87, 128, 137, 196, 200, 225
  parallel implementation, 399
weighted Sobolev spaces, 44, 248
WENO scheme, 233
wiggles, 50, 368
wire-frame information, 185

Yosida method, 299

zero-equation model, 19